SHORT WAVELENGTH COHERENT RADIATION: GENERATION AND APPLICATIONS

AMERICAN INSTITUTE OF PHYSICS
CONFERENCE PROCEEDINGS NO. 147
NEW YORK 1986

OPTICAL SCIENCE AND ENGINEERING SERIES 7

SERIES EDITOR: RITA G. LERNER

SHORT WAVELENGTH COHERENT RADIATION: GENERATION AND APPLICATIONS

MONTEREY, CA 1986

EDITORS:
D. T. ATTWOOD
LAWRENCE BERKELEY LABORATORY
J. BOKOR
AT&T BELL LABORATORIES

L.C. Catalog Card No. 86-71674
ISBN 0-88318-346-3
DOE CONF-860316

Printed in the United States of America

CONTENTS

Preface

The Topical Meeting on Short Wavelength Coherent Radiation: Generation and Applications, held March 24–26, 1986 in Monterey, California brought together over 150 scientists and engineers involved in the development and application of sources of coherent radiation in the extreme-ultraviolet and soft-x-ray spectral regions, as well as those interested in the closely related areas of multiphoton phenomena, soft-x-ray optics, and laser-produced plasma radiation sources. This conference was the third in a series formerly entitled "Laser Techniques in the Extreme Ultraviolet". The name change was intended to reflect the rapid advancement of the technology for coherent radiation sources into the x-ray spectral region, the diversification of technical approaches for source development, and as source technology has matured, the growing interest in scientific and technological applications for coherent short wavelength radiation. This volume contains most of the papers presented at the conference, and the exciting and novel research represented herein clearly bears out our most optimistic expectations.

We wish to express our appreciation to the many people who helped to bring about the conference and this book. The advisory panel and the technical program committee played the critical role of bringing together researchers from diverse fields to give the conference its distinctly interdisciplinary character. We owe a special thanks to the Optical Society of America, and in particular, to Joan Carlisle and Tanya Hill for their tireless efforts in organizing and managing the conference. Crucial financial support for the conference was generously provided by the Office of Naval Research, the Department of Energy, the Air Force Office of Scientific Research, and the Optical Society of America. We are also grateful to our industrial sponsors, Acton Research Corporation and Questek, Inc.

May 1986

J. Bokor
Holmdel, NJ

D. T. Attwood
Berkeley, CA

Technical Program Committee

D. T. Attwood, *Conference Cochair*
Lawrence Berkeley Laboratory

J. Bokor, *Conference Cochair*
AT&T Bell Laboratories

E. M. Campbell
Lawrence Livermore National Laboratory

R. W. Falcone
University of California, Berkeley

S. E. Harris
Stanford University

S. Kikuta
University of Tokyo
Japan

J. Kirz
SUNY at Stony Brook

T. B. Lucatorto
National Bureau of Standards

J. M. J. Madey
Stanford University

C. Pellegrini
Brookhaven National Laboratory

C. K. Rhodes
University of Illinois at Chicago Circle

K. H. Welge
Universitat Bielefeld
Federal Republic of Germany

Advisory Panel

R. R. Freeman
AT&T Bell Laboratories

S. E. Harris
Stanford University

T. B. Lucatorto
National Bureau of Standards

T. J. McIlrath
University of Maryland

OPTIONS AND ISSUES FOR XUV FREE ELECTRON LASER OSCILLATORS

J. E. La Sala and J. M. J. Madey
Stanford Photon Research Center
Stanford University, Stanford, California 94305

D. A. G. Deacon
Deacon Research
900 Welch Road Suite 203, Palo Alto, CA 94304

ABSTRACT

Advancement of free electron laser (FEL) technology into the XUV brings with it many technical challenges. Both FEL gain and mirror reflectivity degrade at shorter wavelengths. However, recent developments in FEL physics, accelerator, and optical technology point the way toward development of a powerful and flexible new class of XUV light sources using FEL technology. We describe the major issues and options that bear on this development.

INTRODUCTION

Since the first operation of a Free Electron Laser (FEL) at Stanford University[1] ten years ago, the FEL has been operated successfully at a number of university and government research facilities at wavelengths in the microwave through the visible part of the spectrum.[2] The FEL is now finding its way into industrial laboratories, and is being used as a research instrument in other fields of science including materials science, chemistry and medicine. The next great challenge for the FEL community is the extension of the technology into the XUV region, where applications abound. There are a number of issues and options under consideration by researchers making the effort to move the FEL to shorter wavelengths. In this paper we will describe those issues and options we consider central to this effort. Where posssible we indicate directions we have chosen in the design of the Stanford Photon Research Center XUV FEL.

FEL PHYSICS

The most important difference between FEL operation in the IR through UV region and the XUV region is the single pass gain required to achieve oscillation. In the long wavelength region 99% mirror reflectivies are readily available, requiring only a few percent gain for oscillation. Of course higher gain results in shorter turn-on time and higher power levels, but all FEL oscillators currently operating or planned for at visible or longer wavelength operation have gains below 100%. Below 1000 Å, on the other hand, the reflectivity of conventional mirrors drops sharply toward 50% and below,[3] implying a single pass gain of at least $\Delta P/P = 3.0$ for threshold oscillation.

FEL's with single pass gain below 1.0 are considered low gain devices, for which the following small signal gain scaling law applies:[4]

$$G \sim N_w^3 \lambda_w^2 \left(\frac{j}{\gamma^3} \right) K^2 [JJ]^2 . \tag{1}$$

This relation assumes a incident optical plane wave and negligible inhomogeneous broadening of the electron beam. The factor $[JJ]^2$ is a Bessel function correction applied in the case of a linearly polarized wiggler to account for power emitted into higher harmonics. N_w is the number of wiggler periods in the FEL, j is the electron beam peak current density, γ is the electron beam energy in units of rest energy, and K is the wiggler field strength parameter related to the peak magnetic field $\hat{B}$ on axis and the wiggler period λ_w as follows:

$$K = 0.093 \hat{B}\,(\mathrm{kG})\, \lambda_w(\mathrm{cm}) . \tag{2}$$

Note the cubic dependence of gain on the number of wiggle periods. If the wiggler is long enough, and the dominant inhomogeneous broadening effect is electron beam emittance ε, the gain scaling for an optimized low gain FEL including the effects of diffraction becomes[5]

$$G \sim \lambda^{\frac{1}{2}} \lambda_w^{\frac{3}{2}} \left[\frac{I}{\gamma^2 \varepsilon_x \varepsilon_y} \right] \sqrt{1 + \frac{K^2}{2}} , \tag{3}$$

where λ is the radiation wavelength. The number of wiggler periods is now set at an optimum value determined by the emittance of the electron beam. The emittance ε_x (ε_y) is the area of transverse phase space associated with the electron beam width and divergence in the x (y)-direction, and is determined by the accelerator structure producing and/or storing the beam. It is natural to define the quantity in square brackets as the electron beam brightness.

If we increase the brightness of the electron beam sufficiently, and use a sufficiently long wiggler, the single pass gain increases to the point that the scaling laws (1) and (3) no longer apply. Instead, a collective instability develops, resulting in an exponential increase in gain with wiggler length:[6]

$$G \sim e^{g N_w} , \tag{4}$$

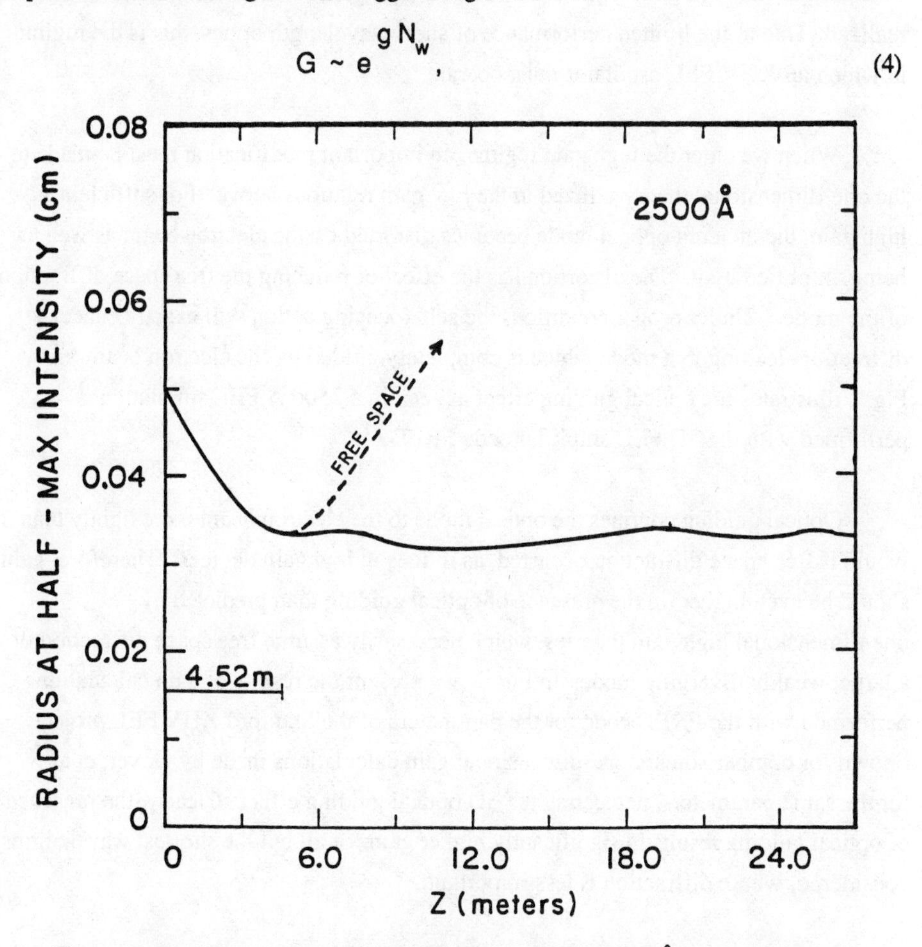

Fig. 1. Radius (measured at half-max intensity) of a 2500 Å optical mode as it passes once through the 27 meter SPRC storage ring FEL. Note the mode is nearly perfectly guided by the electron beam (whose radius is comparable to that of the guided mode) after coming to a focus approximately one Rayliegh range (4.52 m) inside the wiggler. Single pass gain is 132.0.

where

$$g \propto \rho \sim \left(\frac{K^2 \lambda_w^2 j}{\gamma^3} \right)^{\frac{1}{3}} \tag{5}$$

ρ is a gain parameter similar to the "Pierce parameter" of travelling wave tube theory. This defines the high gain regime of an FEL, where gain much greater than 1.0 may be realized. Due to the limited performance of short wavelength optics, this is the regime in which an XUV FEL oscillator must operate.

When we enter the high gain regime, an important modification must be made to the one-dimensional theory utilized in the low gain relations above. For sufficiently high gain, the incident optical mode becomes distorted by the electron beam as well as being amplified by it. The distortion has the effect of reducing the free space diffraction of the mode.[7] Under proper conditions the self-focusing action will exactly cancel diffraction, leading to a mode which is completely guided by the electron beam.[8,9] Fig. 1 illustrates the optical guiding effect as seen in a 2500 Å FEL simulation[10] performed with the 3D FEL amplifier code FRED.[11]

Optical guiding confines the optical mode to the electron beam more tightly than it would if free space diffraction occurred, as it does in low gain devices. Therefore, gain should be even higher in the presence of optical guiding than predicted by one-dimensional high gain theories, which necessarily assume free space diffraction of a large, weakly diverging mode. In Fig. 2 we present the results of gain calculations performed with the FRED code for the parameters of the Stanford XUV FEL project. Shown for comparison are one-dimensional gain calculations made by Gover, et al[12] for the same parameters, neglecting the 3D optical guiding effect. Clearly, the presence of optical guiding results in significantly higher gain for all but the shortest wavelengths considered, where diffraction is less important.

Since we require high gain operation, we must understand optical guiding fully, and determine how best to incorporate its effects in XUV FEL design. Although good progress has been made in the theoretical analysis of high gain operation,[8,9,13] some additional work remains to be done, particularly on transient and saturation effects, as well as the effect of inhomogeneous broadening on guiding, and wiggler field errors.

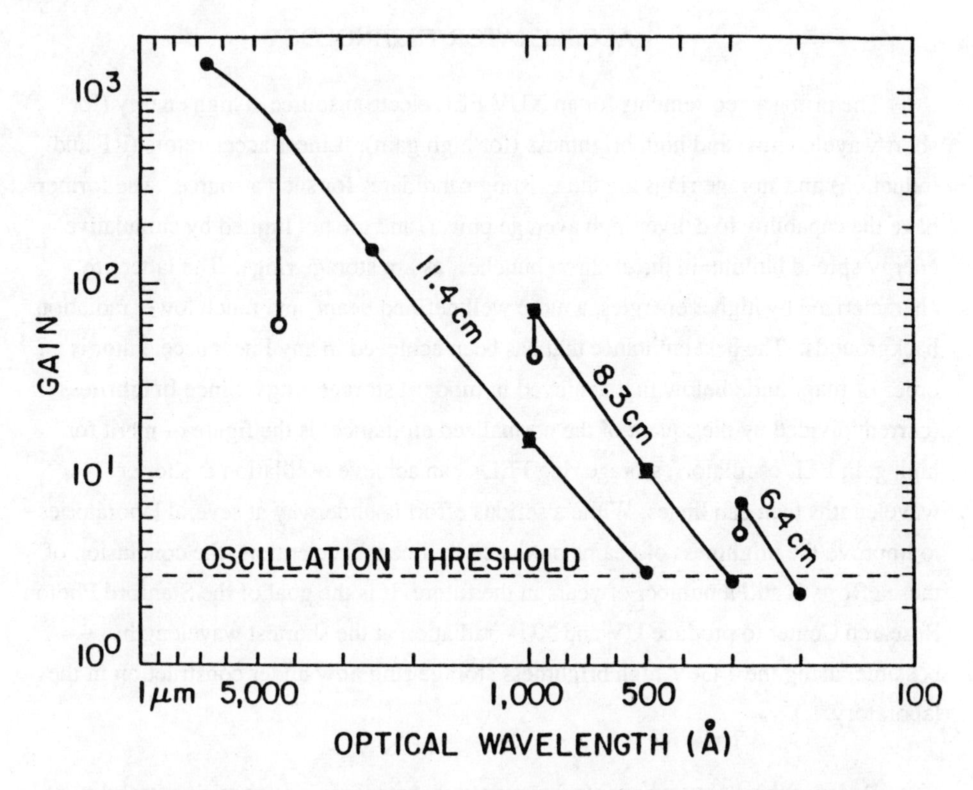

Fig. 2. Solid curves represents single pass gain ($\Delta P/P$) as a fuction of optical wavelength for the 27 meter SPRC storage ring FEL for three different wiggler periods. Oscillation threshold corresponds to effective single cavity mirror reflectivity of 50%. Solid circles are values determined from the FRED 3D simulation code.[11] Empty circles are 1D calculations performed by Gover, et al[12], who assume free space propagation for the amplified mode.

The effects of the electrons' periodic transverse motion on guiding is also under consideration.[14] There is some indication that the degree of guiding varies between laser turn-on and saturation,[15] raising the issue of resonator stability during turn-on. Finally, careful consideration should be given to the development of an experimental data base on these phenomenon using the present generation of FEL amplifiers and oscillators, which are capable – under optimized conditions – of operating at gains approaching 100% per pass.

ACCELERATOR TECHNOLOGY

The primary reqirements for an XUV FEL electron source is high energy (for short wavelengths) and high brightness (for high gain). Linear accelerators (RF and induction) and storage rings are the existing candidates for such a source. The former have the capability to deliver high average power, and are not limited by cumulative energy spread buildup in the electron bunches, as are storage rings. The latter are characterized by higher energies, a more well defined beam, and much lower radiation backgrounds. The best emittance that has been achieved in any linear accelerator is an order of magnitude below that achieved in modern storage rings. Since **brightness** (current divided by the square of the normalized emittance) is the figure of merit for high gain FEL oscillators, storage ring FEL's can achieve oscillation at shorter wavelengths than can linacs. While a serious effort is underway at several laboratories to improve the brightness of beams produced by linear accelerators, the conclusion of these efforts is still a number of years in the future. It is the goal of the Stanford Photon Research Center to produce UV and XUV radiation at the shortest wavelengths possible, using the 1 GeV high brightness storage ring now under construction in the laboratory.

Because the electron bunch(es) circulate repeatedly in a storage ring, stability of beam characteristics is essential. Under normal operation, equilibrium is established by a balance between the effects of excitation and damping due to synchrotron emission in the bending magnets of the ring. An FEL inserted in the ring provides an additional source of electron beam excitation while it is lasing, resulting in a new equilibrium with larger energy spread. The increase in energy spread reduces FEL gain, which contributes to the saturation process. This effect causes the FEL to saturate at lower power than it would in a single pass device (linac).

The major disadvantage of using a storage ring to drive the FEL is the limitation on average laser power due to FEL- induced energy spread in the recirculating electron bunch.[16] For most research applications, however, it is the spectral brilliance of the source, not the average power, which determines the feasibility of new work. Further, because the electron beam is stored for a long period of time (about 1 hour in the SPRC ring when the laser is running), it can be very well characterized, which is a major advantage for studies of FEL physics and for the applications.

In a storage ring there are a number of instabilities to which an electron bunch is susceptible. The major current-limiting instabilities are associated with scattering within the bunch and wake fields created by the charged bunch as it passes through the finite impedence of the vacuum chamber. Modern storage ring technology has been very successful in raising the peak current density at which these instabilities begin to affect beam lifetime. A high current density ring has recently been constructed at SLAC for the SLC project. This damping ring, based on the SPRC ring design, has met design specifications well.[17] Shown in Table 1 below are the operating parameters of the SLAC damping ring, together with those for the SPRC FEL ring. The success of the almost identical SLAC damping ring provides confidence that the SPRC ring will also work well at its rated current density. In Fig. 3 we present a plan view of the SPRC storage ring facility.

Table 1. Comparison of operating parameters for the SLAC damping ring (measured) and the SPRC FEL storage ring (design).

	SLAC Damping Ring[17]	SPRC Ring[18]
Energy (GeV)	1.2	1.0
Circumference (m)	35	107
Bending Radius (m)	2.0	2.0
Momemtum Compaction	0.018	0.009
Damping Time (ms)	2.7	6.8
Beam Lifetime (hrs)	0.13	1.0
Energy Acceptance	±1%	±2.5%
Energy Spread (rms)	0.073%	0.06%
Emittance (m-rad)	$1.8 \times 10^{-8}\ \pi$	$1.7 \times 10^{-8}\ \pi$
Average Current (mA)	136	1000 (10 bunches)
Peak Current (A)	130	270
Brightness (A/cm^2)	3×10^7	9×10^7

Workers at Orsay, France, in collabortion with the Stanford group, successfully operated a (low gain) visible wavelength FEL oscillator in 1983 on the storage ring AC),[19], and are preparing a new series of FEL experiments at UV wavelengths on the storage ring Super-ACO.[20] The ACO storage ring provided the means to verify key theoretical issues of storage ring FEL operation, and demonstrated the possibility of short wavelength FEL's, but was limited by the short straight section and the relatively low beam brightness. The Stanford ring will provide the means to probe the physics of the high gain storage ring FEL, and demonstrate the feasiblity of very short wavelength operation. Experience gained with the SPRC ring will also provide a base of knowlege invaluable for the design and operation of the new national synchrotron radiation facility, the Advanced Light Source (ALS), proposed for construction at LBL.[21] The ALS ring is to operate at 1-2 GeV with an average current of 400 ma and beam quality comparable to the SPRC ring. In addition to 12 short undulator sections for the production of synchrotron radiation, the plan includes a 20 meter bypass for the eventual insertion of an FEL.

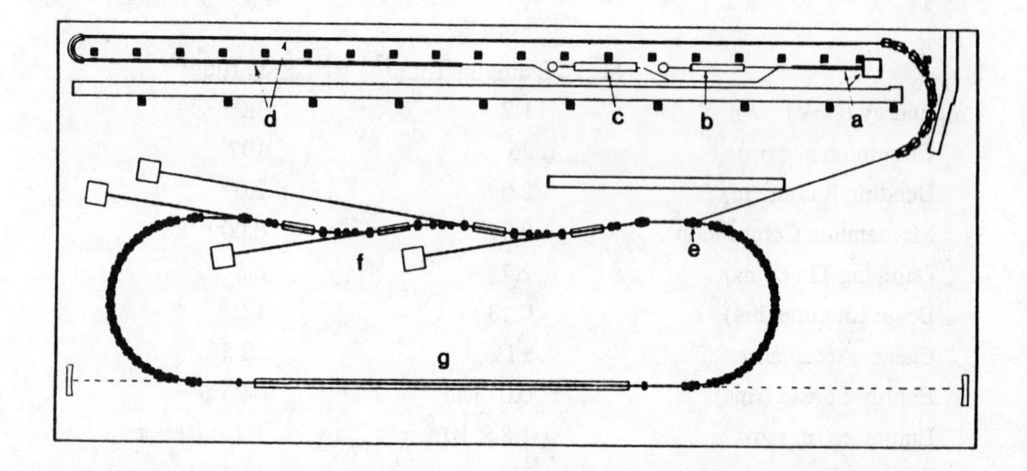

Fig. 3. Plan view of the SPRC storage ring facility under construction at Stanford Univesity. Overall dimensions are 65 × 29 m. a) Microwave electron gun and first accelerating section (45 MeV), b) Mark III infrared FEL, c) Rockwell infrared FEL, d) remainder of Mark III accelerating sections (total 1 GeV), e) ring injection point, f) four sychrotron radiation undulators, beam lines and experimental stations, g) XUV FEL wiggler and resonator.

WIGGLER MAGNET DESIGN

There are a multitude of options available in selecting a design for an FEL wiggler magnet. These include electromagnetic versus permanent magnets, helical versus linear polarization, and tapered versus constant parameter wigglers. For an XUV FEL the list rapidly shortens, and for a gap-tunable FEL the options narrow further.

The oscillation wavelength is related to other FEL parameters through the relation

$$\lambda = \frac{\lambda_w}{2\gamma^2}\left(1 + \frac{K^2}{2}\right) . \qquad (6)$$

With other parameters fixed, the oscillation wavelength can be tuned by varying K, which is proportional to the magnetic field on axis. By using a small wiggler period we can operate a short wavelength FEL with a relatively large value of K, thereby obtaining the highest gain possible. Since electromagnetic wigglers cannot achieve high fields and short periods simultaneously, we are led to permanent magnet structures.[22] Tunability in K is obtained by mounting the (linearly polarized) permanent magnet wiggler to movable jaws for gap scanning. A permanent magnet helical wiggler capable of being gap scanned would be difficult and expensive to construct.

Most permanent magnet wigglers constructed for FELs follow the designs of Halbach,[22] who employs Rare Earth Cobalt (REC) magnets in a variety of geometries. Pure REC magnets have the advantage of compactness and simplicity in construction, and allow linear superposition of external magnetic fields for guiding and focusing of the electron beam inside the wiggler. An effective alternative is to combine REC magnets with steel poles in a hybrid structure. Higher fields are possible in this configuration, and since steel is much easier to machine than the brittle REC components, shaping of higher order field components is fairly simple. In the hybrid design, half as many REC magnets are used for the same wiggler period.

In order to obtain sufficient single pass gain for oscillation in the XUV we must use very long wigglers. Multi-section wigglers containing many hundreds of periods will be needed in the 27 meter SPRC storage ring FEL. The wiggler field must be homogeneous to very high tolerances so that the cumulative trajectory and phase errors stay small. The electron trajectory deviations disrupt the overlap between the light beam and the electron beam, thereby lowering gain. In the high gain regime, where optical guiding is important, these deviations could also weaken self-focusing effects with further loss of gain. Phase errors (accumulated longitudinal displacements of the electrons) contribute to an effective energy spread, also lowering gain. The tolerances grow smaller as the number of periods is increased.

The measurements made on the one-meter Mark III (hybrid) wiggler at Stanford[23] illustrate these problems. In Fig. 4 we present a computer generated electron trajectory through the Mark III wiggler using measured field values. The electron path deviation is much less than the mode radius throughout the central portion of this 47 period wiggler, but the trajectory errors result in noticeable phase shifts at the beginning and end of the trajectory. This wiggler also shows variations in the trajectory as a function of the magnetic gap. As longer wigglers are constructed for the new high gain, short wavelength free electron lasers, we will need to improve our ability to compensate for these errors.

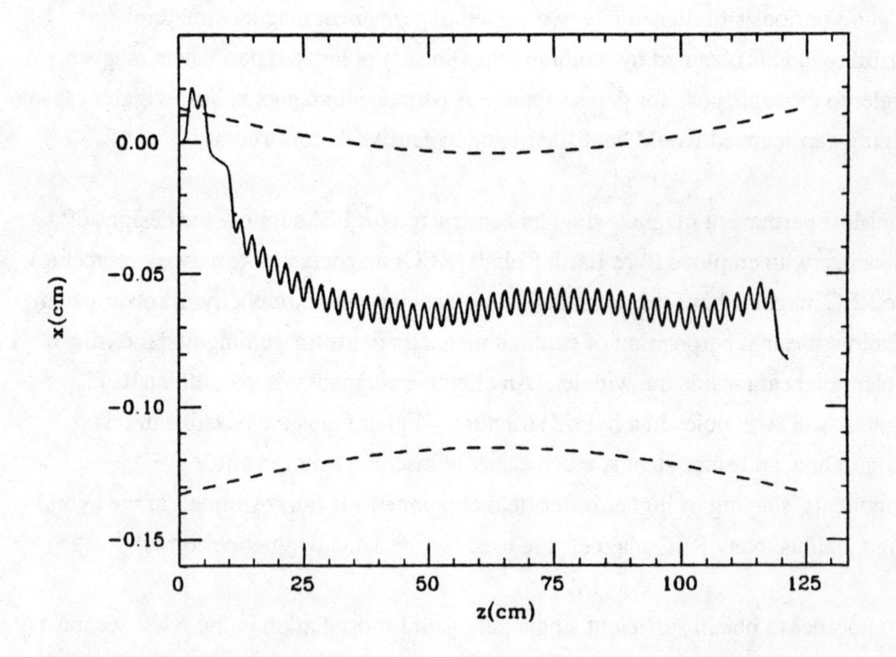

Fig. 4. Computed electron trajectory through the Stanford Mark III infrared FEL wiggler, gap adjusted for K=1.35.[23] The wiggler is 1.08 m in length, and contains 47 periods of hybrid $SmCo_5$/Vanadium Permendur magnets. Dashed lines represent the 1/e intensity waist of the 3.1 μm optical mode.

While it is not obvious, the central portion of the wiggler trajectory, shown in Fig. 4, may be good enough for operation in the XUV if appropriate matching between segments of the multi-section wiggler is provided. Although the trajectory was calculated for a 43 MeV electron beam and mode waist shown is that of a 3.1 μm optical beam, the scaling of the wavelength with energy is such that the curves would look identical (with a new vertical scale) for a 244 MeV electron beam and a 960 Å optical beam. The optical waist scales as the square root of the optical wavelength. From Eqn. (6) one can see that the wavelength scales as the inverse square of the energy. The size of the optical mode therefore scales inversely as the electron beam energy. The deviation of the trajectory from a straight line also scales inversely with energy. The relative size of the deviations and the optical mode are therefore invariant. One can conclude that the Mark III wiggler would be sufficient to keep the electron beam well within an optical mode with a Rayliegh range of 73 cm (as in the figure), which is shorter than will occur even in a gain-focused FEL.

The second issue which must be considered is the phase errors which accumulate between the electrons and the light. For a large K wiggler and constant relative field errors $\delta B/B$, the phase error grows as the 3/2 power of N_w.[24] The Mark III wiggler could not be extended to a single 250 period device without further work to reduce the cumulative field errors. This is the reason for our interest in a multiple section wiggler. If a means can be devised to adjust the electron phase between wiggler segments, the effect of the field errors would no longer grow as fast as the 3/2 power of N_w. We believe considerable progress can be made in the multi-component configuration, but the compensation of the structure as a function of gap has yet to be addressed.

XUV OPTICS

Earlier we made the point that an XUV FEL must operate in the high gain regime because of the poor quality of optical resonators at wavelengths below 1000 Å. In this section we consider more fully this issue and address related problems of mirror durability and resonator stability.

At wavelengths between 1000 Å and 500 Å the primary candidate for resonator mirrors is silicon carbide, which reflects about 50% of this light at normal incidence.[3] Below 500 Å there is currently no single material with comparable reflectivity. Recently Barbee[25] and others have produced multi-layer structures exhibiting reflectivities of 50-60% at 170 Å, and there is hope that greater than 10% reflectivity can be obtained at longer wavelengths. Newnam and co-workers[26] have proposed multi-faceted metal mirrors in a ring resonator configuration for achieving 50% reflectivity over a broad range of wavelengths in this difficult region of the spectrum. For all these approaches, the problems of oxidation and carbon contamination can severely reduce the performance. While the problem of poor reflectivity in the XUV is far from solved, real progress is being made.

An additional issue is that of thermal loading of resonator mirrors, which absorb large fractions of the incident light. The spontaneous radiation from a storage ring XUV FEL, such as the Stanford storage ring device, will illuminate a large portion of the downstream cavity mirror with as much as 180 Watts/cm^2 average intensity over an area 14 cm in diameter. Since much of this power is absorbed, rather advanced methods must be employed to dissipate the thermal energy. Use of cooled apertures inside the resonator cavity will reduce the total load on the mirror structures, but there is, of course, no way to shield the central portion of the mirror. The mirror itself must, therefore, be cooled efficiently without inducing excessive vibration. If necessary, the forward mirror can be moved somewhat further away from the wiggler to reduce the thermal load, which would require the use of an asymmetric cavity.

The lasing mode is much narrower than the spontaneous beam, and presents far more intense radiation at the mirror center. For a 1000 Å oscillator we estimate an intensity of 1.5 kW/cm^2 in a spot 1mm in diameter at the end of the wiggler. This diffraction limited beam will not diverge as rapidly as the spontaneous radiation, so the intensity reduction with distance is limited. Measures must be taken to diffuse this heat rapidly away form the laser spot into the bulk of the mirror, where it can be transferred to the coolant, so that the surface temperature does not become too high.

Thermal loading also has potential effects on the resonator stability. As power builds during turn-on, mirror figure variations will occur. At XUV wavelengths, a figure change of 50 Å can introduce large modal aberrations. Distortion of a few hundred angstroms over the scale of the mode size can easily double the effective radius of curvature of long cavity mirrors, which would have the effect of distorting the laser mode and reducing the round trip gain in the cavity, or even of driving the resonator unstable. Active temperature stabilization or stressing of mirrors are options in dealing with these problems.

PULSED AND CW OPERATION

A storage ring FEL can be operated in either of two extreme modes. In the pulsed mode, the laser is repeatedly Q-switched off to permit the ring to damp the FEL-induced energy spread. When switched back on after a few damping times, the laser power builds to saturation in only a few passes of the electron bunch through the resonator, resulting in very high peak power levels. In the CW mode, the laser operates continuously at lower peak power, with an electron beam whose equilibrium energy spread is determined by ring parameters, the FEL (power-dependent) excitation, and the total charge in the bunch. Both of these modes of operation have been demonstrated at Orsay.[20]

We have performed calculations,[27] with the assistance of the FRED simulation code, of peak and average laser power expected from the SPRC ring FEL operating at 960 Å. The results of these calculations for 10 circulating bunches are summarized below:

	CW Mode	Slow Pulsed Mode
Peak Power	2×10^3 Watts	1.3×10^7 Watts
Pulse Rate	2.8 MHz (natural)	50 Hz (Q-switched)
Average Power	5.6 Watts	0.67 Watts

The CW and pulsed modes described above are extremes in a continuum of operational modes available. The switching rate could, in priciple, be varied to produce a power and time structure suited to experimental needs.

CONCLUSIONS

Owing to the severe drop in mirror reflectivities at wavelengths below 1000 Å, operation of a free electron laser oscillator in the XUV range will likely occur in the high gain regime of the FEL interaction. In this regime interesting new physics will appear; optical guiding offers the prospect of extremely high gain, and creates new challenges to insure the stability of both the storage ring and resonator. The storage ring produces the brightest electron beam of existing technologies and, therefore, will support oscillation at the shortest wavelengths. High gain operation requires a wiggler hundreds of periods in length, and severe thermal loading on mirror surfaces requires placement of mirrors far away from the ends of the wiggler. The progress on these issues in FEL physics, wiggler design, and optics performance will ultimately determine how far into the XUV region it will be possible to operatie the free electron laser.

REFERENCES

1. D.A.G. Deacon, L. R. Elias, J. M. J. Madey, G. J. Ramain, H.A. Schwettman and T. I. Smith, Phys. Rev Lett. 38 (1977) 892.
2. Proceedings of the Workshop on Applications of Free Electron Lasers, Castelgondolfo (Rome), Italy, September 10-12, 1984. See Preface, p. vi.
3. D. T. Attwood, *et al*, Free Electron Generators of Extreme Ultraviolet Coherent Radiation, eds. J. M. J. Madey and C. Pellegrini, American Institute of Physics, Conf. Proc. Vol. 118 (1984) 294.
4. W. B. Colson, Phys. Lett. 64A (1977) 190.
5. T.I. Smith and J.M.J. Madey, J. Appl. Phys. B27 (1982) 195.
6. R. Bonifacio and C. Pellegrini, Optics Comm. 50 (1984) 373.
7. C. M. Tang and P. Sprangle, in: Physics of Quantum Electronics, vol 9. eds, S. F. Jacobs, G. T. Moore, H. S. Piloff, M. Sargent III, M. O. Scully and R. Spitzer (Addison-Wesley, Reasing, Mass, 1982), p.627.
8. G. T. Moore, Opt. Commun. 52 (1984) 46.
9. E. T. Scharlemann, A. M. Sessler and J. S. Wurtele, Phys. Rev. Lett. 54 (1985) 1925.
10. J. E. LaSala, D. A. G. Deacon, E. T. Scharlemann, "Optical Guiding Simulations for High Gain Short Wavelength FEL's", Proceedings of the Seventh International Free Electron Laser Conference, Granlibaaken, CA (1985).
11. W. M. Fawley, D. Prosnitz, E. T. Scharlemann, Phys. Rev. A30 (1984) 2472.
12. A. Gover, E. Jerby, J. LaSala, D. A. G. Deacon, "Feasibility of High Gain Free Electron X-ray Lasers", Proceedings of the Seventh International Free Electron Laser Conference, Granlibaaken, CA (1985).
13. M. Xie, D. A. G. Deacon, " Theoretical Study of FEL Active Guiding in the Small Signal Regime", Proceedings of the Seventh International Free Electron Laser Conference, Granlibaaken, CA (1985).

14. R. Pantell, "Some Concerns About Optical Guiding", Proceedings of the Seventh International Free Electron Laser Conference, Granlibaaken, CA (1985).

15. J. E. LaSala, D. A. G. Deacon, J. M. J. Madey, "Options for the Development of FEL Oscillators from 200 to 1000 Angstroms", International Conference on Insertion Devices for Sychrotron Sources, Proc. SPIE 582 (1985).

16. G. Dattoli, A. Renieri, Nuovo Cimento 59B (1980) 1.

17. G. E. Fischer, W. Davies-White, T. Fieguth, H. Wiedemann, 12th International Conference on High Energy Accelerators, FNRL (1983).

18a. H. Wiedemann, Proceedings of the International Free Electron Laser Conference, Bendor, France (1982) 201.

18b. H. Wiedemann, HEPL Technical Note (1982), Stanford University, CA 94305.

19. C. Bazin, M. Billardon, D. A. G. Deacon, P. Elleaume, Y. Farge, J. M. J. Madey, J. M. Ortega, Y. Petroff, K. E. Robinson, M. Velge; "Results of the First Phase of the ACO Storage Ring Laser Experiment", in Physics of Quantum Electronics, vol 8. eds, S. F. Jacobs, G. T. Moore, H. S. Piloff, M. Sargent III, M. O. Scully and R. Spitzer (Addison-Wesley, Reading, Mass, 1982), p.89.

20. M. Billardon, P. Elleaume, J. M. Ortega, C. Bazin, M. Bergher, M. Velghe, D. A. G. Deacon, Y. Petroff, IEEE J. Quantum Electron. QE-21 (1985) 805.

21. B. Crasemann, W. D. Grobman, D. T. Atwood, "Report of the Workshop on an Advanced Soft X-Ray and Ultraviolet Synchrotron Source:, Applications to Science and Technology", Berkeley, California (November 13-15, 1985).

22. K. Halbach, Proceedings of the International Free Electron Laser Conference, Bendor, France (1982) 211.

23. G. Stolovy, W. Wadensweiler, J. M. J. Madey, S. Benson, " High Quality Hybrid Wiggler for Infrared FEL and Coherent Harmonic Generation", International Conference on Insertion Devices for Sychrotron Sources, Proc. SPIE 582 (1985).

24. B. M. Kincaid, "Laser Harmonic Generation Using the Optical Klystron, and the Effects of Wiggler Errors", Proceedings of the Seventh International Free Electron Laser Conference, Granlibaaken, CA (1985).

25a. T. W. Barbee Jr., Free Electron Generators of Extreme Ultraviolet Coherent Radiation, eds. J. M. J. Madey and C. Pellegrini, American Institute of Physics, Conf. Proc. Vol. 118 (1984) 53.

25b. T. W. Barbee Jr., S. Mrowka, M. J. Hettrick, Appl. Optics 24 (1985) 883.

25c. T. W. Barbee, Jr., "Multilayer Optics for Short Wavelength Lasers", these Proceedings.

26. B. E. Newnam, M. L. Scott, P. N. Arendt, "Multifacet Metal Mirror Design for Extreme-Ultraviolet Applications", these Proceedings.

27. J. E. LaSala, D. A. G. Deacon, J. M. J. Madey, "Performance of an XUV FEL Oscillator on the Stanford Storage Ring", Proceedings of the Seventh International Free Electron Laser Conference, Granlibaaken, CA (1985).

HIGH-GAIN FREE ELECTRON LASERS AS GENERATORS OF SHORT WAVELENGTH COHERENT RADIATION*

Kwang-Je Kim
Lawrence Berkeley Laboratory
University of California,
Berkeley, California 94720

Claudio Pellegrini
Brookhaven National Laboratory
Upton, New York 11973

ABSTRACT

The development of coherent radiation in high-gain free electron lasers, either from initial noise or from low-power input radiation, is analyzed in terms of three-dimensional Maxwell-Klimontovich equations. Exponential growth and saturation, transverse radiation profiles, transverse coherence and spectral features are discussed. Two possible systems of high-gain free electron lasers, one based on a storage ring and by-pass, another based on a linac and damping ring, are considered for the generation of 400 Å radiation.

I. INTRODUCTION

High-gain free electron lasers (FELs) are potentially important as generators of intense, coherent radiation at wavelengths shorter than 1000 Å,[1] a region difficult for atomic lasers. Two modes of operations are possible: In the amplifier mode, coherent input radiation, if available at the desired wavelengths, is amplified to high intensity. In the so-called self-amplified spontaneous emission (SASE) mode, the noise present in the beam is self-amplified to produce high-power coherent radiation. The high-gain FELs thus eliminate the need for the optical cavities required by FEL oscillators. High-gain FELs will deliver higher peak power with lower repetition rate and thus complementary to FEL oscillators. At microwave wavelengths, the principle of the high-gain FELs has been tested experimentally at Lawrence Livermore National Laboratory.[2] In this paper, we discuss general characteristics of high-gain FELs, in particular the properties of SASE. We also evaluate possible high-gain FEL systems as sources of short wavelength coherent radiation.

---

*This work was supported by the Director, Office of Energy Research, Office of High Energy and Nuclear Physics, High Energy Physics Division, U.S. Department of Energy under Contract No. DE-AC03-76SF00098, and, in part, under Brookhaven National Laboratory DOE Contract No. DE-AC02-76CH00016.

The noise in the electron current is due to the discreteness of electrons. The radiation generated by the noise current in undulators, familiar as the undulator radiation in synchrotron radiation research,[3] is analogous to the spontaneous emission in atomic systems. The name SASE derives from this analogy. The characteristics of the undulator radiation is well understood. Due to the self-reinforcing nature of the amplification process, the characteristics of SASE are expected to differ from those of undulator radiation.

In section (II), we give an outline of a consistent three-dimensional analysis of the high-gain process. The characteristics of high-gain FELs derived from this analysis are summarized in the following two sections, in section (III) transverse characteristics and intensity characteristics in section (IV). Section (V) discusses a high-gain FEL for 400 Å based on an electron storage ring and bypass. In section (VI), an alternative system based on a linac and a damping ring is considered.

II. METHOD OF ANALYSIS

The high-gain behavior of FEL amplifier in one-dimensional theory is well-known.[4] Important aspects of SASE have been derived from intuitive analysis.[1] Recently, a one-dimensional theory of SASE was developed based on the coupled Maxwell-Klimontovich equations.[5] The theory was further extended to include three-dimensional effects,[6] which are important for a correct understanding of high-gain properties. Here we give an outline of the analysis given in Ref. (6).

We choose z, the distance from the undulator entrance, as the independent variable. The transverse coordinates are given by a two-dimensional vector $\mathbf{x}$. The dynamical variables desçribing the electron motion are the phase θ and the relative energy deviation η. Here θ is roughly the electron coordinates with respect to beam center in unit of $\lambda_1/2\pi$ where λ_1 is the radiation wavelength. These variables satisfy the well-known pendulum equations.[7]

To properly analyze SASE, it is important to account for the discreteness of the electrons. This is achieved by utilizing the Klimontovich distribution function, $\hat{f}$,[8] which is a sum of two parts, the smooth background $\bar{f}$ and the part containing the high frequency fluctuations and modulations, $\tilde{f}$. The continuity equation for $\hat{f}$ can likewise be separated into two equations: An equation describing the response of the high-frequency part $\tilde{f}$ to the radiation field, and an equation describing the slow, nonlinear evolution of the background distribution $\bar{f}$.

The radiation field is represented by a complex amplitude $a(\nu,\mathbf{x};z)$, which is the slowly varying part of the full amplitude. is the normalized frequency ω/ω_1, where ω_1 is the resonance frequency given by $2ck_u\gamma_o{}^2/(1 + K^2/2)$. Here γ_o is the average beam energy in units of mc^2 (m = electron mass, c = velocity of light), $k_u = 2\pi/\lambda_u$, λ_u is the undulator period length, and K is the magnetic deflection parameter.[3] The wavelength and wave number corresponding to the frequency ω_1 are denoted by λ_1 and k_1, respectively. It is a good approximation to assume that ν is close to an odd integer, which in the following is taken to be $2\ell + 1$. Thus, $\Delta\nu = \nu - 2\ell - 1$ is a small quantity. The amplitude satisfies the three-dimensional Maxwell's equation, in which the source term is specified by $\bar{f}$.

The continuity equation for $\bar{f}$ and the Maxwell's equation form a coupled set. Replacing the slowly varying function $\bar{f}$ by its initial value at $z = 0$, these equations become linear, and can be solved explicitly by the method developed by Van Kampen.[9] The result for the radiation amplitude in the high-gain limit is

$$a(\nu, \mathbf{x}; z) \sim \frac{A(\mathbf{x})\, e^{-2i\mu k_u \rho z}}{\int d^2y\, A(\mathbf{y})\, (1 + U(\mathbf{y})\, dZ(\mu)/d\mu)}$$

$$\times \left[\int d^2y\, A(\mathbf{y})\, a(\nu, \mathbf{y}; 0) - i\, \kappa \sum_i \frac{e^{i\theta_i \nu} A(\mathbf{x}_i)}{\mu + \nu\eta_i/\rho} \right] . \tag{1}$$

Here, the summation over i is over all electrons, $\kappa = e K_\ell / 8\sqrt{2\pi}\, \gamma_o\, \epsilon_o\, c\, k_u\, \rho$, ϵ_o = vacuum dielectric consant, $K_\ell = (-1)^\ell K\, [JJ]$, $[JJ]$ is a shorthand expression involving the difference of two Bessel's functions,

$$Z(\mu) = \rho \int d\eta\, \frac{dV}{d\eta}\, \frac{1}{\mu + \nu\eta/\rho} \tag{2}$$

The mode function A and the complex number μ are, respectively, the eigenfunction and the eigenvalue of

$$\left[\mu + \frac{\Delta\nu}{2\rho} - \frac{1}{2\rho k_u k_1 \nu} \frac{\partial}{\partial \mathbf{x}^2} + U(\mathbf{x})\, Z(\mu) \right] A(\mathbf{x}) = 0 \quad . \tag{3}$$

The functions V and U are, respectively, the initial electron distribution in momentum and in transverse coordinates $\mathbf{x}$ (normalization: $\int d\eta V(\eta) = 1$, $U(0) = 1$). The dimensionless parameter ρ in the above plays an important role in characterizing the high-gain behavior and is defined by[1]

$$\rho = \left[\frac{n\, e^2\, K_\ell^2}{32 \sigma_A\, \gamma_o^3\, k_u^2\, mc^2\, \epsilon_o} \right]^{1/3} . \tag{4}$$

Here n is the line density of electrons, σ_A is the cross-sectional area of the electron beam, $\sigma_A = \int d^2x\, U(\mathbf{x})$. Equation (3) is essentially that derived and studied earlier by Moore,[10] except that the effects of momentum spread are included. In one dimension it becomes the dispersion relation studied by several authors.[4] There are, in general, a discrete set of complex eigenvalues, as well as a continuum of real ones. However, the behavior in the high gain limit is governed by eigenvalue μ with the largest positive imaginary part. In Eq. (1), the term containing the input amplitude a(;0) describes amplifier mode and represents the solution of the initial-value problem in three-dimensional FEL amplification.[10] The term containing the initial electron phases $e^{i\theta_{i\nu}}$ gives gives SASE.

The above results are valid when all electron trajectories are parallel. Taking into account electrons' angular spread, the eigenvalue equation becomes more complicated.[6] Although the new eigenvalue equation has not been studied in detail, the basic conclusions of this paper, which depend mainly on the fact that there exists an eigenvalue with a positive imaginary part and the associated mode function, will not change upon this generalization.

III. TRANSVERSE CHARACTERISTICS

The transverse characteristics of radiation from a high-gain FEL is determined from the structure of Eq. (1). Since the radiation amplitude is specified completely by a single mode function $A(\mathbf{x})$, it follows that the radiation in high gain FEL is guided, as discussed recently.[10,11] The guiding phenomena is important in maintaining high-gain amplification in long undulators; otherwise, the spread of radiation by diffraction would decrease the gain.

It also follows from the explicit form of the solution that the radiation is fully coherent transversely, both in the SASE mode and in the amplifier mode. This result is somewhat surprising for SASE, in comparison with the coherence property of undulator radiation. The latter is, in general, partially coherent transversely, the degree of coherence being determined by the ratio of electron beam phase space area, called the emittance, to the radiation phase-space area.[12]

The optimum amplifier configuration is obtained when the input radiation amplitude is the complex conjugate of the output mode function $A(\mathbf{x})$. This means in particular that the curvature of the input phase front is of the same magnitude as that of the output but of opposite sign.[13] The symmetry in the phase front geometry is somewhat surprising in view of the extreme asymmetry in the intensity levels.

IV. INTENSITY CHARACTERISTICS

The power in the radiation is proportional to the ensemble average of $|a|^2$. The interference between the terms representing SASE and the radiation produced in the amplifier mode clearly vanishes, and the ensemble average of the SASE term can readily be performed assuming that electrons are not correlated initially. The intensity growth and the spectral characteristics are determined mainly by the imaginary part μ_I of μ. In this way, one obtains the power spectrum,

$$\frac{dP}{d\omega} = e^{\tau}\ S(\Delta\omega/\omega_m) \left[g_A \left(\frac{dP}{d\omega}\right)_o + g_s \frac{\rho E_o}{2\pi} \right] , \qquad (5)$$

where $\tau = 8\ \pi\mu_I{}^m\rho N$, μ_I^m = the maximum the value of μ_I as a function of $\Delta\nu$, $\Delta\omega = \omega - \omega_m$, ω_m = the frequency at maximum growth, $S(x) = \exp(-x^2/2\sigma_\Delta{}^2)$, and g_A and g_S are quantities of order unity. The first term in Eq. (5) gives the power spectrum for an FEL operating in the amplifier mode, and one finds the growth of the input power spectrum

$(dP/d\omega)_o$ to be exponential. The power spectrum for SASE is given by the second term, which exhibits the same exponential growth, with the input replaced by the effective noise power spectrum

$$\left(\frac{dP}{d\omega}\right)_{0,SASE} = \frac{\rho E_o}{2\pi} . \tag{6}$$

where E_o is the average beam energy. The function S describes the frequency dependence of the gain for coherent amplification, as well as the spectral shape of the SASE radiation. In one dimension, for zero momentum spread, one obtains $g_A = g_S = 1/9$ and the bandwidth

$$\sigma_\Delta = (9\rho/2\pi \sqrt{3} N)^{1/2} . \tag{7}$$

For momentum spread much larger than ρ, the eigenvalue μ is real and there is no exponential growth. Therefore, for exponential growth to occur,

$$\sigma_\eta \le \rho , \tag{8}$$

where σ_η is the relative momentum spread in the electron beam. The total SASE power, obtained by integrating over the frequency, is

$$P_{SASE} = \rho P_{beam} g_S e^\tau / N_{\ell c} \tag{9}$$

where P_{beam} is the kinetic power in the beam (equal to $E_o I/e$, where I = beam current) and $N_{\ell c} = n\lambda_1 (2\pi)^{-1/2}/\sigma_\Delta$ is the number of electrons in one coherence length.

The slow variation of $\bar{f}$ with respect to z is determined by a procedure known as the quasi-linear approximation in plasma physics.[14] From the resulting nonlinear Fokker-Planck equation, one finds that the average value of η must decrease so as to conserve the total energy of the radiation-beam system. In addition the rms spread of η, σ_η, is found to increase as $\sigma_\eta^2 \approx \rho^2 g_s e^\tau/N_{\ell c}$. Since the growth rate becomes negligible when $\sigma_\eta >> \rho$, the exponential growth will stop when the factor $g_S e^\tau$ becomes about $N_{\ell c}$.[15] In view of Eq. (9) the power at saturation becomes.

$$P_{sat} \sim \rho P_{beam} \tag{10}$$

The saturation occurs when $\rho N_1 \approx \ln(N_{\ell c}/g_s)/8\pi\mu_I^m$. For parameters considered here this becomes[1]

$$N \sim 1/\rho \tag{11}$$

In view of Eq. (7), the bandwidth at saturation is $\omega/\Delta\omega \sim N$, which is the same as the bandwidth of the spontaneous radiation from an undulator with the same N. The crucial importance of the parameter ρ in determining high-gain behavior should be clear by now: It enters into the growth rate [Eq. (5)], the condition for efficiency [Eq. (8)], the effective noise power [Eq. (6)], the bandwidth [Eq. (7)], the saturation condition [Eq. (11)], and the saturated power [Eq. (10)].

Figure 1 shows data from the microwave FEL experiment at Lawrence Livermore National Laboratory.[2] For this experiment, the one-dimensional theory is appropriate, and the growth rate is calculated to be 42.1 dB/m. The observed growth rate was 35 dB/m; the discrepancy may be due to space-charge effects. Taking the observed growth rate and computing the coefficient in Eq. (9), one obtains the dotted line in Fig. 1. The agreement is encouraging.

V. A HIGH-GAIN FEL ON A BY-PASS OF A STORAGE RING

Modern storage rings provide the high-density electron beams required for high-gain FEL operation. For efficient interaction, the undulator must be long and have a narrow gap. Such a device, if placed in a normal section of a storage ring, would severely limit the acceptance of the ring and thus reduce the beam lifetime due to scattering with the residual gas. Moreover, the interaction of the beam with an FEL undulator is disruptive to the beam itself in terms of energy loss and increased momentum spread. To avoid these problems, the undulator can be placed in a special by-pass section,[16] as shown schematically in Fig. 2. The electron beam normally circulates in the storage ring without passing through the FEL undulator. Once every damping time, the beam is directed into the by-pass section, where the interaction with the undulator takes place, generating intense, coherent radiation. As the beam leaves the undulator, it is deflected back into the storage ring, where synchrotron radiation damping reduces the induced energy spread. After one damping time, the beam is ready to be injected into the by-pass again.

To relate the parameter ρ to the electron beam and storage ring parameters, we assume that the electron beams in the undulator are focused with the effective β nction given by

$$\beta_x = \beta_y = \frac{\lambda_u \gamma_o}{\pi K} \tag{12}$$

Equation (4) can then be written as follows:[17]

$$\rho = \left(\frac{1}{16\pi} \frac{r_e}{ec} \frac{K^3 [JJ]^2}{2(1 + K^2/2)} \frac{\lambda_1}{\gamma_o^2} \frac{I_p}{\sqrt{\epsilon_x \epsilon_y}} \right)^{1/3} . \tag{13}$$

Here, we assume operation at the fundamental harmonic, $\ell = 1$. In the above, r_e is the classical electron radius, I_p is the peak current, and ϵ_x and ϵ_y are, respectively, the electron emittances in the horizontal and the vertical directions.

Equation (13) gives the general criteria for design optimization. The peak current should be large, the emittances small. Low electron energy is preferable for a given wavelength λ_1. Also, a large K is preferred, leading to an undulator of small magnet gap.

A feasibility study of a storage ring for a high gain FEL operation at 400 Å was recently carried out at Lawrence Berkeley Laboratory.[18] Important aspects of storage ring issues, such as collective

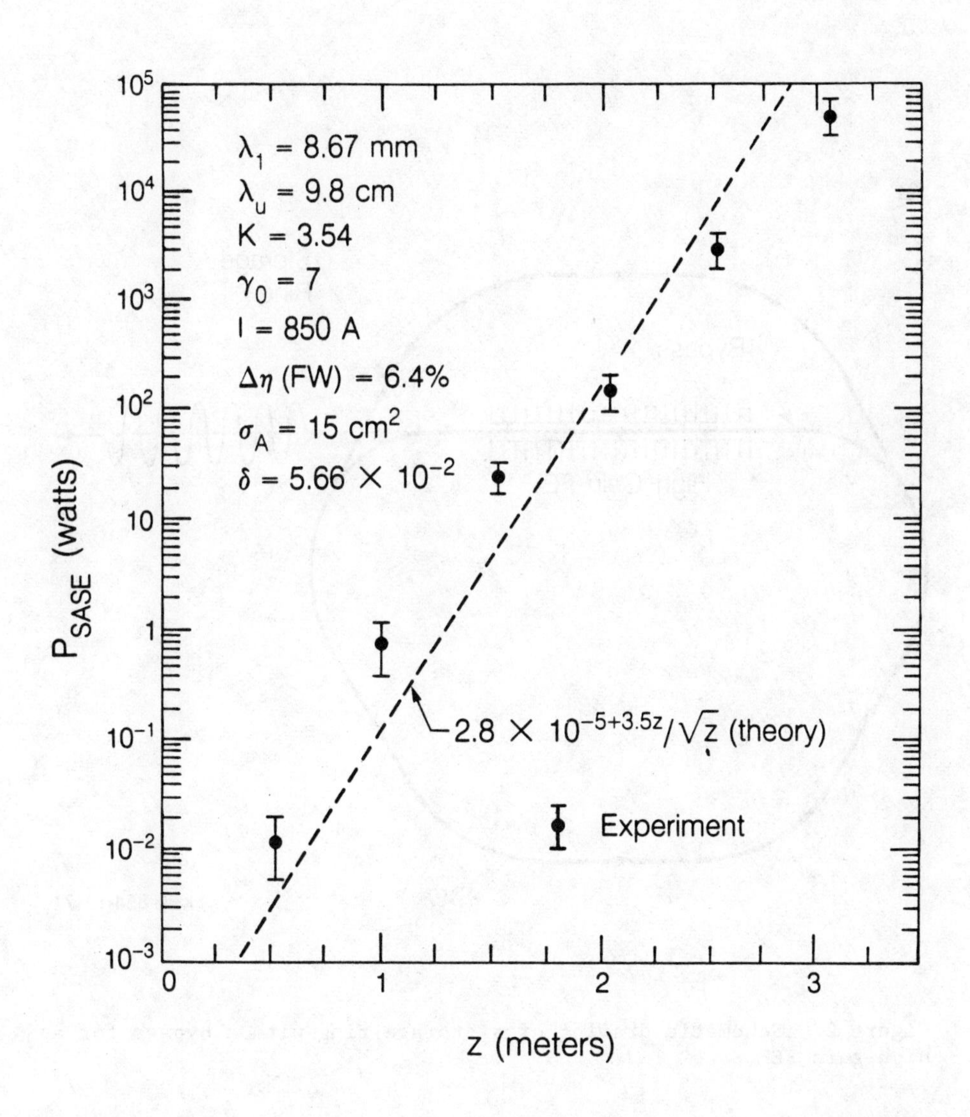

Figure 1. Data from Ref. 2 (courtesy of T. Orzechowski) compared with the prediction.

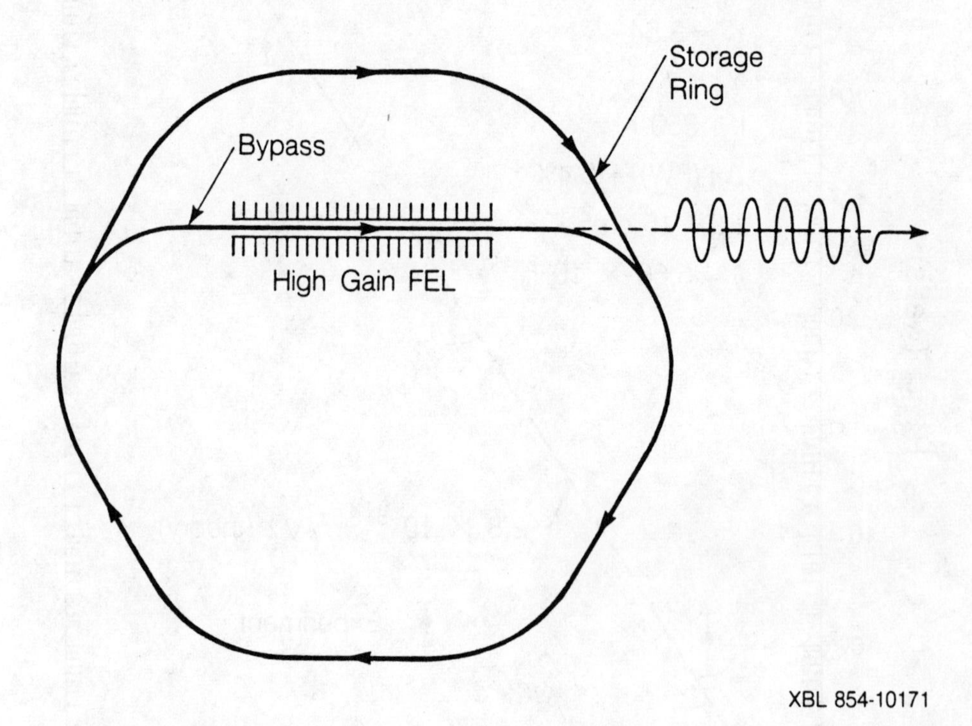

Figure 2. Schematic drawing of a storage ring with a bypass for a high-gain FEL.

instabilities and lattice optimization, by-pass considerations, and operational requirements were studied. The general conclusions were that the high beam quality demanded for these high-gain FEL can be achieved with presently available accelerator technology. Among five different storage rings considered, one with a circumference of 144 m, based on the combined-function lattice, looks most promising. Running at 750 MeV, the horizontal emittance is 1×10^{-8} m-rad, with a peak current of 200 A and a momentum spread of 0.2%. The undulator in the by-pass is of the permanent magnet-steel hybrid type,[19] 20 m in length, and has a period of 2.29 cm with a 3-mm magnet gap. Such a system will produce tens of megawatts of 400 Å radiation, fully coherent transversely, with about a 0.1% relative bandwidth. The radiation is generated in pulses about 100 ps in duration with a repetition rate of about 20 Hz.

VI. HIGH-GAIN FELS BASED ON LINACS

Linacs are more compact than storage rings. However, the emittances and energy spread of electron beams in linacs tend to be larger than those in storage rings. The drawback is removed if the electron beams from a linac are routed to a damping ring, which is basically a compact storage ring whose sole function is to decrease the beam emittances and the energy spread by radiation damping. A system consisting of a linac, a damping ring, and a long undulator offers another approach to a high-gain FEL. Such a system is currently being studied at Brookhaven National Laboratory. A 750-MeV linac has a total length about 50 to 60 m, which is also about the circumference of the damping ring.

REFERENCES

1. B. Bonifacio, N. Narducci, and C. Pellegrini, Opt. Comm. 50, 373 (1984).

2. T.J. Orzechowski et al., Phys. Rev. Lett. 54, 889 (1985).

3. See for example, S. Krinsky, IEEE Trans. Nucl. Sci., NS-30, 3078 (1983).

4. N.M. Kroll and W.A. McMullin, Phys. Rev. A17, 300 (1978); A. Gover, and Z. Livini, Opt. Comm. 26, 375 (1978).

5. K.-J. Kim, Proc. 7th Int. FEL Conf., Lake Tahoe, CA. Sept. 1985. See also J.M. Wang and L.H. Yu, ibid.

6. K.-J. Kim, Three dimensional analysis of coherent amplification and Self-Amplified Spontaneous Emission in Free Electron Lasers, LBL-21426 (April 1986), submitted to Phys. Rev. Lett. SASE for electron beams of infinite transverse extent was recently studied by L.H. Yu and S. Krinsky, these proceedings.

7. For a review see, W.B. Colson, IEEE J. Quant. Elect. QE17, 1417 (1981).

8. Y.L. Klimontovich, Soviet Phys. JETP 6, 753 (1958). See also S. Ichimaru, Basic Principle of Plasma Physics (W.A. Bejamin, Inc., 1973). Appendix A.

9. N.G. Van Kampen, Physica 21, 949 (1951); K.M. Case, Ann. Phys. 7, 349 (1959).

10. G.T. Moore, Opt. Comm. 54, 121 (1985).

11. E.T. Scharlemann, A.M. Sessler, J.S. Wurtele, Phys. Rev. Lett. 54, 1925 (1985).

12. K.J. Kim, preprint LBL-20181 (July, 1985), to be published in Nucl. Instr. Method.

13. This was independently noticed by G.T. Moore, University of New Mexico preprint.

14. See for example, G. Schmidt, Physics of High Temperature Plasma (Academic Press, New York, 1979), section 9-2. The method was applied to 1-D FEL amplifier problem by T. Taguchi, K. Mima and T. Mochizuki, Phys. Rev. Lett., 46, 824 (1981).

15. This result was first derived by S. Krinsky using the fluctuation dissipation theorem (BNL-preprint).

16. J. Murphy and C. Pellegrini, J. Opt. Soc. Am. 1B, 530 (1984).

17. K-J. Kim, J.J. Bisognano, A.A. Garren, K. Halbach, and J.M. Peterson, Nucl. Instr. Methods, A239, 54 (1985).

18. J.J. Bisognano et al., preprint LBL-19771 (March, 1985), submitted to Part. Acc.

19. K. Halbach, Journal de Physique Colloque, C1, 44, 211 (1983).

Low Relativistic Electron Beams Traversing a Superlattice as a X-ray Source

C. T. Law and A. E. Kaplan
School of Electrical Engineering
Purdue University, West Lafayette, IN 47907

Abstract

Soft X-ray can be generated by low relativistic electron beams traversing through a medium with short spatial period. The amount of radiation generated is reduced owing to photon absorption and electron scattering. We investigate the influence of photon absorption and electron scattering on the amount of radiation, the critical electron energy and the total length of the superlattice. The results indicate that a heavy medium may be a better radiator for our system.

Introduction

When fast electrons pass through a periodic medium, electromagnetic wave is generated (resonant transition radiation) [1-3]. The resonant angle ϑ_n with respect to the electron beam and wavelength λ for this radiation is related by so called resonant condition

$$\sqrt{\bar{\varepsilon}}\cos\vartheta_n = c/v - n\lambda/l \tag{1}$$

where $\bar{\varepsilon}$ is the average dielectric constant, v is the velocity of electron, n is an integer and l is the spatial period of interfaces. This is a condition of constructive interference of electromagnetic field at a distance point. Usually $l \gg \lambda$, so that ultra-relativistic electron are required to satisfy Eqn (1) and to emit radiation. With the progress in technology, we can construct superlattice with short spatial period l. It was shown previously that a system composed of thin layers ($l \leq 500A^o$) enables one to reduce the critical energy for satisfying Eqn (1) to a few tens Kev or lower [4,5] The fortunate factor is that the layer structures with $l < 100A^o$ and with broad spectrum of materials are well known and are widely explored as X-ray mirrors for the same range of frequency [6,7]. In the proposal [4,5], the electron beam is supposed to pass through such a solid-state structure. However, photon absorption and electron scattering prevent one from using too low electron energy. It turns out that a slightly increase of the energy of electron beam overcomes this problem while this energy is still low. In this paper, taking photon absorption and scattering electrons into consideration, we obtain the optimal range of electron energy E_0 (which appears to be between a few hundred Kev and a few Mev), the total thickness L of the superlattice and maximal radiation for a given frequency of radiation ω.

In this paper, we show that usually the effect of photon absorption is dominant over electron scattering. Indeed, the photon absorption length [8] defined as $1/\bar{\mu}$ is of the order of micron at a wavelength about 10 A^o where $\bar{\mu}$ is the average photon absorption coefficient. On the other hand, the critical length for scattering electron [9] defined as $E_o\overline{(dE/dx)}^{-1}$ (where $\overline{dE/dx}$ is average energy loss per unit path length) is usually of the order of millimeter for electron energy E_o of 1 Mev. Permittivity of the material also strongly affect the optimal radiation and the shape of the radiation spectrum. Usually [10,11,13], a light medium with small atomic number is used as a radiator and air is employed as a spacer in a system utilizing ultra-relativistic electrons. However, a heavy medium with atomic number may be an appropriate candidate as a radiator (and a light medium as a spacer) for producing strong radiation with narrow spectral width in our system since we have to use a solid-state structure.

Emission in a Periodic Absorbing Media

The differential cross section for transition radiation including photon absorption in a multilayer system can be expressed in the following form [11,12,13]

$$\frac{d^2N}{d\Omega d\omega} = F_1 F_2 F_3 \tag{2}$$

where F_1 is the differential cross section for one single interface, F_2 denotes the coherence interference of radiation in a single plate and F_3 represents the coherence summation of radiation from each layer (definition of F_1, F_2 and F_3 is given below). The unit of the differential cross section is number of photon per unit solid angle in steradian per unit photon energy in ev per electron. One of the characteristics of transition radiation is the requirement of minimum distance for significant generation of radiation. The distance is known as formation length [3,13]

$$Z_i = \pi\lambda(1/\beta - \sqrt{\varepsilon_i - \sin\vartheta})^{-1} \tag{3}$$

where $Z_i (i=1,2)$ are formation length of the two media, $\varepsilon_i (i=1,2)$ are the permittivity of the two media, $\beta = v/c$, v is the speed of the electron, c is the speed of light and ϑ is the angle of emission with respect to the path of electron. Usually [2,3,10,11,13] the permittivity is calculated with the plasma frequencies $\omega_i (i=1,2)$ for the two media

$$\varepsilon_i = 1 - (\omega_i/\omega)^2 \tag{4}$$

where plasma frequencies $\omega_i (i=1,2) = 2c\sqrt{\pi r_o N_i Z_{a_i}}$, classical electron radius $r_o = 2.82\times10^{-13}$ cm, $N_i (i=1,2)$ are the number of atom per cm^3 in each medium and $Z_{a_i} (i=1,2)$ are atomic numbers (for sufficiently high frequency). However, ω_i cannot be used in Eqn (4) for most of the material in the soft X-ray photon energy (0.1 - 2 Kev) [8]. Instead of plasma frequency ω_i in Eqn (4), we must use parameters ω_{a_i} with tabulated data [8] to calculate permittivity where $\omega_{a_i} (i=1,2) = 2c\sqrt{\pi r_o N_i f_{R_i}}$ and $f_{R_i} (i=1,2)$ are the real part of atomic scattering factors f_i [8]. This modification affects the spectral shape. This point will be emphasized later.

The differential cross section for single interface is given by [3]

$$F_1 = \frac{\alpha(\Delta\varepsilon\,\beta)^2}{\pi^2\omega}\,|\,G(\beta,\varepsilon_1,\varepsilon_2,\vartheta)\,|^2 \tag{5}$$

where $\Delta\varepsilon=\varepsilon_1-\varepsilon_2$, fine structure constant $\alpha=1/137$ and

$$G(\beta,\varepsilon_1,\varepsilon_2,\vartheta) = \frac{\sin\vartheta\cos\vartheta(1-\beta^2\varepsilon_2-\beta\sqrt{\varepsilon_1-\varepsilon_2\sin^2\vartheta})}{(1-\beta\varepsilon_2^2\cos^2\vartheta)(1-\beta\sqrt{\varepsilon_1-\varepsilon_2\sin^2\vartheta})(\varepsilon_1\cos\vartheta+\sqrt{\varepsilon_1\varepsilon_2-\varepsilon_1^2\sin^2\vartheta})}$$

According to the above, formula the differential cross section F_1 is proportional to $(\beta\Delta\varepsilon)^2$ and depends on angle of emission in a complicated fashion. A simple picture of transition radiation ensues when speed of electron and frequency of radiation are low so that one of the media behaves like metal. Under these conditions, we have dipole radiation pattern. As electron energy increase, the radiation concentrates in a narrow cone.

F_2 accounting for coherent summation of radiation in two neighboring interfaces assumes the interference pattern of two sources

$$F_2 = 4\sin^2(l_2/\,Z_2) \tag{6}$$

where l_2 is the thickness of the denser medium. According to this expression, radiation in a single plate is four times as large as that of single interference when interference is constructive.

For a single electron traversing $2M$ layers (M layer of medium 1 and M layers of medium 2), the factor for coherent summation from each layer with photon absorption is [11,12]

$$F_3 = \frac{1-e^{-\bar{\mu}L}}{\bar{\mu}} \tag{7}$$

where the average X-ray absorption coefficient $\bar{\mu}=(\mu_1 l_1+\mu_2 l_2)/\,l$, the spatial period $l=l_1+l_2$, $l_i\,(i=1,2)$ are the thickness of layers for the two media and total length $L=Ml$. For small absorption, F_3 is proportional to total length L or M.

With assumptions that the interaction between electrons inside the beam is negligible, the average energy loss of electrons is owing to the decrease in number of electrons with distance travelled and M is sufficiently large, we obtain the formula for the number of photon generated per unit photon energy in ev per electron for a small angle $\Delta\vartheta$ about the resonant angle ϑ_n, Eqn (1), as

$$\frac{dN}{d\omega} = \begin{cases} (dN/\,d\omega)_{cr}((1-e^{-\bar{\mu}L})/\,\Gamma-(\bar{\mu}L+e^{-\bar{\mu}L}-1)/\,\Gamma^2) & \text{for } L \le L_{cr} \\ (dN/\,d\omega)_{cr}(e^{\Gamma}-1-\Gamma)e^{-\bar{\mu}L}/\,\Gamma^2 & \text{for} L > L_{cr} \end{cases} \tag{8}$$

where $(dN/\,d\omega)_{cr} = \dfrac{16\alpha(\Delta\varepsilon\,\beta)^2}{\pi\omega}\,|\,G(\beta,\varepsilon_1,\varepsilon_2,\vartheta_n)\,|^2\,\dfrac{L_{cr}}{l}\sin^2(l_2/\,Z_2)$, critical length for electron energy loss $L_{cr}=E_o(\overline{dE/dx})^{-1}$, angular width of the radiation $\Delta\vartheta=1/M$, average electron energy loss per unit path length $\overline{dE/dx}=((dE/dx)_1 l_1+(dE/dx)_2 l_2)/l$ and $\Gamma=\bar{\mu}L$. There are numerous ways to calculate $\overline{dE/dx}$ [14]. For simplicity, we use the collision loss formula [9]

$$(dE/dx)_i=\frac{2\pi N_A r_o^2 m_o c^2 Z_{a_i}}{\beta^2 A_i}\rho_i[\ln((\gamma-1)^2(\gamma+1)m_o^2c^4/(2I_i^2)) + ((\gamma-1)^2/8+(2(\gamma-1)+1))\ln 2/\gamma^2+1-\beta^2] \quad (9)$$

where m_o is electron mass, ionization potential of the atom $I_i(i=1,2)=9.73Z_{a_i}+58.8Z_{a_i}^{-0.79}$ ev (for atomic number $Z_{a_i}>13$), $\rho_i(i=1,2)$ are densities of the media, $A_i\ (i=1,2)$ are atomic mass and Avogadro number $N_A=6.024\times10^{-23}$. With Eqn (9), we find the optimal total length L_{opt} of the superlattice required to obtain a maximum possible power of radiation

$$L_{opt}=\ln((1+\Gamma)/\bar{\mu}) \quad (10)$$

For given media and electron energy E_o, the maximum radiation can be found by maximizing $G(\beta,\varepsilon_1,\varepsilon_2,\vartheta_n)$ with respect to emission angle ϑ_n for a chosen n. And the optimal spatial period l and ratio $r=l_2/l$ are found by the resonance condition Eqn (1) and the condition $l_2/Z_2=2\pi m+\pi/2$ where m is an integer. Together with L_{opt} from Eqn (11), we find the optimal radiation with n and m set to one. Some of the results are tabulated in table I.

a) $E_o=1Mev$	Photon energy = 0.157 Kev (80.4°)				
Material	$l(A^o)$	ratio r	ϑ_n	$L(um)$	η
Be/Ba	563	.48	21.1°	.53	1.32×10^{-7}
Be/Ce	529	.46	21.9°	.62	1.17×10^{-8}
Be/Ta	555	.48	21.3°	.41	2.08×10^{-7}
Be/Ge	490	.49	20.8°	.52	1.67×10^{-8}
Be/Eu	570	.48	21.0°	.54	6.75×10^{-8}

b) $E_o=1Mev$	Photon energy = 0.6985 Kev (18 A°)				
Material	$l(A^o)$	ratio r	ϑ_n	$L(um)$	η
Be/Ba	151	.500	19.3°	11.37	2.16×10^{-8}
Be/Ce	150	.499	19.3°	6.07	4.88×10^{-9}
Be/Ta	148	.494	19.5°	1.44	1.13×10^{-7}
Be/Ge	150	.498	19.3°	6.21	4.58×10^{-8}
Be/Eu	150	.498	19.3°	6.19	1.11×10^{-8}

c) $E_o=0.5Mev$	Photon energy = 0.157 Kev (80 A°)				
Material	$l(A^o)$	ratio r	ϑ_n	$L(um)$	η
Be/Ba	281	.498	28.0°	.50	4.11×10^{-7}
Be/Ce	274	.496	28.4°	.57	3.77×10^{-7}
Be/Ta	280	.486	28.1°	.37	6.47×10^{-8}
Be/Ge	284	.493	27.8°	.48	5.11×10^{-9}
Be/Eu	283	.491	27.8°	.50	2.08×10^{-8}

d) $E_o = 0.5 Mev$		Photon energy = 0.6895 Kev (18 A°)			
Material	$l(A^o)$	ratio r	ϑ_n	$L(\mu m)$	η
Be/Ba	67.3	.500	27.2°	10.20	5.86×10^{-9}
Be/Ce	67.2	.499	27.2°	5.50	1.33×10^{-9}
Be/Ta	66.8	.497	27.3°	1.31	3.11×10^{-8}
Be/Ge	67.2	.499	27.3°	5.80	1.25×10^{-8}
Be/Eu	67.2	.499	27.3°	5.70	3.02×10^{-9}

Table 1 Values of l, r, ϑ_n, L and η for various materials, E_o and photon energy.

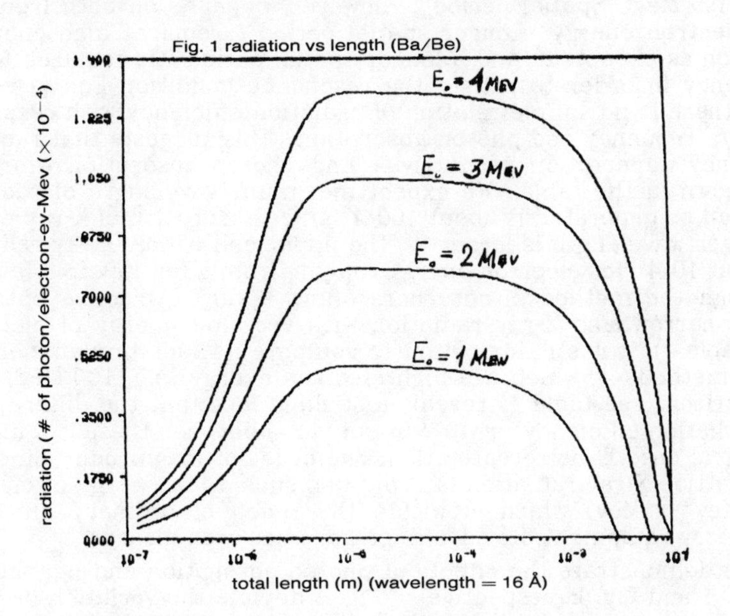

Fig. 1 radiation vs length (Ba/Be)

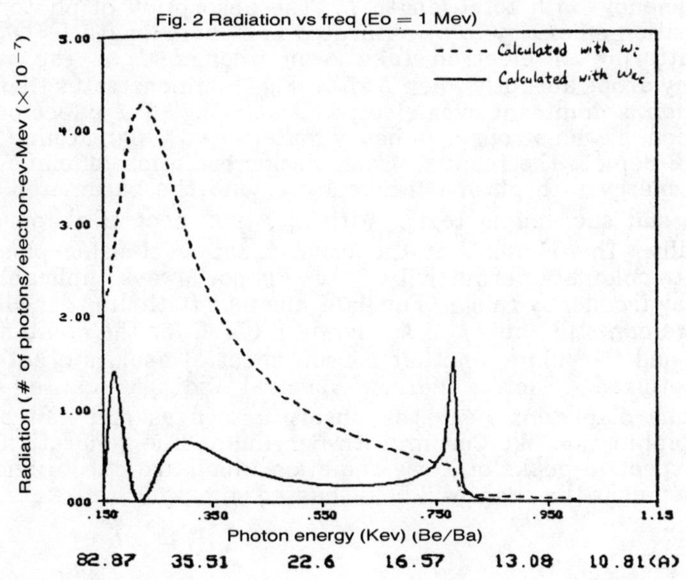

Fig. 2 Radiation vs freq (Eo = 1 Mev)

For comparison purpose, we introduce radiation efficiency η $(dN/d\omega)E_0^{-1}$ which has a unit of number of photon per photon energy in ev per electron per electron energy in Mev. Table 1 demonstrates that optimal length is a function of frequency, e.g. $L \approx 0.5\mu m$ at a wavelength of $80A^o$ and $L \approx 5\mu m$ at a wavelength of 18 A^o. Since L_{opt} depends on photon absorption $\overline{\mu}$ and collision loss $\overline{dE/dx}$, L_{opt} should depend on material. On the other hand, the angle of emission ϑ_n depends on electron energy, e.g. $\vartheta_n \approx 20^o$ at E_0 of 1 Mev and $\vartheta_n \approx 27^o$ at E_0 of 0.5 Mev. Spatial period l, however, depends on both frequency and electron energy. Longer spatial period l requires high energy of electron as expected. And a longer spatial period l is required for low frequency in order to satisfy the resonance condition Eqn (1). However, there is no simple relation of radiation efficiency with respect to energy, frequency and photon absorption. This suggests that radiation efficiency depends on permittivity and photon absorption which are not shown in the table. We expect maximum wavelength of radiation that can be generated is about 100 A^o since absorption of X-ray photon at longer wavelength is larger [8]. The upper end of soft X-ray radiation is about 10 A^o for electron energy ranging from a few Kev to a few Mev. Although the method discuss here offers a very attractive option to obtain narrowband X-ray radiation with very low energy of electrons (see table 1), it is still important to compare its radiation efficiency to other methods [13] which use high electron energy (e.g. 100 Mev). This comparison (see table 1) reveals a striking fact that the difference in the radiation efficiency ($\eta \approx 10^{-7}$ in Ref [11]) between these two method in typical case is not greater than one order of magnitude. Since the superlattice X-ray radiation is exploiting such a low energy of electron (200 Kev - 1Mev) which is within the reach of ordinary university laboratory equipment, its advantage becomes even clearer.

We demonstrate the effects of photon absorption and permittivity on Fig. 1 and Fig. 2. respectively. Fig. 1 depicts the variation of radiation efficiency with total length L. The absorption of photon makes the radiation efficiency to be saturated in a short length $L \approx 1/\overline{\mu}$. Then the scattering of electron take over when $L \approx L_{cr}$. The radiation efficiency drops abruptly when $L > L_{cr}$. Fig. 1 demonstrates that photon absorption is dominant over electron scattering. The effect of photon absorption is even stronger in heavy material with high atomic number [8]. Fig. 2 depicts the results of calculating radiation efficiency versus photon energy with plasma frequency ω_i and the parameter ω_{a_i} [see Eqn (4) and succeeding text]. With ω_{a_i}, the spectral shape changes drastically. This shows that the usual practice of using plasma frequency to calculate permittivity [3,10,12,13] is not always applicable to the soft X-ray frequency range. For light element (Lithium, Beryllium and Born), we can still apply ω_i. As shown in Fig. 2 for the combination of Barium and Beryllium, another narrow spectral peak is created when ω_{a_i} is utilized. Such a narrow spectral width is caused by the anomalous dispersion near the absorption edges and can occur in other combination like Cerium with Beryllium. The generation of very narrow spectral peaks of X-ray radiation which are due to anomalous dispersion near the edges will be discussed somewhere.

Conclusions

We have considered the role of photon absorption and electron scattering in X-ray emission by electrons with low energy (few Kev to few Mev) which are sufficient to produce resonant transition radiation in a superlattice with short spatial period (50 A^o to 500 A^o) proposed earlier [4,5]. We have formulated a numerical procedure to calculate maximum radiation with the electron energy, combination of material, and desired frequency specified. During the calculation process, the spatial period, optimal total length, ratio r of the thicknesses of each medium, and resonant angle are evaluated. This completely specifies the design of the system. For example, we obtain $l \approx 150A^o$, $\vartheta_n \approx 20^o$, $r \approx 0.5$ and $L_{opt} \approx 5\mu m$ for a chosen $E_0 = 1Mev$ and desired wavelength about $18A^o$. We further compare our system to other systems [13] employing higher electron energy (100 Mev). It turns out that the difference of the two system is about one order of magnitude for most cases. This makes our system (low electron energy plus superlattice) preferable as an inexpensive narrowband X-ray source.

We appreciate discussion with S. Datta. This work was supported by AFOSR.

References

1. V. L. Ginzvurg and I. M. Frank, Zh. Eksp. Theor. Fiz. 36, 15 (1946) (in Russian).
2. G. M. Garibyan, Sov. Phys. JETP 33, 23 (1971).
3. M. L. Ter-Mikaelian, High Energy Electromagnetic Processes in Condensed Media (Wiley Interscience, New York, 1972).
4. A. E. Kaplan and S. Datta, Appl. Phys. Lett. 44, 661 (1984); A. E. Kaplan and S. Datta, AIP Conf. Proc. 119, 304 (1984).
5. S. Datta and A. E. Kaplan, Phys. Rev. A31, 790 (1985).
6. T. W. Barbee and D. C. Keith, Stanford Synchrotron Radiation Laboratory 78/04, III-36 (May 1978).
7. E. Spiller, AIP Conf. Proc. 75, 124 (1981).
8. B. L. Henke, P. Lee, T. J, Tanaka, R. L. Shimabukuro and B. K. Fujikawa, Atomic Data and Nuclear Data Tables 27, No. 1 (1982).
9. L. Page, E. Bertel, H. Joffre and L. Sklavenitis, Atomic data 4, No. 1 (1972).
10. M. L. Cherry, G. Hartmann, D. Muller and T. A. Prince, Phys. Rev. D10, 3594 (1974).
11. P. J. Ebert, M. J. Moran B. A. Dahling, B. L. Berman, M. A. Piestrup, J. O. Kephart, H. Park, R. K. Klein and R. H. Pantell, Phys. Rev. Lett. 54, 843 (1985).
12. C. W. Fabjan and W. Struczinkski, Phys. Lett. 57B, 183 (1975).
13. M. A. Piestrup, P. F. Finman, A. N. Chu, T. W. Barbee, Jr., R. H. Pantell, R. A. Gearhart, F. R. Buskirk, IEEE J. Quan. Elect. QE-19, 1771 (1983).
14. B. Jouffrey, Electron Microscopy in Materials Science Part IV, 1405 (1975).

Transition Radiation as a Coherent Soft X-ray Source

M. A. Moran, B. A. Dahling
Lawrence Livermore National Laboratory, University of California
Livermore, California 94550

and

M. A. Piestrup
Adelphi Technologies
Woodside, California 94062

and

B. L. Berman
Department of Physics, George Washington University
Washington, D.C. 20052

and

J. O. Kephart
Department of Electrical Engineering, Stanford University
Stanford, California 94305

A series of experiments using 54 MeV electrons to irradiate thin foil targets has demonstrated the spatially coherent nature of soft x-ray transition radiation.

An ongoing series of experiments at the Lawrence Livermore National Laboratory electron positron linear accelerator has studied transition radiation produced by 25- and 54-MeV electrons traversing targets consisting of thin low-z foils. Our results have been consistent with theoretical predictions of the absolute intensity and angular and spectral distributions of photon emissions in the soft x-ray energy range.[1,2] Energy-integrated measurements of the angular distribution have demonstrated the transition radiation spatial coherence and point the way to application of these experiments to the study of x-ray properties of materials and to the development of new kinds of coherent photon sources.

Figure 1 shows the experimental arrangement that was used for the first series of measurements. Here, a flow-type proportional counter using a mixture of 90% neon 10% isobutane was remotely scanned across the beam line to measure the angular distribution, with 1 mrad resolution, of x rays with energies from 0.3 to 6 keV.

The targets consisted of from one to twenty foils, approximately 1 μm-thick, of beryllium, carbon, mylar, magnesium, aluminum, silicon, and titanium. After traversing the target, the electron beam was deflected into a dump-hole in the floor where a scintillation detector was used to monitor the beam current. Backgrounds were due mostly to bremsstrahlung, were generally small compared to the transition radiation, and were measured by inserting a 0.127 mm-thick aluminum foil between the target and the detector.[1,2]

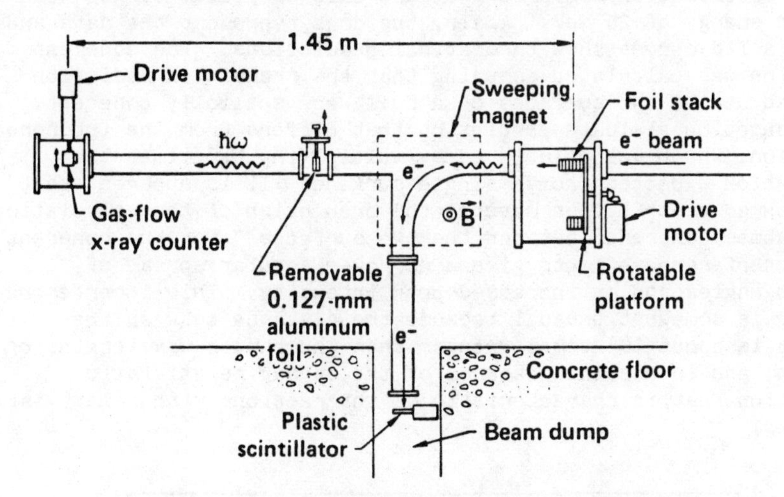

Figure 1: Schematic Diagram of Basic Experimental Arrangement

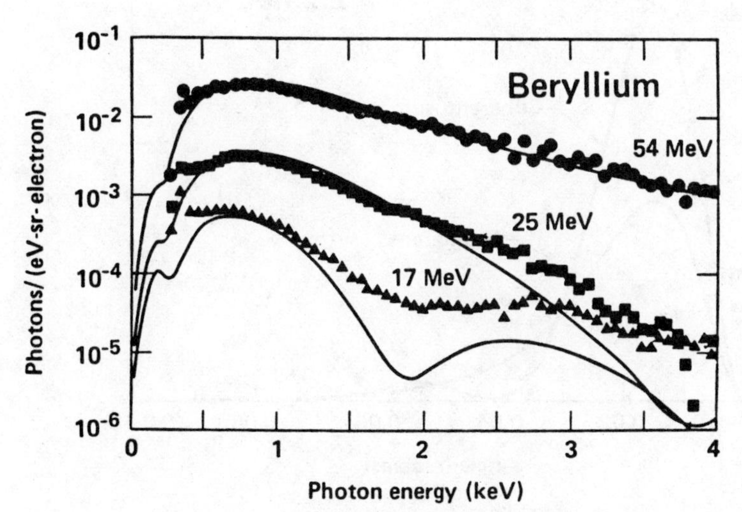

Figure 2: Transition X-Ray Spectra From Beryllium Target

Figure 2 shows the transition x-ray spectrum from a target consisting of eighteen 1 μm-thick beryllium foils for three different incident electron energies. The dots represent data from the experiment and the curve was calculated from a simplified theoretical description of transition radiation.[3] The agreement between experiment and theory is excellent and is typical of the results obtained with foils of the other materials mentioned above.

Figure 3 shows the energy-integrated transition radiation x-ray angular distribution measured from the same beryllium target with an electron energy of 25 MeV. Again, the dots represent the data and the two solid curves show theoretical predictions. The coherent prediction was calculated assuming that the transition radiation generated at the two surfaces of a foil were spatially coherent. This assumption yields a prediction that differs from the incoherent prediction, which was calculated by multiplying two times the distribution expected from a single surface. It is apparent that our measured distribution matches the prediction of the calculation that assumes coherence between the two surfaces. For the coherent form, interference effects give a narrower angular spread of emission angles and an increased peak intensity. This interference behavior is somewhat unusual because the distance between the surfaces is about 10^3 times greater than the photon wavelengths of interest, and is possible because of the strong relativistic contraction that is characteristic of interactions with relativistic particles.

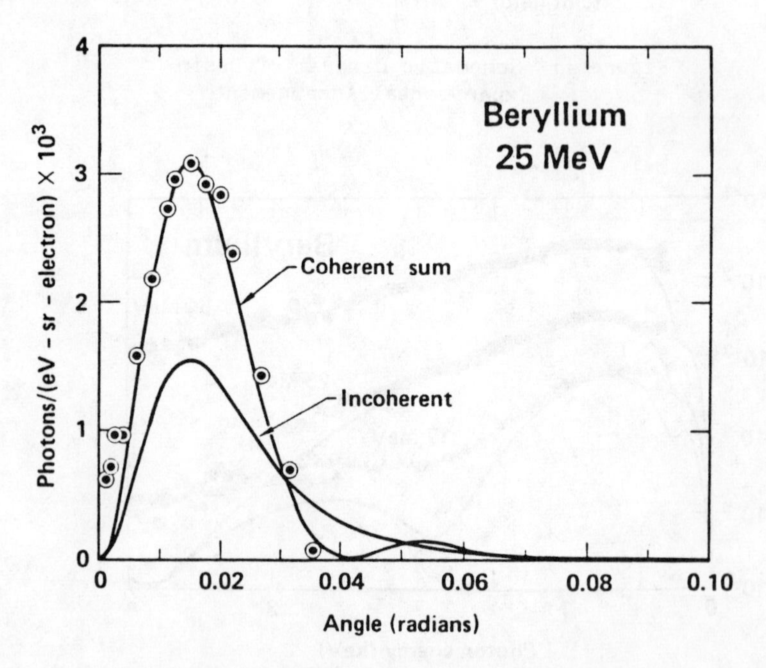

Figure 3: Transition Radiation (Energy-Integrated) Angular Distribution with Beryllium Target

The single foil coherence motivated an attempt to observe interference between the individual foils in a target. Inter-foil interference is not apparent in the data in Fig. 3 because the inter-foil spacing for this target (approximately 3000 μm) results in an interference pattern with structure that is much smaller than the 1-mrad resolution of the detector.

Figure 4 shows results from later experiments that studied angular distributions (Fig. 4(a) and 4(b)) and spectra (Fig. 4(c) and 4(d)) from two different targets exposed to a beam of 54 MeV electrons. Each target used two 0.55 μm-thick polypropylene foils; but one target had an inter-foil spacing of 3000 μm, while the second had a 50-μm spacing. The differences between the two angular distributions are obvious. The oscillations in the angular distribution for the 50 μm-spaced target are caused by interference effects between the two foils. Again, the data (dots) is compared with corresponding predictions (solid curves). The sharp oscillations in the prediction for the 3000 μm spacing were not resolvable by the detector. But for the 50 μm spacing the interference oscillations are very broad in angle and the data and prediction agree remarkably well. In these cases, the actual source strengths are greater than the plots indicate because the detector efficiency (approximately 25%) was included in the calculations.

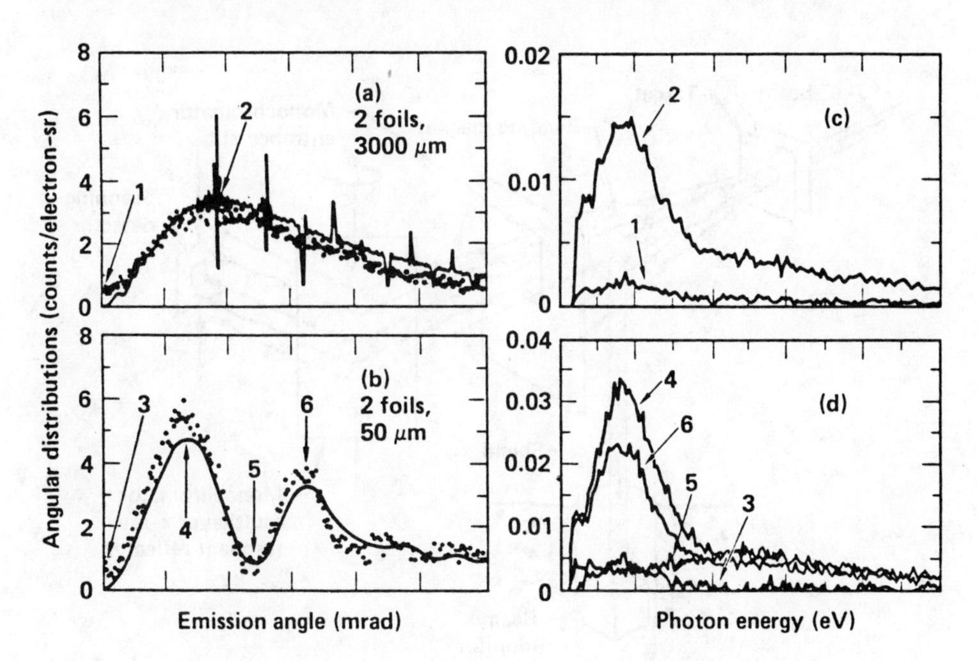

Figure 4: Transition Radiation (Energy-Integrated) from Polypropylene 2-foil Targets

The spectra in Fig. 4(c) and 4(d) were recorded at the angles indicated on the corresponding angular distributions. The strong peak near 150 eV that dominate the spectra are associated with the polypropylene foil thickness. The spectrum in Fig. 4(c) is typical of spectra recorded over a wide range of emission angles for the 3000 μm-spaced target. As Fig. 4(d) shows, the spectra recorded for the 50 μm-spaced target show wide variations, depending on the particular emission angle chosen.

The results in Fig. 4 demonstrate interference effects with photons having wavelength of about 83 A between foils having a separation of 50 m. In this case, the interference occurs in a structure that is nearly 10^4 times longer than the wavelength of interest. This behavior demonstrates that each transition photon is a coherent response to the entire target structure. This is true for our results because the experiments were always conducted with low average beam currents (below 10^{-10}A) where there was very rarely more than one photon at a time within the target.

In order to study these effects more carefully we recently performed energy-resolved measurements of the transition radiation angular distribution. Figure 5 shows the experimental arrangement

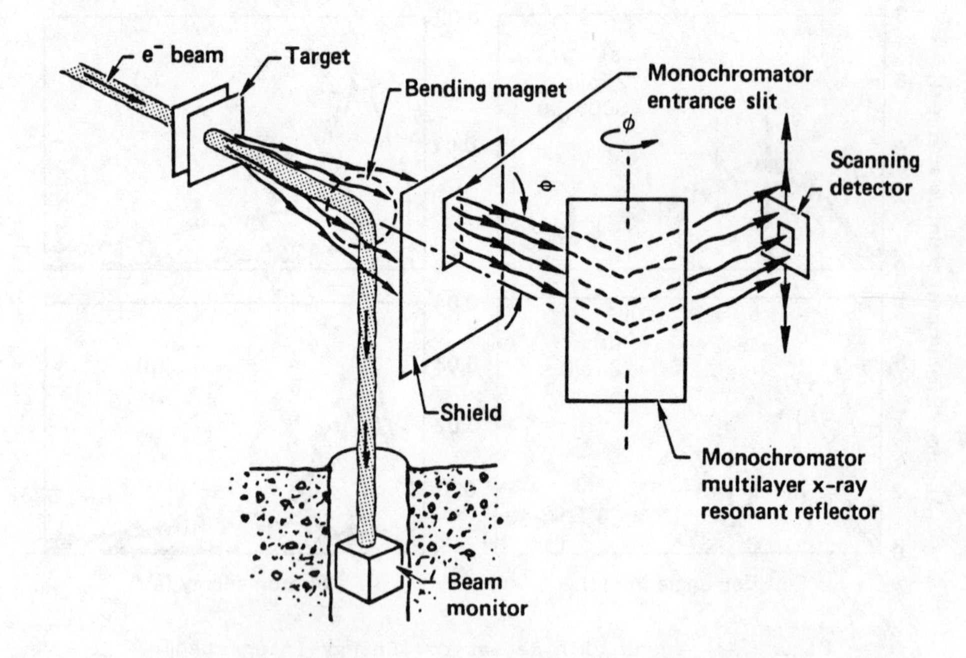

Figure 5: Schematic Diagram for Energy-Resolved Measurements

for these measurements. A Si/W multilayer resonant reflector with a 2d period of 149.2 A and measuring 1" by 2.75" was placed in the beam line and used as the basis of a simple x-ray monochromator. Then the reflector was rotated to an appropriate Bragg angle and the detector was placed at a corresponding position to measure the reflected x-ray angular distribution.

Figure 6 shows angular distributions that were measured in this fashion. Figure 6(a) shows the measured (dots) angular distribution of 180 eV photons generated when 54 MeV traversed the same two-foil target with the nominal 50-μm spacing described previously. Unfortunately, in this case, the portion of the data file for emission angles less than 4 mrad have been lost, but the remainder of the data clearly shows the interference pattern that is characteristic of this particular situation. Figure 6(b) shows similar data for a photon energy of 250 eV. The calculations (solid curves) show excellent qualitative agreement with the form of the measured patterns. In order to obtain the best possible comparison, the calculations have been multiplied by a normalization factor that corresponds to the reflectivity of the multilayer reflector at the appropriate wavelength. We found that in order to achieve good agreement between the measured and calculated distributions, the foil thickness and separation had to be changed from their nominal values of 0.55 and 50 μm, respectively, to 0.60 μm for the foil thickness and 63μm for the separation. This agreement could not be achieved for both energies by adjusting either the foil thickness or spacing alone.

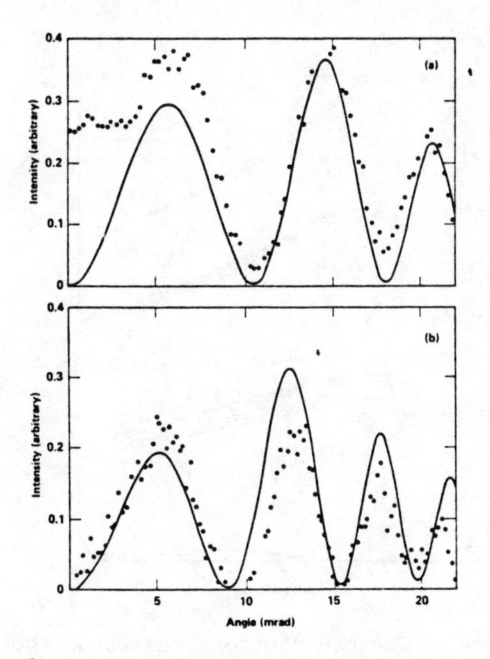

Figure 6: Energy-Resolved Angular Distributions

The foil thickness had been measured by several different techniques, and the nominal value of 0.55 m was only accurate to within $\pm$10%. Thus, the thickness that was inferred from the calculations shown in Figure 6 is consistent with the possible range of foil thicknesses.

The foil separation had not been measured, but was inferred from the target assembly technique. The foils had been glued to the surface of 50 μm-thick one cm square silicon frames having a 7 mm square clear aperture. Thus, the 50-μm value was the minimum value that could have been expected from this structure. The inferred value of 63 μm is consistent with the presence of a thin layer of glue that slightly increased the foil separation.

The calculations in Figure 6 have been used to infer the foil thickness and separation. In order to do this, the calculations assumed that the frequency-dependent dielectric constant, $\varepsilon(\omega)$, of the polypropylene was given by the Drude free-electron approximation: $\varepsilon(\omega)=1-(\omega_p/\omega)^2$, where ω_p is the plasma frequency of polypropylene and ω is the photon frequency. Now, if independent means can be used to characterize the target structure, then the analysis demonstrated above can be turned around and used to measure the frequency-dependent x-ray dielectric constant of the foil material: for each monochromator photon energy, ε could be determined.

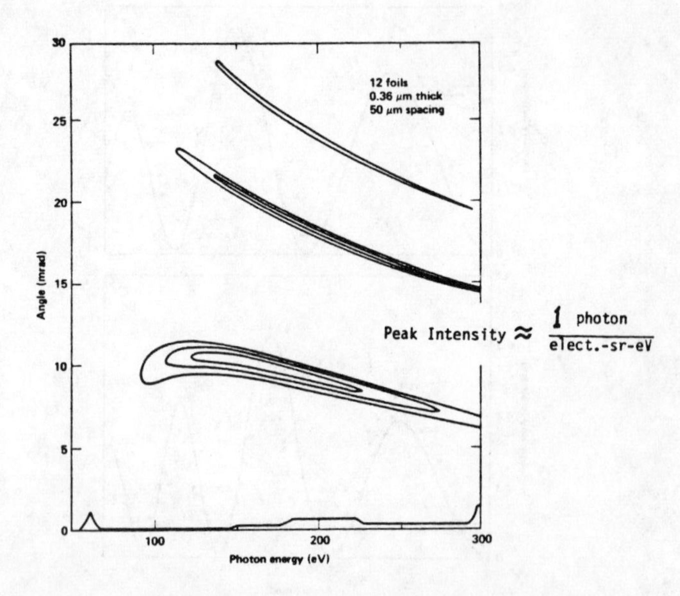

Figure 7: Contour Plot of Transition Photon Energy vs. Angle of Emission

Another application of our results might be the development of a tuneable, spatially coherent x-ray source driven by 50 MeV electrons. In this case, the number of foils in the target would be increased and the transition photon energies would be determined by the emission angle. For a given angle of emission the photon energy, E, would be defined with a precision approximated by: $\Delta E/E=1/N$, where N is the number of foils in the target. However, the number of foils would be limited to give a total thickness not much larger than the x-ray absorption length for the photon energies of interest. Figure 7 shows a contour plot of transition photon energy vs. angle of emission. The contours show that the photon energies would very in a predictable way with angle of emission. The peak emission rate is about 1 photon/(elect-sr-ev).

In summary, we have shown the results of measurements and calculations that demonstrate both the intra- and inter-foil coherence of transition in multiple-foil targets. Coherent behavior results in distincitive angular distributions that are accurately predicted by theoretical calculations. Two possible applications of this work are the study of the x-ray optical constants of thin foils and the development of a spatially coherent, tuneable soft x-ray source.

The authors thank R. H. Pantell for the loan of electronic equipment. This work was performed at the Lawrence Livermore National Laboratory under the auspices of the U.S. Department of Energy under Contract No. W-7405-Eng-48.

References:

[1]M.A. Piestrup, J.O. Kephart, H. Park, R.K. Klein, R.H. Pantell, P.J. Ebert, M.J. Moran, B.A. Dahling, and B.L. Berman, Phys. Rev. A 32, 917 (1985)

[2]P.J. Ebert, M.J. Moran, B.A. Dahling, B.L.Berman, M.A. Piestrup, J.O. Kephart, H. Park, R.K. Kline, and R.H. Pantell, Phys. Rev. Lett. 54, 893 (1985).

[3]M.J. Moran, B.A. Dahling, P.J. Ebert, M.A. Piestrup, B.L. Berman, and J.O. Kephart (to be published)

X-RAY MICROSCOPY EXPERIMENTS WITH SYNCHROTRON RADIATION

D.Rudolph, G.Schmahl, B.Niemann and W.Meyer-Ilse

Forschungsgruppe Röntgenmikroskopie
Universität Göttingen
Geismarlandstraße 11, D-3400 Göttingen
Fed.Rep. of Germany

Abstract

The x-ray microscope at the electron storage ring BESSY in Berlin is described.An example of biological specimens imaged with 4.5 nm x-radiation is given. Aspects and future developments are discused.

Introduction

It is well known that microscopy with soft x-rays between 2.3 and 4.5 nm wavelength offers the possibility to investigate biological specimens wet and unstained in their natural For detailed discussion of the possibilities of x-ray microscopy also in comparison to electron and light microscopy see [1,2,3,4,5,6].

X-ray microscopy requires high resolution x-ray optics and intense x-ray sources of high spectral -brilliance. Concerning x-ray optics there are two classes of optical systems: Condenser optics to illuminate the object with quasi-monochromatic x-radiation. Up to now for condenser optics zone plates [7,8] or a combination of grazing inci-dence optics and zone plates [5] were used. In future in addition multilayered mirrors [13,14] possibly in combination with zone plates and/or gratings may be applied. As micro optics with high resolution up to now only micro zone plates are used [7,9].

Suitable x-ray sources have to be of high spectral brilliance. Tunability to different wavelengths may be advantageous though not necessary in any case. Bending magnet synchrotron radiation from 0.5 - 1.0 GeV electrons in dedicated storage rings is currently the most widely used radiation [1,6].

The Beamlines and the X-Ray Microscope at BESSY

At the electron storage ring BESSY in Berlin two beamlines are dedicated to x-ray microscopy. Figure 1 shows the arrangement of the beamlines of the x-ray microscopic area.

Both beamlines emerge from from the teefitting behind the valve V1 with which the beamlines can be separated from the outlet of the storage ring. The second line is dedi-

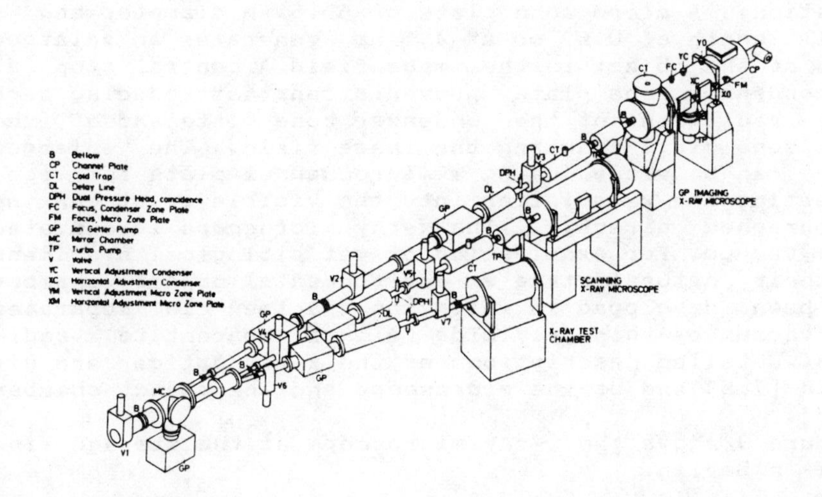

Fig.1: The arrangement of the beamlines of the microscopic area at the BESSY storage ring

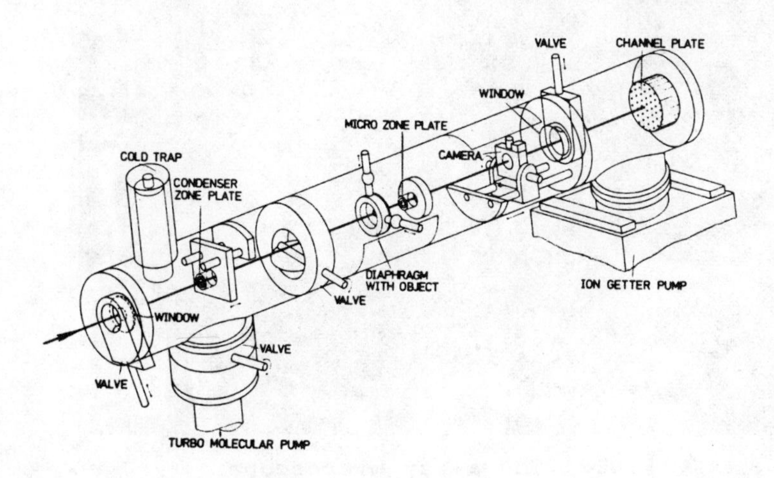

Fig.2: The schematic arrangement of the x-ray microscope

cated to the scanning x-ray microscope [12]. A third beam can be directed into the x-ray test chamber by reflection from a grazing incidence grating monochromator installed in the chamber MC. This chamber is used for testing x-ray optical elements and x-ray detector devices etc.

The first beamline ends up in the x-ray microscope, figure 2 shows the schematic arrangment. The polychromatic x-radiation meets the condenser zone plate at a distance of 15 m of the source which is the tangent point to the electron storage ring. The condenser zone plate with a diameter of 9 mm and a focal length of 304 mm for 4.5 nm radiation. A micro zone plate of 55.6 µm diameter and a focal length of 0.69 mm at 4.5 nm generates an enlarged image of the object in the image field.A central stop at the condenser zone plate prevents contrast reducing zero order radiation of the condenser zone plate and of the micro zone plate reaching the image field. The enlarged image can be viewed using a microchannel plate for con-converting the x-radiation into the visible or it can be photographed directly. The x-ray microscope is operated under vacuum. For examination of wet biological specimens in their natural state an environmental object chamber has been developed in which the specimen is separated from vaccum by thin polyimide foil transparent to x-radiation. Detailed descriptions of the x-ray optics are given in [7,9] and of the microscope and the object chamber in [6].

Figure 3 shows the x-ray microscope at the storage ring BESSY in Berlin.

Fig.3: The x-ray microscope

Results

Figure 4 shows an image of a part of a human fibroblast made with the x-ray microscope with an x-ray magnification of 330 x. The cells were critical point dried without any staining. At the left a part of the nucleus can be seen. Images of wet cells, e.g. a human epithelial cheek cell and spores of the Australian moss Dawsonia superba were made [15] and investigations of viability of those spores under x-irradiation were started [16].

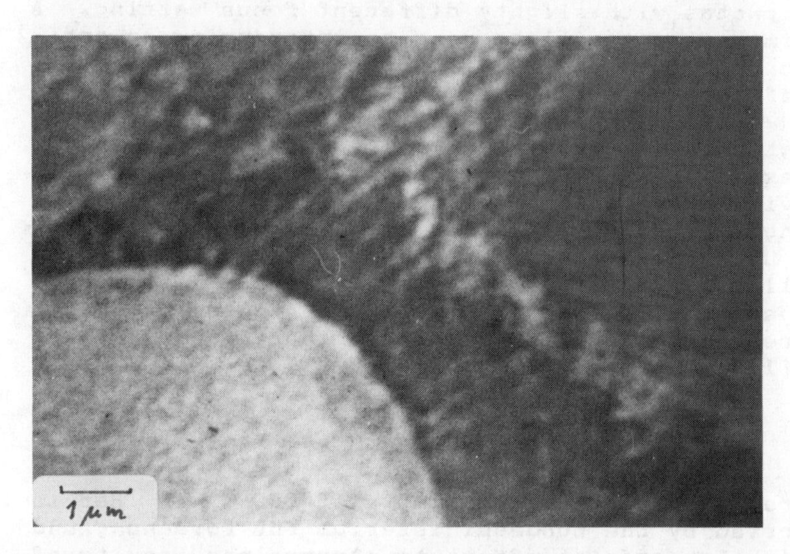

Fig.4: Part of a human fibroblast, critical point dried, imaged with 4.5 nm radiation, x-ray magnification 330 x

Future Aspects

Concerning x-ray sources in the near future a major advance in the spectral brilliance is to be expected through the use of undulators in 1 - 5 GeV electron storage rings. As laboratory sources Plasma Focus sources [10] and Laser Produced Plasma sources [11] will be available. In the future x-ray lasers may become important x-ray sources of high brilliance operating at single spectral lines.

Up to now for x-ray imaging amplitude zone plates are used with a theoretical maximum first order diffraction efficiency of 10%; in practice values of about 5% are achieved. Higher values of diffraction efficiency can be obtained by using phase zone plates (up to about 20% for laminar phase structures). Phase zone plates with laminar phase structures are under development.

Zone plates with higher diffraction efficiency enable shorter exposure times, in addition a larger amount of the x-radiation which loads the object is used for image generation, leading to a reduced radiation dose applied to the specimen.

Up to now a coarse focussing of the object is done by looking at the x-radiation converted to the visible by the microchannel plate and subsequently taking a few fine focussing photos with slighty different focus setting. A further improvement will be done by introducing an optical microscope into the x-ray microscope for prefocussing. Prefocussing with visible light will reduce the x-radiation load to the object to the amount necessary for taking just only one x-ray image.

In the existing x-ray microscope at BESSY the image is directly viewed by converting the x-radiation to the visible using a microchannel plate or it is directly photographed. Using semiconductor imaging devices such as CCD's will provide higher detective quantum efficiency and the possibility of electronic image processing. Such devices are under development; preliminary results are given in [17].

Acknowledgements

The x-ray microscopic experiments at BESSY in Berlin are supported by the Bundesministerium für Forschung und Technologie. For a joint effort to compare the results of dif- ferent microscopic methods Dr.P.C.Cheng, IBM Th.J.Watson Research Center, Yorktown Heights, NY, USA, provided us with critical point dried human fibroblasts (see figure 4).

References

1. Howells, M., J.Kirz, D.Sayre and G.Schmahl,Soft X-ray microscopes, physics today, 1985, Vol. 38, No.8, p. 22 - 32

2. Niemann, B., G.Schmahl, D.Rudolph, X-ray microscopy: recent developments and practical applications, in: SPIE Vol.368-Techniques and Capabilities (Sira)(1982) p. 2 - 8

3. Kirz, J. and D.Sayre, Soft X-ray microscopy of biological specimens, in: Synchrotron radiation research, H. Winnick and S.Doniach, eds., Plenum Press, 1980, p. 277 - 322

4. Duke, P.J., X-ray microscopy: Recent developments and future prospects, Proceedings RMS, Vol.16/3, 1981, p. 186 - 192

5. Schmahl, G., D.Rudolph, B.Niemann and O.Christ, X-Ray Microscopy of Biological Specimens with a Zone Plate Microscope, in: Ultrasoft X-Ray Microscopy:Its Application to Biological and Physical Sciences, D.P. Parsons, ed.,Annals of the New York Academy of Sciences, Vol.342, 1980, p. 368 - 386

6. Rudolph, D., B. Niemann, G. Schmahl and O. Christ, Göttingen X-Ray Microscope and X-Ray Microscopy Experiments at the BESSY Storage Ring,in: X-Ray Microscopy, G.Schmahl and D.Rudolph, eds., Springer Series optical Sciences, Vol.43, Springer-Verlag Heidelberg, 1984, p. 192 - 202

7. Schmahl, G., D.Rudolph,P.Guttmann and O.Christ, Zone Plates for X-Ray Microscopy, ibid., p. 63 - 74

8. Thieme, J., Construction of Condenser Zone Plates for a Scanning X-Ray Microscope, ibid., p. 91 - 96

9. Guttmann, P., Construction of a Micro Zone Plate and Evaluation of Imaging Properties, ibid., p. 75 - 90

10. Herziger, G., X-Ray Emission from a 1 KJ Plasma Fokus ibid., p. 19 - 24

11. Ginter,M.L., Laser Produced Plasma VUV and Soft X-Ray Light Sources, ibid., p. 25 -29

12. Niemann, B., The Göttingen Scanning X-Ray Microscope, ibid., p. 217 - 225

13. Barbee, T.W., Sputtered layered synthetic microstrucre (LSM) dispersion elements,in:Low Energy X-Ray Diagnostics - 1981, D.T.Attwood and B.L.Henke, eds., AIP Conference Proceedings No.75, 1981, p. 131 - 145

14. Dhez,P. and J.M.Esteva, Les multicouches metalliques, Direction des Recherches,Etudes et Techniques,Rapport AEPA 80/043, 1980

15. Schmahl, G., B. Niemann, D. Rudolph, P. Guttmann and V.Sarafis.1985. X-Ray Microscopy:experimental results with the Göttingen X-ray microscope at the electron storage ring BESSY in Berlin, Journal of Microscopy, Vol.138, p. 279 - 284

16. Niemann, B., V.Sarafis, D.Rudolph, G.Schmahl,W.Meyer-Ilse and P. Guttmann. 1986. X-Ray Microscopy with Synchrotron Radiation at the Electron Storage Ring BESSY in Berlin, Proceedings of the "International Conference X-Ray and VUV Synchrotron Radiation Intation, Stanford 1985, Nuclear Instruments and Methods, in press

17. Germer,R. and W.Meyer-Ilse. 1986. X-Ray TV-Camera at 4.5 nm, Rev.Sci.Instrum., in press

SOFT X-RAY INTERFEROMETRY AND HOLOGRAPHY

S. Aoki
Institute of Applied Physics, University of Tsukuba,
Sakura, Ibaraki 305, Japan

S. Kikuta
Department of Applied Physics, Faculty of Engineering,
University of Tokyo, Hongo, Bunkyo-ku, Tokyo 113, Japan

ABSTRACT

Four types of soft X-ray interferometers are proposed, and two of them, Lloyd's mirror and Young's experiment are examined. Phase shifts of refracting objects are observed with these interferometers. Two types of X-ray holograms are taken. Gabor in-line X-ray holograms are recorded by using undulator radiation on X-ray resists. Two-demensional lensless Fourier-transform X-ray holograms are recorded and reconstructed with visible light.

INTRODUCTION

More than ten years ago we demonstrated that the X-ray holograms obtained by partially coherent X-rays could be successfully reconstructed with visible light.[1, 2, 3] On account of the low intensity of the X-ray source, resolution was limited to several microns at that time. The situation has been much improved with the developments of synchrotron radiation sources and new X-ray optical elements since the late 1970's. These developments have been solving the problem of high resolution X-ray measurements.

Both soft X-ray interferometry and holography are closely connected with coherence of X-rays. Recent successes of producing soft X-ray lasers[4] and new advances in the generation of undulator radiation[5] have encouraged us to investigate the high resolution coherent X-ray optics.

In this paper we first review the optical systems of soft X-ray interferometry and holography, and then give the recent results of our own experiments.

OPTICS FOR INTERFEROMETRY

Two kinds of coherence properties of X-ray source are required for X-ray interferometry. One is temporal coherence and the other is spatial coherence.[6] The former depends on the spectral bandwidth of the source, and the latter depends on the size of the source. Temporal coherence is defined with the expression of the coherence length, $\Delta\ell=\lambda^2/\Delta\lambda$. Line spectra of conventional X-ray sources can be used for the experiments on X-ray interferometry. The continnum spectra of synchrotron radiation source must be monochromatized with a proper monochromator. Quasi-monochromatic undulator radiation can be used with or without a monochromator, and it depends on the spectral bandwidth required in the experiment. Spatial coherence is defined with the expression of the degree of coherence. The coherent area of the quasi- monochromatic source is given by $\Delta x=0.16R\lambda/\rho$ where R is a distance between the source and the point on an illuminated plane and ρ is the radius of the circular area of the source.

Interference of soft X-rays is essentially similar to that of visible light. Phase shifts are introduced into the interferometer by the refracting object. For X-rays the refractive index is written by complex number n:[7]

$$\begin{aligned} n &= 1 - \delta - i\beta \\ &= 1 - (N r_0 \lambda^2/2\pi)(f_1 + i f_2) \end{aligned}$$

Here N is the number density of atoms, r_0 the calssical electron radius and λ the wavelength. f_1+if_2 is the complex atomic scattering factor. The soft X-ray interferometer offers a very direct way of measuring refractive indices.

No soft X-ray interferometer has been constructed so far except those of Kellstrom.[8] Various types of soft X-ray interferometers will be shown as follows. There are two general methods of obtaining beams for interference experiments. One is the method based on division of wave-front. And the other is the method based on division of amplitude. The typical examples of the former are shown in Fig. 1. Figure 1(a) shows the soft X-ray interferometer which utilizes Young's double slits arrangement. In order to obtain the phase information, the phase object is put in front of one of the apertures. And then the shift of the interference fringes will

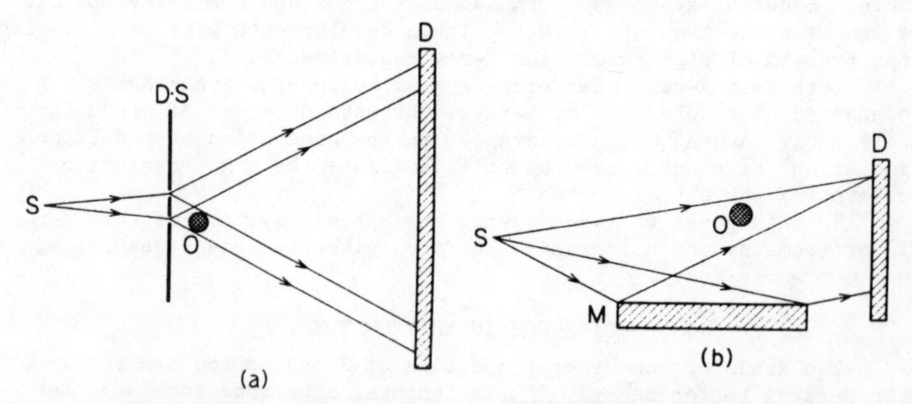

Fig. 1. Soft X-ray interferometers. The method based on division of wave-front.
(a) Young's experiment. (b) Lloyd's mirror.
S: X-ray source, D·S: double slits, M: Mirror, O: object, D: film.

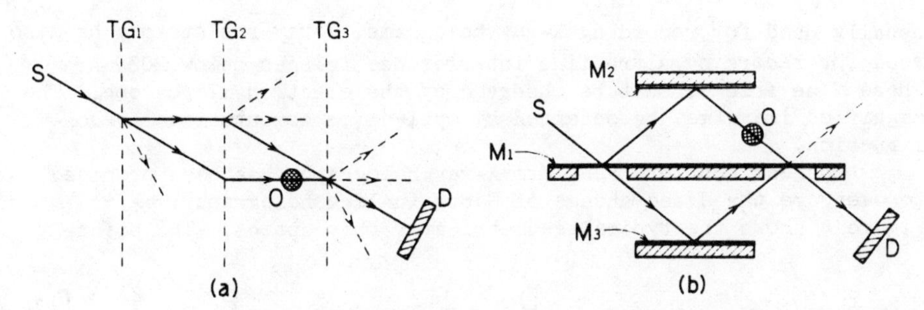

Fig. 2. Soft X-ray interferometers. The method based on division of amplitude. (a) Triple Laue case (LLL) utilizing three transmission gratings (TG_1, TG_2, TG_3).
(b) Triple Bragg case (BBB) utilizing three multilayered mirrors (M_1, M_2, M_3).

be observed. Figure 1(b) shows the soft X-ray interferometer which utilizes Lloyd's mirror. The direct wave and the reflected wave are superposed on the photographic film. The phase object is inserted into one of the optical paths. These methods are easy to construct optical systems. The typical examples of the other method are shown in Fig. 2. Figure 2(a) shows the soft X-ray interferometer which is analogous to the so-called triple Laue case (LLL) X-ray interferometer.[9] Three transmission gratings are used instead of crystals. The transmission grating with high spatial frequencies below 1000 Å may be very useful for the soft X-ray interferometer. Figure 2(b) shows the soft X-ray interferometer which is analogous to the triple Bragg case (BBB) X-ray interferometer. Three multilayered mirrors are used instead of crystals. Very thin multilayers M_1, M_2, and M_3 serve as a beam splitter, a mirror and an analyzer, respectively. Multilayers made by the Langmuir-Blodgett's technique or layered synthetic microstructures (LSM) can be used.

OPTICS FOR HOLOGRAPHY

With the invention of the optical laser, holography has become one of the most important techniques for highly precise measurements. Partially coherent and very intense X-rays which are emitted from the undulator will bring a great impact on X-ray holography. In X-ray holography, holograms are made with X-rays and the reconstruction of the magnified image is given by visible light. Two types of X-ray holography will be given. Figure 3 shows the typical geometries of recording Gabor in-line X-ray holograms. The resolution is limited by the size of the source and the resolving power of the film. A microfocus X-ray source can be used without any pinholes as is shown in Fig. 3(a). A pinhole which will guarantee the coherence area may be introduced into the optical path (Fig. 3(b)). Newly developed optical elements, such as zone plates, Wolter mirrors and spherical multilayered mirrors, will be usefull for focusing coherent X-rays (Fig. 3(c)). Although high resolution photographic plates are usually used for recording X-ray holograms, X-ray resists may be also used for recording ultra fine interference fringes below 2000 Å. These fine fringes must be observed by the electron microscope. The magnified image may be obtained by optical or computational reconstruction.

Lensless Fourier-transform X-ray holography has been proposed to overcome the disadvantage of Gabor in-line holography.[10, 11] Figure 4 shows the typical geometries of this optics. The point

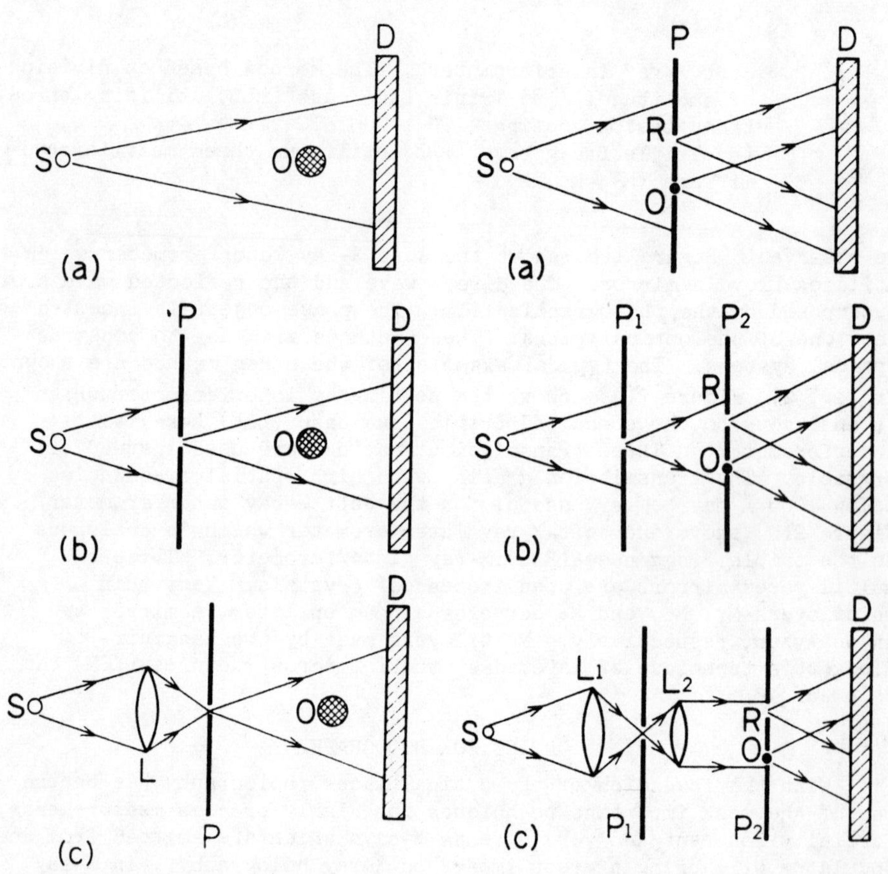

Fig. 3. Geometry for Gabor in-line X-ray holography.
(a) A point source.
(b) A pinhole (P) for coherent illumination.
(c) High resolution optics for a very small X-ray source.
L: mirror or zone plate.

Fig. 4. Geometry for lensless Fourier-transform X-ray holography.
(a) A point source.
(b) A pinhole (P_1) for coherent illumination.
(c) High resolution optics.
R: reference, O: Object,
L_1, L_2: mirror or zone plate

reference source and the object are set in the same plane which is perpendicular to the optical axis. Both of them must be coherently illuminated by a point X-ray source (Fig. 4(a)). If the X-ray source is partially coherent, a pinhole must be used to make a spatially coherent source (Fig. 4(b)). The resolution is mainly limited by the size of the reference source. In order to obtain a high resolution X-ray image a very small reference source less than 1000 Å in diameter will be required. In such a case efficient optical elements must be used to collect or focus coherent X-rays (Fig. 4(c)).

X-RAY SOURCES

Experiments were performed with a conventional microfocus X-ray generator and the undulator which is inserted in PF storage ring at Tsukuba. A fine focus less than 1μm in diameter can be obtained with the former generator. The intensity, however, is so small that the application of this source is limited. The undulator has the following paramteters:[5]

i) Number of the period 60
ii) Length of the period 6cm
iii) Spectral range (first harmonic) 13 Å – 30 Å
iv) Bandwidth $\Delta\lambda/\lambda=1/20$
v) Angular divergence $\sigma_x=0.534$ mrad, $\sigma_y=0.065$ mrad
vi) Brightness $\sim 10^{14}$ photons/sec·mA·mrad^2·1%BW

Although the spectral bandwidth was not narrow enough, we used no monochromator for the experiment. Undulator radiation was collimated by a 1 mm-dia. pinhole and deflected by a flat mirror made of platinum-coated silicon carbide. As the grazing angle was 2°, X-rays shorter than 4.7 Å were cut off.

EXPERIMENTAL RESULTS

Figure 5 shows the interferogram produced by the interferometer of Lloyd's mirror type. The phase object was a polyethylene-terephthalate fiber of 10μm in diameter and the X-ray was AlKα radiation (λ=8.34 Å). The film used was the Agfa 10E56 holographic plate whose resolution is about 2000 lines/mm. Shift of the fringes caused by the phase object is clearly seen in the figure. We can calculate the value of δ from the amount of the phase shift. In this case $\delta=1.4\times10^{-4}$.

Figure 6 shows the phase shift of the Young's interference fringes. A 1.6μm thick carbon foil was put on the part of one slit. The spacing of the double slits was 60μm. The wavelength of undulator radiation used was 20 Å. The narrow interference fringe in the middle part of the figure is due to the second harmonic of undulator radiation. The film used was the Agfa 10E56 plate. The exposure time was less than a second with an electron current of 155 mA at 2.5 GeV. We can calculate the value of δ to be 1.1×10^{-3}. The anomalous dispersion effect will be directly observed by using

Fig. 5. The interferogram produced by the Lloyd's mirror.

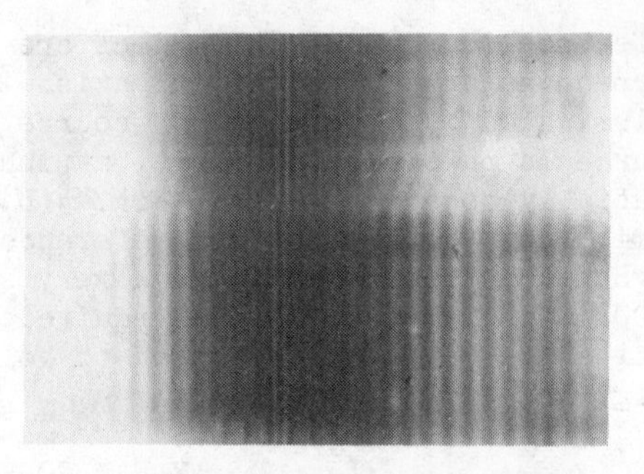

Fig. 6. The interferogram produced by the Young's experiment.

these simple interferometers.

The high resolution X-ray resist PMMA (polymethyl methacrylate) was used to record the Gabor in-line X-ray hologram. PMMA has the resolution better than 1000 Å. Figure 7 shows the X-ray hologram of chemical fibers which were put on a copper mesh. The wavelength of undulator radiation used was 15.5 Å. The distance from the sample to the resist was 22.3 mm. The exposure time was less than one second with an electron current of 80 mA. Reconstruction of the image from the X-ray hologram is under investigation.

The two-dimensional lensless Fourier-transform X-ray hologram was obtained for the first time. A schematic diagram of the experimental arrangement is shown in Figure 8. A 1.2μm thick aluminum foil was used to cut the visible and the vacuum-ultraviolet light. A 5μm-dia. pinhole (S) was used to obtain the coherent X-ray source. As is shown in Fig. 9(a), the two-dimensional model object (O), that is, an array of six pinholes and the reference pinhole (R) are arranged in a nickel foil of thickness 6μm, X-ray holograms were recorded by using 15.5 Å (λ_1) undulator radiation. The distance from the pinhole (S) to the sample and that from the sample plane to

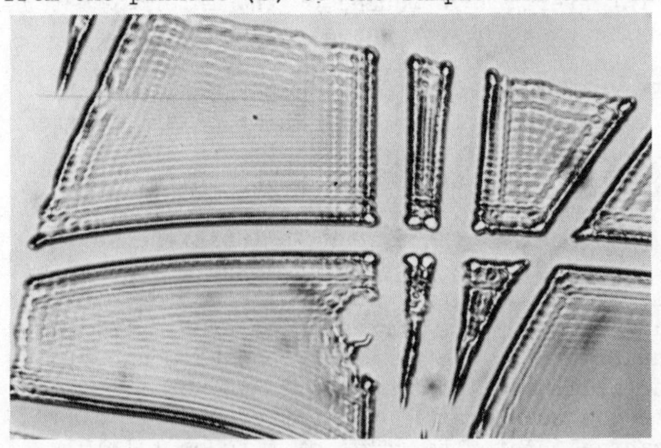

Fig. 7. The Gabor in-line X-ray hologram of chemical fibers on PMMA.

the hologram (Z_1) were 65 cm and 76 cm, respectively. Figure 9(b) shows the X-ray hologram. The film used was the Agfa 10E56 plate. The exposure time was 5 minute with an electron current of 133 mA. In the reconstruction process, the hologram was illuminated by a plane wave of the He-Ne laser (λ_2=6328 Å). The distance from the hologram to the image plane (Z_2) was 25.3 cm. The reconstructed images I and J, and the noise image K are shown in Fig. 9(c). The magnification (M) of the images is calculated to be 136 from an equation $M=\lambda_2 Z_2/\lambda_1 Z_1$, being in good agreement with the experiment.

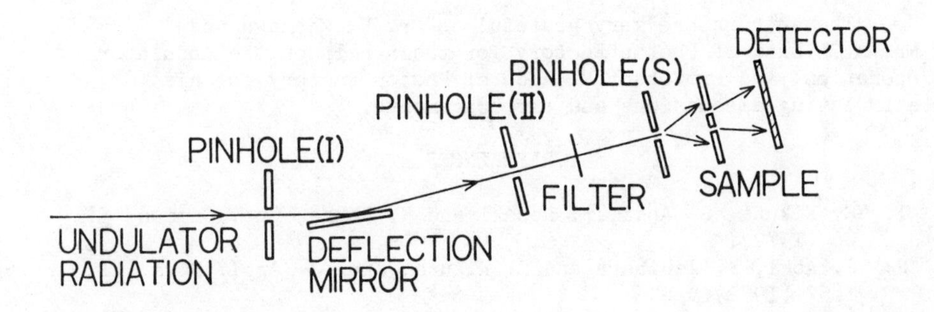

Fig. 8. Schematic diagram of experimental arrangement for lensless Fourier-transform X-ray holography at PF.

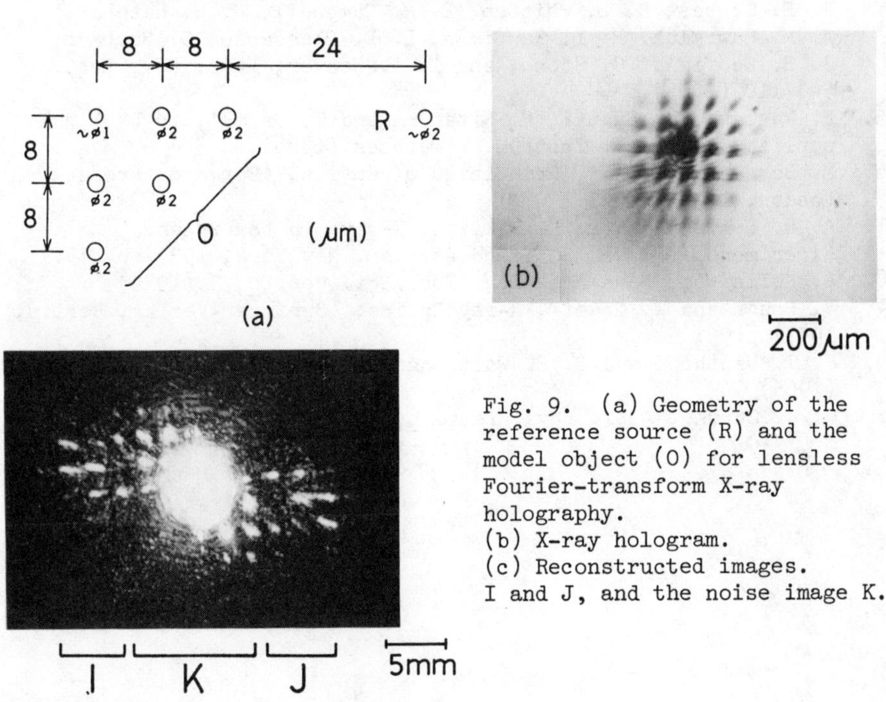

Fig. 9. (a) Geometry of the reference source (R) and the model object (O) for lensless Fourier-transform X-ray holography.
(b) X-ray hologram.
(c) Reconstructed images.
I and J, and the noise image K.

CONCLUSION

We have demonstrated that simple optical systems for soft X-ray interferometry were useful for determining the refractive index. We have also reported some results of our recent experiments on making X-ray holograms with undulator radiation. Although the resolutions are not so high enough, the improvement of the size of the reference source will give a much better resolution than 1000 Å.

ACKNOWLEDGEMENTS

The authors are very grateful to Dr. H. Maezawa and Mr. A. Mikuni of Photon Factory for their help on the undulator operation, and Professor M. Ando of Photon Factory for his stimulating discussions and encouragements.

REFERENCES

1. S. Kikuta, S. Aoki, S. Kosaki and K. Kohra, Opt. Commun. 5, 86 (1972).
2. S. Aoki, Y. Ichihara and S. Kikuta, Jpn. J. Appl. Phys. 11, 1857 (1972).
3. S. Aoki and S. Kikuta, Jpn. J. Appl. Phys. 13, 1358 (1974).
4. D. L. Matthews, P. L. Hagelstein, M. D. Rosen, M. J. Eckart, N. M. Ceglio, A. U. Hazi, H. Medecki, B. J. MacGowan, J. E. Trebes, B. L. Whitten, E. M. Campbell, C. W. Hatcher, A. M. Hawryluk, R. L. Kauffman, L. D. Pleasance, G. Rambach, J. H. Scofield, G. Stone, and T. A. Weaver, Phys. Rev. Lett. 54, 110 (1985).
5. H. Maezawa, Y. Suzuki, H. Kitamura and T. Sasaki, to be published in Nucl. Instrum. & Methods (1986).
6. M. Born and E. Wolf, Principles of Optics, (Pergamon Press, London, 1970) p.491.
7. A. H. Compton and S. K. Allison, X-rays in theory and Experiment, (D. Van Nostrand Co. Inc, New York, 1935) p.279.
8. G. Kellstrom, Nova Acta Reg. Soc. Sci. Ups. 8, 5 (1932)
9. V. Bonse and W. Graeff, X-ray Optics, (Springer-Verlag, Berlin, 1977) p.93.
10. J. T. Winthrop and C. R. Worthington, Phys, Lett. 15, 124 (1965).
11. G. W. Stroke, Appl. Phys. Lett. 6, 201 (1965).

SCANNING SOFT X-RAY MICROSCOPY AT THE NATIONAL SYNCHROTRON LIGHT SOURCE

C. Jacobsen, J. Kirz[@], I. McNulty, and R. J. Rosser,
Physics Department, SUNY at Stony Brook, Stony Brook, N.Y., 11794[*]

C. Buckley, R. E. Burge, M. T. Browne, R. Cave,
P. Charalambous, P. J.Duke, J. M. Kenney, A. G. Michette, and G. Morrison
Physics Department, King's College, London, England[+]

F. Cinotti and H. Rarback
National Synchrotron Light Source, Brookhaven National Laboratory, Upton, N.Y. 11973,[&] and

J. Pine,
California Institute of Technology, Pasadena, Ca. 91125[#]

ABSTRACT

In scanning soft X-ray microscopy a quantitative map of specimen absorptivity is obtained. Sharp changes in the transmission spectrum at absorption edges provide contrast and are the basis for elemental analysis by comparison of images taken at differing wavelengths. We have developed an instrument that uses monochromatized synchrotron radiation as its source. A collimated portion of this source is focused with a Fresnel zone plate to a submicron spot, across which the specimen is scanned under computer control. To illustrate the microscope's capabilities, we have imaged whole cultured cells in a fixed but wet and unstained state, and measured the distribution of calcium in thin sections of human bone tissue.

I. INTRODUCTION.

The development of X-ray microscopes has been largely motivated by the needs of the biologist. One of the prime requirements is to view intact (wet, unstained) biological

@ J. S. Guggenheim Fellow.
* Supported in part by the National Science Foundation.
+ Supported in part by the Science and Engineering Research Council.
& Supported by the Department of Energy.
# Supported in part by an NIH Biomedical Research Support Grant.

material at a resolution unattainable with the optical microscope. In addition there is considerable interest in measuring the distribution of various elements within the specimen at high resolution. The motivation, and various approaches to the problem have been reviewed recently [1-3]. Important new results have been presented at this Conference [4,5], and at recent meetings[6,7].

We have chosen to build a scanning microscope. The advantage of this instrument is that it lends itself naturally to electronic detectors, and thereby to both a quantitative, computer-based imaging system, and to a system that minimizes radiation exposure to the specimen[8].

The microscope has been operating at the VUV ring of the National Synchrotron Light Source. It is the first instrument to use a zone plate to form a microprobe of soft x rays. Prior to this, either a microfocus X-ray tube[9] or a pinhole[10] was used to form the probe. An alternative approach, being developed by Spiller[11], makes use of mulilayer mirrors at normal incidence. Other zone plate based scanning microscopes are under development at King's College, for use at Daresbury[12], in Gottingen for use at BESSY[13], and in Hefei [14].

The zone plates used in our microscope were electron beam fabricated structures. The first version, made in 1982 at IBM by Kern and coworkers[15] consists of gold rings supported on a silicon nitride membrane. The finest rings of this zone plate have a width of 150 nm. It has been in use for over three years, and has maintained its good performance throughout. All previously reported results from the Stony Brook microscope[8,16,17] were obtained with this zone plate.

During the past year we have started to use zone plates fabricated at King's College, London[18]. These zone plates were produced by contamination writing onto thin carbon substrates, and have therefore a carbonaceous structure. (See the paper by Buckley et al.[19] for progress and improvements). The finest rings have a width of 75 nm. In this paper we show results obtained by both the IBM and the King's College zone plates.

The diffraction limited resolution of the ideal zone plate is given in first order by the width of the finest ring multiplied by 1.22. The size of the microprobe formed by the zone plate becomes diffraction limited if the illumination is coherent. Since we wish to approach this diffraction limited spot size using an incoherent source, the necessary coherence is generated by monochromatization (temporal coherence) and collimation (spatial coherence).

II. THE MICROSCOPE.

The electron beam circulating in the storage ring forms the entrance slit for the toroidal grating monochromator (TGM). This simple device operating at near grazing incidence (5.7 degree total deflection) refocuses the beam and selects the wavelength with a resolving power of 300 near 400 eV. Near the exit slit of the monochromator a small pinhole (20 μm - 100 μm in diameter) collimates the radiation. This pinhole is the fixed source that is demagnified by the zone plate to form the microprobe. The demagnification factor is 120 for the IBM zone plate, about 600 for the King's College zone plate. (Dimensions are summarized in Table I).

Table I. Zone plate parameters.

	IBM zone plate	King's C. zone pl.
Outer diameter (μm)	89	43
Apodized diameter (μm)	48	17
Total number of zones	140	138
Focal length at 3.2 nm	4.7 mm	1.05 mm
Outer zone width	150 nm	75 nm
Substrate	silicon nitride	carbon
Zone rings	gold	carbon

Located between the pinhole and the zone plate is a 200 μm diameter, 0.12 μm thick silicon nitride window through which the radiation escapes the vacuum into the atmoshere. The specimen is mounted on a stage that can be moved into the plane where the microprobe is focused. The space between the vacuum window and the proportional counter is flooded with helium to reduce losses due to absorption in the air. The zone plates are fabricated with an opaque central region. The microprobe is formed in the shadow cast by this opaque disk. A collimator, slightly smaller than the opaque disk, is placed between the zone plate and the specimen to eliminate unfocused radiation.

The stage scans the specimen in this plane under computer control using piezoelectric transducers. X rays transmitted by the specimen are detected by a flow proportional counter. These components can be seen in Figure 1. Counts collected during a fixed time period are used as a measure of the specimen absorptivity at a given point, and are used to form the image. This image is displayed in real time and stored on disk for off-line analysis.

III. PERFORMANCE.

Neurons and fibroblasts were grown at the California Institute of Technology on finder grids coated with about 100 nm of nitrocellulose. The finder grid facilitates correlation of optical and x-ray micrographs, and rapid positioning of the interesting area to be imaged in our instrument. The cultures were fixed before examination.

To image these cells, a culture was drained until only the medium that adhered to the cells was left. It was covered by another nitrocellulose film supported on a

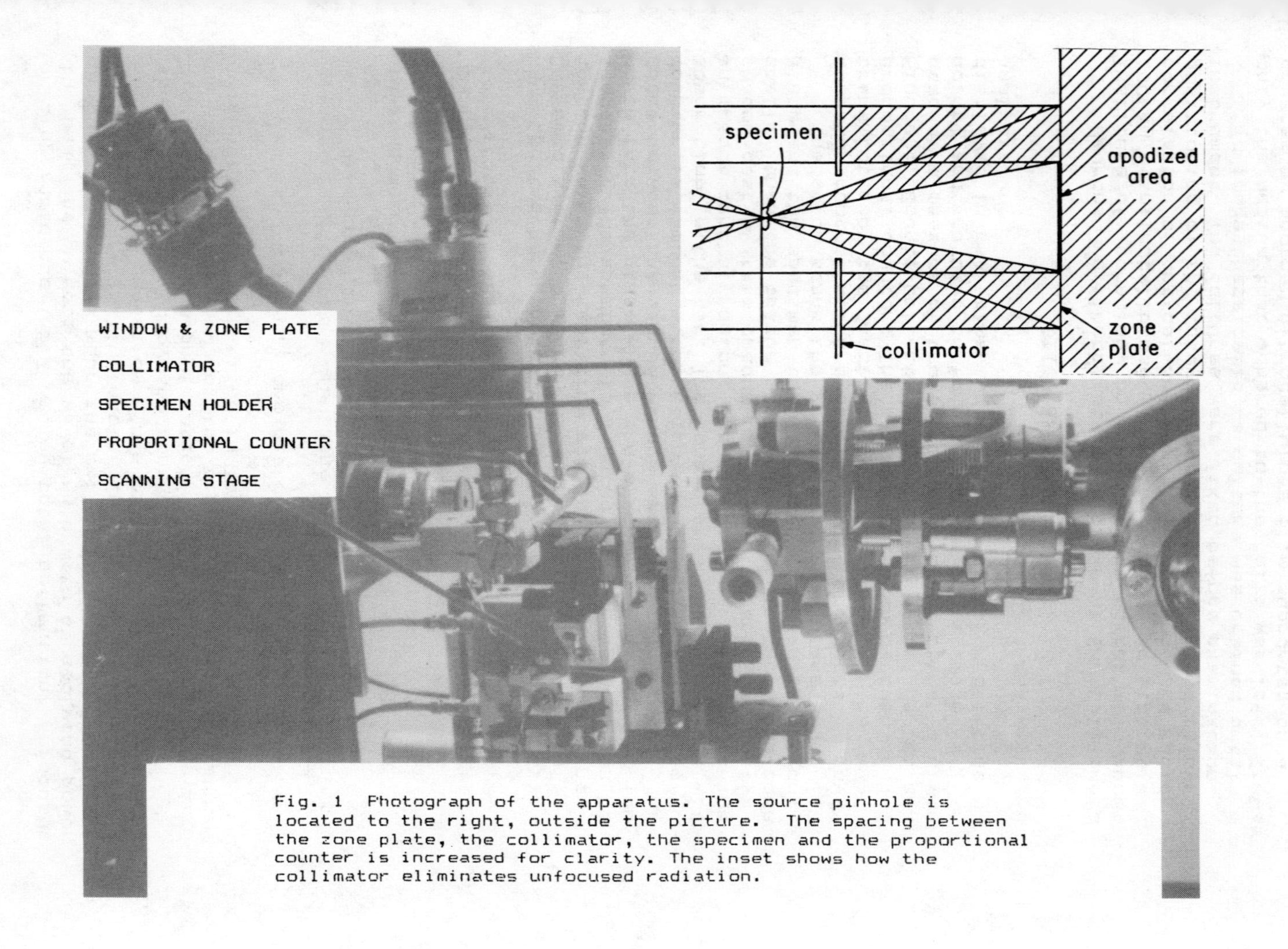

Fig. 1 Photograph of the apparatus. The source pinhole is located to the right, outside the picture. The spacing between the zone plate, the collimator, the specimen and the proportional counter is increased for clarity. The inset shows how the collimator eliminates unfocused radiation.

plastic ring to avoid dehydration. This sandwich was then mounted on the stage for imaging. Phase contrast optical micrographs were taken before and after imaging with X rays to look for gross damage due to radiation or dehydration. In most cases, where the culture was exposed to the X-ray beam for about 10 hours, no visible damage was found.

Figure 2 is that of a glial cell from a chick ciliary ganglion after two days in culture. It was obtained with the King's college zone plate at a wavelength of 3.2 nm. This optical element was shown to be capable of image formation with 0.14 μm resolution, and a collection efficiency of about 1%.

To demonstrate the ability of the microscope to map the spatial distribution of a chemical element, we examined sections of bone tissue. The IBM zone plate was used in this investigation. Pairs of X-ray micrographs were taken at closely spaced wavelengths (3.55 nm and 3.58 nm) near the L III edge of calcium, where the absorption coefficient for that element undergoes dramatic change. The logarithmic difference of the images, appropriately scaled, provides a measure of the amount of calcium present at each point[17,20]. Such information is of interest in the study of a variety of bone diseases, such as osteoporosis.

Pathological specimens of human iliac bone tissue were generously supplied by M. C. Voisin. These were fixed and sectioned prior to imaging. Differences in calcium distribution in specimens from individuals suffering from different disorders is clearly seen. Two examples are shown, one from a young woman (Fig.3) with a relatively mild case of osteoporosis, the other from an older man (Fig. 4) with a severe case of the desease. Detailed analysis of this series of images is in progress, and will be published elsewhere.

IV PLANS FOR THE FUTURE.

Due to the limited amount of coherent flux available from the beamline used in these experiments, the time required to form the images shown was on the order of an hour. For a practical instrument of biological research that is too long. To overcome this problem, an undulator beamline is under construction on the X-ray ring at the National Synchrotron Light Source[21]. This new beamline is designed to provide at least three orders of magnitude more coherent flux. It should make it possible to collect high resolution images in a minute or less.

To reach higher resolutions, finer zoneplates will be required. Such optical elements are being fabricated at King's College, and at IBM, in collaboration with the Lawrence Berkeley Laboratory Center for X Ray Optics[22].

We are also interested in developing the capability to image cultured cells that at least at the outset are

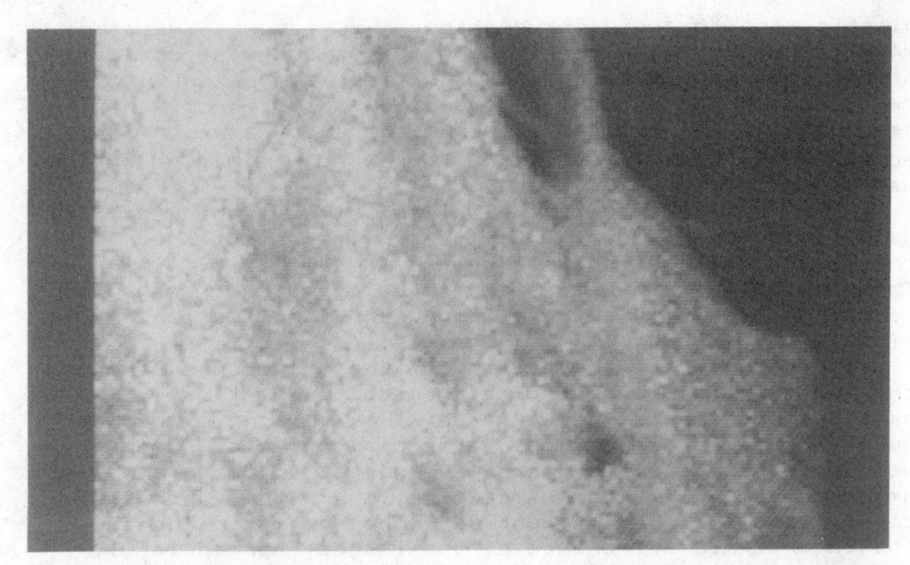

Fig. 3. Calcium distribution map from a section of iliac bone from a young woman. Lighter shades correspond to higher concentrations of calcium. The edge of the section is seen in the upper right. Image area is $48 \times 48\ \mu m^2$, with 120×120 pixels.

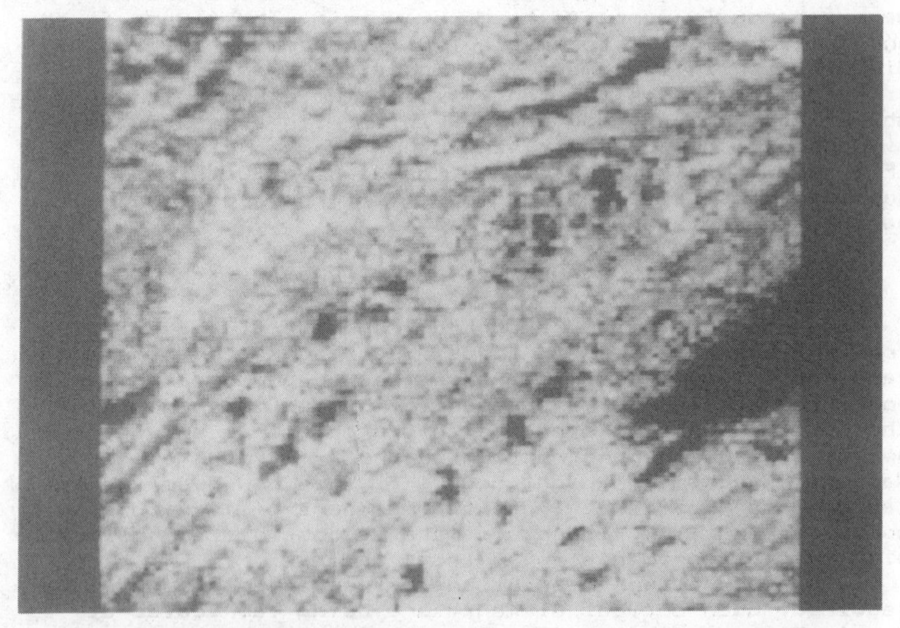

FIG. 4. Calcium distribution map from a section of iliac bone of an older man with severe osteoporosis. Reduced calcium concentration relative to Figure 3 is seen. Image area is $72 \times 72\ \mu m^2$, with 120×120 pixels.

alive. To that end we have installed an incubator and sterile hood near the microscope, that is available for investigators who wish to make use of the instrument.

V. ACKNOWLEDGMENTS.

We are grateful to the many people who have contributed to this work. In particular we thank R. Feder, D. Kern and D. Sayre from IBM; W. Thomlinson from Brookhaven; M. Howells from the Lawrence Berkeley Laboratory; B. X. Yang from Stony Brook; M. C. Voisin from Creteil; and D. Deeds, H. Rayburn and T. Stevens from Caltech for their participation, help, and advice. This work was performed in part at the National Synchrotron Light Source, supported by the U. S. Department of Energy under Contract DEAC0276CH00016.

REFERENCES.

1. J. Kirz and H. Rarback, Rev. Sci. Inststrum. 56, 1 (1985).
2. E. Spiller, in Handbook on Synchrotron Radiation, Vol. 1, ed. by E. E. Koch, D. E. Eastman, and Y. Farge. (North Holland, Amstrdam, 1983)
3. M. Howells, J. Kirz, D. Sayre, and G. Schmahl, Phys. Today 38, 22 (August 1985).
4. G. Schmahl et al., these proceedings.
5. H. Aritome et al., these proceedings.
6. Proc. Int. Conf. X-Ray and VUV Synchrotron Radiation Instrumentation, Stanford, Ca. Nucl Instr. and Meth. (in press).
7. L. Beese, R. Feder, and D. Sayre, Biophys. J. (to be publ.)
8. C. Jacobsen et al. (submitted for publ.)
9. H. H. Pattee, Jr., J. Opt. Soc. Am., 43, 61 (1953).
10. P. Horowitz and J. A. Howell, Science 178, 608 (1972).
11. E. Spiller, in X-Ray Microscopy, Ed. by G. Schmahl and D. Rudolph (Springer, Berlin, 1984) p 226.
12. P. J. Duke, ibid. p. 232.
13. B. Niemann, ibid. p. 217.
14. X.-s. Xie et al. in Ref. 6.
15. D. P. Kern et al., J. Vac. Sci. Technol. B1, 1096 (1983).
16. H. Rarback et al., in X-Ray Microscopy, op. cit. p. 203.
17. J. M. Kenney et al., J. Microsc. 138, 321 (1985).
18. A. G. Michette et al., in X-Ray Microscopy, op. cit. p. 109.
19. C. J. Buckley et al., these proceedings.
20. J. M. Kenney, Ph. D. Dissertation, SUNY at Stony Brook, 1985.
21. H. Rarback et al. in Ref. 6.
22. Y. Vladimirsky, priv. comm.

STABILITY OF MULTILAYERS FOR SHORT-WAVELENGTH OPTICS

E. Ziegler, Y. Lepetre, Ivan K. Schuller, J. Viccaro
Argonne National Laboratory
Materials Science & Technology Division
Argonne, IL 60439

E. Spiller
IBM T. J. Watson Research Center
Yorktown Heights, NY 10598

ABSTRACT

A variety of multilayer mirrors with transition metal absorber layers (W-C, WRe-C, Co-C, and Cr-C) have been fabricated and tested up to 1000 C using standard θ-2θ x-ray diffraction, Debye-Scherrer scattering and microcleavage transmission electron microscopy.

The θ-2θ x-ray diffraction during annealing shows the Bragg peak position to shift toward lower angles with increasing temperature. This irreversible shift starts at around 300 C and is equivalent to as much as 12% expansion of the multilayer period with a temperature change from ambient to 750 C.

In all cases a crystallization occurs in the metal component between 650-750 C. The different types of crystalline compounds formed have been identified by the Debye-Scherrer technique. As a consequence of this crystallization abrupt changes occur in the multilayer structure. Electron microscopy shows that the surface roughness increases by formation of hillocks and the layered structure is destroyed. Moreover the x-ray reflectivity decreases considerably.

The expansion and crystallization are of great importance in cases where a precise multilayer period is required or in devices intended for high x-ray flux applications.

INTRODUCTION

Metallic multilayers are expected to fill the lack of reflectors between the natural crystals x-rays applications domain and the noble heavy metals U.V. reflecting materials.[1] In addition, to be useful, high thermal or radiation load capabilities are required. This is the case for applications such as synchrotron radiation,[2] and free electron laser.[3]

To the best of our knowledge, no high flux-long time experiments have been performed. A short time-high flux experiment (700 MW/cm^2 in a 20 nsec pulse) reports[4] that the W-C multilayer reflecting surface placed at 15 cm was destroyed during the reflection of a 1 keV radiation pulse peak. On the other hand, W-C radiation exposure tests[5] at SSRL for 2.5 months and at ACO for 300 hours showed no change at CuKα energy (8 keV). Earlier annealing studies[6] at 400°C for 4 hours in vacuum for ReW/C, Cr/C, V/C, and Ru/C multilayers showed no appreciable changes in X-ray reflectivity, while Ni/C, Co/C and Fe/C multilayers crystallized and lost their reflectivity. A volume expansion of the layers with temperature has been also reported.[6,7]. Expansions as large as 5% have been observed when the samples were heated to 400°C.[6] This interpretation is based on the observation of a shift of the small angle X-ray peaks to lower angles.

The purpose of this paper is to study the behavior of metallic multilayers under thermal load, which as a first approximation is expected to lead to similar consequences as under radiation load.

MULTILAYER FABRICATION

A variety of multilayers mirrors with transition metal absorber layers (W-C, WRe-C, Cr-C, and Co-C) have been prepared on Si substrates in a multisource evaporation system.[8,9] Here, two electron beam guns produce evaporation at typical rates of 0.5 Å/sec in a vacuum of ~ 10^{-6} Torr. Some W-C multilayers were also prepared using a multisource triode sputtering system. In this case the substrates rotate on a planetarium and the sputtering targets are alternately commuted.

EXPERIMENTAL SET-UPS

X-ray diffraction measurements on the multilayers were performed with a Rigaku D/max diffractometer at CuKα line. Each sample, placed inside a controlled atmosphere chamber, was heated by electric resistances surrounding the multilayer and its holder. A temperature controller was able either to maintain a given temperature (between ambient temperature and 1000°C) or to perform a given heat-cool thermal cycle. A large Al foil angular window and the possibility of sample alignment allowed in-situ X-ray reflectrometry measurements. Transmission Electron Microscopy (TEM) and electron diffraction using a microcleavage preparation technique described earlier[10] were made on a Jeol 100 CX microscope using a hot stage grid holder. Debye-Scherrer scattering was also performed on these multilayers before and after annealing.

EXPERIMENTS

We have measured the first Bragg peak position as a function of temperature for different ReW-C multilayers. Figure 1 shows the relative variation of the $\lambda/\sin\theta$ parameter from the $\lambda/\sin\theta_o$ value at the initial ambient temperature. All the samples exhibit few changes to 300°C followed by a 4% increase of $\lambda/\sin\theta$ to 500°C, and then depending on the sample (period, number of layers, atmosphere) a variation of 5% to 12% between 600 and 750°C. This variation is due to the multilayer period (d) variation ($2d = \lambda/\sin\theta$). In order to understand the underlying physical mechanism of this expansion, a variety of other studies have been performed.[11]

The irreversibility of the expansion has been studied by performing a series of cyclic thermal annealing experiments. Figure 2 shows the evolution of the first Bragg peak position (2θ) for a ReW-C stack versus "Temperature Log (time)" as is customary in amorphous glasses which exhibit irreversible behavior.[12] The black symbols in figure 2 show that the peak position decreases with increasing temperature. The white symbols show that the peak position does not return to its initial value with decreasing temperature: i.e. the expansion remains at the last value attained and even continues to increase slightly. Since a heat-cool-heat cycle is not instantaneous after the second heating step the peak position is always lower (black symbol).

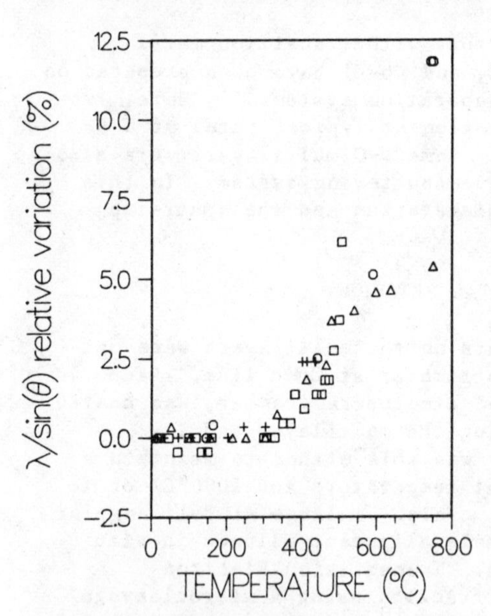

Fig. 1 $\lambda/\sin\theta$ relative variation of the first order Bragg peak value versus temperature for ReW-C multilayers. The change from the value at ambient temperature is given in percent.

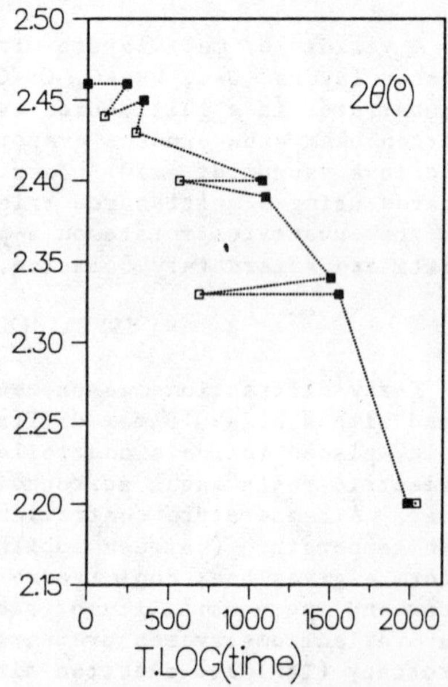

Fig. 2 Evolution of the first order Bragg peak position (2θ) versus temperature x log(time) for a 100-layer ReW-C stack (period = 36Å).

Together with the study of the X-ray peak position (Fig. 1) we have followed the evolution of the first Bragg peak intensity versus temperature (figure 3). Care was taken that the observed effects are not due to temperature induced changes in diffractometer alignment. Moreover, most samples were annealed in an Ar environment in order to avoid oxygen contamination. In this case, a correction was applied to account for changes in the absorption due to the Ar gas. Some samples were also annealed in air (curve xxx Fig. 3) in order to study their behavior in a room environment. Figure 3 shows that the intensity of the first order

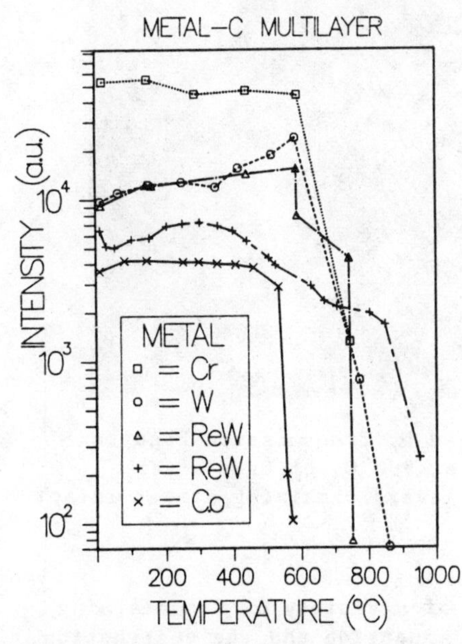

Fig. 3 Intensity of the first order Bragg peak as a function of temperature.

Bragg peak in all cases is relatively constant and then precipitously decreases at 600 to 750C (500C in air). A top TEM view after this step of a W-C (Fig. 4a) and a Cr-C (Fig. 4b) multilayer clearly shows the formation of large grains. Interestingly the C layers remain almost intact as observed in the upper corner of fig. 4b for Cr-C. Electron Diffraction (Fig. 5) shows crystalline rings which were not present at the initial stage. In order to study the nature of the crystallites we performed Debye-Scherrer diffraction on sample scrapped off the substrate. These studies indicate that the abrupt decrease of the first Bragg peak intensity corresponds to the crystallization of the metallic layer which destroys the layered structure. Table I summarizes the results and shows the crystalline material formed in each case. Notice that i) the Debye-Scherrer diffraction before annealing shows the multilayers to be amorphous or very disordered (Fig. 6a); ii) crystallization of W or Re at ~ 750°C (Fig. 6b) causes the destruction of the layers, iii) this occurs before the formation of carbides at higher temperatures.

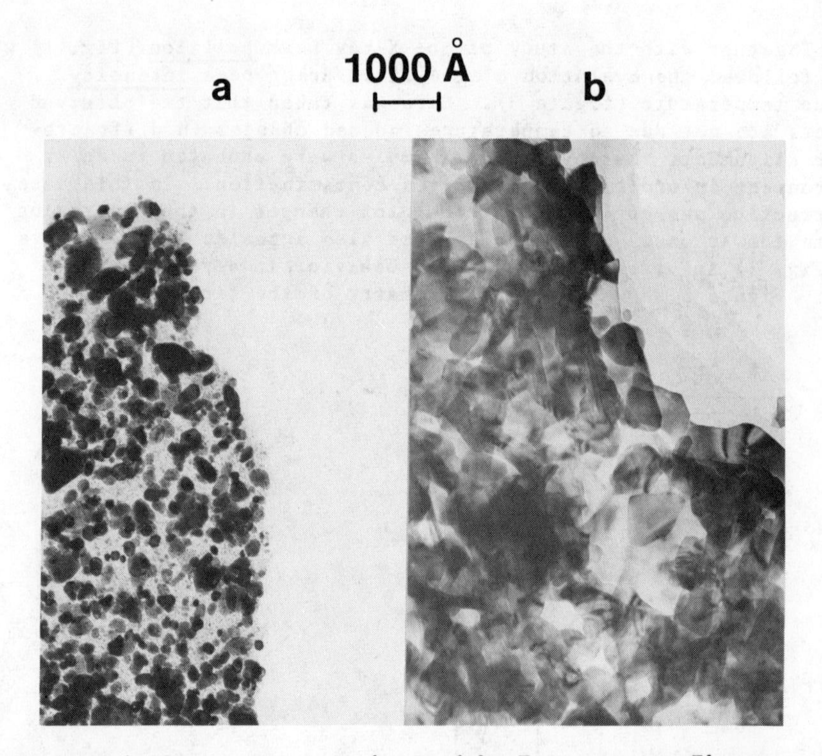

Fig. 4 Multilayer top view obtained by Transmission Electron Microscopy a) W-C after annealing at 900°C, b) Cr-C after annealing at 745°C. Notice the C layers remaining almost intact (upper corner).

CONCLUSION

X-ray diffraction measurements of a variety of C containing multilayers show an anomalous layer expansion and the destruction of the layered structure around 700°C. The electron microscopy confirms that the destruction is due to crystallization, and the Debye-Scherrer gives an unambiguous determination of the crystallites formed.

Because this expansion and crystallization seemingly occurs for all metal-carbon multilayers, care must be taken to include these changes in design consideration where a precise multilayer period is assumed. In the case of high X-ray fluxes efficient cooling is necessary to limit the optics temperature. We hope that these type of studies will lead to the development of more stable multilayered structures in the near future.

We wish to thank P. Dhez, T. Morrison and H. Homma for useful conversations and R. Rivoira for some W-C multilayer samples. This work was supported by the US-DOE, BES-Materials Sciences, under contract #W-31-109-ENG-38.

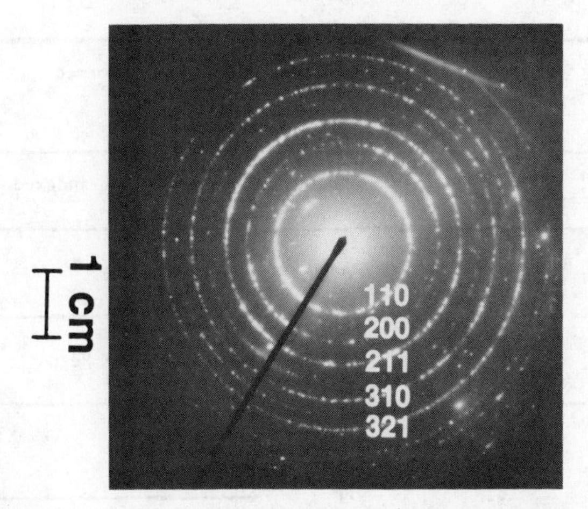

Fig. 5 Electron Diffraction of a W-C multilayer after crystallization showing seven distinct tungsten rings.

Fig. 6 a) Debye-Scherrer picture of a ReW-C multilayer after fabrication showing only the presence of silicon substrate.
b) Corresponding Debye-Scherrer picture after annealing at 750°C showing a crystalline Re pattern.

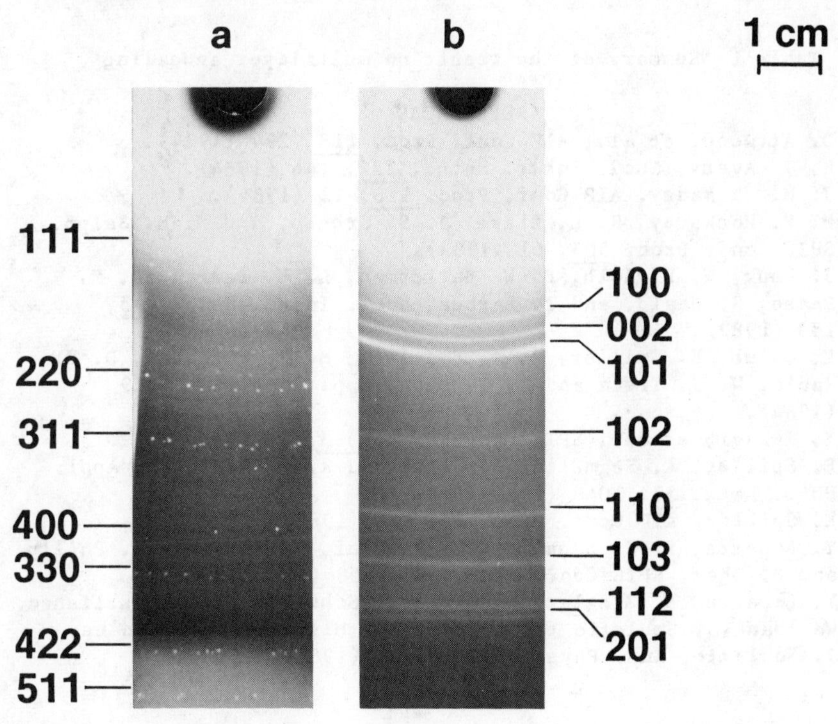

High Z material (d = period)	Number of layers	Gas environment	Crystal formed	Annealing temperature
ReW (d = 77 Å)	34	Ar	hexagonal not indexed	950°C
ReW (d = 36 Å)	110	Ar	Re	750
		Air	WO_3 + Re	500
W (d = 32 Å)	51	Ar	W	770
			WO_3	630
Co (d = 32 Å)	128	Ar	not indexed	575
Cr (d = 71 Å)	27	Ar	Cr_2O_3	745

Table 1 Summary of the result on multilayer annealing

REFERENCES

1. D. Attwood, et al., AIP Conf. Proc. 118, 294 (1984).
2. R. T. Avery, Nucl. Instr. Meth., 222, 146 (1984).
3. J. M. J. Madey, AIP Conf. Proc. 118, 12 (1984).
4. M. P. Hockaday, R. L. Blake, J. S. Grosso, and M. M. Selph, SPIE Conf. Proc. 563, 61 (1985).
5. J. Wong, W. L. Roth, B. W. Batterman, L. E. Berman, D. M. Pease, S. Heald, and T. Barbee, Nucl. Instr. Meth. 195, 133 (1982).
6. L. Golub, E. Spiller, R. J. Bartlett, M. P. Hockaday, D. R. Kania, W. J. Trela and R. Tatchyn, Appl. Opt. 23, 3529 (1984).
7. Y. Takagi, et al., SPIE Conf. Proc. 563, 66 (1985).
8. E. Spiller, A. Segmüller, J. Rife and R. P. Haelbich, Appl. Phys. Lett. 37, 1048 (1980).
9. E. Spiller, AIP Conf. Proc. 75, 124 (1981).
10. Y. Lepetre, I. K. Schuller, G. Rasigni, R. Rivoira, R. Philip and P. Dhez, SPIE Conf. Proc. 563, 258 (1985).
11. Y. Lepetre, E. Ziegler, and Ivan K. Schuller, to be published.
12. We thank J. Souletie for suggesting this experiment to us. J. Souletie, Ann. Phys. Fr., 10, 69 (1985).

X-RAY FOURIER OPTICS

V.V.Aristov, A.I.Erko, V.V.Martynov

Institute of Problems of Microelectronics Technology and Superpure Materials, USSR Academy of Sciences 142432 Chernogolovka, USSR

ABSTRACT

The paper deals with the coherent and incoherent production of Fourier images. Modelling experiments on obtaining Fourier images in total incoherent radiation of the visible range have carried out. The results of the experimental observation of Fourier images of two-dimensional periodic objects in the range of ultra-soft X-ray radiation are reported. An undulator at 2.5 nm wavelength was used as a coherent X-ray radiation source. Diffraction images up to order 16 of the Fourier-plane with a minimum linewidth 0.2 μm have been obtained.

INTRODUCTION

The soft X-ray and vacuum UV radiation wavelength range has become the subject of intensive studies thanks to the advent of potent short-wave radiation sources, focusing instrumentation and focusing techniques, X-ray beam collimation. One of principal benefits of this wavelength range lies in the fact that resolution of optical systems reaches the maximum at radiation wavelength ~ 5 nm and is restricted only to the manufacturing process of optical elements. At greater wavelengths the resolution is confined to classical focusing limit and for short wavelengths to a weak radiation interaction with matter. The problems of coping with soft X-ray range presents difficulties, for in this case one fails to use optical analogs of focusing elements - lenses based on the radiation refraction effect. A high absorption coefficient of basically all materials in the X-ray radiation wavelength range permits fairly efficient X-ray lenses to be designed. Radiation focusing, deviation or separation of X-ray beams are feasible only through the use of grazing incidence or coherent optical diffraction effects.

Among X-ray diffraction elements can be included periodic and quasi-periodic structures fabricated either by thin-film technology (multilayer mirrors)[1-4] or planar technology (diffraction gratings, Fresnel focusing lenses). Modern microelectronics methods permit fabricating zone structure with resolution up to 100 nm[5]. This dictated limiting resolution potentialities and is likely to be a record result nowadays. However, zone plates with an absorbing pattern are found to have a fairly low diffraction efficiency. Zone optics efficiency may be increased while employing phase materials in the X-ray spectrum area that allows a better diffraction efficiency to the first order up to 30-40%. Another possible image transmission method with high resolution is the use of coherent optics techniques, the Talbot effect, for example, i.e. the projection structures, so called Fourier optics[6-10]. The paper offers researches into the principles of the progress of X-ray Fourier optics based on the Talbot effect and experimental results on high resolution image transmission in soft X-rays.

COHERENT IMAGE TRANSMISSION

Fig. 1 exhibits the scheme of coherent image transmission. For simplicity, without any loss of generality, we shall discuss a one-dimensional case. Let $\psi(X)$ be the amplitude distribution in the image plane, $\gamma(\xi_1 - \xi_2)$ the spatial coherency radiation function in the object plane, $g(\xi)$ the amplitude distribution function in the object plane. N is the number of the periods, d the period size, $E(x-\xi) = \frac{1}{r} \exp\left\{-\frac{ik}{2R}(x-\xi)^2\right\}$ the function of radiation propagation in the Fresnel approximation (further on the factor 1/r and constant factors before the integral will be omitted for simplicity). Then for monochromatic radiation

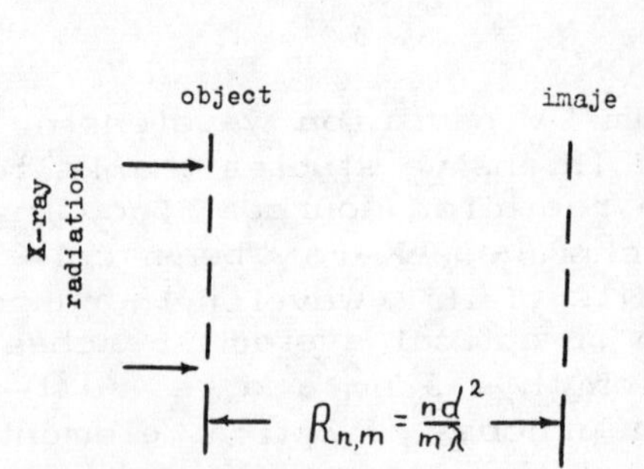

Fig. 1. Scheme of coherent image transmission.

$$|\psi(x)|^2 = \int_0^{Nd}\!\!\int \gamma(\xi_1 - \xi_2)\, g(\xi_1)\, g^*(\xi_2)\, E(x-\xi_1)\, E^*(x-\xi_2)\, d\xi_1\, d\xi_2 \qquad (1)$$

If the size of the source is sufficiently small so that the half-width of the spatial coherency function is larger than the size of the object mask, $\Delta > Nd$, then $\gamma(\xi_1 - \xi_2) \approx 1$ and

$$|\psi(x)|^2 = \int_0^{Nd} g(\xi_1)E(x-\xi_1)d\xi_1 \int_0^{Nd} g^*(\xi_2)E^*(x-\xi_2)d\xi_2 = \left|\int_0^{Nd} g(\xi)h(x-\xi)d\xi\right|^2 \quad (2)$$

where $h(x) = \sin\left(\frac{\pi N d x}{R\lambda}\right) / \left(\frac{\pi N d x}{R\lambda}\right)$ is the response dependence of the given image transmission system[8]. That is, in the imaging plane there are summarized amplitudes from all the object points and coherent image transmission takes place. An image of a specific period is transmitted in the same way as it would be done using an objective with Nd in diameter and the focal length $f'_{n,m} = \frac{nd^2}{m\lambda}$ (n, m = 1, 2,...; n is the number of the self-reproduction plane, m is the number of the period multiplication) in spatially coherent radiation[8].

The Talbot effect realization in the X-ray wavelength range presents some difficulties because of the lack of coherent radiation sources. The first experiement was performed with a purpose to obtain a diffraction grating with a small period[11], Fourier image of one-dimensional rectangular grating being produced. The first Fourier image of two-dimensional structures in X-ray radiation was given in [9]. The experiment was carried out using an undulator installed at the Photon Factory of the National Laboratory for High Energy Physics, KEK, Tsukuba, Japan. The undulator radiation source parameters were as follows: the number of periods was 38, period length 6 cm, the spectral range (λ) 1.0-3.0 nm, band width $\Delta\lambda/\lambda$ =0.11, the angular source dimensions σ_x (mrad)=0.534, σ_y (mrad)=0.065. The microphotograph of the object which was a polyimide membrane transparent to X-rays with a periodic gold pattern of a minimum size 0.3 μm is shown in Fig.2a. The pattern was fabricated by electron-beam lithography and the lift-off process.

At these radiation parameters and a 10 m source-object distance the image resolution to be estimated reaches 0.5 μm according to the object-registrator distance. X-ray sensitive PMMA polymer on a silicon wafer was used as a registrator. Fig.2a shows a microphotograph of a gold X-ray mask on a polyimide membrane transparent to X-rays. The two-dimensional structure period was 5 μm, the frame linewidth 0.3 μm.

The Fourier image of the mask structure obtained at a distance of 10.1 mm with radiation wavelength 2.5 nm is presented in Fig.2b. The image resolution is not worse than 0.6 μm.

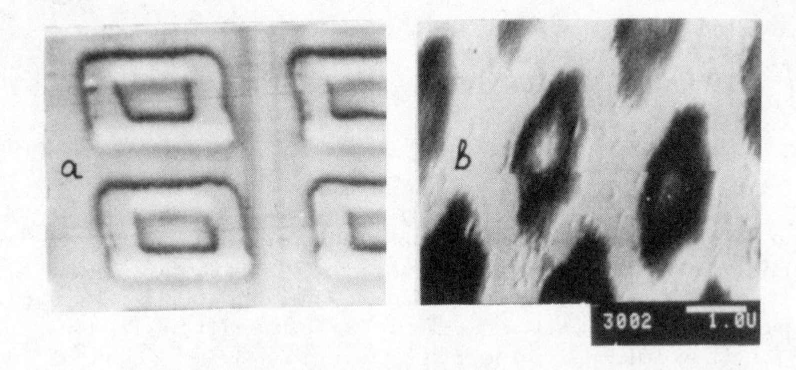

Fig.2. (a) Microphotograph of the X-ray mask. Structure period 5 μm, linewidth 0.3 μm. (b) Mask Fourier image at the 10.1 mm distance with a PMMA resist.

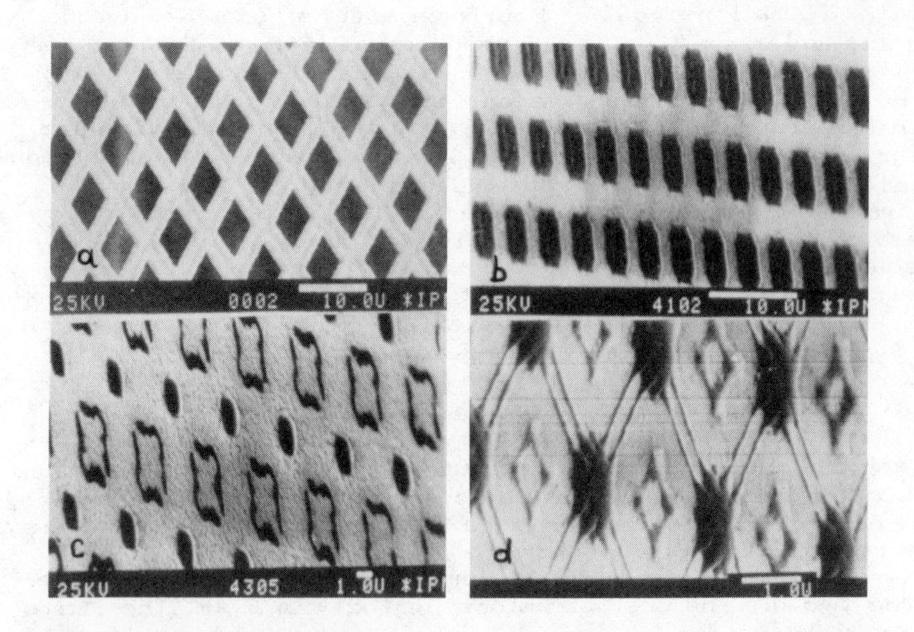

Fig.3. (a) Microphotograph of the copper grid of a 12 μm period. (b), (c), (d) Fourier images at 35, 17 and 4 mm distances, respectively.

High coherence of the undulator radiation enabled us to obtain a Fourier image of a copper grid with a period of 12 μm (Fig.3a) up to order 16 of n=1 (m=16). Fig.3b–d exhibit the images for m=2,4,8 and 16 at the distances of 35 mm, 17, 4 mm, respectively. The resolution of the Fourier image of the order of 16 is 0.2 μm.

The experimental data obtained prove that the Talbot effect can be applied to transmit high resolution images in soft X–ray radiation. An interesting characteristic of the method under consideration is the possibility of the diffraction synthesis of complex structure images with very small elements. The initial object structure therewith can be of an element size which is by an order of magnitude larger than required in the image. The experimental diffraction images, for instance, are of 0.2 μm linewidth while the smallest copper grid was 4.7 μm in size.

NONMONOCHROMATICITY OF RADIATION

Now let radiation be nonmonochromatic. Then for the given distances R_1 and R_2 in radiation with a wavelength satisfying the equation $1/R_1 + 1/R_2 = m\lambda/nd^2$, ψ (x) will satisfy equation (2) and for other wavelengths the image will be defocused. The computed effective radiation spectrum contributing to the image is given in Fig.4a.

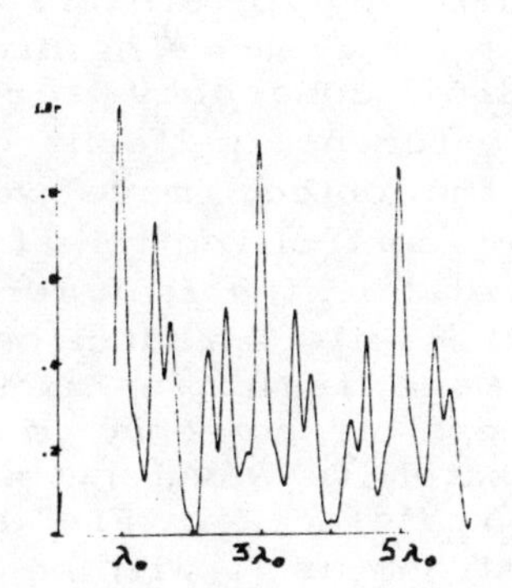

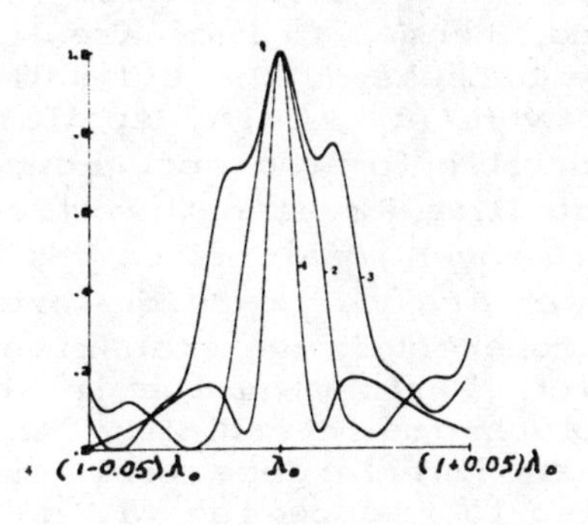

Fig.4. (a) The area of the effective radiation spectrum contributing to the image at coherent transmission. (b) The view of an individual peak of the effective spectrum for various duration ratios of periodic object structure at coherent transmission (1) b/d= 0.05 (2) b/d= 0.1 (3) b/d= 0.15

For the case of an incident plane wave, for example, self-reproduction planes are separated by a distance of $R_{n,m} = \frac{nd^2}{m\lambda} = \frac{n}{m} R_0$, images in the planes with even and uneven n being shifted by a half-period, i.e. they are not superimposed (if the size of any element of the periodic object structure $b < d/2$). Since R and λ enter into the expression of radiation distribution $E(x-\xi)$ in the same manner, then the effective radiation spectrum in Fig.4a is similar to distances from self-reproduction planes $\lambda_{n,m} = \frac{nd^2}{mR} = \frac{n}{m}\lambda_0$. Here the images for even and uneven n are shifted by a half-period, that is, they are nonoverlapping and produce images independently. The imaging contrast deteriorates only with increasing the spectrum width ofthe radiation incident on the object to $\Delta\lambda = 2\lambda_0$, then it is unaffected. A better contrast can be gained with increasing the duration ratio of the object structure (i.e. decreasing the b/d ratio).Images of each peak of the spectrum are overlapping not blurring each other. That is, one is always able to observe Fourier images in polychromatic radiation, but due to the final width $\Delta\lambda$ these images are blurred[8]. As a result of numerical computations we have ascertained a dependence between the peak width of the effective spectrum and the maximum size of a separate element of the periodic object structure b, $\Delta\lambda/\lambda \sim (b/d)^2$. Fig.4b presents computed plots of individual peaks for different duration ratios (b/d). It is seen that a decrease in the peak width and, hence, an increase in resolution are obtainable only by decreasing the b/d ratio that is far from being always convenient. Again, spatial radiation coherency is responsible for the occurrence of interference patterns on satellite wavelengths which make the object image worse. In conventional optics the means of controlling this type of interference image distortions consist in the transfer to incoherent image transmission in the self-reproduction effect. Performing the incoherent image transmission in polychromatic radiation makes it possible not only to eliminate interference distortions on satellite wavelengths but also to reduce the width of a separate peak in Fig.4a down to quasi-monochromatic radiation (as it will be demonstrated below).

TOTAL INCOHERENT IMAGE TRANSMISSION

Fig.5 gives a scheme of incoherent image transmission[8]. Substituting $\gamma(\xi_1-\xi_2) = \sum_{n=0}^{N-1}\delta(|\xi_1-\xi_2|+nd)$ in (1), that is valid if the size of the source mask is $\gg Nd$, where $\delta(x)$ is the delta-function, we obtain

$$|\Psi(x)|^2=\int_0^{Nd} g(\xi_1)E(x-\xi_1)\left\{\sum_{n_2=-INT(\xi_1/d)}^{N-INT(\xi_1/d)-1} g(\xi_1+n_2 d)E[x-(\xi_1+n_2 d)]\right\}^* d\xi_1 \qquad (3)$$

where $INT(\xi_1/d)$ is integer ξ_1/d. Having considered that

$g(\xi)=\int_0^d g(\eta)\left\{\sum_{n=0}^{N-1}\delta(\xi-(\eta+nd))\right\}d\eta$ we get

$$|\Psi(x)|^2=\int_0^d g^2(\eta)\int_0^{Nd}\left\{\sum_{n_1=0}^{N-1}\delta(\xi+n_1 d-\eta)E[x-(\xi+n_1 d-\eta)][\sum_{n_2=-n_1}^{N-n_1-1}\delta(\xi+n_1 d+n_2 d-\eta)\cdot \right.$$
$$\left.\cdot E^*(x-(\xi+n_1 d+n_2 d-\eta))]\right\}d\xi d\eta=\int_0^d g^2(\eta)\,h^2(x-\eta)\,d\eta \qquad (4)$$

where $h(x)=\left\{\sum_{n=0}^{N-1}\delta(x+nd)\right\}*\frac{\sin(\pi N d x/R\lambda)}{(\pi N d x/R\lambda)}$ is the response function of the given optical system, since the expression in braces is the Fourier-image intensity of the δ-functional grating.

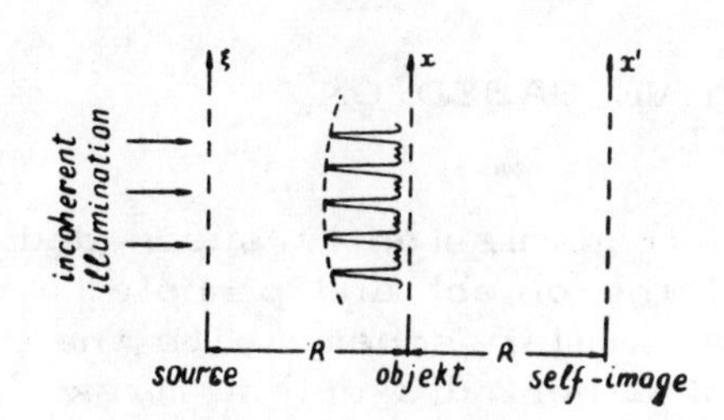

Fig.5. Scheme of incoherent image transmission.

That is, the image is transmitted in the same manner as it is done using an objective Nd in diameter and focal length $f_{n,m}=\frac{nd^2}{m\lambda}$ but in spatially incoherent radiation. An objective of the kind differs from a common one in the transmission procedure when the image is transmitted by point gratings with a period d/m rather than by a single point.

In polychromatic radiation on satellite wavelengths, images of point gratings are to be defocused and then formed again for all points of the object η (see (4)). That is why no interference patterns distorting the object image develop there during coherent image transmission. Since the peak width in the effective radiation spectrum forming an image is $\Delta\lambda/\lambda\sim(b/d)^2$ and we have b=0 (image transmission occurs by point gratings) then $\Delta\lambda/\lambda\to 0$, i.e. each maximum constitutes quasi-monochromatic radiation (if the number of periods N$\gg$1). The computed effective radiation spectrum at incoherent image transmission is given in Fig.6a. The maximum width is practically independent of the duration ratio of the object structure (Fig.6b).

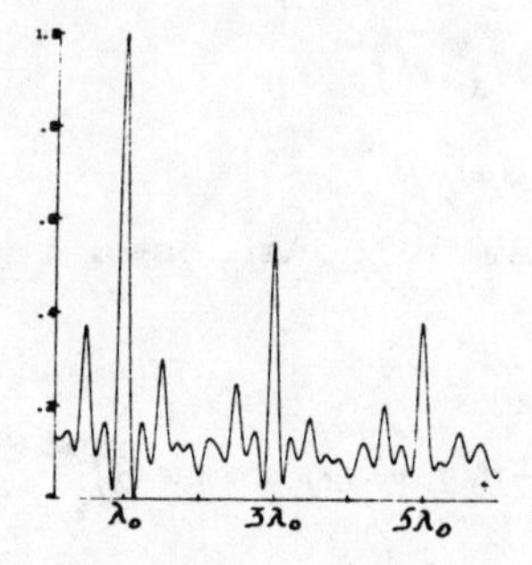

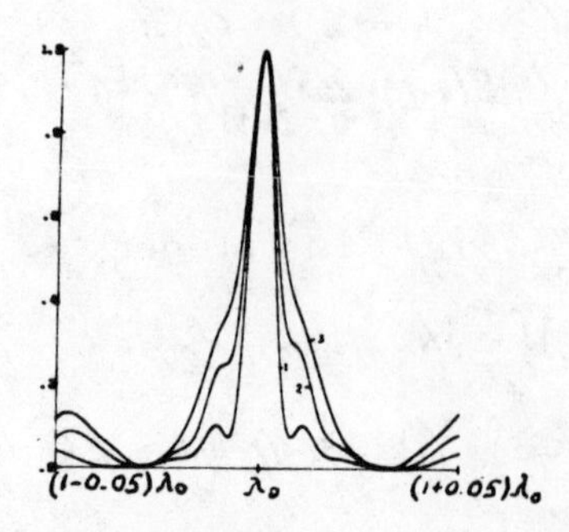

Fig.6. (a) The area of the effective radiation spectrum contributing to the image at incoherent transmission; (b) the view of an individual peak of the effective spectrum for various duration ratios of a periodic object structure; at incoherent transmission: (1) b/d= 0.05 (2) b/d= 0.1, (3) b/d= 0.15

THE DIFFRACTION OBJECTIVE BASED ON THE TALBOT EFFECT

A natural incoherent image transmission is succeeded by interchanging the position of the object and pinhole masks in the scheme in Fig.5. A similar scheme with the use of a single object instead of a periodic object mask in monochromatic radiation has been suggested in[12]. Each point of the object located in the source plane will produce the Fourier image of the pinhole mask in the image plane. These Fourier images overlap incoherently those of masks of pinholes from other object points. In this case, the effective radiation spectrum creating the image is similar to that in Fig.6 and interference patterns in satellite wavelength radiation will be incoherently blurred.

In fact, the pinhole mask in such a scheme provides a diffraction objective with a whole set of focal lengths $f_{n,m} = \frac{nd^2}{m\lambda}$ (n,m=1,2...; n is the number of a Fourier plane without period multiplication, m is the number of period multiplication in the given Fourier plane). The expression relating the distance between the object and objective to that between the objective and image is similar to the lens formula

$$1/R_1 + 1/R_2 = 1/f_{n,m} \qquad (5)$$

In the case of the object as a single periodic structure, R_1 and R_2 can vary continuously and an increase in the size of the object $k=R_2/R_1$ changes in the same manner from 0 to ∞. If the object presents a periodic structure, then besides the condition (7) there is an additional condition to be fulfilled for Fourier images of the pinhole mask of the objective spaced by a period to coincide. Therefore, distances R_1 and R_2, and hence, an increase in $k=R_2/R_1$ vary discretely, the relative aperture therewith growing.

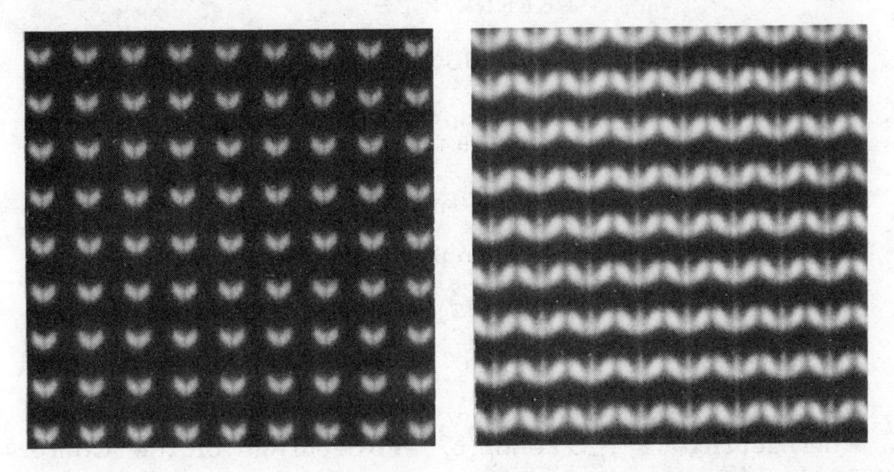

Fig. 7. The image of a two-dimensional object (d=400 m) in polychromatic radiation (a visible range): (a) coherent image transmission; (b) incoherent image transmission using a diffraction objective.

Fig. 7 displays the experimental images of a two-dimensional periodic structure in white light (a visible wavelength range) during coherent (a) and incoherent, using a diffraction objective, (b) image transmission. Interference patterns are well observed on satellite wavelengths during coherent image transmission and disappear at incoherent one.

CONCLUSIONS

The paper deals with the action of nonmonochromatic radiation on the Talbot effect. At a sufficient spatial radiation coherency, Fourier images are shown to occur in radiation with spectrum of any width, the image contrast deteriorating when the spectrum width enlarges up to $\Delta\lambda=2\lambda_0$ and then it is constant and improves only with the duration ratio of the object structure (a decrease in the ratio: a maximum size of a separate element to the period). In order to get rid of false images (interference patterns on satellite wavelengths) and to reduce the effective radiation spectrum contributing to the image up to the quasi-monochromatic one, it is proposed to change from coherent image transmission to incoherent one in the process.

REFERENCES

1. A.V.Vinogradov, B.Ya.Zeldovich, Appl. Opt. 16, 89 (1977).
2. A.D.Akhsakhalyan, S.V.Gaponov et al. Zh. Tekh. Fiz. (USSR) 54, 4, 747 (1984).
3. S.V.Gaponov, F.V.Garin, S.L.Gusev et al. Nuclear Instr. and Methods. 208,227 (1983).
4. S.V.Gaponov, V.M.Genkin et al. Pisma Zh. Eksp. Teor. Fiz. 41, 2, 53 (1985), English Translation.
5. H.Aritome, K.Nagata, S.Namba. "Microelectronic Engineering". 3, 459 (1985).
6. V.V.Aristov, G.A.Bashkina, L.V.Dorozhkina, A.I.Erko, V.V.Martynov. Poverkhnost. Fizika, khimiya, mekhanika. (USSR) 12, 113 (1983).
7. V.V.Aristov, A.I.Erko, Ch.V.Kopezky. "Microcircuit Engineering-82", Grenoble, Proceedings of the Conference, 137.
8. V.V.Aristov, A.I.Erko, V.V.Martynov, Opt. Communs. 53, 159 (1985).
9. V.V.Aristov, S.Aoki, A.I.Erko, S.Kikuta, V.V.Martynov, Opt. Communs., 56, 4, 223 (1985).
10. V.V.Aristov, A.I.Erko, V.A.Kudryashov. "Microelectronic Engineering". 3, 589 (1985).
11. D.C.Flanders, A.M.Hawryluk, H.I.Smith, J. Vac. Sci. Techn. 16, 1949 (1979).
12. Bryngdahl Olaf, JOSA, 63, 4, 416 (1973).

PROPOSAL FOR A FEMTOSECOND X-RAY LIGHT SOURCE

R. W. Falcone and M. M. Murnane
Physics Department
University of California, Berkeley, CA. 94720

ABSTRACT

We propose a new type of laser produced plasma which uses intense, sub-picosecond laser pulses to create highly ionized solids with electron temperatures of several hundred electron volts. The plasma will emit short wavelength radiation on a sub-picosecond time scale for applications including flash x-ray scattering from non-equilibrium condensed matter systems and pumping x-ray lasers.

INTRODUCTION

Intense x-ray sources are commonly produced by focussing high power laser pulses onto solid metal targets[1]. At laser pulse durations greater than about 10 picoseconds the intense light vaporizes solid material, generating an expanding coronal plasma above the surface. A gradient in plasma density develops and the laser is primarily absorbed by the so-called critical surface where the plasma frequency equals the optical frequency. Short wavelength radiation is emitted from this hot, dense region by line emission and recombination and bremsstrahlung radiation.

In this paper we propose the creation of a new type of plasma by irradiation of metal targets with intense subpicosecond laser pulses. This plasma has several important differences with conventional laser produced plasmas. <u>First</u>, with short laser pulses absorption and electron heating occur within an optical skin depth and there is little expansion. The particle density is consequently two to three orders of magnitude greater, resulting in rapid heating, collisional excitation and ionization. <u>Second</u>, laser light is absorbed and only partially reflected from the dense plasma, even though the plasma frequency exceeds the laser frequency, because the electron motion is highly damped by rapid electron-ion collisions. <u>Third</u>, the actual thickness of the heated plasma region may be several optical skin depths and is determined by both rapid heat transfer into the solid and plasma ablation. This relatively thin volume is rapidly cooled by conduction, ablation and radiation. It is key to this proposal that rapid cooling processes and reduced density cause the plasma temperature and x-ray emission to follow the subpicosecond time scale of the laser heating pulse.

We expect the plasma to be a novel source of incoherent, short pulse duration, soft and hard x-rays for applications involving flash x-ray scattering from materials and pumping of x-ray lasers. Additional interest in this work comes from a capability to study the strongly coupled plasmas[2] produced at somewhat lower temperatures by reducing the focused laser intensity.

PLASMA ANALYSIS

As with conventional laser produced plasma x-ray sources it is expected[1] that the conversion efficiency of incident laser energy to short wavelength radiation will be greatest for high atomic number targets. We have analyzed a gold (Au) plasma produced by a 100 femtosecond laser pulse at a power density of 10^{16} watts/cm^2 and a wavelength of 308 nanometers. A laser system capable of producing this intense flux is currently being built based on a colliding pulse laser oscillator, and dye and excimer laser amplifiers.

In our model, absorption of laser light is calculated assuming inverse bremsstrahlung absorption as in a normal free electron metal. As the electrons heat, the optical properties of the material rapidly approach those of an ionized plasma as electronic structure effects become less important. The dielectric constant of the medium is calculated using the electron-ion collision frequency of an ionized plasma according to Spitzer.[3] To account for spatial variations we numerically solve a one dimensional finite difference heat flow equation. We include electron lattice interactions,[4] heat conduction,[1] and spatially varying energy deposition due to laser attenuation in the metal. A simple rate equation model relates the electron temperature and average ionization stage of the Au ions.

To first approximation (ignoring heat losses) the energy balance equation is simply

$$n_e \frac{dE}{dt} = I(1-R)\alpha \tag{1}$$

where n_e is the number density of electrons, E is the average electron energy, I is the incident laser power density, R is the reflectivity from the plasma surface and α is the absorption coefficient. In the regime of interest we have

$$\omega_p > \omega > \nu_c \tag{2}$$

where ω_p is the plasma frequency, ω is the laser optical frequency, and ν_c is the electron-ion collision frequency. The dielectric function then yields the approximate electron heating rate

$$\frac{dE}{dt} \cong \frac{4I\nu_c}{n_e c} \tag{3}$$

where electron heating (and thus high plasma temperatures and efficient x-ray emission) results from high electron-ion collision rates. Although the actual dielectric constant is a function of electron temperature, ionization state, density and shielding effects, we have found that over the wide range of plasma conditions which occur during laser heating of Au the fractional energy deposited per unit length is

$$(1-R)\alpha \cong 4 \times 10^{5} \text{cm}^{-1} , \tag{4}$$

a value which is remarkably within a factor of two of the room temperature value for Au. The assumptions of this plasma model[3], including several electrons per Debye sphere, are met under the rapid heating conditions of our proposed experiment. We also believe that linear absorption will dominate at our applied power density.[5]

Electron heating occurs at an average rate of 10 eV/fsec and the time to ionize sequential stages averages about 5 fsec. The result of our calculation is that we predict electron temperatures of several hundred eV and ionization stages approaching Au^{20+} just inside the surface. A typical computer calculation is shown in Figure 1 in which the electron temperature at the surface is displayed versus time (here we neglect radiation and ablation cooling).

Thermal conductivity of 10^5 watts/M / K is assumed and this results in a doubling of the extent of the heated plasma beyond the optical skin depth of 100 Å. Ions remain relatively cool (<5 eV) over the duration of the laser heating pulse.

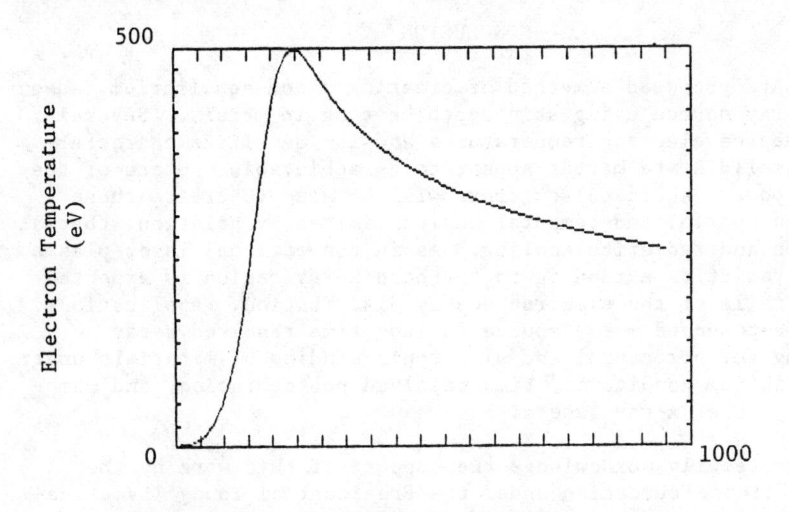

Fig. 1. Electron temperature at a gold surface illuminated with a 100 femtosecond long pulse at 308 namometers at a power density of 10^{16} watts/cm^2.

We believe that plasma ablation will occur in two stages. The first stage is a rapid escape of free electrons from the Au surface. These electrons may be due to multiphoton photoelectric emission or thermionic emission. In any case, space charge limits the density and spatial extent of this non-neutral plasma so that it will not prevent the laser from reaching the Au surface. (This may not be true at longer laser wavelengths, however.) The second stage of expansion occurs as the neutral plasma ablates with a velocity[1]

$$v \cong \sqrt{\frac{zE}{m}} \tag{5}$$

where z is the charge state of the ions and m is the ion mass. For our conditions this velocity will be approximately 10^7 cm/sec, implying an expansion on the order of 100 Å in 100 fsec when the plasma is at peak temperature.

Radiation processes are very difficult to calculate. Using standard formulas[3] for bremsstrahlung and recombination radiation we find that they contribute less than 10% of the energy loss due to thermal conduction and ablation cooling. Line radiation from this marginally optically thick plasma appears to be an important cooling mechanism and needs to be evaluated more precisely both for the heating model and to determine the actual spectral distribution of emitted light. Typically, laser plasma light x-ray sources operate at efficiencies in excess of several percent.[1]

CONCLUSION

We have proposed a method of creating a non-equilibrium dense plasma x-ray source using skin depth heating in metals. Several hundred degree electron temperatures and ion densities characteristic of solid state matter appear to be achievable. State of the art high power, short pulse lasers will be used to create these plasmas on spatial and temporal scales limited by ablation, thermal conduction and radiative cooling. As in conventional laser plasma sources, radiation extending to the hard x-ray region is expected from the tails of the electron energy distribution. Applications of this femtosecond x-ray source include time resolved x-ray scattering for structural and electronic studies of materials under non-equilibrium conditions,[6] time resolved photoemission, and pumping new types of x-ray lasers.

We gratefully acknowledge the support of this work by the National Science Foundation under the Presidential Young Investigator Program, Grant No. 83-51689, and additional support under this program from AT&T Bell Laboratories, E. I. du Pont de Nemours and Company, Schlumberger-Doll Research and Shell Companies Foundation.

REFERENCES

1. C. E. Max, "Physics of the Coronal Plasma in Laser Fusion Targets" in Laser Plasma Interaction, R. Balian and J. C. Adam, eds., North Holland (1982).
2. R. More, "Plasma Processes in Non-Ideal Plasmas," to appear in Laser Plasma Interactions, Vol. 3, Mary B. Hooper, ed., Plenum Press.
3. L. Spitzer, Physics of Fully Ionized Gases (Interscience Publishers, N.Y., 1956).
4. M. I. Kaganov, I. M. Lifshitz, and L. V. Tanatarov, Sov. Phys. JETP 4, 173 (1957).
5. Y. Shima and H. Yatom, Phys. Rev. A12, 2106 (1975).
6. K. Murakami, et al., Phys. Rev. Lett. 56, 655 (1986).

ELLIPSOIDAL FOCUSING OF SOFT X-RAYS FOR LONGITUDINALLY PUMPING SHORT WAVELENGTH LASERS

J. F. Young, J. J. Macklin, and S. E. Harris
Edward L. Ginzton Laboratory, Stanford University
Stanford, CA 94305

ABSTRACT

We propose using a prolate ellipsoid for collecting and focusing soft x-ray radiation from a laser-produced plasma to longitudinally pump short wavelength lasers. Total collection efficiencies greater than 20% into gain regions with aspect ratios of more than 50 are possible. This combination results in significant improvements relative to present transverse multiple-spot geometries.

SUMMARY

We propose a new geometry for collecting and focusing soft x-ray radiation from a laser produced plasma which offers collection efficiencies greater than 20% and aspect ratios of gain length to diameter of more than 50. This combination typically results in (power density) x (length) improvements of 30 relative to present transverse multiple-spot geometries.[1] The high aspect ratios significantly reduce laser excitation problems associated with amplified spontaneous emission and high atomic number densities. Removal of the laser plasma target from the active medium simplifies cell design, permits spectral filtering of the pumping radiation, and may lead to improved conversion efficiency.

The basic geometry is shown in Fig. 1; a section of a prolate ellipsoid is used to reflect, at grazing incidence, the radiation from a laser-produced plasma located at one focus to the active atomic species located at the other focus. Surfaces of revolution generated by ellipses, parabolas, and hyperboloids, singly and in combination, have been used previously to construct x-ray microscopes and telescopes. Resolution, field of view, and very short wavelength requirements, however, have generally resulted in designs with low efficiency. Laser pumping has no such restrictions: gain depends only on the product of power density and length. In addition, the low spatial resolution requirement of laser pumping significantly reduces the importance of small angle scattering and eases surface finish and figure tolerances. Thus we have based the design of the ellipsoidal reflector on a non-imaging ray tracing analysis of its collection and concentrating performance.

The virtues of the longitudinal pumping geometry could also be achieved using normal incidence optics in spectral regions where high reflectivities are available, but even with unity reflectivity the f-number of the optical system would have to be less than 2 to obtain the efficiencies calculated for the ellipsoid.

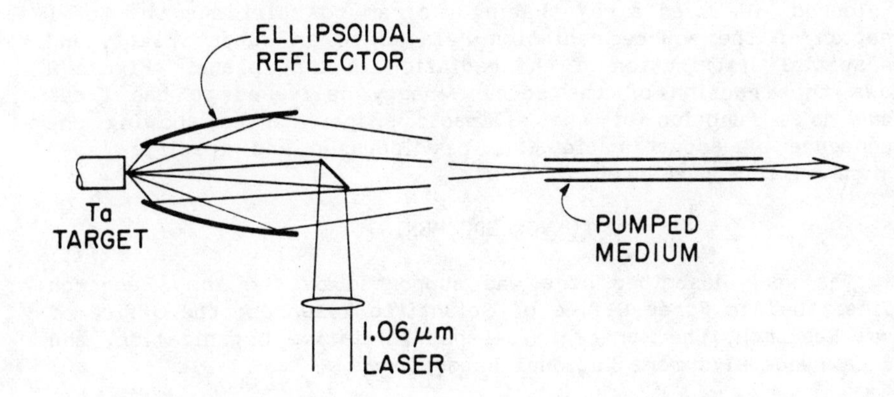

Fig. 1. Schematic of ellipsoidal focusing of x-rays.

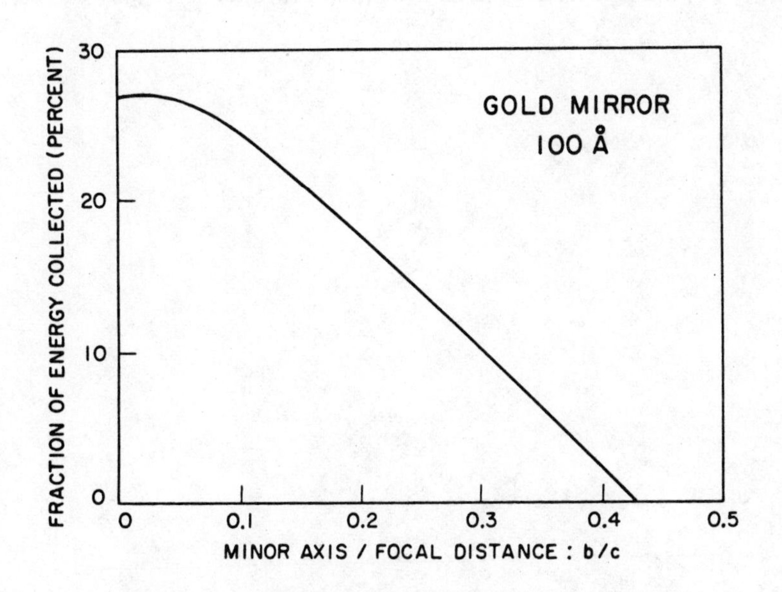

Fig. 2. Energy collected vs. ellipse shape.

Our calculations assume an angular distribution function for the radiation emitted by the target which is proportional to the cosine of the angle from the target normal. The reflectivity of the ellipsoid is modeled by a simple straight-line approximation to the measurements of Malina and Cash[2] for a gold surface at 10.4 nm. The reflectivity is taken as 1.0 for grazing angles of incidence less than 3 deg, 0.0 for angles greater than 25 deg, and a linear interpolation is used for angles between those limits. Both ideal point sources and extended disk-shaped sources were considered. We used a ray tracing program to calculate the total fraction of the source radiation relayed to an output plane, and the spatial distribution of the radiation in that plane. Figure 2 shows the fraction of the source energy delivered to the focal plane as a function of the ellipsoid shape; curves showing the dependence of aspect ratio will be discussed and specific laser designs will be presented.

ACKNOWLEDGEMENTS

The work described here was supported by the Army Research Office, the Air Force Office of Scientific Research, the Office of Naval Research, the Strategic Defense Initiative Organization, and the Lawrence Livermore National Laboratory.

REFERENCES

1. R. G. Caro, J. C. Wang, J. F. Young, and S. E. Harris, Phys. Rev. A 30, 1407 (1984).
2. Roger F. Malina and Webster Cash, Appl. Opt. 17, 3309 (1978).

Suppression of multiphoton ionization with circularly polarized light

T.J. McIlrath[a], *P.H. Bucksbaum, M. Bashkansky*[b], *R. R. Freeman*
AT&T Bell Laboratories
Murray Hill, New Jersey 07974

L. F. DiMauro
Department of Physics and Astronomy, Louisiana State University
Baton Rouge, Louisiana 70803

ABSTRACT

Multiphoton ionization in xenon has been studied using 1064 nm circularly polarized coherent light, producing electrons with energies up to 15 eV. A strong suppression of the cross section is observed for electrons below 4 eV, but not for linearly polarized light at the same intensity. We show that if the electrons are emitted in single step nonresonant coherent processes, rather than stepwise excitation, the centrifugal potential barrier in the atomic Hamiltonian produces this suppression.

Studies of multiphoton ionization (MPI) at focused intensities of $10^{13} - 10^{15}$ W/cm^2 reveal a number of dramatic and unexpected features[1]. Chief among these is above-threshold-ionization (ATI), in which electrons are not only produced near the ionization threshold, but at a series of energies separated by the photon energy $h\nu$. Several theoretical papers have attempted to explain the qualitative behavior of these higher order peaks, appealing either to step-wise transitions in the continuum, or direct coherent processes from the ground state[2].

Experiments reported previously have used linearly polarized light. Here we show that the use of circularly polarized radiation at 1064 nm for MPI in xenon dramatically changes the energy distribution of the final state electrons, suppressing all electrons from threshold to > 4 eV. Although neither the number of high energy photoelectrons nor the total ionization rate is significantly reduced, the average photoelectron energy moves much higher. This phenomenon is wavelength dependent; for 532 nm radiation only the first ionization peak is suppressed. A simple model, based on angular momentum absorption and suppression of the overlap with low energy final states by centrifugal repulsion, agrees well with these observations, and predicts this effect independently from the detailed structure of the atom. Thus the effect should be a general property shared by virtually all atoms. These results indicate that ATI photoelectrons are produced in single step, non-resonant coherent processes directly from the ground

[a] On leave from IPST, University of Maryland
[b] Physics Department, Columbia University

state.

The apparatus used in these experiments has been described in another paper at this meeting and will not be repeated here[3]. Figure 1a is a 10,000 shot electron spectrum for MPI in xenon using 1.1mj ($\approx 10^{13}$ W/cm^2) pulses of 532 nm linearly polarized light. The vertical scale is normalized to number of counts per shot, per milliTorr of xenon pressure, per meV energy bin. The laser polarization is along the detection axis. The MPI electron angular distributions are generally peaked along the laser polarization for linearly polarized light. At least six photons must be absorbed in order to ionize xenon, which has series limits of 12.13 and 13.43 eV, corresponding to the $^2P_{3/2}$ and $^2P_{1/2}$ fine structure levels, respectively, of the $5s^25p^5$ ground configuration of Xe^+. The two lowest energy peaks in the figure represent six photon ionization to these two final states. The higher peaks are examples of ATI. Figure 1b shows a spectrum at similar intensity, but with circular polarization.

The electron spectra are repeated in figures 1c and 1d, but for 1064 nm laser pulses at $\approx 1.5\times10^{13}$ W/cm^2. Eleven photons are required to ionize to the lowest state of Xe^+, and 12 photons to reach the higher $^2P_{1/2}$ state. There is no sign of a peak in the spectrum corresponding to either of these transitions. The lowest energy spectral feature for linear polarization occurs in the vicinity of ATI processes representing 12 and 13 photon absorption to the two fine structure states. The relative suppression of the threshold electrons in xenon for linear polarization MPI has been observed previously[4]. Figure 1d is a spectrum taken at the same 1064 nm laser intensity as Fig. 1c, but with circular polarization. The lowest MPI electron peaks are missing as before, but in addition, the next four ATI peaks in the spectrum are also absent. This striking effect appears to be independent of laser intensity, over a range of at least an order of magnitude. The small number of low energy electrons which were detected are consistent with background, plus small imperfections in the circular polarization. We have repeated these measurements under varying conditions of gas pressure and laser intensity, but have never succeeded in observing low energy electron peaks with circularly polarized 1064 nm light.

The disappearance of the low energy electrons can be explained by a severe reduction of the transition matrix elements for low energy ATI electron states due to the high orbital angular momentum of the final continuum states. For xenon, absorption of n photons with n units of angular momentum requires transitions into the $L \geq n-1$ continuum. To understand how this suppresses low energy electrons, consider the transition matrix element M for ATI in atomic hydrogen. The general features of this calculation can be extended easily to ATI transitions in other atoms since the hydrogenic wavefunctions are good descriptions of high angular momentum states of complex atoms. M is given by:

$$M = (eE)^n \sum_{i_1} \cdots \sum_{i_{n-1}} \left(\frac{\langle \psi_f | d^+ | i_{n-1} \rangle \cdots \langle i_1 | d^+ | \psi_0 \rangle}{\delta E_{n-1} \cdots \delta E_1} \right) \quad (1)$$

In this expression, the wavefunctions $|i_m\rangle$ represent intermediate states. The initial and final states are $|\psi_0\rangle$ and $|\psi_f\rangle$, respectively; and the detuning associated with the intermediate state $|i_m\rangle$ is $\delta E_i = E_i - E_0 - m\hbar\omega$. E is the electromagnetic field

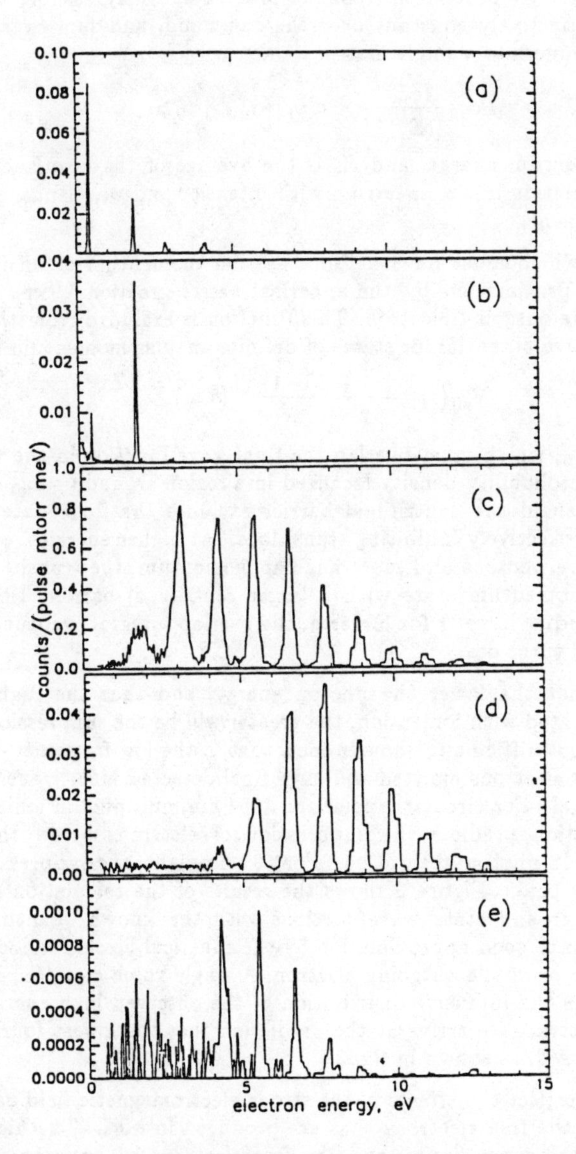

1. Photoelectron energy spectrum for (a) 1.1 mJ, 532nm, linearly polarized; (b) 1.5 mJ, 532 nm, circularly polarized; (c) 11 mJ, 1064 nm, linearly polarized; (d) 11 mJ, 1064 nm, circularly polarized; (e) 6.3 mJ, 1064 nm, circularly polarized radiation. Collection solid angle $\approx 2.5\times10^{-3}$ster along direction of linear polarization and in plane of circular polarization. Anisotropy of linear polarization signal enhances collection efficiency by $\geq$ 12 over circular polarization.

strength. In this high order non-resonant process one may assume that the energy denominator is effectively constant over the summand, and sum over the numerator alone[5]. M then reduces to a single matrix element,

$$M \propto \frac{(eE)^n}{(\delta E)^{n-1}} \times < \psi_f | (d^+)^n | \psi_0 > \tag{2}$$

where e is the electron charge, and δE is the average of the resonance energies. We can study the variation in the spectrum with intensity and wavelength as predicted by the r^n matrix element.

The final state, which must have definite angular momentum of $n\hbar$ for hydrogen, is given to good approximation by the spherical bessel function $j_n(kr)$, where k is the wave vector of the outgoing electron. This function is excluded from the center of the atom by an effective potential for states of definite angular momentum l:

$$V_{eff}(r) = -\frac{1}{r} + \frac{l(l+1)}{2r^2}. \quad \text{(a.u.)} \tag{3}$$

The function $r^n \psi_0$, which must overlap the final wave function in the matrix element, has a maximum probability density localized in a region around $r = na_0$ as shown in figure 2. Near threshold the centrifugal barrier excludes the final state wave function from this region, effectively inhibiting transitions. At higher energies, corresponding to absorption of more photons and more angular momentum, the transition is made to a higher angular momentum state with a larger centrifugal barrier. However, because $r^n \psi_0$ is also peaked at larger r for larger n, the overlap integral eventually becomes significant, and MPI turns on.

It is apparent that the lower the photon energy, and thus the higher the angular momentum associated with ionization, the greater will be the suppression of low energy electrons. (It is not difficult to show in fact, that in the low frequency--classical--limit, the lowest energy electrons emitted will have final detected kinetic energy equal to the ionization potential.) For circularly polarized 1064 nm multiphoton ionization of hydrogen, this calculation predicts the suppression of electrons below the 5.0 ev peak corresponding to 16 photon absorption. For 532 nm light, the suppression only affects the lowest energy peak. Figure 3 shows the results of the calculation for xenon using Bates-Damgaard ground state wavefunctions with the known quantum defect. The final state is given to good approximation by the spherical bessel function $j_n(kr)$, where k is the wave vector of the outgoing electron. A single value of $E/\delta E$, introduced into Eq. (2) reproduces the intensity distribution of the observed high energy peaks in the experimental spectra. We arrive at the prediction that the lowest four electron peaks should be suppressed, as shown in fig. 3.

These estimates neglect the effects of the strong electromagnetic field on the ionization potential and on the free electrons that are produced in MPI. The highly polarizable Rydberg states, the series limits, and the free electrons all undergo sizable shifts in their potential energy due to coherent field effects generally referred to as "ponderomotive forces."[6] These impose an extra potential energy, due to the quiver of the electrons in the oscillating field, equal to $e^2 < E^2 > /(4m\omega^2)$. The observable effects of this ponderomotive potential term are difficult to calculate in these experiments, because

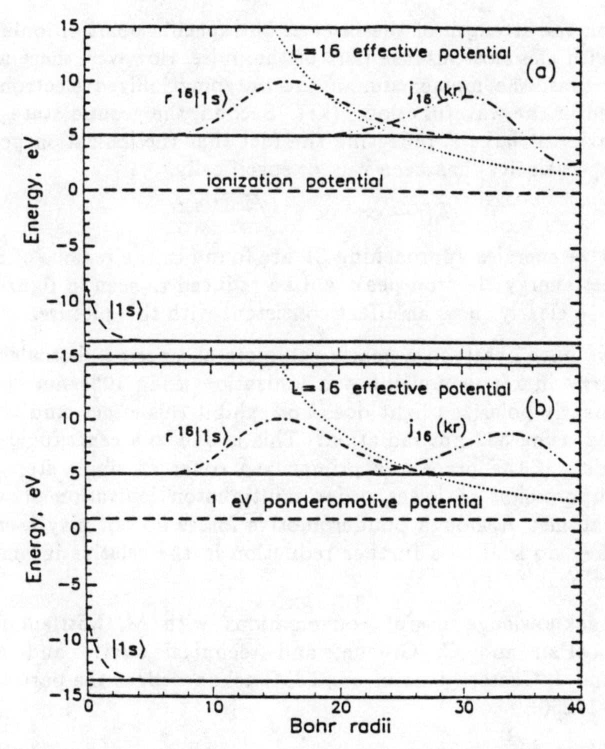

2. (a). Effective potential V_{eff} and continuum wavefunctions $j_{16}(kr)$ for angular momentum l=16, superimposed on 1s ground state wavefunction and $r^{16}|1s\rangle$ for hydrogen; (b). Same as (a) but with a 1 eV ponderomotive potential.

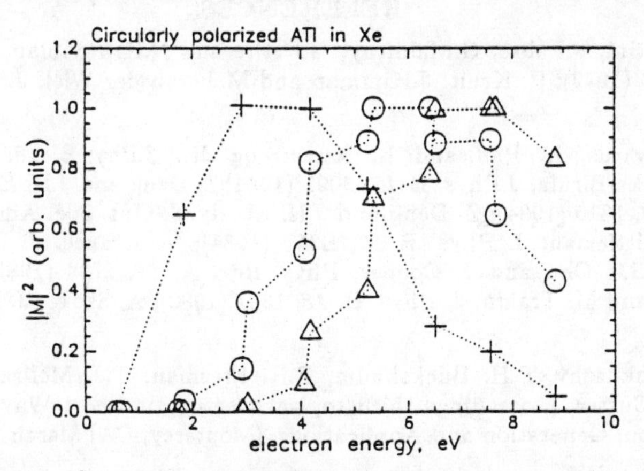

3. Variation of $|M|^2$ between the initial $|5p\rangle$ state and final $|E_n\rangle$ continuum state in xenon, corresponding to n photon absorption. $E_n = nh\nu - E_i$ is the free electron energy where E_i is the ionization energy. Angular momentum of continuum state is l=n+1, and E_i includes both Xe^+ final states. (+): $h\nu$ = 532 nm; (o): $h\nu$ = 1064 nm; (Δ): $h\nu$ = 1064 nm, with a ponderomotive potential of 1 eV.

they depend on the strength of the laser at the specific point of ionization, which may not coincide with the most intense part of the pulse. However, there are two important consequences: first, the momentum of the outgoing ionized electron is reduced, and this is reflected in the wavefunction $j_l(kr)$. Second, the ground state wavefunction has different behavior at large r, reflecting the fact that the ionization potential E′ (classically, the escape velocity) has been raised: specifically,

$$\psi_0(r \rightarrow \infty) \propto e^{-(\sqrt{2mE'/\hbar})r}$$

If ponderomotive energies approaching $h\nu$ are found in the regions of significant ionization, the lowest energy electron peak will be reduced as seen in figures 2b and 3. The data in figure 1e clearly show an effect consistent with this picture.

In summary we have observed a nearly complete absence of photoelectrons below 4 eV of kinetic energy in xenon multiphoton ionization using 1064 nm circularly polarized laser light. Linearly polarized light does not exhibit this effect, and the phenomenon is greatly reduced using 532 nm radiation. This is due to a centrifugal barrier to ionization which exists if the process is primarily a coherent single step from the ground state. The suppression of lower order multiphoton ionization should be a general feature of all atoms. Although ponderomotive forces do not play a crucial role in this suppression they do lead to a further reduction in the relative intensity of the lowest visible electron peak at 4 eV.

We wish to acknowledge useful conversations with M. Mittleman, A. Szoke, L. Armstrong, L. Pan and C. Greene, and technical advice and assistance by D. Schumacher and J. Custer. One of us (TJM) acknowledges the partial support of NSF grant CPE 81-19250.

REFERENCES

1. P. Agostini, F.Fabre, G.Mainfray, G.Petite and N.K. Rahman, Phys. Rev. Lett. *42*, 1127 (1979); P. Kruit, J.Kimman and M.J. van der Wiel, J. Phys. B *14*, L597 (1981).

2. M. Edwards, L. Pan and L. Armstrong Jr., J.Phys.B *18*, 1927 (1985); Z. Bialynicka-Birula, J.Phys. B *17*, 3091 (1984);Z. Deng and J.H. Eberly, Phys. Rev. Lett. *53*, 1810 (1984); Z. Deng and J.H. Eberly, J. Opt. Soc. Am. B *2*, 486 (1985); M.H. Mittleman, J. Phys. B *17*, L351 (1984); M. Crance, J. Phys. B *17*, L355 (1984); S.-I Chu and J. Cooper, Phys. Rev. A, *32*, 2769 (1985); Y'Gontier, M. Poirier and M. Trahin, J. Phys. B. *13*, 1381 (1980); A. Szoke, J. Phys. B, *18*, L427 (1985).

3. M. Bashknashy, P.H. Bucksbaum, R.R. Freeman, T.J. McIlrath L.F. DiMauro and J. Custer, proceedings third topical meeting on Short Wavelength Coherent Radiation: Generation and Applications, (Monterey, CA) March 24-26, 1986.

4. P. Kruit, J. Kimman, H.G. Muller and M.J. van der Wiel, Phys. Rev. A *28*, 248 (1983); L. A. Lompre, A.L'Huillier, G. Mainfray and C. Manus, J. Opt. Soc. Am. B *2*, 1906 (1985).

5. H.B. Bebb and A. Gold, Phys. Rev. 43,1 (1966). For circularly polarized light, the nth order contribution to M includes only states with principal quantum number $\geq$n. These states lie within Ry/n^2 of the ionization energy E_i; for non-resonant processes all of the energy denominators for discrete intermediate states can be approximated by $E_i - n\hbar\omega$. Using closure on the first (m-1) contributions reduces M to integrals over the continuum. For the higher order contributions the intermediate states all have angular momentum greater than m$\hbar$ and the centrifugal barrier reduces the penetration of the wavefunction so that for low energy states the radial overlap with the groundstate vanishes. The only contribution is from a narrow range of states with sufficient penetration to overlap the initial state. Thus the energy denominator is approximately constant and positive and can be removed from the summand.

6. T.W.B. Kibble, Phys. Rev. *150*, 1060 (1966); P. Avan, C. Cohen-Tannoudji, J. Dupont-Roc and C. Fabre, J. de Physique *37*, 993 (1976); J.H. Eberly, Progress in Optics, 7, 359 (1969).

OPTICAL GAIN AT 185 NM IN A LASER ABLATED, INNER-SHELL PHOTOIONIZATION-PUMPED INDIUM PLASMA

R. A. Lacy and R. L. Byer
Stanford University, Department of Applied Physics, Stanford, California 94305

W. T. Silfvast and O. R. Wood II
AT&T Bell Laboratories, Holmdel, New Jersey 07733

S. Svanberg
Lund Institute of Technology, Department of Physics, Lund, Sweden

ABSTRACT

Inner-shell photoionization of In^+ ground state ions by a laser-produced plasma broad-band soft X-ray pump source has created a population inversion on the 185 nm $4d^9\,5s^2\ ^2D_{5/2} \rightarrow 4d^{10}\,5p\ ^2P^\circ_{3/2}$ transition in In^{++}. A small signal intensity gain of exp(0.38) was measured in a ~4 mm path length, yielding a gain coefficient of $\alpha \sim 0.95\ cm^{-1}$. The initial population of $\sim 10^{16}\ cm^{-3}$ In^+ ground state ions is prepared by ablating a liquid indium target with a 10 nsec, 100 millijoule (mJ), 532 nm laser pulse yielding a highly ionized indium plasma. The expanding plasma cools and recombines to give the required $In^+\ 4d^{10}\,5s^2$ population, at which time a 70-100 psec, ~50 mJ, 1.06 μm laser pulse, incident on a nearby tantalum target, generates a hot plasma radiating broad-band soft X-rays. The soft X-rays photoionize inner-shell 4d electrons from the In^+ ions, populating the $4d^9\,5s^2$ In^{++} state, which is inverted with respect to the nearly empty $4d^{10}\,5p$ state.

INTRODUCTION

In 1967, Duguay and Rentzepis[1] first pointed out the possibility of producing population inversions by photoionization of inner-shell electrons from atoms using broad-band X-rays. Many atomic species have significantly larger photoionization cross-sections for inner-shell electrons than that for the outermost electrons. Also, the inner-shell photoionization cross-sections are usually quite broad in energy, even to the point of matching well with black-body distributions[2]. Thus broad-band X-ray sources, for instance laser-produced plasmas, can efficiently pump inversions by inner-shell photoionization.

In 1983, Silfvast and co-workers[3] demonstrated the first inner-shell photoionization-pumped laser. They used a laser-produced plasma to generate broad-band soft X-rays, which photoionized a 4d electron from neutral Cd atoms. This populated the $4d^9\,5s^2$ state which was inverted with respect to the empty $4d^{10}\,5p$ state below it. With a mirror resonator, the inversion led to lasing at 441.6 and 325.0 nm on the $^2D_{5/2} \rightarrow {}^2P^\circ_{3/2}$ and $^2D_{3/2} \rightarrow {}^2P^\circ_{1/2}$ transitions, respectively.

The indium work described here was an extension of the Cd experiment. Our strategy was to extend inner-shell photoionization-pumped lasers to shorter wave - lengths by the isoelectronic scaling of the known $4d^9\,5s^2 \rightarrow 4d^{10}\,5p$ transition. However, the indium experiment is quite different from the previous Cd work, as ions are the specie to be photoionization-pumped rather than neutral atoms.

In^+ is isoelectronic to neutral Cd. The removal of a 4d electron from ground state In^+ yields $In^{++}\,4d^9\,5s^2$, which radiates at[4] 185.0 nm on the transition $4d^9\,5s^2\ {}^2D_{5/2} \rightarrow 4d^{10}\,5p\ {}^2P^{\circ}_{3/2}$ (see Figure 1). The demon - stration of gain on this tran - sition shows the usefulness of isoelectronic scaling in the search for shorter wavelength lasers.

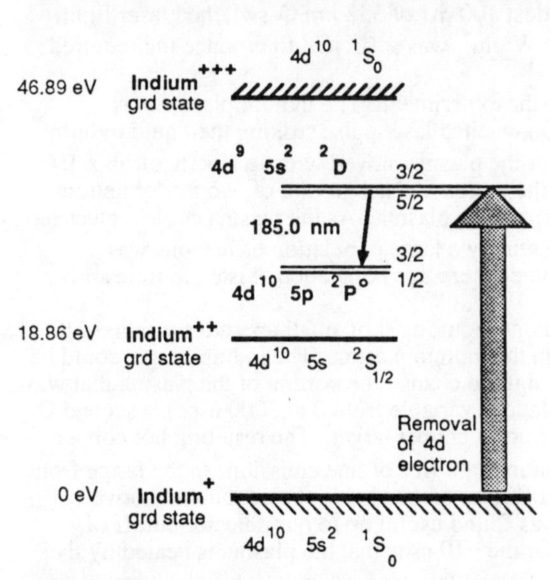

Fig. 1. Energy levels of singly and doubly ionized indium showing photoionization-pumping to produce inversion on the 185 nm transition.

This scheme required a large density (~ 10^{16} cm^{-3}) of ground state In^+ ions prior to photoionization pumping. Laser ablation of a liquid indium target provided the ions needed. We also had to determine the densities of the ionic species produced by ablation as a function of time and space. This was achieved using a laser-produced copper plasma as a VUV continuum light source for time-resolved absorption spectroscopy of the indium plasma. Finally, we had to photoionize inner-shell 4d electrons from ground state In^+ ions and measure the resulting optical gain at 185.0 nm. A laser-produced tantalum plasma served as the source of X-rays for the photoionization process. The gain measurements were made using a mirror to reflect light from the excited indium plasma back through the gain region.

PRODUCTION AND CHARACTERIZATION OF AN INDIUM PLASMA

Based on the earlier experiments[3] with Cd, and on Nilsson's[5] computer calculations[6] of the Einstein A coefficient (A ~ 1.3×10^7 sec^{-1}) of the $4d^9\,5s^2\ {}^2D_{5/2} \rightarrow 4d^{10}\,5p\ {}^2P^{\circ}_{3/2}$ transition in In^{++}, it appeared that a density-length product of ~10^{16} cm^{-2} of ground state In^+ would be needed prior to photoionization-pumping. This population must be in a path length less than 1 cm long to be pumped efficiently by the X-rays from a point source. Conventional ion sources such as hollow cathode discharges are not adequate to this task. A heat pipe was also impractical: producing 10^{16} cm^{-3} of indium neutrals would require 1300 °C, and those neutrals must be ionized into the In^+ ground state by some means.

A high intensity laser pulse will create a plasma from a solid target. The literature suggested that laser ablation of a metal target could achieve the high ion densities we sought[7]. Because solid targets are rapidly damaged, we ablate from a liquid indium target, which is self-healing and provides consistent results over tens of thousands of laser shots. A modest 100 mJ of 532 nm Q-switched laser light, focused to an intensity of 4.5×10^8 W/cm^2, was sufficient to produce the required density of In^+ ground state ions.

Figure 2 shows a schematic of the experiment. The indium plasma was produced with laser ablation by a Q-switched laser pulse striking the liquid indium target. We observed that the front of the plasma moved with a velocity of $\sim 6 \times 10^6$ cm/sec at a distance of 3 mm from the target. A background of two torr of helium gas slowed down and cooled the expanding plasma. As the plasma cooled, electron-ion recombination occurred and eventually a large population of In^+ ions was present. It was necessary to determine where this population existed in time and space.

Because the plasma was inhomogeneous, a set of small apertures was used to define a narrow line of sight through the indium plasma. The indium target could be moved perpendicular to this line of sight to change the portion of the plasma that was viewed. At a selected time after ablation (variable from 0 to 2000 nsec), a second Q-switched Nd:YAG laser fired and struck a copper target. The resulting hot copper plasma radiated strong VUV continuum light free of line emission[8] in the range from 50 to 160 nm. There were some Cu II lines embedded in the continuum above 160 nm, but this radiation source was found useful up to our detector cutoff of 350 nm. This radiation lasts only for the ~10 nsec that the plasma is heated by the laser pulse[9], and was used to probe the rapidly moving indium plasma with fine time

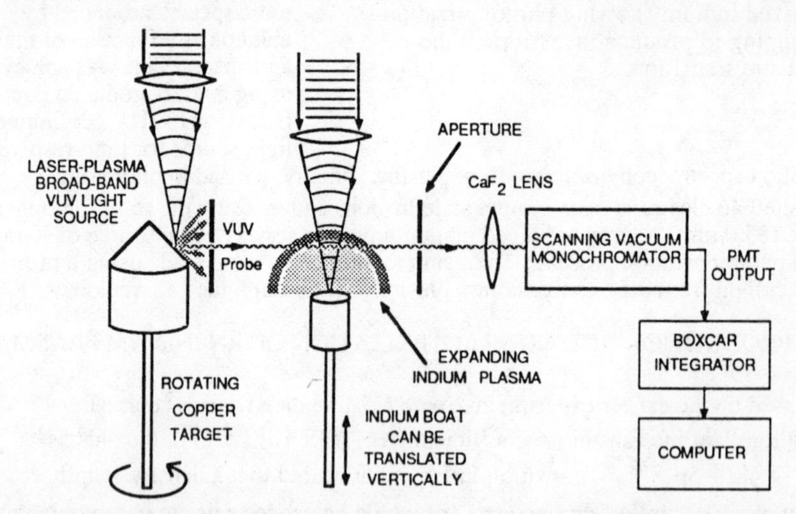

Fig. 2. Schematic of apparatus for producing and characterizing an indium plasma. Q-switched laser pulse striking molten indium creates a plasma. Second laser shot strikes copper target at a later time, creating a hot plasma that radiates continuum VUV for ~10 nsec. This radiation is the probe for space- and time-resolved absorption spectroscopy of the indium plasma.

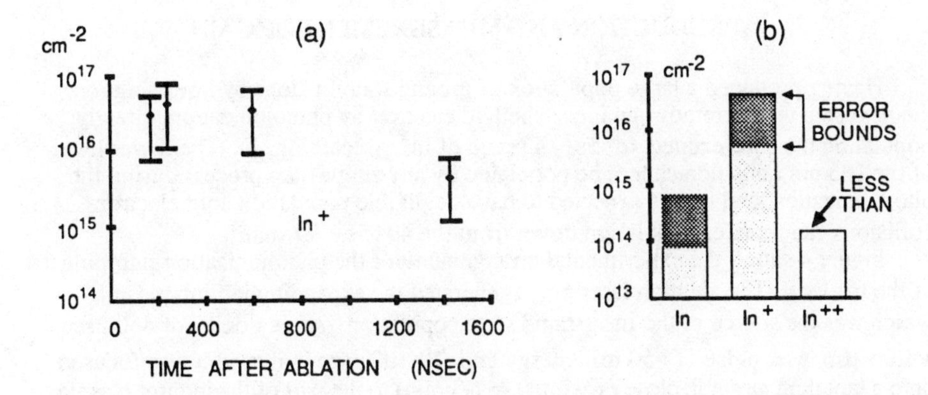

Fig. 3. Indium ion density-length product measurements. Fig. 3a shows the time evolution of ground state singly ionized In at 3 mm above the In target. Fig. 3b shows the amounts of ground state neutral, singly, and doubly ionized In at 240 nsec after ablation and 3 mm above the In target.

resolution. As the broad-band probe light passed through the indium plasma, a scanning vacuum monochromator monitored absorption on the resonance lines of indium neutrals and ions. A computer recorded the absorption data, which was processed to yield the line integral of the product of the ion density and the path length for each ion stage. Given a good hydrodynamic model of the expanding plasma, it might be possible to specify the ion densities themselves as a function of position within the plasma. However, the density-length products alone were sufficient to suggest that the laser-ablated indium plasma was suitable for our purposes, as noted below.

Some results of the indium ion density-length-product measurements are shown in Figure 3. Figure 3a shows the time evolution of the ground state of In$^+$ at a distance of 3 mm from the ablation target. At 3 mm above the target, the density-length of ground state In$^+$ is greatest between 200 and 400 nsec after ablation. Figure 3b shows the density-length of three ground state species at 3 mm from the ablation target and at 240 nsec after ablation. This is near the peak in time and space of the In$^+$ ground state population. At this time and position, the path probed through the plasma is 10 mm long, but the ion density is not evenly distributed[10] Models of the most extreme possible distributions were used in analyzing the absorption data in order to establish the error bounds shown in Figure 3.

From Figure 3a we observe that a sufficient number of ground state In$^+$ ions are produced by laser ablation to meet our original density-length goal. Figure 3b leads to a second important observation: the In$^+$ ground state population is an order of magnitude larger than that of the neutral specie, and is at least two orders of magnitude greater than that of the doubly-ionized specie. Both conditions are important in the production of a population inversion by photoionization pumping. The absence of In^{++} before photoionization pumping means that initially there will be no population in the lower level of the gain transition. The absence of neutrals removes an undesirable source of absorption of the X-rays needed to pump the In$^+$ ions into the upper level of the gain transition.

PRODUCTION AND MEASUREMENT OF GAIN

Having produced a large population of ground state In^+ ions by laser ablation, the next step was to remove an inner-shell 4d electron by photoionization, directly populating the core-excited $4d^9\ 5s^2\ ^2D$ state of In^{++} (recall Fig. 1). The lower level of the 185 nm transition cannot be populated by any single-step process during the photoionization, and is thus expected to have negligible population until electron collisions can transfer population down from the $4d^9\ 5s^2\ ^2D$ state.

Figure 4 shows the experimental arrangement for the photoionization-pumping of the In^+ ions. The ablation laser pulse generated the expanding indium plasma which was the source of the In^+ ground state population. After a delay of 460 nsec, a 1.06 μm laser pulse of ~50 mJ energy and 70-100 psec pulsewidth was focused onto a tantalum target in close proximity (~1-3 mm) to the part of the indium plasma where the In^+ ground state density was high. The preferred delay is believed to be ~240 nsec, but the laser timing circuits did not permit this. The hot tantalum plasma radiated soft X-rays which photoionized the In^+ ions, populating the core-excited $4d^9\ 5s^2\ ^2D$ state of In^{++}.

Figure 4 also shows the method used to determine the gain of the plasma at 185 nm. A set of 4 mm diameter apertures defined the region of the indium plasma to be probed. A mirror reflected light from the indium plasma back through the plasma. By blocking and unblocking the space between the plasma and the mirror and measuring the change in intensity observed at the monochromator, the single-pass gain of the plasma was measured. Several checks on the accuracy of this method have been performed. Silfvast and coworkers[11] have used this method to measure gain in photoionization-pumped argon, and have shown that it agrees with probe beam measurements using a CW argon-ion laser. Also, experimental measurements in the indium experiment on transparent lines (expected to show neither gain nor loss) agreed with the theoretically predicted ratio for intensities when blocking and unblocking the mirror. In addition, we performed the gain measurements using two mirrors of different reflectivities, obtaining similar values for the gain coefficient with either mirror. The reflectivities were measured before and after the gain measurements, and no evidence of change due to coating by target debris or other problems was encountered. Thus we have full confidence in this method of measuring the gain.

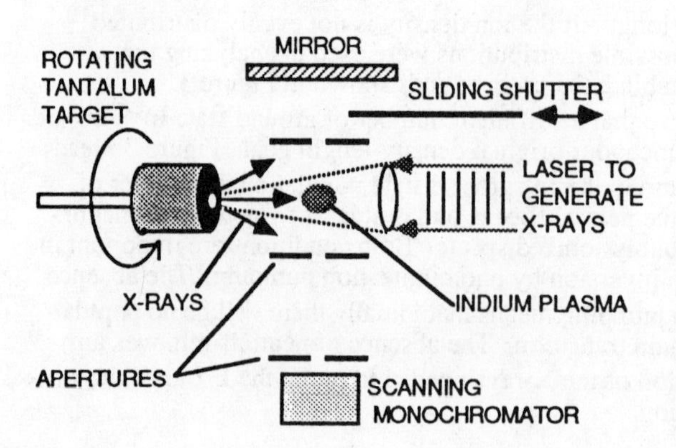

Fig. 4. Schematic of apparatus for photoionization-pumping of the indium plasma, and the measurement of the gain produced at 185 nm.

The results of the gain measurements for four different trials are shown in Figure 5. The largest single-pass intensity gain measured was 1.46, for a ~4 mm path length through the plasma. If the excited state population were evenly distributed over the path length, this would imply an exponential gain coefficient of 0.95 cm^{-1}. Because the distribution was not even, the gain coefficient in part of the plasma was in fact greater than 0.95 cm^{-1}.

In all trials, the indium was ablated with 67 mJ of 532 nm light at an intensity of ~1×10^8 W/cm^2. The X-rays were generated from a tantalum target using between 43 and 55 mJ of 1.06 μm light in a 70 to 100 psec pulse. For trial 1, a 50 cm focal length lens focused the light onto the tantalum target, which was ~5 mm from the gain region. In all other trials, a 25 cm lens was used, and the tantalum target was ~2 mm from the gain region. For each trial, the focused spot size was adjusted for maximum gain. The gain was only weakly dependent on the focusing conditions, as has been found in similar experiments[12]. The explanation is that a tight focus on the tantalum target produces a hot plasma with small emitting surface area; a loose focus yields a cooler, less bright plasma but with a larger emitting surface area. Near the optimum spot size, the total useful flux of X-rays is a slowly varying function of the spot size due to the counterbalancing of plasma temperature (and hence brightness) and the plasma surface area.

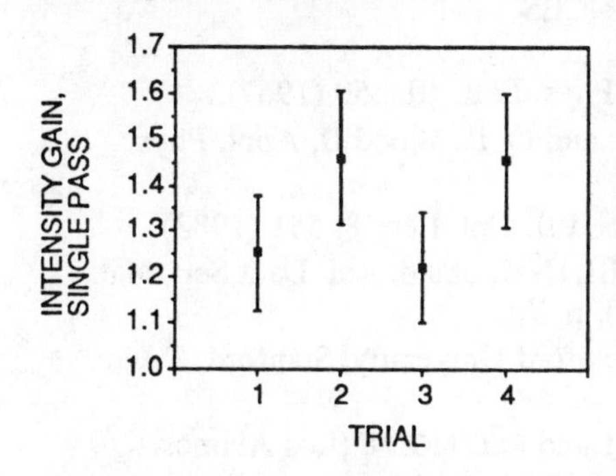

Fig. 5. Gain measurements at 185 nm. The pump laser energy for X-ray production was: 55 mJ for Trial 1, 48 mJ for Trials 2 and 3, and 43 mJ for Trial 4. (See text for other conditions.)

DISCUSSION

While the method of laser ablation has been found useful for producing high densities of ions, one should be warned of some of its drawbacks. One difficulty is the lack of a hydrodynamic computer code for predicting the distribution of ions in space and time. Without an *a priori* model of the conditions in the ablated indium plasma, it is difficult to determine how best to scale up the gain to make a super-fluorescent laser at 185 nm.

A further complication of laser ablation is that the amount of material evaporated is very nonlinearly dependent on the laser intensity reaching the target. One must be careful to ensure reproducibility with this method. We found that small changes in laser energy or in the distance from target to focusing lens would drastically change the ion densities produced. However, we worked close to the intensity threshold at which indium ions, rather than neutral atoms and clusters, were produced. This was done to prevent splashing of the liquid indium target due to the high pressures generated at the target surface by high ablation intensities. Reproducibility should be less of a problem at higher intensities, away from the threshold of ionization.

CONCLUSION

Based on the existence of known laser transitions in Cd^+, and on the ability to pump those transitions by photoionization of inner-shell electrons with broad-band soft X-rays, the authors predicted that gain at 185 nm might be obtained with the same pumping method on the same transition in In^{++}. In this paper, we report that a gain of at least 0.95 cm^{-1} at 185 nm has been produced by photoionization-pumping of an indium plasma.

Furthermore, laser ablation was shown to generate the large In^+ ground state ion density required for this scheme. A laser-produced plasma VUV continuum light source was used to make time-resolved density-length measurements of three ion species present in the laser ablated indium plasma. This VUV light source proved to be easy and reliable to operate.

ACKNOWLEDGEMENTS

This work was funded by the Air Force Office of Scientific Research. The authors thank Spectra-Physics for the generous loan of a Quanta-Ray Nd:YAG laser.

REFERENCES

1 M. A. Duguay and P. M. Rentzepis, Appl. Phys. Lett. 10, 350 (1967).

2 H. Lundberg, J. J. Macklin, W. T. Silfvast, and O. R. Wood II, Appl. Phys. Lett. 45, 335 (1984).

3 W. T. Silfvast, J. J. Macklin, and O. R. Wood II, Opt. Lett. 8, 551 (1983).

4 C. E. Moore, Atomic Energy Levels, Vol. III, (Nat. Stand. Ref. Data Ser., Nat. Bur. Stand., Washington D. C., 1971), p. 70.

5 A. C. Nilsson, Dept. of Applied Physics, Stanford University, Stanford, CA 94305. Unpublished work.

6 R. D. Cowan's computer codes RCN MOD 31 and RCG MOD 8 (Los Alamos Scientific Laboratory, Los Alamos, New Mexico).

7 K. Dittrich and R. Wennrich, Prog. Analyt. Atom. Spectrosc. 7, 139 (1984).

8 C. G. Mahajan, E. A. M. Baker, and D. D. Burgess, Opt. Lett. 4, 283 (1979).

9 A. W. Ehler and G. L. Weissler, Appl. Phys. Lett. 8, 89 (1966).

10 N. G. Basov, V. A. Gribkov, O. N. Krokhin, and G. V. Sklizkov, Sov. Phys. JETP 27, 575 (1968).

11 W. T. Silfvast, O. R. Wood II, and D. Y. Al-Salameh, presented paper TuB1 at the Topical Meeting on Short Wavelength Radiation, Monterey, California, March 1986.

12 W. T. Silfvast, O. R. Wood II, J. J. Macklin and H. Lundberg, in Laser Techniques in the Extreme Ultraviolet, edited by S. E. Harris and T. B. Lucatorto (AIP Conference Proceedings No. 119, 1984) pp. 427-436.

LASER-SYNCHROTRON STUDIES OF THE DYNAMICS OF UV-PHOTON-STIMULATED DESORPTION IN ALKALI HALIDES

R. F. Haglund, Jr. and N. H. Tolk
Department of Physics and Astronomy, Vanderbilt University, Nashville, TN 37235

ABSTRACT

Laser-synchrotron studies of neutral alkali emission from alkali halide crystals are yielding new insights into the dynamics of energy absorption, energy localization and bond-breaking in photon-stimulated desorption. The ground-state neutral desorption is triggered by thermal diffusion of photon-induced electronic defects; however, the excited-state neutral alkalis are formed in a surface-specific process on an extremely short time scale. In addition, there is new evidence for a surface overlayer which retards substrate desorption, thus suggesting a new approach to the optical damage problem at ultraviolet wavelengths.

INTRODUCTION

In recent years, there has been a growing interest in electronically-induced changes at surfaces and in the near-surface bulk of optical materials. Among the key scientific questions are the nature of charge-transfer processes in electron, ion and neutral atom interactions with surfaces; the dynamics of bond- breaking; the role played by defect formation in the near-surface bulk, and the relative importance of surface states in the absorption, localization and rechanneling of the incident electronic energy.

In this paper, we shall discuss laser-synchrotron studies of desorption induced by electronic transitions (DIET) which touch on various aspects of these questions. We shall describe the experimental protocol used in our studies of photon-stimulated desorption; present new results which give new insights into the dynamics of surface bond-breaking *via* electronic mechanisms; and consider evidence for the existence of a desorption-retarding overlayer, which may point the way to new techniques for protecting ultraviolet optical materials from photon-induced damage.

DESORPTION AS A PROBE OF SURFACE DYNAMICS

There are two principal means of studying dynamical processes on surfaces: *scattering* experiments, in which both the initial and final states of probe particles or photons are measured before and after the interaction with a well-characterized surface; and *desorption* experiments, in which a particle or photon in a known state transfers energy and possibly (as in sputtering) momentum to the surface, but final state measurements are made on atoms or molecules ejected from the surface. Desorption induced by electronic transitions (DIET) is particularly interesting because the excitation of the surface and near-surface bulk atoms leading to desorption occurs efficiently even without momentum transfer, and because the deposition profile for the incident electronic energy is relatively well understood.

0094-243X/86/1470103-7$3.00

DIET has been an object of great interest since the pioneering work of Menzel, Gomer and Redhead,[1] in which they applied two-body potential theory from gas-phase physical chemistry to the problem of ions desorbed by electron impact from metal-adsorbate systems. Subsequently, Knotek and Feibelman[2] successfully linked a model for the ion-desorption process to specific properties of certain materials. This work received further impetus with the elucidation of the relationship between angular distributions of desorbing ions and bond angles on surfaces by Madey and Yates,[3] providing detailed information on surface geometrical and electronic structure.

However, with the discovery that the yields of ground-state and excited-state atoms from both electron- and photon-stimulated desorption (ESD/PSD) in the alkali halides exceed ion yields by as much as five orders of magnitude,[4,5] it became clear that *neutral-species* desorption holds the key to the *dynamics* of electronic processes on many non-metallic surfaces. As the focus in DIET studies has shifted to neutral-species desorption, optical spectroscopic techniques, such as laser-induced fluorescence (LIF),[6] have assumed increasing importance. By measuring relative yields, velocity distributions, and the time, photon-energy and surface-temperature dependence of the neutral atom yields in PSD experiments, for example, it is possible to determine in detail the mechanisms responsible for the desorption and to correlate these mechanisms to specific materials properties.[7]

EXPERIMENTAL APPARATUS AND PROTOCOL

In the experiments described here, an arrangement shown schematically in Fig. 1 was used to produce and detect neutral atoms desorbed by ultraviolet photon irradiation from the Tantalus synchrotron storage ring at the University of Wisconsin. The photon beam was incident along the surface normal of a clean, optically

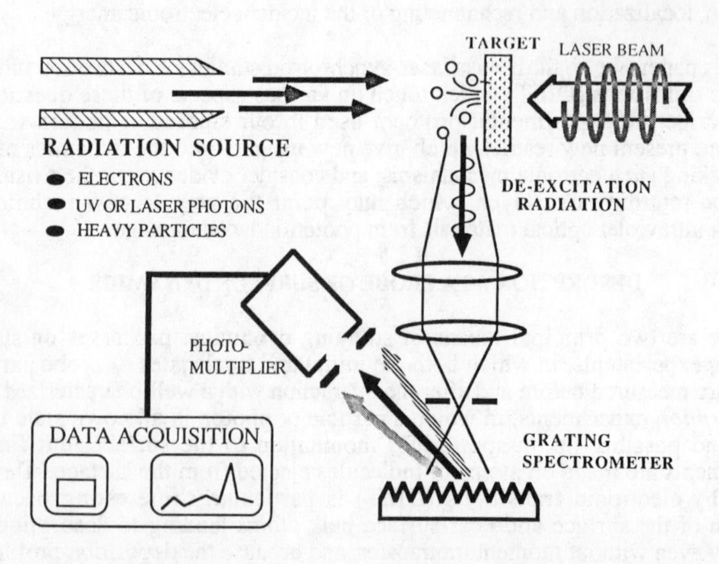

FIG. 1. Schematic of the experimental layout used for PSD measurements.

transparent alkali halide sample maintained under ultrahigh vacuum (base pressure $4x10^{-10}$ Torr). The (100) surface of the single-crystal target was cleaved and polished prior to its introduction into the UHV system and then cleaned by heating to 400° C. The sample temperature was varied by resistive heating of the copper target holder.

Fluorescence decay signals from desorbing ground-state and excited-state atoms were detected by a spectrometer-photomultiplier combination arranged to view a small region out on front of the surface through an appropriate imaging system. The location of the focal volume with respect to the target surface was varied by translating along the horizontal axes of the target manipulator. Fluorescence count rates ranged from a few kHz to a few hundred kHz.

One fine-structure component of the first optical resonance line (589 or 671 nm, respectively, for atomic sodium and lithium) was used to identify the desorbing alkalis. Ground-state neutral atoms were detected from laser-induced fluorescence (LIF) radiation produced by the light from a frequency-stabilized tunable dye laser, transported to a quartz entrance window of the UHV chamber through a single-mode, polarization-preserving optical fiber. To count neutral excited-state atoms, on the other hand, the laser was turned off and the desorbing atoms were detected by setting the spectrometer to the characteristic free-atom wavelength.

To measure relative densities of desorbing alkalis, the laser was injected parallel to and about 1 cm away from the surface, in a direction perpendicular both to the incident beam and to the spectrometer line-of-sight. For velocity distribution or desorption time history measurements, the laser was injected along the rear surface normal of the alkali halide crystal, anti-parallel to and collinear with the photon beam. Velocity distributions of the desorbing ground-state atoms were obtained by scanning the laser through the Doppler-shifted wavelength band resonantly absorbed by the alkali atoms.

EXPERIMENTAL RESULTS

Figure 2 shows a typical yield measurement for ground-state neutral atoms as a function of photon energy. Apart from a slight shoulder in the curve at a photon energy corresponding to the Na 2p core-level excitation, there is a notable absence of structure, suggesting that the excitation mechanism is not a resonant one. It should be noted, however, that the yield is relatively high throughout the entire photon energy range spanned by the measurement, indicating that the PSD process is relatively efficient over a broad range of uv photon energies.

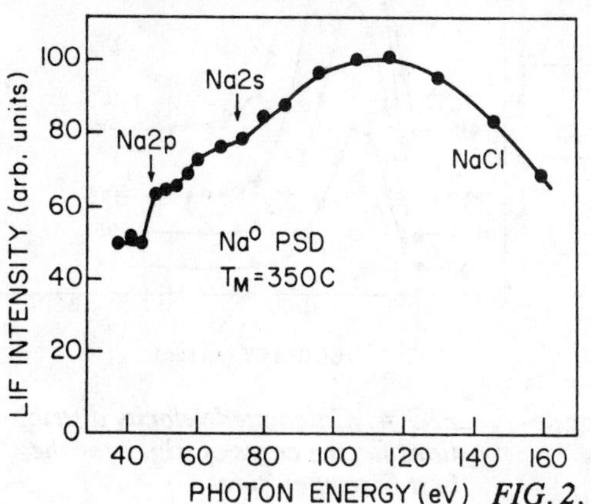

FIG. 2. LIF intensity as a function of photon energy for ground-state neutral Na desorbed from single-crystal NaCl.

Significant clues to the mechanism underlying ground-state neutral desorption were furnished by velocity-distribution measurements. The number *vs.* velocity distribution of the desorbing alkali atoms is given by

$$dn/dv = C\ v^2 \exp(-mv^2/2kT), \tag{1}$$

where the velocity v is related to the frequency ν of the emitted photons by

$$\nu = \nu_o\,[1 + (v/c)\cos\theta]. \tag{2}$$

The parameter ν_o in the first-order Doppler formula is the central frequency for stationary atoms and θ is 0 in the geometry of these experiments.

Figure 3 shows the measured yield *vs.* laser-frequency from PSD experiments on various sodium halide crystals, while Fig. 4 shows calculated velocity distributions for desorbed sodium atoms at the temperature best fitting the observed data.[8] The velocity distribution is definitely Maxwell-Boltzmann. The linearly increasing difference between the fit temperature T_f and the measured temperature T_m (see inset) suggests a systematic error, probably arising from a temperature gradient between the thermocouple mount and the crystal surface. Measurements of PSD yield as a function of surface temperature [not shown here] also display an Arrhenius thermal desorption characteristic as a function of increasing surface temperature.

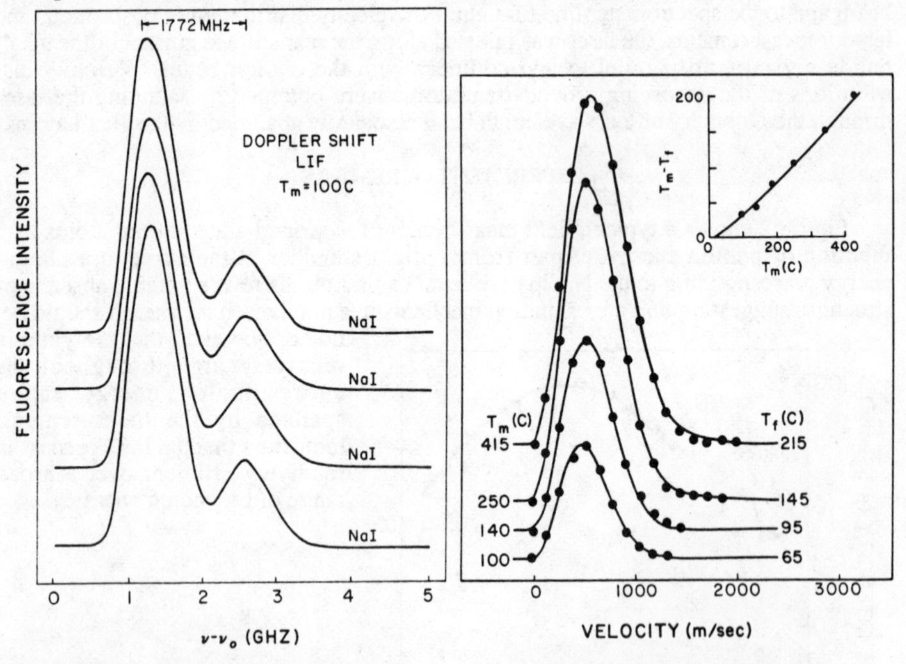

FIG. 3. Measured hyperfine spectra for desorbing Na atoms from several sodium halides.

FIG. 4. Calculated velocity distributions for the cases of Fig. 3 for the best-fit temperatures.

We have also measured the time history of ground-state and excited-state lithium desorbing from lithium fluoride crystals. In the case of ground-state lithium (Fig. 5), the fluorescence yield showed an initial sharp drop after the beam was turned off, followed by a slowly-decaying emission lasting many seconds. For excited-state atoms, time histories were measured by using the time structure of the synchrotron beam to provide a nanosecond excitation pulse, and routing the fluorescence signals through a time-to-amplitude converter. These data, shown in Fig. 6, indicate that the *excited-state* neutral lithium turns on and off with the exciting photon beam. Thus, while the ground-state desorption is consistent with a model in which thermally-driven diffusion transports radiation-induced defects from the near-surface bulk to the surface where thermal desorption occurs,[7] excited-state desorption appears to be a surface-specific process, probably involving direct excitation of surface states.

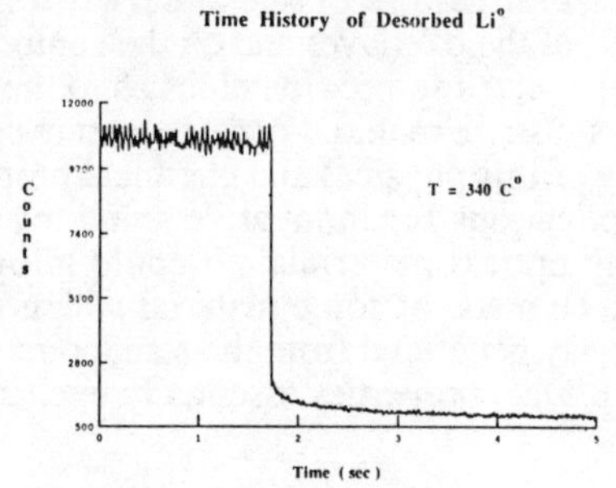

FIG. 5. Time history of Li° desorption from LiF following photon irradiation.

Time History of Desorbed Li* and Li°
Stimulated by UV Photons
31.293 ns
Li* + Bulk Fluorescence
Bulk Fluorescence
Synchrotron pulse
Li°
Counts
Time (ns)

FIG. 6 Time history of Li desorbed from LiF by uv photons.*

The data of Fig. 2 indicate a rather broad range of photon energies over which the alkalis are desorbed with significant efficiency from alkali halides by PSD. However, the critical question of energy threshold for neutral ground-state desorption is not answered by these experiments. A recent experiment with filters has confirmed earlier conjectures that these thresholds are astonishingly low -- and in the case of LiF, below the band-gap energy. Figure 7 exhibits the results of a recent experiment in which the ground-state yields of Li desorbed by uv photons from LiF were measured by passing white light from the storage ring through filters in order to measure the relative desorption efficiencies in various regions of the spectrum. After correcting for the monochromator efficiency, it can be seen that nearly twenty-five *per cent* of the total neutral yield results from photons below 13.2 eV -- the band gap for the LiF filter. (Indeed, the threshold may be even lower, since this particular filter showed obvious signs of significant color-center formation.) This means that such desorption is likely to take place efficiently even at excimer-laser photon energies, perhaps even without invoking multiphoton transitions.

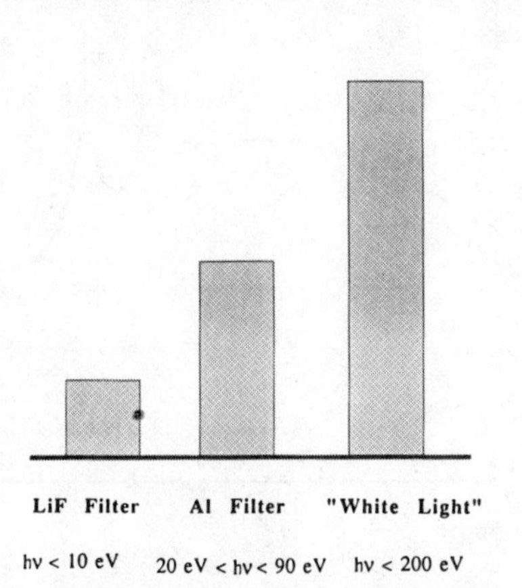

FIG. 7. Relative PSD yields of Li° from LiF for the photon-energy ranges indicated.

From these and related ESD experiments, a picture is emerging which clearly demonstrates that electronic mechanisms play a leading role both in *microscopic* desorption processes and in *macroscopic* erosion and surface damage in optical materials of interest in the ultraviolet region of the spectrum. However, we are also seeing evidence indicating that a hydrogen-rich overlayer of as yet undetermined structure and composition may play a role in retarding the onset of desorption from substrates.

Figure 8 shows a spectrum of excited-state neutral atom desorption in a PSD experiment with white light on LiF. The ground-state Li^{o} peak is much in evidence -- but so also are the various hydrogen lines and OH in the spectrum. Similar results have been seen in ESD experiments in which comparative yields from different alkali halides were measured. It is clear that the *relative* loss of substrate, *vis a vis* the overlayer, material depends both on the details of the overlayer and on the composition of the substrate. Further measurements are expected to provide clues as to the exact role of this "protective overlayer" in binding substrate material and preventing surface bond-breaking. More detailed understanding of this physical and chemical properties of the overlayer may have far-reaching implications for innovative solutions to the problem of damage in ultraviolet and X-ray optical materials. It could allow, for example, high-damage-threshold coatings to be made of some material which would change the surface *electronic* properties in a way beneficial from the standpoint of uv-photon-induced desorption, while leaving its *optical* properties essentially unchanged.

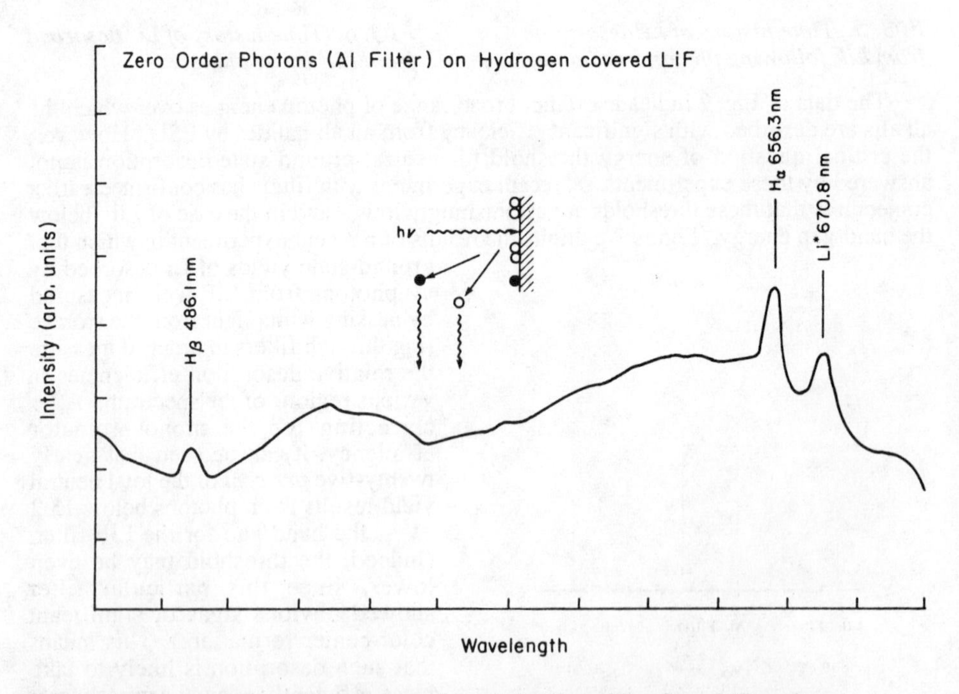

FIG. 8. Spectrum of excited-state neutral particles desorbing from LiF under ultraviolet photon irradiation. The identifiable atomic lines are indicated; the broad continuum background arises from surface and near-surface-bulk luminescence.

CONCLUSIONS

The PSD experiments described here have produced striking new information building on previous evidence that surface *dynamics* in PSD is determined primarily by the neutral desorption channels. The time-resolved measurements have pinpointed the critical role played by surface states in the desorption of excited-state neutral Li from PSD on LiF, while showing that the role of diffusive defect-induced desorption is central to understanding ground-state neutral PSD. The low photon-energy thresholds measured for efficient PSD means that this process must now be considered important in excimer-laser-induced material-damage phenomena, where it was formerly thought to be of little consequence. And the demonstration of a mechanism for retarding the onset of substrate damage clearly has fascinating implications for the optical damage question, since it suggests that there may be ways of treating optical surfaces chemically in order to retard desorption, surface modification and ultimately, catastrophic erosion and damage.

ACKNOWLEDGEMENT

It is a pleasure to acknowledge the collaborative efforts of the other principal participants in the experimental work reported here: Richard Rosenberg, Synchrotron Radiation Center, University of Wisconsin; Guillermo Loubriel, Sandia National Laboratories; and Royal Albridge, Vanderbilt University.

REFERENCES

1. D. Menzel and R. Gomer, J. Chem. Phys. **41** (1964) 3311. P. A. Redhead, Can. J. Phys. **42** (1964) 886.

2. M. L. Knotek and P. J. Feibelman, Phys. Rev. Lett. **40** (1978) 964.

3. T. E. Madey and J. T. Yates, Jr., Surface Sci. **63**, 203 (1977).

4. N. H. Tolk, L. C. Feldman, J. S. Kraus, R. J. Morris, M. M. Traum and J. C. Tully, Phys. Rev. Lett. **46** (1981) 134.

5. N. H. Tolk, M. M. Traum, J. S. Kraus, T. R. Pian, and W. E. Collins, Phys. Rev. Lett. **49** (1982) 812.

6. N. H. Tolk, R. F. Haglund, Jr., M. H. Mendenhall, E. Taglauer and N. G. Stoffel, in: Desorption Induced by Electronic Transitions: DIET II, eds. W. Brenig and D. Menzel (Springer-Verlag, Berlin, 1985), p. 152.

7. R. F. Haglund, Jr., R. G. Albridge, D. W. Cherry, R. K. Cole, M. H. Mendenhall, W. C. B. Peatman, N. H. Tolk, D. Niles, G. Margaritondo, N. G. Stoffel and E. Taglauer, Nucl. Instrum. Meth. in Phys. Research **B13**, 525 (1986).

8. N. G. Stoffel, R. Riedel, E. Colavita, G. Margaritondo, R. F. Haglund, E. Taglauer and N. H. Tolk, Phys. Rev. B **32**, 6805 (1985).

PROGRESS IN THE ANALYSIS OF SELENIUM X-RAY LASER TARGETS*

M. D. Rosen and P. L. Hagelstein
University of California
Lawrence Livermore National Laboratory
Livermore, California 94550

ABSTRACT

We review progress in the modeling of Ne-like-Se XRLs. Dielectronic recombination plays an important role in the level kinetics as well as in ionization balance. Refraction becomes important at target lengths greater than 2cm by reducing signal at 0° view, and by having much larger signals emitted at a 10-20 mrad view. We predict success in scaling these systems to lower λ with higher Z targets, but at great cost in required driver power.

INTRODUCTION

A previous experimental series on the Novette laser, using a specially designed exploding selenium foil,[1] succeeded in demonstrating[2] laser amplification at 206 and 209 Å. The amplification was attributed to J = 2 to J = 1, 3p to 3s transitions in neon-like selenium. However, the theoretical foundation for predicted gain resulted in fundamental differences between data and theory.[1,3]

In this article, we review progress during 1985 that has advanced our understanding of the selenium x-ray lasing system. We also predict some potential successes and problems as we pursue the goal of demonstrating x-ray lasing at shorter wavelengths.

ATOMIC PHYSICS ISSUES

The major discrepancy between theory and data is the low gain observed on the J = 0 to J = 1, 3p to 3s transition compared to the predicted gain of about 10 cm^{-1}. In addition, the observed J = 2 to J = 1 gain was about 6 cm^{-1}, while the predicted value was 3 to 4 cm^{-1}.

The theory and predictions made last year contained at least one significant shortcoming. While it was widely recognized that dielectronic recombination could play an important role in atomic-level population kinetics, such recombination was only

*Work performed under the auspices of the U. S. Department of Energy by the Lawrence Livermore National Laboratory under contract number W-7405-ENG-48.

modeled in a crude manner. We used a reasonable guess as to the total rate, but connected that total rate from ground state to ground state of iso-electronic sequences. This rate was apparently correct because it reproduced gross features of n = 3 to n = 2 x-ray spectra that gave us an indication of ionization balance (approximately 40% fluorine-like and 20% neon-like selenium). However, in dielectronic recombination (e.g., from a fluorine-like ion), a ground-state electron is elevated to an n = 3 level or higher, while a free electron decreases in energy by an equal amount, from the continuum into a high-lying level. While autoionization is the inverse process and can undo the doubly excited neon-like state, the ion can stabilize in other ways: in particular, by radiative stabilization. That process can leave the neon-like ion in a singly excited state. Cascade processes that follow favor populating the J = 2 levels over the J = 0 levels because of their higher (by a factor of 5) statistical weight. In addition, the 3s levels can be populated in this manner, thus lowering the gain on the J = 0 to J = 1, 3p to 3s transition. When we crudely modeled dielectronic processes by proposing that the flow from fluorine-like to neon-like is directly through their ground states, rather than (correctly) through excited states, we may have correctly predicted ionization balance, but eliminated the possibility of lowering J = 0 gain and raising J = 2 gains.

As a result of such deficiencies in previous theory, we have begun to include all the doubly excited states resulting from dielectronic recombination in the atomic physics input-data files. These files are read [along with hydrodynamic quantities from a two-dimensional (2-D) LASNEX simulation] into the XRASER computer code. With these new files, it is of interest to compare our new results for gain prediction with those of last year.

The 2-D LASNEX simulation models our nominal target, which consists of a 750-Å coating of selenium on 1500 Å of a plastic called formvar. A 2ω 2-beam Novette laser impinges on the foil with a line focus that is 200 μm in height at an intensity of 4×10^{13} W/cm^2 in a 450-ps, full-width-at-half-maximum (FWHM) Gaussian pulse.

The 2-D results do not differ fundamentally from those obtained by 1-D simulations[4]. However, the peak electron temperature for the 1-D simulation is somewhat higher than that predicted by the 2-D simulation. The temperature, density, and velocity history is fed into the XRASER calculation of gain, along with the detailed atomic physics model that includes most of the relevant doubly excited states formed by dielectronic recombination. For simplicity, we consider the conditions for a central computational zone that describes a typical part of the main lasing region.

The first result of interest from our calculation is that the ionization balance of 40% fluorine-like and 20% neon-like selenium is reproduced by the XRASER calculation, thus lending credence to our previous estimates. In Figs. a and b, we plot the prediction for gain vs time for the J = 0 to J = 1 and J = 2 to J = 1 lasers, respectively. Times early with respect to the peak of the driving laser pulse are largely irrelevant because the foil has not fully exploded. Thus, the initial high density and steep density gradients of the foil deny the laser system full access to gain down its full length, and refraction bends the laser out of the high-gain medium.[1,3] Times much later than the laser peak are uninteresting because the rapidly falling density leads to falling gain. The time of greatest interest is roughly the 100 ps immediately after the driving laser peak, which optimizes gain vs a nonrefractory density gradient.

At +100 ps in Fig. 1, the gain on J = 0 is about 7 cm^{-1}; the gain on the upper 209-Å J = 2 line shown in Fig. 2 is about 6.5 cm^{-1}. (The 206-Å J = 2 line has a similar gain.) Compared with our previous model, the J = 0 gain is slightly reduced due, in part, to a slightly higher 3s population. The J = 2 gain is nearly doubled, due to our directly populating the upper J = 2, 3p levels via the decay of the doubly excited states. Nevertheless, the predicted J = 0 gain remains too high with respect to the data. Although we have, as yet, no complete explanation for the discrepancy, we now have obtained a predicted gain for J = 0 that is comparable to that for J = 2, and the J = 2 predicted gain is now quite close to the observed value.

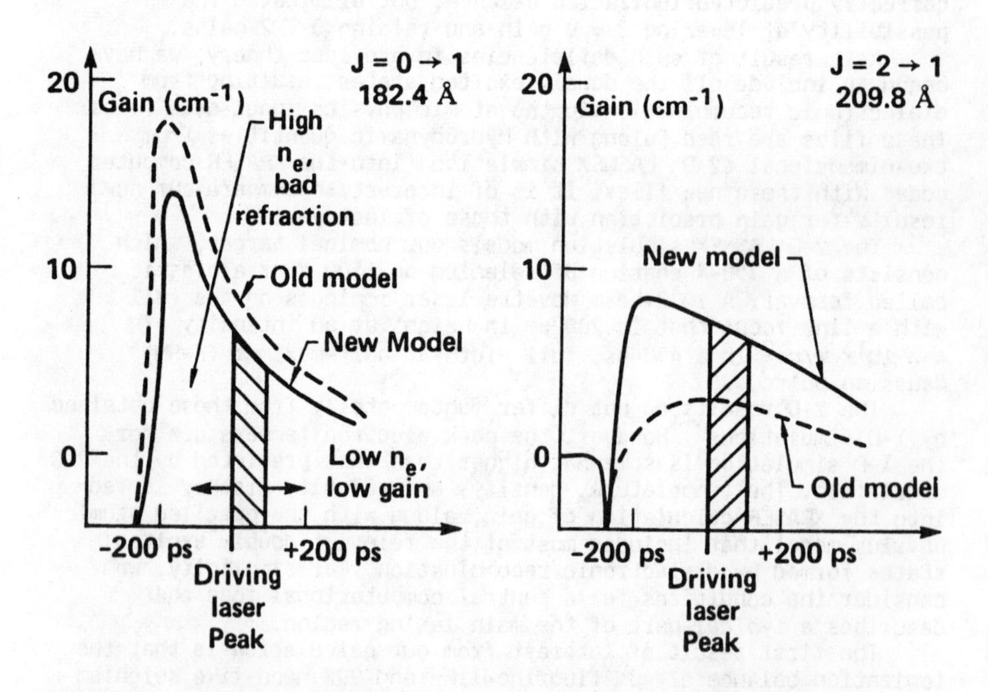

Fig. 1a. Predicated gain vs time for J = 0 to J = 1, 3p to 3s (183-Å) transitions in neon-like selenium.

Fig. 1b. Predicated gain vs time for J = 2 to J = 1, 3p to 3s (209-Å) transitions neon-like selenium.

Another issue arising from our data is the apparent lack of lasing lines from fluorine-like states. Even with our recent and more complete modeling, we continue to predict a gain of about 2.5 cm^{-1} for fluorine-like lasing at 204 Å. While the value is significantly less than the gain of 6 cm^{-1} observed for J = 2 neon-like lines, it is curious that the lines remain unobserved given the large fraction of ions in the fluorine-like state.

HYDRODYNAMIC ISSUES

As another topic, we have performed studies of the laser medium under conditions that can be characterized as nonoptimal, i.e., arising from a subtle hydrodynamic effect. The nominal line focus of 1 cm is not uniform along the full 1-cm length. The laser intensity at the extreme edges of the 1-cm length is lower by a factor of 2 than that in the central part of the line of focus. This observation is based on both equivalent-plane optical measurements and on x-ray pinhole pictures. We have processed the LASNEX simulations at various intensities and created an x-ray pinhole calibration curve. From that curve and the experimental pinhole picture, we have confirmed that the laser intensity at the edge is one-half that at the center.

Figure 2 shows the density profile of an exploding foil 100 ps after the peak of a 220-ps, FWHM Gaussian driving laser pulse. The nominal intensity I_0 is 7 X 10^{13} W/cm^2. The profile for the intensity $I_0/2$ represents the density profile seen by the laser as it tries to exit the 1-cm-long tube (the intensity at the edge is reduced by a factor of 2). The outcome is not highly deleterious with respect to refraction. However, for an off-optimal shot, the central intensity may be $I_0/2$. In this case, the intensity at the edge is $I_0/4$, and the density profile at that edge shown in Fig. 3 is much steeper (more density) than the profile in the center of the lasing tube. The result can lead to deleterious refraction effects as the laser beam leaves the laser tube. This is a specific example of a general phenomenon: namely, off-optimal targets are increasingly sensitive to the I/2 edge effect

BEAM PROPAGATION ISSUES

The Nova II target chamber will allow for two-sided illumination with a line focus of up to 5 cm in length. As we move to shorter-wavelength x-ray lasing schemes, our first priority is to reach saturation on our selenium scheme. Figure 3(a) shows the trajectories of x-ray laser beams as they traverse the length of the exploding foil for distances greater than 1 cm. The onset of severe degradation beyond 2 cm is quantified in Fig. 3(b).

From the trajectories of Fig. 3a another consequence emerges. Mentally, extend the target one more cm past 3cm. Clearly a zero degree (straight down the Z axis) view will collect little extra signal vis a vis a 2 or 3cm target due to refracton. However a view from 10 or 20 milliradians off axis can see orders of magnitude more

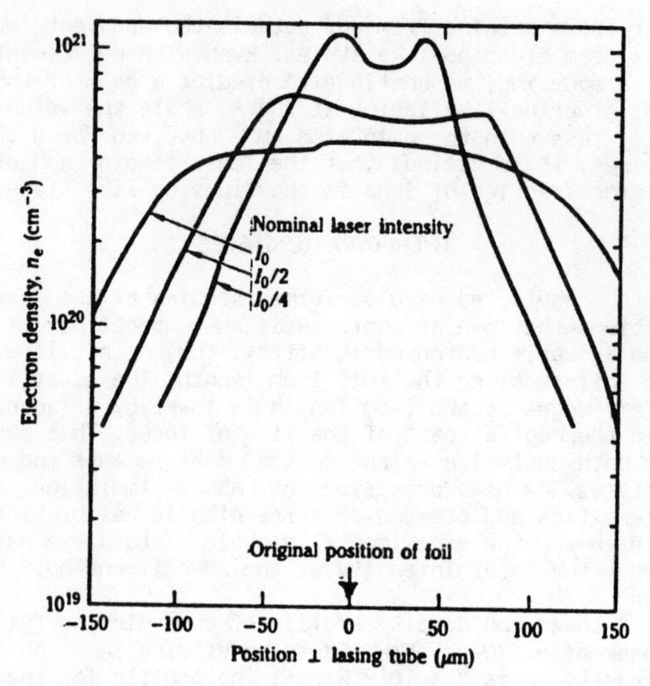

Fig. 2. Electron density vs position perpendicular to the lasing tube axis for three incident laser intensities.

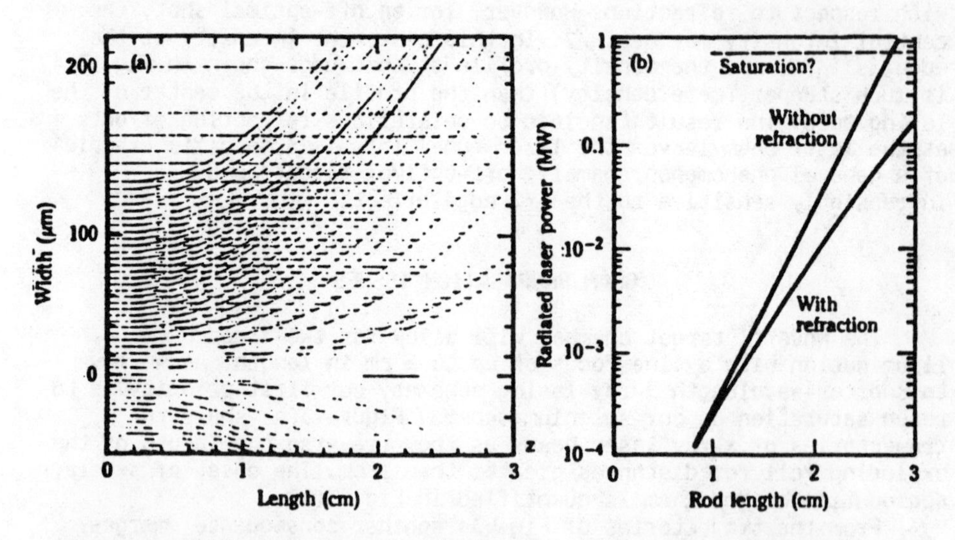

Fig. 3a. Trajectories of x-ray laser beam viewed from above an exploding foil and propagating over a distance of 3 cm.

Fig. 3b. Radiated laser power vs rod length calculated with and without refraction effects.

signal. To demonstrate this, reflect Fig. 4a around Z = 0, from -2cm to 2cm. Then the "inverted rainbow" of rays (emerging at 10-20 mrad) all get a full 4cm worth of exponentiation. Thus during this time, there is far more signal at 10-20 mrad than there is at 0 mrad.

We are attempting to circumvent refraction limitations to reaching saturation by inventing new target geometries for the x-ray laser. One scheme involves exploding foils that face one another and explode toward each other. This scheme creates a temporary trench that traps the x-ray laser beam in a region of high gain and allows propagation over distances much longer than 1 cm. At 3 cm, we predict a degradation of only 10% compared to the 90% degradation for the single foil shown in Fig. 3(b). Preliminary hydroexperiments at KMS Fusion on this scheme support our optimism. However, the new target forms a confining trench in only one dimension. We are therefore pursuing cylindrical exploding foils as a 2-D analog of the trench targets. Other targets with tamping from thick formvar substrates also show promise as nonrefractory lasing media.

SHORTER WAVELENGTH LASING ISSUES

Shorter-wavelength schemes using nickel-like analogs to our neon-like system or rapid-cooling recombination schemes are currently under study and will be presented elsewhere. Here we present calculations for the gain of a higher-Z, neon-like system: namely molybdenum (Z = 42).

The nominal target is 1000 Å of molybdenum on 1000 Å of formvar. The target is irradiated by a 2ω laser with a total irradiance of 3.5×10^{14} W/cm^2 for 500 ps. At the peak of the driving laser, electron temperatures reach 2 keV, densities reach 5×10^{20} cm^{-3}, and the scalelength is about 200 μm. Lasing with a gain of about 3 cm^{-1} is predicted for both the J = 2 and J = 0 transitions.

An interesting change takes place in proceeding to higher Z. For selenium, the J = 0 was at 182 Å, shorter than the two J = 2 transitions at 206 and 209 Å. For molybdenum, the J = 0 at 139 Å is a longer wavelength than either of the two J = 2 transitions at 131 and 133 Å. The crossover occurs near yttrium (Z = 39). Moreover, the other J = 0 line (at 167 Å in Se) which was predicted to have a low gain in Se, is predicted to have gain of order 3cm^{-1} for Mo, and to be at about 105 Å. Reaching laser wavelengths in the "water-window" below 40Å will require laser powers far greater than currently available if we are to rely strictly on J = 2 Ne-like 3p-3s lasing. Using scaling laws for exploding foil targets[4], we find (details to be published elsewhere) the required driver power scaling as (XRL-wavelength)$^{-4}$. Ways of circumventing this obstacle are currently being studied.

REFERENCES

1. M. D. Rosen et al., Phys. Rev. Lett. 55, 44 (1985).
2. D. L. Matthews et al.. Phys. Rev. Lett. 55, 48 (1985).
3. Laser Program Annual Report 1984, Lawrence Livermore National Laboratory, Livermore, Calif., UCRL-50021-84 (1985), pages 3-10 to 3-14.
4. R. A. London and M. D. Rosen, submitted to Phys. Fluids. Available as LLNL Report No. UCRL-94085, Feb. 1986.

STATUS OF THE NOVA X-RAY LASER EXPERIMENTS*

D. Matthews, S. Brown, M. Eckart, B. MacGowan,
D. Nilson, M. Rosen, G. Shimkaveg, R. Stewart,
J. Trebes, and J. Woodworth

University of California
Lawrence Livermore National Laboratory
Livermore, CA 94550

ABSTRACT

We review the progress of the x-ray laser experiments done at the Nova Laser facility. To date we have achieved lasing at wavelengths as short as 10.5 nm and output powers to ~1-10 MW in a pulsewidth of 175 psec FWHM. We are experiencing considerable x-ray laser beam breakup for exploding foil amplifiers >30 mm in length. We also describe our plans for achieving single-mode fully-saturated lasers at wavelengths <4.4 nm.

Since our original description[1] and demonstration[2] of optical laser driven exploding foil x-ray amplifiers, we have made considerable progress in their understanding, design and demonstration. In the previous talk, M. Rosen and P. Hagelstein described our progress in the modeling and design of x-ray lasers. We will focus on our new experimental capabilities and results.

Our primary activity, since September 1984, has been to construct and activate the new Nova 2-beam laser facility. This new facility gives us considerably more capability for performing x-ray laser experiments than was possible at Novette. Table 1 summarizes the most important differences while Fig. 1 shows a schematic of the basic Nova laser facility. Eventually, the Nova 10-beam laser will also be configured with 10 cylindrical lens doublets. This will provide us with five times the current power capability for a given amplification length.

---

*This work was performed under the auspices of the U.S. Department of Energy by the Lawrence Livermore National Laboratory under contract No. W-7405-Eng-48.

Table 1. Novette vs. Nova 2-beam characteristics

	Novette	Nova
Power/beam @ 0.53 μ	2.4 TW	>5 TW
0.32 μ	No	Yes
Line focus	0.1x11 mm	0.1x0.1→52 mm variable
Timing fidu	No	Yes
shaped pulses	No	Yes

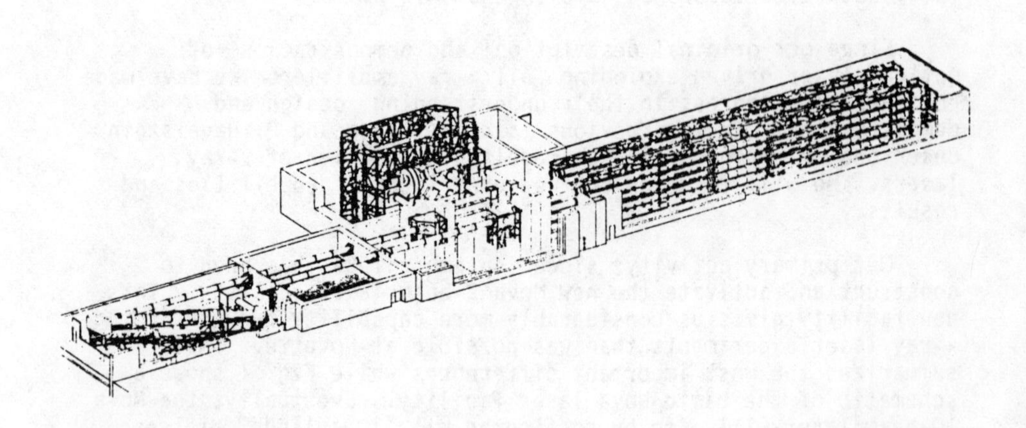

Figure 1. The Nova Laser Facility

We have also spent considerable effort improving our diagnostics package. Figure 2 illustrates the placement and identification of our most important diagnostics. The principle diagnostics for monitoring the output of the x-ray amplifier are the McPigs II and III gated microchannel plate intensified grating spectrometers,[3] and the TGSS or transmission grating streaked spectrograph.[3] McPigs II monitors the spontaneous emission from the plasma amplifier while McPigs III collects a portion of the axial or amplified spectrum. The transmission grating spectrograph time-resolves the spectral emission with a time resolution of ~20 psec FWHM. The XCSS is a streaked x-ray crystal spectrometer, the PHC is an x-ray pinhole camera and the HENWAY is a conventional x-ray spectrograph which uses film to record data. A new, important diagnostic is the near field camera (NFD), which is used to photograph the x-ray laser beam profile at a distance of ~160 cm from the exit aperture of the x-ray amplifier. Figure 3 shows a cutaway drawing of this diagnostic. Its primary components are a carbon mirror operated at 2θ = 11° and an aluminum filtered filmpack With this high-pass, L-edge filter combination we can view the x-ray laser beam with ~20% bandpass and a background rejection ratio of 25/1, assuming 100 eV blackbody level of emission from out of the plasma amplifier. Figure 4 shows a portion of the x-ray laser beam viewed with this instrument. It shows diverging beamlets (shown as large regions of white in this photo) emanating from a region where the x-ray laser target foil was originally laser beam camera, NFD located

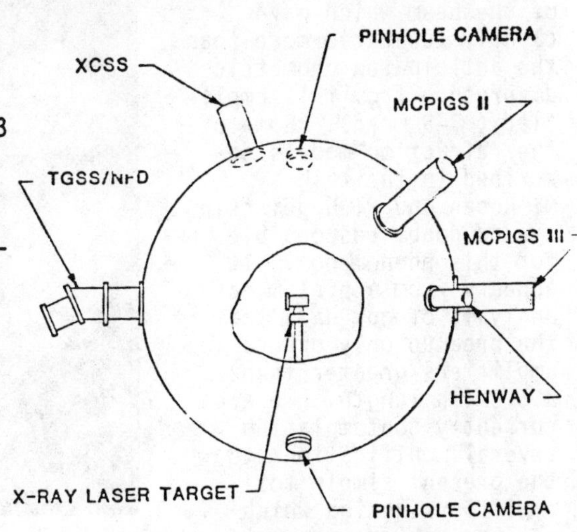

Figure 2. Experimental Setup

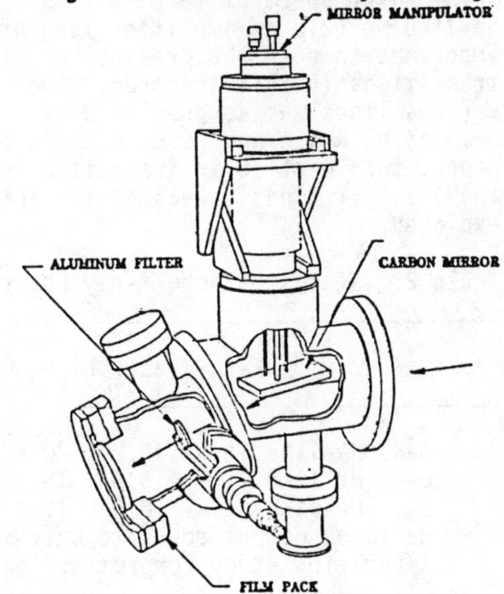

Figure 3. Schematic of xray near field diagnostic

(shown as the dark vertical band in Fig.4). The fine strips and dark horizontal band on the photo are for pointing checks and film calibration, respectively. Within the limited field of view of the instrument (~17 mrad full angle) we see two portions of the beam which have clearly separated more than the anticipated geometric divergence from this amplifier (~7-8 mrad). Some of the refraction models described in the talk by M. Rosen and P. Hagelstein are no doubt responsible for this phenomenon. It appears from a preliminary analysis of our data that the breakup only occurs for amplifiers greater than 2.5 cm in length. We are currently contemplating several modifications to the present simple foil amplifiers design which will be tested to see if this problem can be cured.

Figure 4. X-ray photograph of Se xray laser beam at a distance of 16- cm from the source

We present the status of our experiments in Table 2. A representative x-ray laser spectrum for Ne-like Se (2.2 cm amplifier length is shown in Fig. 5). It is important to note the presence of five anticipated laser transitions in this spectrum. The relative intensities of the various lines was as predicted except for the J=0 to 1 at 18.24 nm. As M. Rosen and P. Hagelstein discussed in the previous paper, this particular transition is predicted to be as strong as the J=2 to 1 lines. Reasons for this anomaly are still being explored.

Table 2. Status of Nova X-ray Laser Experiments

X-ray Laser Lines Observed to Date						
o	Ne-like Se --	18.24	20.69	20.96	22.13	26.3 nm
o	Ne-like Y --	15.5	15.7	16.5	21.8	
o	Ne-like Mo --	10.5	13.1	13.3	13.9	14.2

Measured output power to date is at least 1 MW @ 20.6 20.9 nm
Saturation study compromised by beam breakup for L>30 mnm

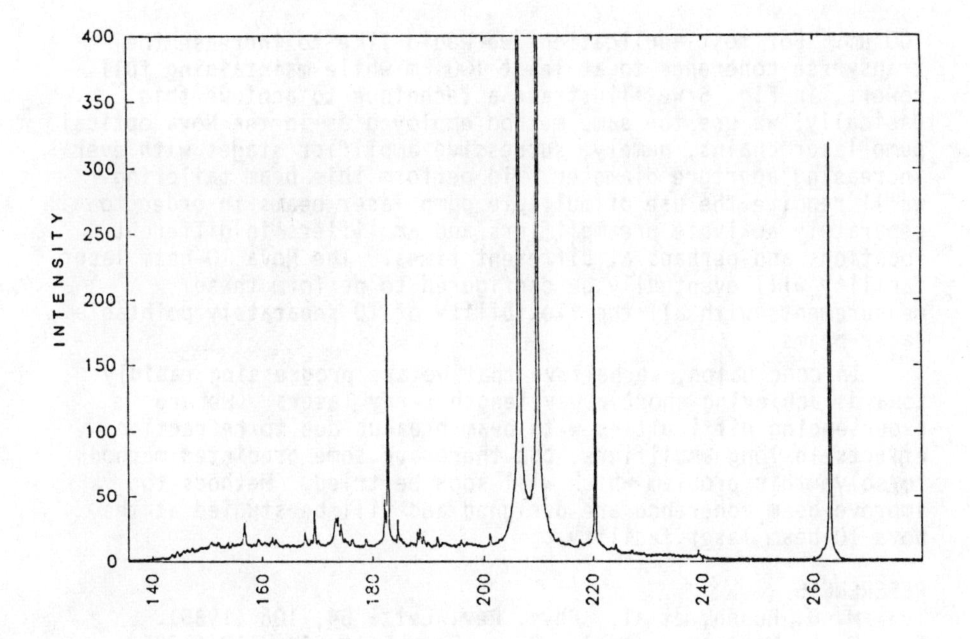

Figure 5. Se x-ray laser spectrum with 20.6 and 20.9 nm deliberately off-scale to emphasize weaker lines.

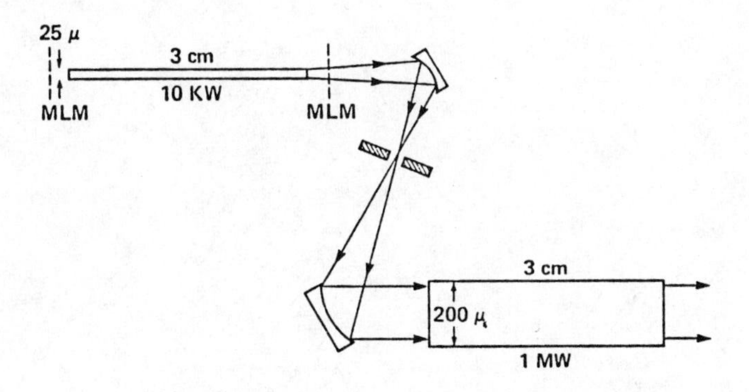

Figure 6. Method to improve x-ray laser beam coherency.

In order to successfully use the x-ray laser, we must not only improve the beam divergence, but also the transverse coherence. We calculate the transverse coherence length to be $L\lambda/D = 3\ \mu m$ at the output end of a 5 cm long amplifier, whereas the longitudinal coherence length is $\lambda^2/\Delta\lambda =$

100 μm. For most applications we would like to increase the transverse coherence to at least 100 μm while maintaining full power. In Fig. 6 we illustrate a technique to achieve this. Basically, we use the same method employed as in the Nova optical pump laser chains, namely, successive amplifier stages with ever increasing aperture diameter. To perform this beam tailoring we'll require the use of multiple pump laser beams in order to separately activate preamplifiers and amplifiers in different locations and perhaps at different times. The Nova 10-beam laser facility will eventually be configured to perform these measurements with all the flexibility of 10 separately pointable laser beams.

In conclusion, we believe that we are progressing rapidly towards achieving shorter wavelength x-ray lasers. We are experiencing difficulties with beam breakup due to refraction effects in long amplifiers, but there are some predicted methods to solve this problem which will soon be tried. Methods to improve beam coherence are designed and will be studied at the Nova 10 beam laser facility

REFERENCES

1. M. D. Rosen, et al., Phys. Rev. Lett. 54, 106 (1985).
2. D. L. Matthews, et al., Phys. Rev. Lett. 54, 110 (1985).
3. For more information on these diagnostics and other aspects of the x-ray laser program at Lawrence Livermore National Laboratory, see the publication Energy and Technology Review, November, 1985, available from Lawrence Livermore National Laboratory's Technical Library, report number UCRL-5200-85-11.

RECENT PROGRESS IN SOFT X-RAY LASER DEVELOPMENT AT PRINCETON

S. Suckewer, C. H. Skinner, D. Kim,
E. Valeo, D. Voorhees, and A. Wouters
Plasma Physics Laboratory, Princeton University
Princeton, New Jersey 08544

Abstract

Recent advances in research on soft X-ray lasers at Princeton are described. A one-dimensional code has been constructed which is in good agreement with the measured radial dependence of soft X-ray gain at 182 Å in a magnetically confined recombining plasma. Multichannel detectors have been installed in the diagnostic spectrometers and spectra of the line emission in axial and transverse directions are presented. Initial measurements of the relative divergence and, very recently, absolute divergence measurements of the axial 182 Å beam have been made by scanning the axial spectrometer across the beam. The absolute divergence was measured to be in the range 5-10 mrad, depending on experimental conditions and the maximum power of soft X-ray beam was ∿ 100 kW. Finally, a new two laser approach to create gain at wavelengths below 100 Å is briefly described.

I. INTRODUCTION

Progress in the development of soft X-ray lasers has been recently reported by several laboratories.[1] At Princeton, an approach based on a magnetically confined recombining plasma column, cooled by radiation losses, has generated amplification of stimulated emission of ∿100 (a one pass gain length of $k\ell \approx 6.5$).[2] In this experiment a commercially available 1 kJ TEA CO_2 laser (duration 80 nsec, maximum gain was obtained at an energy 300 J) was incident on a carbon disc target in a strong (90 kG) magnetic field. Rapid recombination, after the laser pulse, created a population inversion between levels 3 and 2 in hydrogen-like carbon, (CVI). Installation of a soft X-ray mirror[3] in a double pass arrangement provided an additional demonstration of the amplification of stimulated emission. With a measured normal incidence reflectivity of 12% at 182 Å, a 120% increase in axial stimulated emission was observed.[2] In work with axially oriented, thick (35μ) carbon fiber targets coated with a thin layer of aluminum, gains of up to 6 cm^{-1} were also generated on the CVI 182 Å transition[4].

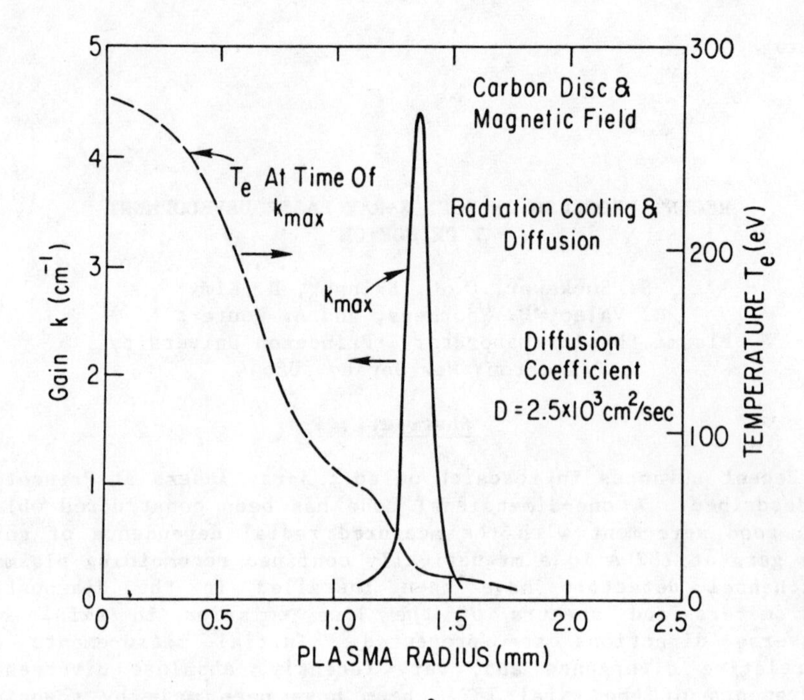

Fig. 1 Radial profiles of CVI 182 Å gain, k_{max}, and electron temperature, T_e, versus radius in the plasma column as predicted by a 1D code.

In this article, we will focus on recent progress in the theoretical interpretation of our earlier results, measurements of the relatively small divergence of the soft X-ray laser beam at 182 Å with power ∿100 kW, and on the development of a system (two-laser approach) for the study of conditions for lasing action significantly below 100 Å.

II. COMPARISON OF GAIN MEASUREMENTS WITH A ONE DIMENSIONAL MODEL

A key element in the achievement of relatively large amplification of stimulated emission was the realization, from measurements of the radial profiles of the CVI line radiation, that the most favorable conditions for maximum gain should exist in the off-axis regions of the plasma column. Gain was measured by recording the amplification (enhancement), E, of the axial CVI 182 Å emission; that is the ratio of the 182 Å stimulated plus spontaneous emission in the axial direction to the mostly spontaneous 182 Å emission in the transverse direction. The enhancement is related to the one pass gain, G, by:

$$E = (\exp G - 1)/G.$$

By varying the position of the laser focus on the carbon disc target with respect to the observation volume of the axial and transverse monochromators it was possible to measure the gain as a function of radius r in the plasma column. The results of one such scan were presented earlier.[5] Although in that experiment the CO_2 laser energy was not optimal, there was a rapid rise of the enhancement in the region r = 1.5 to 2.5 mm off axis. Further experiments[2] with optimal plasma conditions led to measurements of an amplification of $E \approx 100$ near r = 1.3 mm.

A one-dimensional hydrodynamic plus atomic physics model has been developed to aid understanding these results.[6] In this model a single mean flow velocity was used to describe the ion mass motion. After solution of time-dependent equations for the ion density, momentum and electron energy, the gain was calculated by a post processor code from the electron density and temperature and the number density of the ground state populations of fully stripped and hydrogen like carbon. Because the laser pulse length is longer than the compressional Alfven transit time, radial pressure balance is quickly established in the plasma. Strong heating on the cylindrical axis of symmetry (at the laser focus) leads to a centrally peaked temperature profile with a corresponding electron density minimum. On the other hand, off axis, strong radiative cooling by CIV leads to low temperature, high density conditions conducive to a fast recombination rate and high 182 Å gain. With the introduction of an ion diffusion rate an order of magnitude greater than the classical value, totally stripped ions were transported from the center to the cold off axis region where fast recombination generated high gain.

Figure 1 shows the predicted gain versus radius and it can be seen that high gain occurs in a narrow ∿50-100μ wide annulus at a radius of 1.4 mm. This is in excellent agreement with the experimental results.

III. AXIAL AND TRANSVERSE EMISSION SPECTRA

Recently, multichannel detectors, based on microchannel/reticon arrays, were installed in the axial and transverse soft X-ray spectrometers. These permit the recording of emission spectra in the axial and transverse directions in a single laser shot. One example is shown in Fig. 2. In the transverse spectrum, the spontaneous CVI 182 Å emission is weak compared to the strongest line in the spectrum, OVI 173 Å. However in the axial direction, the stimulated 182 Å emission dominates the spectrum.

Some exciting results obtained very recently were made by scanning the position of the axial spectrometer in the transverse direction (perpendicular to the axis of the plasma column). The axial emissions are imaged by a grazing incidence mirror onto the entrance slit of the axial spectrometer (see Ref. 2 for experimental arrangement). The mirror is constructed by bending a glass strip therefore the optical quality of the system is not ideal. Hence a transverse scan of the axial spectrometer gives information of the relative divergence of the stimulated

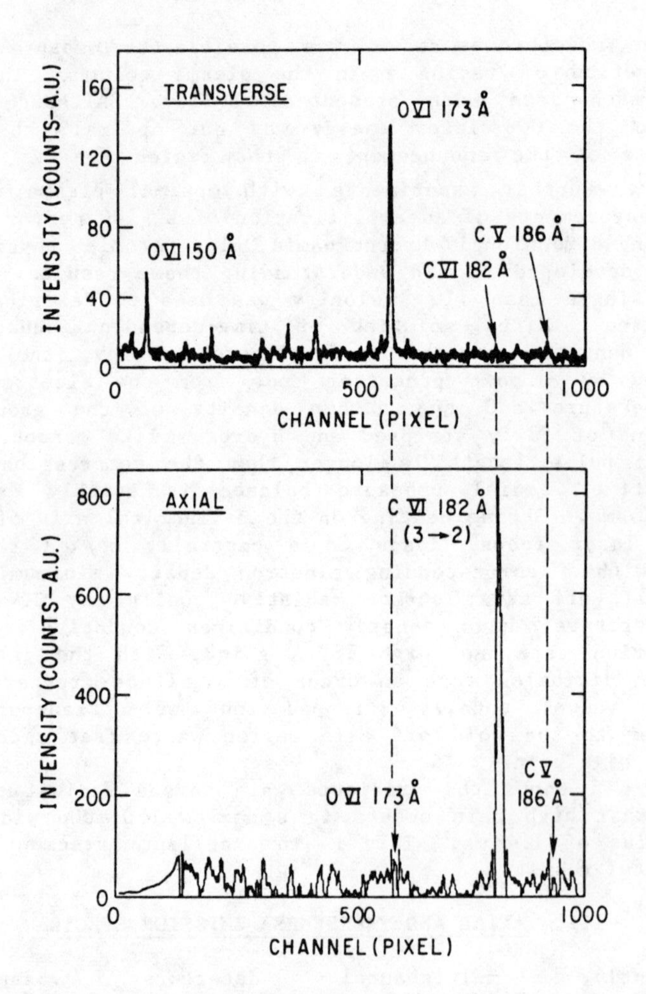

Fig. 2 Transverse and axial spectra in the region near 182 Å from a carbon disc target with four carbon blades. The laser energy was 500 J.

182 Å emission in comparison to non-lasing lines. As shown in Fig. 3 the 182 Å emission was so intense on axis that the detector was saturated whereas the CVI 186 Å and OVI 173 Å lines remained weak with a flat spatial profile.

The axial spectra presented in Fig. 2 were obtained at a position of the spectrometer corresponding to x = 250μ in Fig. 3. This horizontal position of spectrometer was established for all earlier measurements[2] by alignment of the system i.e. CO_2 laser focusing mirror, target slot, and the spectrometer entrance slit using a He-Ne laser beam. However, from Fig. 3 we can see that maximum gain is near x = 200μ with a corresponding amplification of 182 Å radiation about factor of 5 larger than

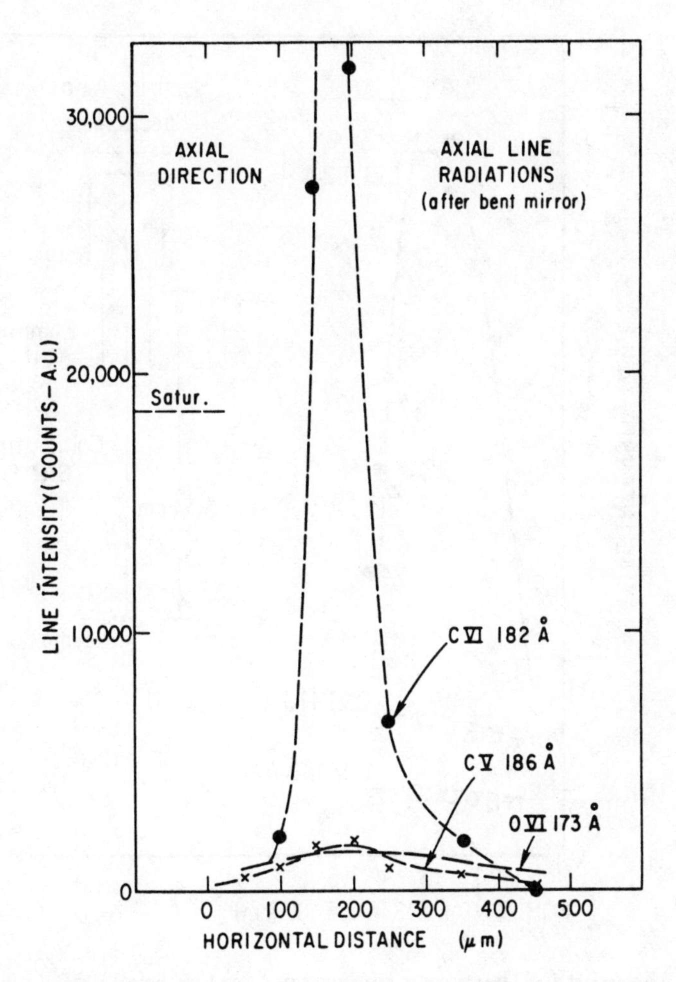

Fig. 3 Transverse scan of CVI 182 Å, CV 186 Å, and OVI 173 Å emission showing a strong central peak for the CVI 182 Å stimulated emission. In this figure, "Satur." indicates the level above which the detector is saturated. Intensities above this level were obtained by comparing the nonsaturated region of the 182 Å spectral profiles.

presented earlier[2]. The sensitivity of axial instrument for the data of Fig. 3 was higher than in Fig. 2 in order to measure intensity changes of the non-lasing lines.

To obtain information about the absolute divergence of the soft X-ray laser beam, the grazing incidence bent mirror was removed. In its place a 1 mm wide and 10 mm high collimating slit was installed to block reflections from the walls of the vacuum chamber. Figs. 4 and 5 show the result of a shot-by-shot horizontal scan of both the collimating slit and the axial soft

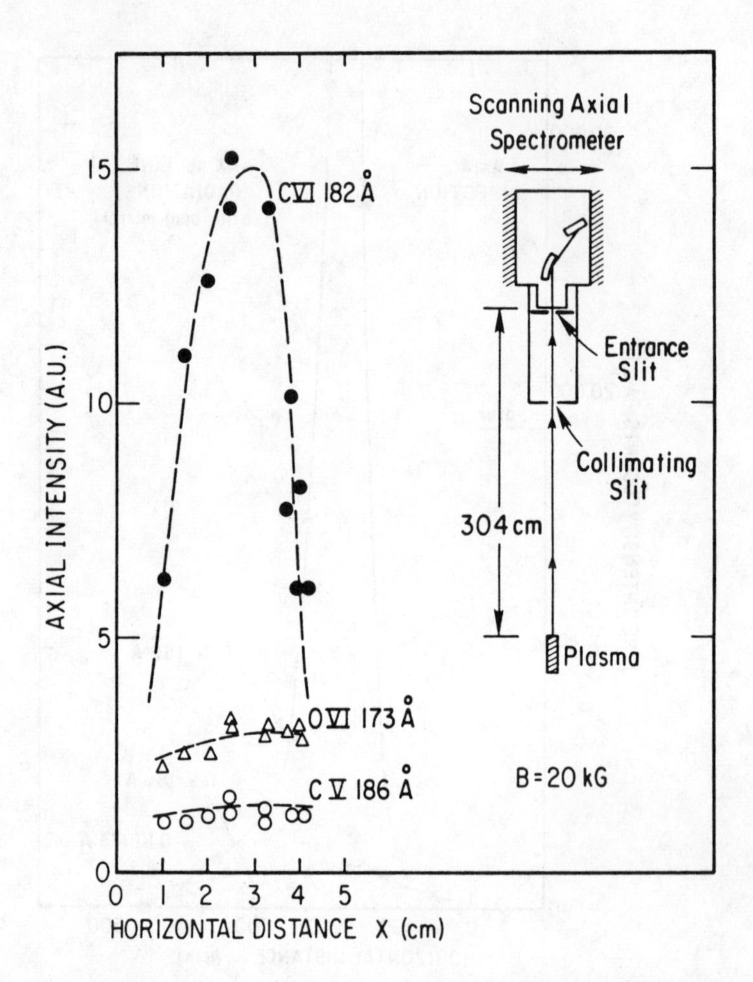

Fig. 4 Absolute divergence measurements (∿9 mrad) of the 182 Å lasing radiation for a magnetic field B = 20kG. For comparison are shown intensities of non-lasing lines OVI 173 Å and CV 186 Å, recorded simultaneously with 182 Å line.

X-ray spectrometer. The principle of the experiment is presented schematically on the right-hand side of Fig. 4, and on the left-hand side is shown the intensity distribution of the CVI 182 Å, CV 186 Å, and OVI 173 Å lines for a magnetic field B = 20 kG. For every shot the intensities of all three lines were recorded simultaneously on the multichannel detector of the axial soft X-ray spectrometer. One may see that the lasing line, CVI 182 Å, is strongly peaked on axis with a FWHM $\approx$ 2.7 cm at distance 304 cm from plasma (target). This corresponds to a horizontal divergence of the beam of ∿9 mrad. At the same time the intensities of the non-lasing lines OVI 173 Å and CV 186 Å are quite constant over the scan region ∿3 cm.

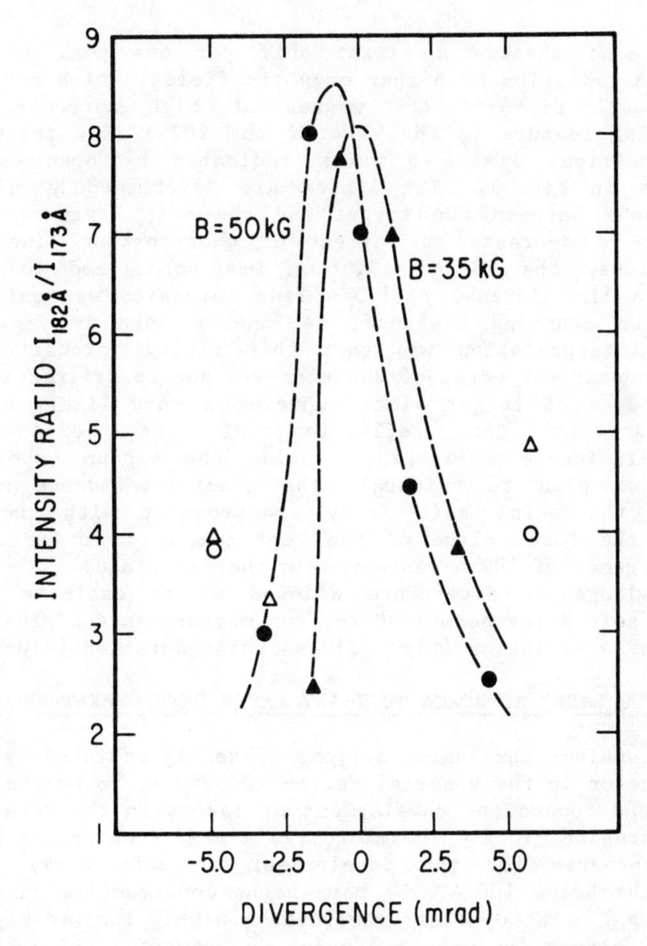

Fig. 5 Divergence measurements (∿5 mrad) of the 182 Å radiation (in relation to OVI 173 Å line) for B = 35kG and 50kG.

With increasing magnetic field (B = 35 kG and 50 kG) we observed further narrowing of soft X-ray laser beam down to ∿5 mrad (Fig. 5). This indicates that with a higher magnetic field maximum gain is created in a more narrow plasma region (less than 50μ transversely). Refraction of the 182 Å emission in the plasma is negligible as the electron density ($n_e \lesssim 10^{19} cm^{-3}$) is too low and thus an estimate of the divergence of the stimulated emission in single pass amplification can be obtained from ray tracing. With an annular width of 50μ as predicted by the one-dimensional code and a plasma length of 1 cm, the angular range of rays that pass through the gain region is 10 mrad. With a peak in the gain profile at a particular radius the divergence will be even less and thus the 1D calculation provide a good understanding of the remarkably low value of the measured divergence.

We also observe a slight shift of the peak intensity of the 182 Å radiation at higher magnetic fields, which may be caused by a small tilt of the magnet at high currents. Another interesting feature is the rise of the 182 Å line intensity near the geometrical limits of scan indicated by open circles and triangles in Fig. 5. The limits are determined by diameter of vacuum tube between the target and the soft X-ray spectrometer. In order to decrease the effect of shot-to-shot line intensity fluctuations, the 182 Å radiation was normalized in Fig. 5 to OVI 173 Å line intensity (173 Å line intensity was quite uniform across the scanning distance, as can be seen in Fig. 4). Our initial interpretation was that this rising intensity of 182 Å radiation near the edges of the scan was due to diffraction effects on the edges of target slot. However a more likely explanation may just be the reflection of the 182 Å radiation (grazing-incidence reflection) inside the vacuum tube. In the future, we plan to reinstall the grazing-incidence mirror and relocate the axial soft X-ray spectrometer with the entrance slit in the focal plane of the bent mirror in order to measure the divergence of 182 Å radiation in the far-field.

Knowledge of divergence allowed us to estimate the total power of soft X-ray beam ∿100 kW from measurements of the intensity of the 182 Å radiation (∿1-3 mJ) and pulse duration (∿10-30 nsec).

IV. TWO LASER APPROACH TO X-RAY LASER DEVELOPMENT BELOW 100 Å

We consider the lasing actions presently achieved by Livermore and Princeton in the spectral region 100-200 Å, to be the beginning of the road toward the development of lasers in the more important spectral region ∿10 Å. Therefore, as a next step we are proceeding with a program for the development of soft X-ray lasers at wavelengths below 100 Å. We have begun constructing an experiment in which a 1.5 kJ CO_2 laser generates a highly ionized magnetically confined plasma column in which a powerful ($I \gtrsim 10^{16}$ W/cm^2) picosecond laser beam will produce a population inversion and gain. The role of the CO_2 laser is to provide access to the high energy, short wavelength transitions of high Z ions which are then excited by the picosecond laser via multiphoton processes.

The feasibility of exciting either one subvalence electron or two valence electrons of an ion through multiphoton processes[7] was studied for the argon and krypton isoelectronic sequences excited by a powerful picosecond KrF* (2480 Å) laser.[8] The excitation of two valence electrons (e.g. $4s^24p^45s^2$ state) is especially attractive because of the faster progression to shorter wavelengths with increasing change of the target ion. One potential lasing transition would be $4s^24p^45s^2 \rightarrow 4s^24p^55s$ and Hartee-Fock values of the wavelength scaling of this transition in the Kr isoelectronic sequence are shown in Fig. 6 from Ref. 8. The potential lasing wavelength for e.g. Cd^{12+} is 89 Å. An additional advantage with the higher ionization stages is that

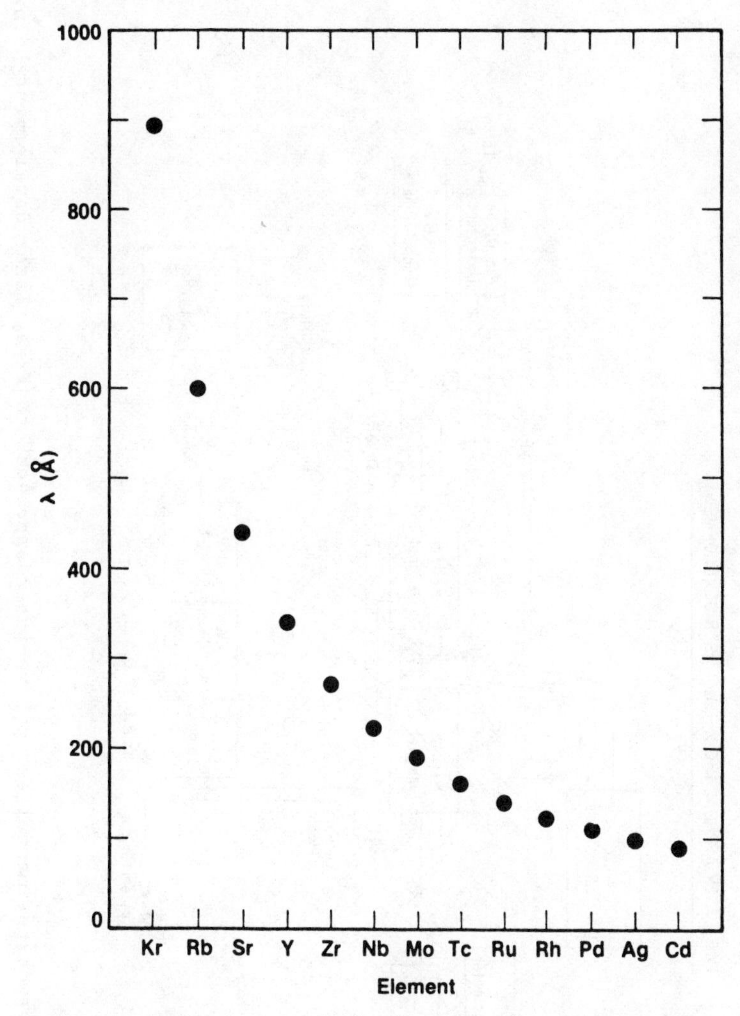

Fig. 6 Wavelength of the $4s^2 4p^4 5s^2$ - $4s^2 4p^5 5s$ transitions in the KrI isoelectronic sequence (from Ref. 8).

competing processes such as photoionization or autoionization are reduced or eliminated. Multiphoton excitation is also expected to significantly increase the population inversion and gain at 182 Å in the current experiment.

As shown in Fig. 7, the new experimental system incorporates two lasers; a 1.5 kJ CO_2 laser to produce the ionized medium and a powerful picosecond laser to generate the population inversion. The target is placed in a solenoidal magnet which will radially confine the plasma and produce the long, thin geometry suitable for laser action. Primary diagnostics will consist of multichannel soft X-ray spectrometers and an X-ray streak camera.

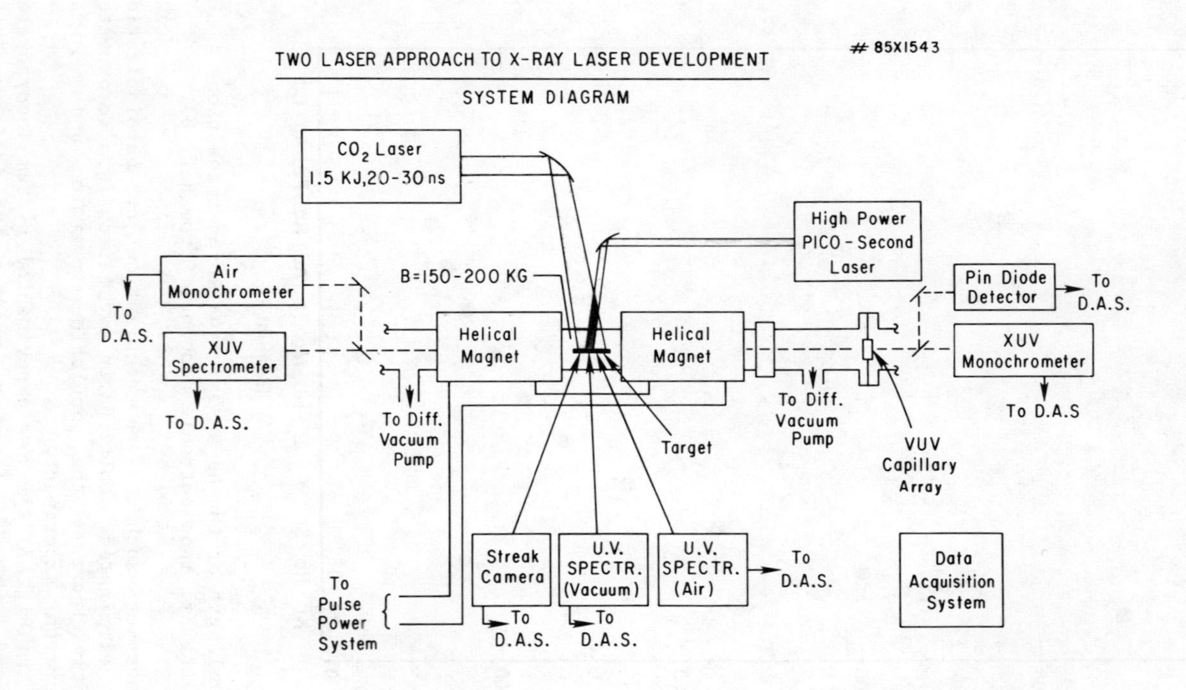

Fig. 7 Diagram of new experimental setup "Two-Laser Approach" to X-ray laser development below 100 Å.

The powerful picosecond laser is expected to generate a 1 J, 1ps, 2480 Å laser pulse. The main oscillator is a YAG-laser pumped dye laser at 6470 Å which amplified in a three stage dye amplifier producing an energy output of a few mJ. This is then frequency doubled and mixed with the 1.06μ YAG-laser to produce a 1 ps, 2480 Å pulse with an energy of a few hundred μJ. In the final stage, two KrF* amplifiers increase the pulse energy to the joule level with an (unfocussed) power in the terawatt range. Focussing by a suitable lens will produce intensities in excess of 10^{16} W/cm^2 and in addition to short wavelengths laser applications, will produce new insights into the interaction of radiation with ions in a regime where the laser field is comparable to the Coulomb field between the electrons and the nucleus.

ACKNOWLEDGMENTS

We would like to acknowledge support and encouragement from H. Furth and J.R. Thompson; assistance with the analysis of the multiphoton excitation scheme by C.W. Clark, M.G. Littman, T.J. McIlrath, and R. Miles; significant contributions by C. Keane, L. Meixler, C.H. Nam, J.L. Schwob, T. Srinivasan, and W. Tighe; and technical assistance from L. Guttadora and J. Robinson.

This work was made possible by financial support by the U.S. Department of Energy Basic Energy Sciences, Contract No. KC-05-01, and the U.S. Air Force Office of Scientific Research, Contract No. AFOSR-86-0025.

REFERENCES

1. See e.g. other papers in this Proceeding.
2. S. Suckewer, C.H. Skinner, H. Milchberg, C. Keane, and D. Voorhees, Phys. Rev. Lett. 55, 1753 (1985).
3. T.W. Barbee, Jr., S. Mrowka, and M.C. Hettrick, Appl. Opt. 24, 883 (1985).
4. H. Milchberg C.H. Skinner, S. Suckewer, and D. Voorhees, Appl. Phys. Lett. 47, 1151 (1985).
5. S. Suckewer, C.H. Skinner, H. Milchberg, C. Keane, and D. Voorhees, PPPL Report 2207 (March 1985); also Proc. Int. Conf. "Laser 85", Las Vegas, NA (Dec. 1985).
6. E.J. Valeo, C. Keane, and R.M. Kulsrud, Bull. Am. Phys. Soc. 30, 1600 (1985); also S. Suckewer et.al.; Proc. Int. Conf. "Laser 85", Las Vegas, NA (Dec. 1985).
7. T.S. Luk, H. Pummer, K. Boyer, M. Shahadi, H. Egger, and C.K. Rhodes, Phys. Rev. Lett., 51, 110 (1983).
8. C.W. Clark, M.G. Littman, R. Miles, T.J. McIlrath, C.H. Skinner, S. Suckewer, and E. Valeo, J. Opt. Soc. Am. March 1986.

DIRECT PHOTOIONIZATION PUMPING OF HIGH GAIN VUV, UV AND VISIBLE INVERSIONS IN HELIUM, CADMIUM AND ARGON VIA TWO-ELECTRON (SHAKEUP) AND OF SODIUM VIA THE OUTPUT FROM THE LLNL SOFT-X-RAY LASER

W. T. Silfvast, O. R. Wood, II and D. Y. Al-Salameh
AT & T Bell Laboratories, Holmdel, New Jersey 07733

ABSTRACT

Picosecond two-electron (shakeup) photoionization due to the soft-x-ray flux from a laser-produced plasma has been used to produce 1.3 cm^{-1} gain in He^+ on the H-like 3 - 2 transition at 164.0 nm, 0.25 cm^{-1} gain in Cd^+ on the $4d^9 5s6s$ - $4d^9 5s5p$ transition at 205.5 nm and 0.9 cm^{-1} gain in Ar^+ on the $3p^4 4p$ - $3p^4 4s$ transitions at 427.7 and 476.5 nm. These results, when taken together with the earlier demonstrations of gain in Cd[1] and Zn[2] via one-electron photoionization, suggest that it should be possible to produce gain in Na on the $2p^5 3s$ - $2p^6$ transition at 37.2 nm by photoionizing sodium vapor with the 20.6 and 20.9 nm outputs from the Lawrence Livermore National Laboratory soft-x-ray laser.[3]

INTRODUCTION

The broadband soft-x-ray flux from a laser-produced plasma has been shown to be an efficient photoionization source to produce metastable states[4] and to directly pump lasers.[1] Large small-signal gains have now been demonstrated in a number of species by inner-shell photoionization (see Fig. 1). Photoionization of Cd vapor (the first inner-shell photoionization laser) has resulted in gains in Cd^+ as high as 40 cm^{-1} at 442 nm and output energies as high as 0.1 mJ/pulse.[5] Photoionization of Zn vapor has produced population inversions in Zn^+ in the near infrared.[2] Photoionization of In^+ ions produced by laser ablation, in an isoelectronic scaling of the first Cd experiment, has recently resulted in population inversions in In^{++} in the VUV.[6] Now, two electron (shakeup) photoionization of He, Cd and Ar has been shown to produce large inversions in He^+ at 164 nm, in Cd^+ at 205 nm and in Ar^+ at 428 and 477 nm. This same mechanism was recently proposed as an excitation mechanism for a VUV laser in lithium.[7] While these two-electron processes are not as efficient as one-electron photoionization of Cd and

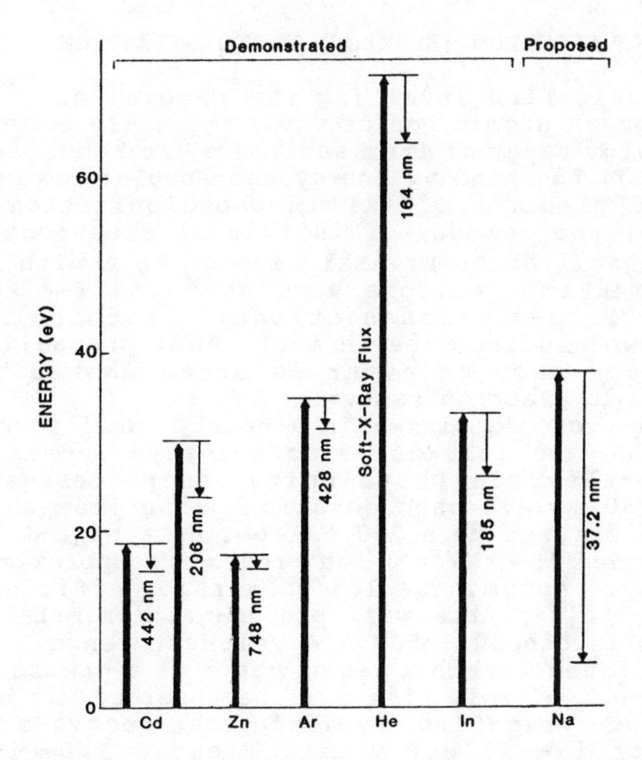

Fig. 1. Partial energy level diagrams for inner-shell photoionization lasers in Cd, Zn, Ar, He, In and Na.

Zn (because the cross sections are lower) the results achieved in He, Cd and Ar indicate that large pumping fluxes are available at short wavelengths from relatively low intensity laser-produced plasmas.

Scaling the results achieved in cadmium to sodium suggest that an intense pulse of amplified spontaneous emission at 37.2 nm should be possible. The stimulated emission cross section in Na is almost identical to that of Cd when taking into account the Cd isotopic distribution. The inner-shell photoionization cross section in Na is approximately 2/3 that of Cd and peaks at a photon energy about 50% higher. This suggests that to produce a laser in Na in the XUV, the only significant changes from the Cd experiment are the use of a higher temperature blackbody or the use of the LLNL soft-x-ray laser as a source and the use of a shorter excitation pulse to minimize electron collisional excitation of the lower laser level.

TWO-ELECTRON (SHAKEUP) PHOTOIONIZATION

Photoionization involving the removal of a single electron from an atomic gas or vapor typically occurs over a photon energy range of from several eV to hundreds of eV depending upon the binding energy and angular momentum of the removed electron. Maximum photoionization cross sections for the removal of individual electrons range from less than 1 Mb to several tens of Mb's with larger angular momentum electrons such as d and f-electrons having the larger cross sections. Photoionization involving two electrons has a much lower probability of occurring and such an event is often masked by the dominant single electron removal.

A laser-produced plasma is a nearly ideal source for producing the high fluxes of soft-x-rays necessary to excite two-electron photoionization processes. For example, a 50 mJ, 70 psec duration pulse from a Nd:YAG laser, when focused to a 250 μm spot on a high-Z target (an intensity of 10^{12} W/cm^2) can produce an approximate 20 eV blackbody. Assuming a 10% conversion efficiency to soft-x-rays, this source will produce approximately 10^{25} photons/sec in the 50 - 500 Å wavelength range. If the target is located within a gas or vapor at a pressure of 1 - 10 Torr, most of this flux will be absorbed via single-electron photoionization (typical cross sections are of the order of 1 - 30 Mb) within 1 cm^3. Two-electron photoionization (typical cross sections are 10^3 to 10^4 times smaller) with such a source can produce two electron states at population densities of the order of 10^{11} to 10^{12} cm^{-3}.

POPULATION INVERSIONS IN CADMIUM VIA SHAKEUP

Strong emission from long lived autoionizing states in Cd^+ was recently observed[8] by combining picosecond excitation using the soft-x-ray flux from a laser-produced plasma with fast detection using a 125 psec risetime microchannel plate detector and a 1 GHz bandwidth oscilloscope. Using a 50 mJ, 70 psec duration soft-x-ray source (as described above) approximately 15 subnanosecond-duration emission lines were observed in the Cd photoionization spectrum between 2000 and 4000 Å. These lines, necessarily autoionizing transitions due to their fast decay (plasma densities were not high enough to account for such a rapid decay by electron collisions), were computer sorted for common upper levels using known $4d^9 5s5p$ Cd^+ levels as possible lower levels. One of the levels found using this technique was identified as the $4d^9 5s6s$ (^1D) $^2D_{5/2}$ state of Cd^+ (see Fig. 2). Five emissions lines located this level, the first identified autoionizing level in Cd^+, at 168041.1 +/- 2.0 cm^{-1} and

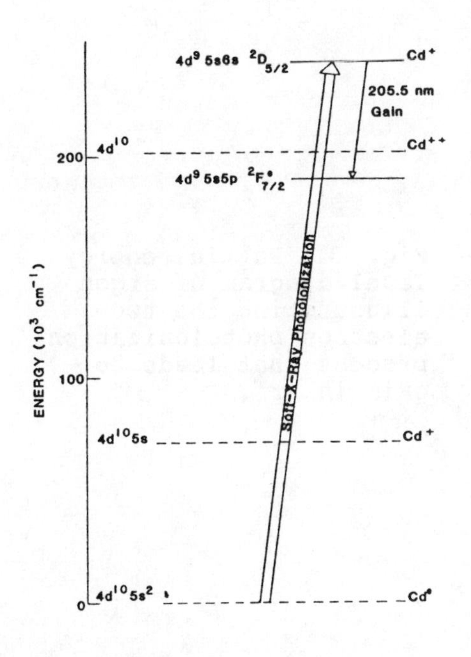

Fig. 2. Partial energy level diagram of cadmium illustrating the two-electron photoionization process that leads to gain in Cd^+.

gave measured intensities in good agreement with the oscillator strengths predicted by the RCN/RCG multiconfiguration Hartree-Fock atomic physics code.[9]

The temporal behavior of the strongest transition from this level, the $4d^9 5s6s\ ^2D_{5/2}$ - $4d^9 5s5p\ ^2F_{7/2}$ transition at 205.5 nm, was recorded with a microchannel plate detector using a time correlated photon-counting technique.[10] The result, indicating a single exponential decay with a decay time of 460 +/- 50 psec, agrees well with the 490 psec autoionizing time predicted by the atomic physics code.[9] By comparing the absolute intensity of this emission line and emission from $4d^9 5s5p\ ^2F_{7/2}$ to nearby lines in other species with known populations and radiative rates, we have shown that the populations of the upper and lower levels of this transition are inverted and that the 205.5 nm transition is characterized by a small-signal gain coefficient of approximately 0.25 cm^{-1}.

POPULATION INVERSIONS IN ARGON VIA SHAKEUP

Two-electron photoionization with intense short wavelength radiation from a laser-produced plasma has recently been used to produce significant population densities in the $3p^4 4p$ state of Ar^+ and population inversions in Ar^+ on transitions from this state to lower lying $3p^4 4s$, 4d states at 427.7 and 476.5 nm with gains of the order of 0.9 cm^{-1}. An energy level diagram of the relevant levels in Ar and Ar^+ is shown in Fig. 3.

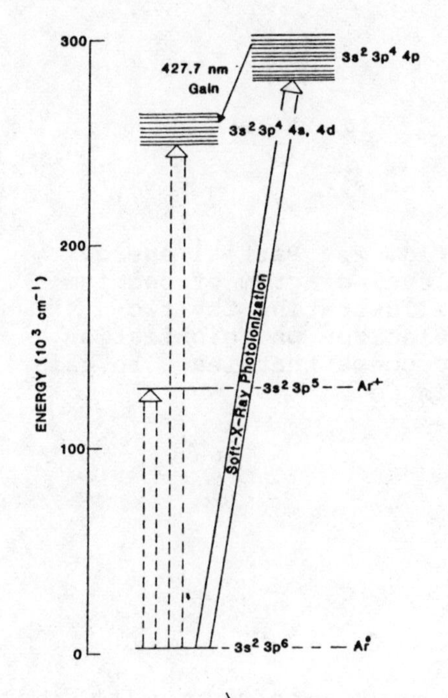

Fig. 3. Partial energy level diagram of argon illustrating the two-electron photoionization process that leads to gain in Ar^+.

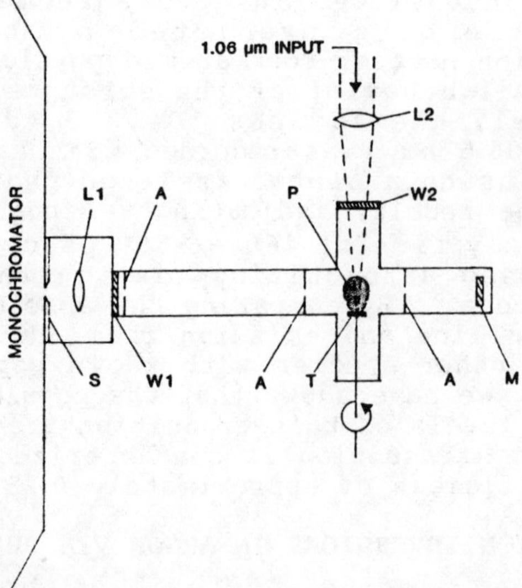

Fig. 4. Experimental arrangement for measuring gain in argon (and helium) photoionized with the soft-x-ray flux from a laser-produced plasma. Legend: A - 3 mm aperture, L1 - 10 cm f.l. LiF lens, L2 - 25 cm f.l. quartz lens, M - flat mirror, P - photoionized argon (helium) plasma, S - monochromator slits, T - tantalum target, W1 - 2 mm thick CaF_2 window and W2 - 2 mm thick quartz window.

The experimental arrangement for producing a photoionized Ar^+ plasma and for measuring the small-signal gain on visible transition in Ar^+ is shown in Fig. 4. The output from a 30 mJ, 70 psec duration 1.06 μm laser was focused with a 25 cm lens onto a rotating tantalum target in a cell filled with Ar. The measured spot size on the target was approximately 0.015 cm (radius). The resulting plasma of photoionized Ar at a pressure of 10 Torr was estimated to have dimensions of the order of 1 cm in a direction normal to the target and 0.5 cm in diameter at a position 0.5 cm above the target. Three apertures, two of which were located approximately 1.5 cm on either side of the plasma and the third was spaced some 30.5 cm in front of the plasma, were used to define a region of observation in the plasma. A flat dielectric coated pyrex mirror, with a reflectivity exceeding 99% in the blue spectral region, was placed 12 cm behind the plasma. The plasma emission was detected with a monochromator equipped with a photomultiplier. The detection system had a risetime of the order of 2.8 nsec.

Immediately after the termination of the fast soft-x-ray pulse, the population densities of 18 individual $3p^4 4p$ Ar^+ levels were measured by calibration with emission from levels of known density and radiative rates. The measured densities, shown in Fig. 5, are consistent with the

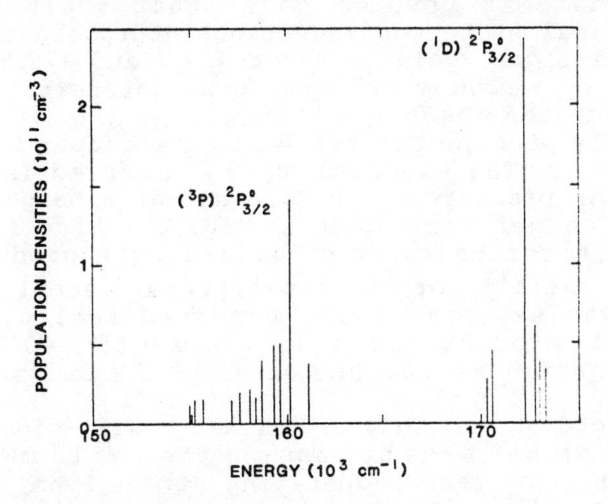

Fig. 5. Population densities of individual $3p^4 4p$ levels in Ar^+ measured immediately after the termination of a 70 psec duration soft-x-ray excitation pulse.

densities that would be expected from a 17 eV blackbody source using the two-electron photoionization cross sections calculated with the RCN/RCG multi-configuration

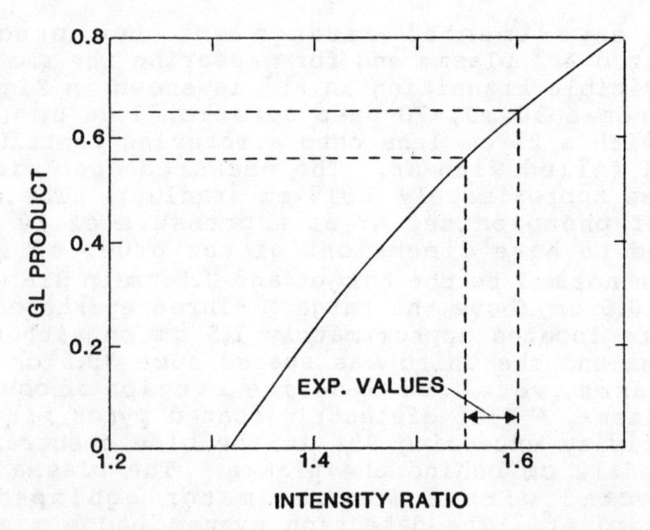

Fig. 6. Calculated GL product versus intensity ratio for experimental arrangement illustrated in Fig. 4 using a 99% reflecting mirror at 477 nm (solid curve) together with largest and smallest measured intensity ratios (dashed vertical lines).

Hartree-Fock atomic physics code[9] that included both initial and final state configuration interactions.

The small-signal gain in Ar at 427.7 and 476.5 nm was obtained both by measuring the ratio of intensities in the presence and in the absence of the mirror shown in Fig. 4 and by directly probing the gain with the output from a cw argon-ion laser. The experimentally observed intensity ratios at an Ar pressure of 10 Torr and at a height of 0.5 cm above the target are shown in Fig. 6. Also shown in Fig. 6 is a theoretical plot of gain-length product (GL) vs. intensity ratio[11] for the experimental geometry shown in Fig. 4. The experimentally observed ratios, 1.55 - 1.60, give GL products ranging from 0.55 - 0.65, and translate to gain coefficients of 0.8 - 0.9 cm^{-1} at 476.5 nm.

An independent measurement of the small-signal gain in Ar at 476.5 nm was made by passing the 1 - 10 mW output from a cw argon-ion laser, operating at 476.5 nm, through the plasma near the soft-x-ray source. The value of gain determined using this technique was in close agreement with the value obtained from the intensity ratio technique described above.

POPULATION INVERSIONS IN HELIUM VIA SHAKEUP

Population inversions and significant small-signal gain (gain coefficients up to 1.3 cm^{-1}) have been produced in H-like (n = 3 - 2) He^+ at 164 nm via two-electron photoionization using the soft-x-ray flux from a laser-produced plasma created by a 30 mJ, 70 psec duration pulse from a Nd:YAG laser. An energy level diagram of the relevant levels in He and He^+ is shown in Fig. 7.

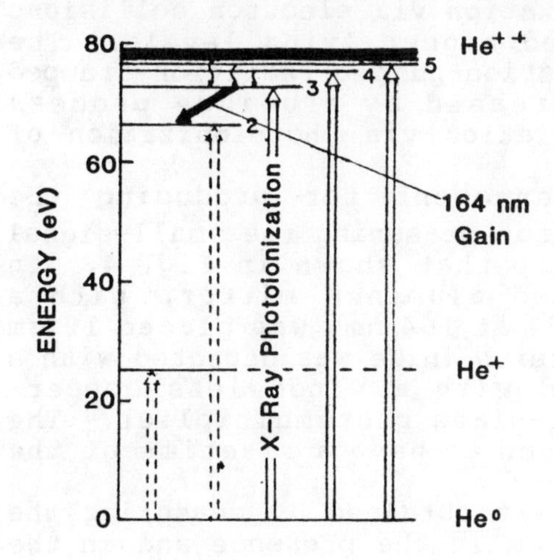

Fig. 7. Partial energy level diagram for helium illustrating the processes leading to gain in He^+.

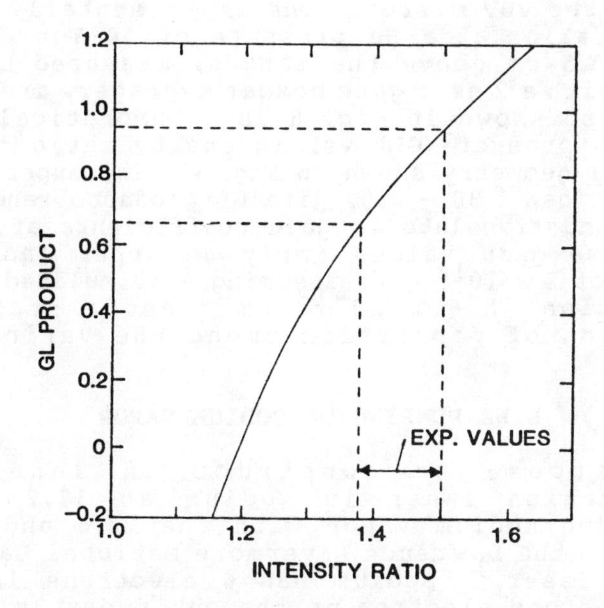

Fig. 8. Calculated GL product versus intensity ratio for experimental arrangement illustrated in Fig. 4 using a 65% reflecting mirror at 164 nm (solid curve) together with largest and smallest measured intensity ratios dashed vertical lines).

Photoionization by soft-x-ray radiation occurs (to some extent) to all of the state of He^+ from the ground state of neutral He. The n = 3 upper laser level is pumped directly by two-electron photoionization from the ground state of neutral helium, and indirectly (in times of the order of 1 - 2 nsec) by relaxation via electron collisions from photoionization pumped higher lying levels. The decay rate of the photoionization-pumped radiation-trapped lower laser level is increased by a unique process involving absorption of radiation via photoionization of ground state He atoms.

The experimental arrangement for producing the photoionized He^+ plasma and for measuring the small-signal gain at 164 nm is similar to that shown in Fig. 4. In this case a MgF_2-overcoated aluminum mirror, with a measured reflectivity of 65% at 164 nm, was placed 12 cm behind the plasma. The emission in He was detected with a VUV monochromator equipped with a windowless copper-iodide-coated microchannel-plate photomultiplier. The detection system was observed to have a risetime of the order of 1 nsec.

The small-signal gain was obtained by measuring the ratio of intensities at 164 nm in the presence and in the absence of the VUV mirror. The experimentally observed intensity ratios at a He pressure of 50 Torr and at a height of 0.5 cm above the target, measured by signal averaging with a 2 nsec gate boxcar averager, are shown in Fig. 8. Also shown in Fig. 8 is a theoretical plot of gain-length product (GL) vs. intensity ratio[11] for the experimental geometry shown in Fig. 4. The experimentally observed ratios, 1.38 - 1.5, give GL products ranging from 0.7 - 0.9, and translate to gain coefficients of 1.0 - 1.3 cm^{-1}. These gain values imply an upper laser level population of 2×10^{12} cm^{-3} assuming a stimulated emission cross section of 6×10^{-13} cm^{-2} and a statistical distribution of population among the various n = 3 sublevels.

LINE PUMPING OF SODIUM VAPOR

We propose to construct an inner-shell photoionization laser in sodium at 37.2 nm[12] by photoionizing sodium vapor with the 20.6 and 20.9 nm outputs from the Lawrence Livermore National Laboratory soft-x-ray laser.[3] Sodium has 6 electrons in the 2p subshell and one electron in the outermost 3s subshell ($2p^6 3s$). Removal of the 3s electron produces a $^1S_0$ ground state ion (lower laser level) and removal of a 2p electron produces a p-state divided into 4 levels including $^3P_{0,1,2}$ and $^1P_1$. The $^1P_1$ level is the primary candidate for the upper laser level because of its high oscillator strength

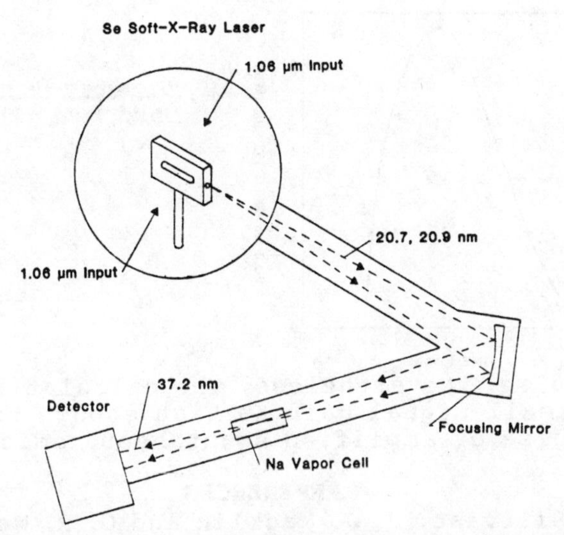

Fig. 9. Experimental arrangement for pumping sodium vapor with the output from the LLNL soft-x-ray selenium laser.

and, consequently, high stimulated emission cross section (4×10^{-14} cm^2) to the ground state. Pumping is to be accomplished via inner-shell photoionization of a 2p electron with the 20.6 and 20.9 nm outputs from the LLNL soft-x-ray Se laser. The probability of removing the 2p electron with these wavelengths is many orders of magnitude higher than for removing an outer 3s electron.

The experimental arrangement for line pumping sodium vapor is shown in Fig. 9. The 1 mJ, 175 psec duration output from the LLNL Se laser at 20.6 and 20.9 nm is focused with a grazing incidence ellipsoidal mirror into a 10 cm long cell filled with sodium vapor at approximately 1 Torr. The amplified spontaneous emission from sodium at 37.2 nm is separated from the 20.6 and 20.9 nm pumping radiation using a grazing incidence spectrograph equipped with a microchannel plate detector gated with an Auston switch.

The primary problem with this scheme is the production of ground state ions (lower laser level) through the collision of photoelectrons with ground state neutral sodium atoms. Although the cross section for this process is high, the mean time for these collisions is of the order of 150 psec at a sodium vapor pressure of 1 Torr. Therefore, if the excitation pulse is fast enough this problem can be avoided. A calculation of the temporal behavior of the GL product at 37.2 nm from 1 Torr of sodium vapor pumped with the 1 mJ, 175 duration, pulse from the LLNL Se laser is shown in Fig. 10. Gain at 37.2 nm is seen to occur only during the first 65 psec of the

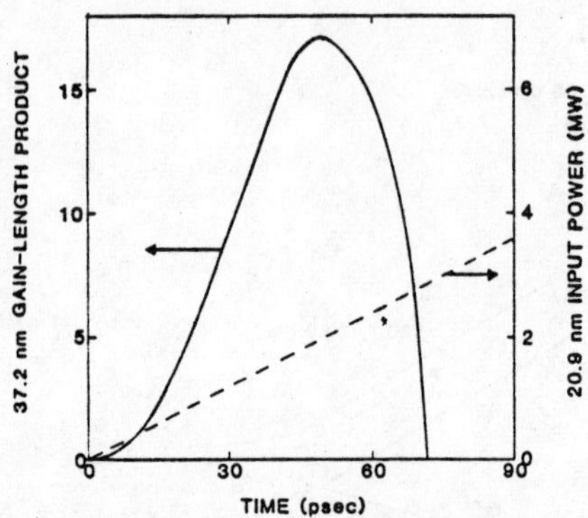

Fig. 10. Temporal behavior of the gain-length product at 37.2 nm (solid curve) produced by pumping sodium vapor with the 20.6 and 20.9 nm output from the LLNL Se laser (dashed curve).

pumping pulse. Nevertheless, the calculations indicate that the small-signal gain is high enough to produce an intense pulse of amplified spontaneous emission at 37.2 nm.

REFERENCES

1. W. T. Silfvast, J. J. Macklin and O. R. Wood, II, Opt. Lett. 8, 551 (1983).
2. H. Lundberg, J. J. Macklin, W. T. Silfvast and O. R. Wood, II, Appl. Phys. Lett. 45, 335 (1984).
3. D. L. Matthews, P. L. Hagelstein, M. D. Rosen , M. J. Eckart, N. M. Ceglio, A. U. Hazi, H. Medecki, B. J. MacGowan, J. E. Trebes, B. L. Whitten, E. M. Campbell, C. W. Hatcher, A. M. Hawryluk, R. L. Kauffman, L. D. Pleasance, G. Rambach, J. H. Scofield, G. Stone and T. A. Weaver, Phys. Rev. Lett. 54, 110 (1985).
4. R. G. Caro, J. C. Wang, R. W. Falcone, J. R. Young and S. E. Harris, Appl. Phys. Lett. 42, 9 (1983).
5. W. T. Silfvast, O. R. Wood, II, J. J. Macklin and H . Lundberg, in Laser Techniques in the Extreme Ultraviolet, S. E. Harris and T. B. Lucatorto, eds. (AIP, N. Y., 1984), p. 427.
6. R. A. Lacy, W. T. Silfvast, S. Svanberg, O. R. Wood, II and R. L. Byer, in Short Wavelength Radiation: Generation and Applications, D. T. Attwood and J. Bokor, eds. (AIP, N. Y., 1986), This volume.
7. S. E. Harris and R. G. Caro, Opt. Lett. 11, 10 (1986).
8. W. T. Silfvast, O. R. Wood, II, J. J. Macklin and D. Y. Al-Salameh, in Laser Spectroscopy VII, T. W. Hansch and Y. R. Shen, eds. (Springer-Verlag, Berlin, 1985), p. 171.
9. R. D. Cowan, The Theory of Atomic Structure and Spectra (UC Press, Berkeley, 1981), Sec. 8-1 and 16-1.
10. S. K. Poultney, in Advances in Electronics and Electron Physics, L. Marton, ed. (Academic, N. Y., 1972), Vol. 31.
11. W. T. Silfvast and J. S. Deech, Appl. Phys. Lett. 11, 97 (1967).
12. M. A. Duguay and P. M. Rentzepis, Appl. Phys. Lett. 10, 350 (1967).

SOFT X-RAY PUMPING OF INNER-SHELL EXCITED LEVELS FOR EXTREME ULTRAVIOLET LASERS

R. G. Caro, P. J. K. Wisoff, G. Y. Yin, D. J. Walker, M. H. Sher, C. P. J. Barty, J. F. Young, and S. E. Harris

Edward L. Ginzton Laboratory, Stanford University
Stanford, CA 94305

ABSTRACT

We report the construction of a facility for investigating x-ray pumping of extreme ultraviolet lasers. Measurements of large excited populations in Li^+ and an investigation of a Li^+ "shake-up" laser are described.

* * *

When a 1.06 μm laser beam is focused onto a high-Z target at intensities of 10^{12} - 10^{14} W cm^{-2}, a plasma is formed from which a burst of soft x-rays is emitted with a conversion efficiency in excess of 10%. The burst of x-rays can be used as an x-ray flashlamp to pump potential extreme ultraviolet (XUV) laser systems.[1-4]

In this work we report the construction of a facility at Stanford University consisting of a Nd:Glass laser system capable of generating a 20 J pulse of 1 ns duration at 1.06 μm. This laser is synchronized with a tunable dye laser which can produce pulses with energies as high as 1 mJ, a duration of approximately 20 ps, and which are tunable from 400 nm to 900 nm. This facility is well suited to the experimental investigation of a variety of x-ray flashlamp-excited XUV laser systems. In particular, it is ideal for the study of XUV lasers produced by the initial excitation of a highly energetic metastable level (by the x-ray flashlamp), followed by rapid transfer of population to a nearby radiating level by means of a tunable dye laser.

A schematic of this system is illustrated in Fig. 1. The heart of the system consists of two mode-locked and Q-switched Quantronix 416 Nd:YAG lasers for which the acousto-optic mode-locker units are driven by a common rf signal. In this way synchronization of the output pulse trains from the two lasers is obtained such that the relative timing of the two pulse trains is constant within ± 50 ps over a period of 4-8 hours. By inserting an etalon in the cavity of one of these lasers, a train of 1 ns pulses is produced. A single one of these pulses is selected by a Pockels cell pulse selector placed external to the cavity, and that pulse is amplified through a series of Nd:YAG and Nd:Glass amplifiers until a pulse of 20 J energy is obtained from the final 33 mm diameter Nd:Glass rod. The configuration of this part of the system is shown in Fig. 2.

0094-243X/86/1470145-12$3.00

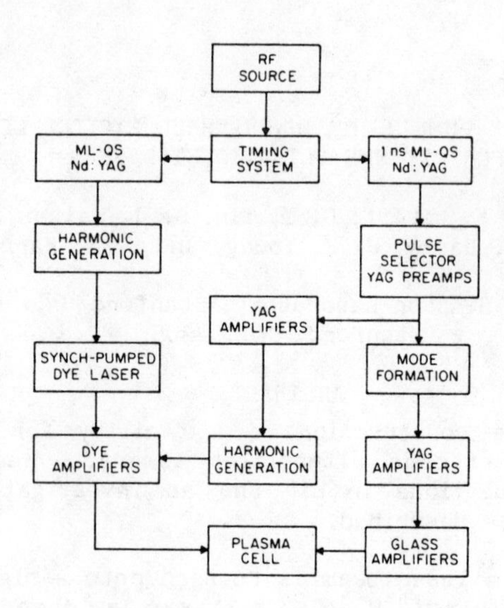

Fig. 1. Schematic of Stanford laser facility.

The output from the second mode-locked and Q-switched Nd:YAG laser is converted to the second, or third, harmonic at 532 nm or 355 nm, and used to synchronously pump a dye laser oscillator,[5] as shown in Fig. 3. This oscillator can produce easily tunable output pulses with 20 ps duration and nearly transform limited linewidth in the spectral region from 400 nm - 900 nm. Output energies range from 20 μJ at 560 nm when pumping with 532 nm light, to 2 μJ at 400 nm when pumping with the lower energy third harmonic of the Nd:YAG laser. The output from this dye oscillator is then amplified in a chain of three dye amplifiers, each pumped longitudinally by a pulse of 1 ns duration at 532 nm or 355 nm. This pump pulse is derived from the 1 ns, 1.06 μm pulse at an early stage before amplification in the Nd:Glass amplifiers. It is then passed through a further three Nd:YAG amplifiers to produce a 200 mJ beam at 1.06 μm that can be converted to the second or third harmonic for pumping the dye amplifier system. A schematic illustration of this dye amplifier system is shown in Fig. 4.

The experiments that have been performed to date with this laser facility have had two objectives. The concept of using x-rays emitted from a laser-produced plasma to excite large densities of energetic excited levels in atoms and ions has been thoroughly experimentally investigated[1-3] using modest, 100 mJ, plasma-producing laser energies. It was necessary, however, to verify that these techniques could be scaled up to the 20 J energies available from the current facility. In order to compare with the original work of Caro, et al.,[1-2] experiments have been performed to investigate the excitation of Li^{+} (1s2s) ions at 60 eV as a result of photoionization by laser-produced x-rays.

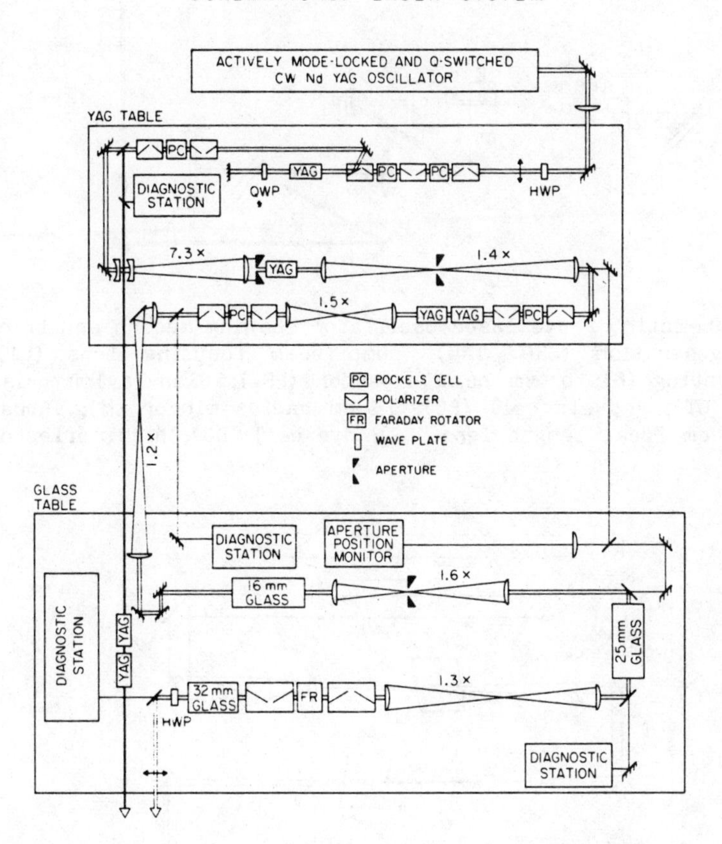

Fig. 2. Schematic of 1.06 μm plasma-producing laser system. HWP, QWP = half wave plate and quarter wave plate.

In addition, the Li^+ vacuum ultraviolet laser system at 165.3 nm, recently proposed by Harris and Caro,[6] has been experimentally investigated. The energy level diagram of this system is shown in Fig. 5. When a (1s) electron is removed from a Li atom by x-ray photoionization, the principal product is the Li^+ (1s2s) ion. However, due to the process of "shake-up," there is a significant probability that the Li^+ (1s3s) ion will be produced.[7-9] Indeed, experiments indicate[9] that 15% of the ions created by (1s) removal should be in the level Li^+ $(1s3s)^3S_1$. In contrast, only 7% should be found in the Li^+ $(1s2p)^3P$ levels and so an inversion should be produced on the $[(1s3s)^3S - (1s2p)^3P]$ transition at 165.3 nm.

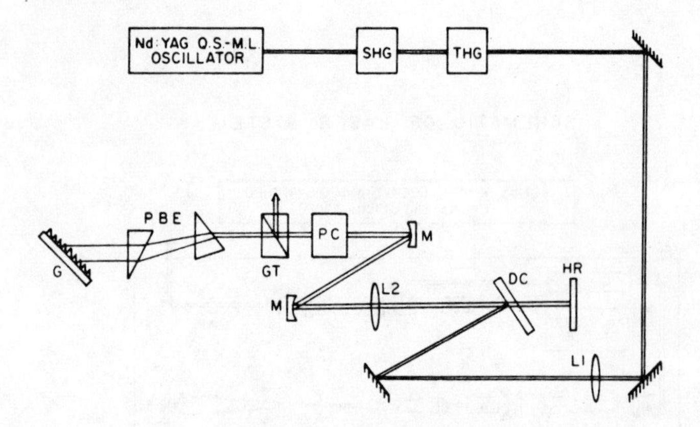

Fig. 3. Schematic of dye laser oscillator showing second and third harmonic generation (SHG, THG); pump beam focusing lens (L1); Littrow grating (G); prism beam expander (PBE); Glan-Taylor prism polarizer (GT); Pockels cell (PC); 15 cm radius mirror (M); intracavity 7.5 cm focal length lens (L2); dye cell (DC); high reflector (HR).

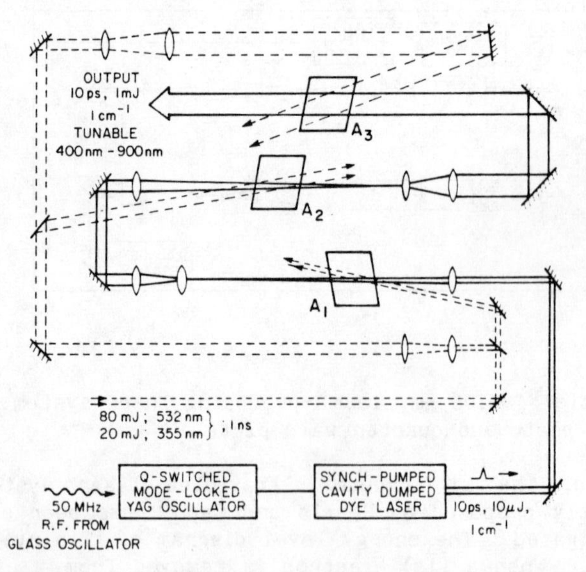

Fig. 4. Schematic of dye amplifier system.

As shown in Fig. 6 and discussed in Ref. 6, modeling of the photoionization process, as well as of a number of competing processes that occur in the photoionized plasma, indicates that gains in excess of exp(10) should be observable at 165.3 nm when an experiment with the geometry shown in Fig. 7 is performed and a 10 J plasma-producing laser is used. As can be seen in Fig. 7, the

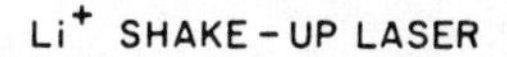

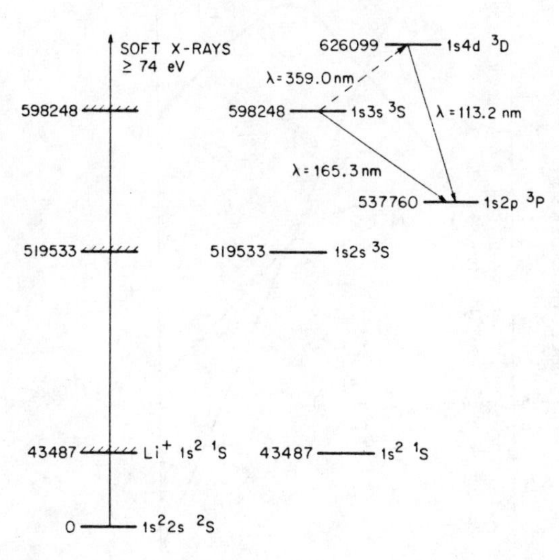

Fig. 5. Li shake-up laser energy levels.

plasma-producing laser beam is split into three beams, each of which is focused onto a separate tantalum target at an intensity of the order of 10^{12} W cm^{-2}. The plasmas that are produced emit soft x-rays that propagate out into the surrounding Li vapor (10^{17} cm^{-3}) and photoionize the atoms to produce Li^+ (1s2s) and Li^+ (1s3s) ions in large quantities.

Although relative magnitudes of the populations in the various Li^+ levels can be deduced by observations of emission, quantitative measurements can more easily be made by means of absorption spectroscopy. In this technique, the wavelength of the beam from the picosecond dye laser is scanned through a transition which has as its lower level, that in which population is to be measured. By analyzing the absorption profile obtained, a measure of the excited level population-length product, N^*L , can be deduced.[1-2] By altering the delay between the dye laser probe beam and the plasma-producing 1.06 µm beam, the temporal behavior of the population can be examined.

In Figs. 8 and 9 are shown the dependence of the N^*L product of the Li^+ $(1s2s)^3S$ ion as a function of time (Fig. 8) and distance from the Ta targets (Fig. 9). For comparison, the predictions of a theoretical model[10] are shown as line A in Figs. 8 and 9. It can be seen that the functional dependence of the data is in good agreement with the theoretical predictions, while the absolute magnitude is low by a factor of approximately 2-3.

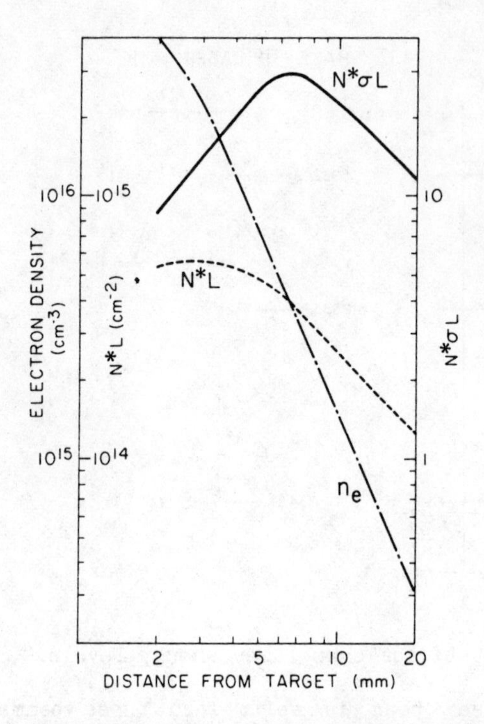

Fig. 6. Model predictions of electron density n_e, effective number density-length product, N^*L, and gain, $N^* \sigma L$ at 165.3 nm as a function of distance from the target. The model assumes three target spots, each irradiated by a 1 ns, 2 J, 1.06 μm laser pulse.

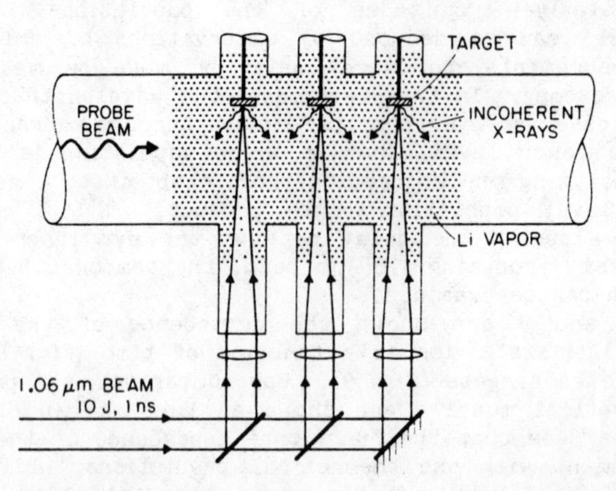

Fig. 7. Schematic of experimental geometry.

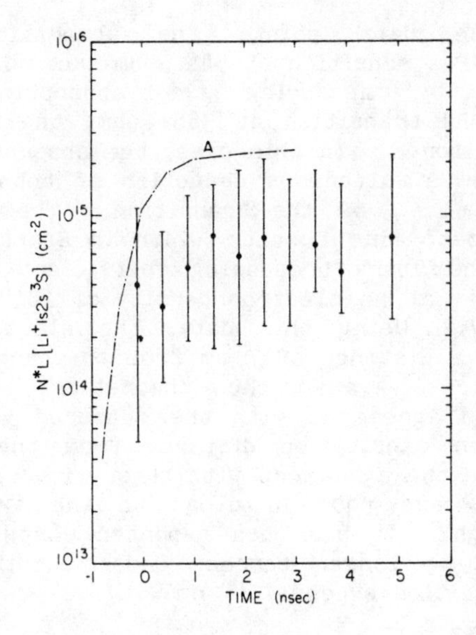

Fig. 8. N^*L [Li^+ (1s2s)^{3}S] as a function of time with respect to the peak of the 1.06 μm laser pulse. Distance from targets = 7 mm; Li density = 10^{17} cm^{-3}; energy incident on each of three targets = 2.3 J.

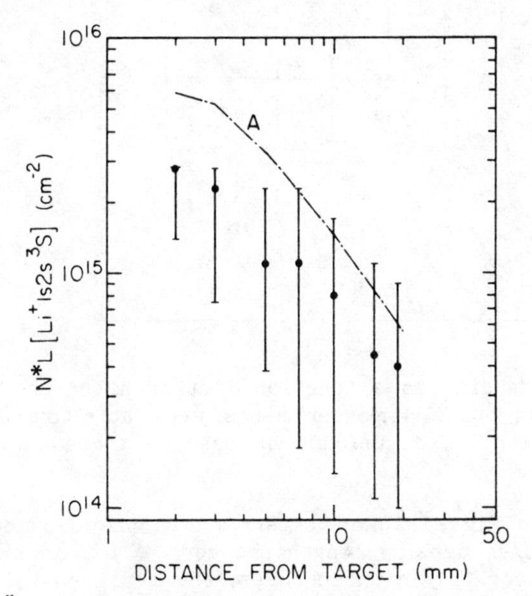

Fig. 9. N^*L [Li^+(1s2s)^{3}S] as a function of distance from the target at a time 1.9 ns after the peak of the 1.06 μm pulse. Li density = 10^{17} cm^{-3}; energy incident = 2.3 J on each of three targets.

To obtain these data points the absorption on the Li^+ [$(1s2s)^3S$ - $(1s2p)^3P$] transition at 548.5 nm was monitored. In a similar fashion, by measuring the absorption on the Li^+ [$(1s3s)^3S$ - $(1s4p)^3P$] transition at 368.4 nm, the Li^+ $(1s3s)^3S$ population can be monitored. In this case, the absorption is in a regime that allows the simultaneous deduction of both N^*L and the Lorentz width, $\delta\omega_L$, of the transition. The predominant mechanism contributing to this Lorentz width is Stark broadening and, from Griem,[11] the Stark broadening coefficient is approximately $\delta\omega_L$ = 2.6 cm^{-1} at an electron density of 10^{16} cm^{-3} and a temperature of 0.5 eV. . Using this data, the electron density dependence on time at a distance of 7 mm from the target has been determined (Fig. 10). Again, the theoretical predictions, curve A , are in good agreement with the measured values. The dependence of electron density on distance from the target is shown in Fig. 11. Here the agreement with theory is only fair, and the model predictions are poor in close to the target where pumping rates are high. It has been reported elsewhere[10] that this simple model is no longer correct under conditions where predicted electron densities exceed 10^{16} cm^{-3}.

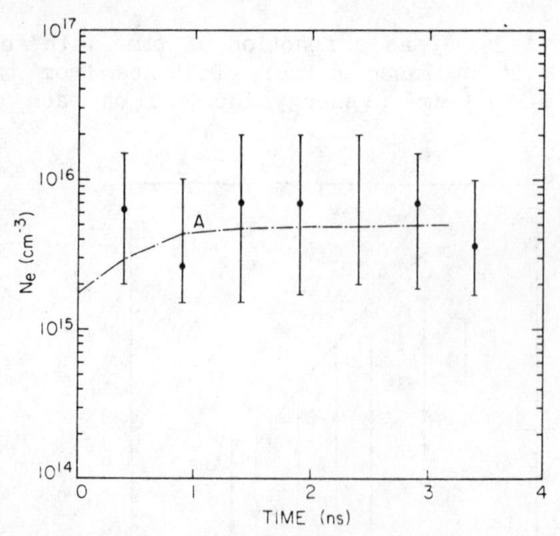

Fig. 10. Electron density as a function of time after the peak of the plasma-producing pulse. Measurements were at 7 mm distance from the targets with 2.7 J incident on each of three targets; Li density = 10^{17} cm^{-3}.

In contrast, in Figs. 12 and 13 are shown the dependence of the Li^+ $(1s3s)^3S$ number density-length product, N^*L , on time, and distance from the target. In these cases, the measured values are lower than those predicted from Fig. 6 by a factor of approximately 30. The gains depicted in Fig. 6 are small signal gains in the absence of amplified stimulated emission (ASE) at 165.3 nm. If the populations of Fig. 6 were being realized, ASE would act to

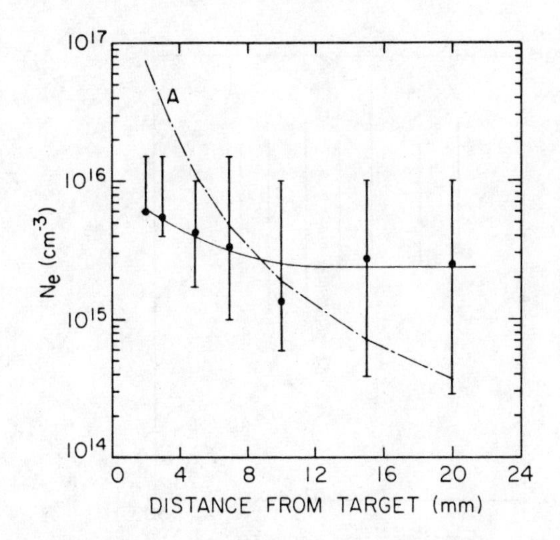

Fig. 11. Electron density as a function of distance from the three targets at a time 1.4 ns after the peak of the 1.06 μm pulse. Li density = 10^{17} cm^{-3}; energy per target = 2.3 J.

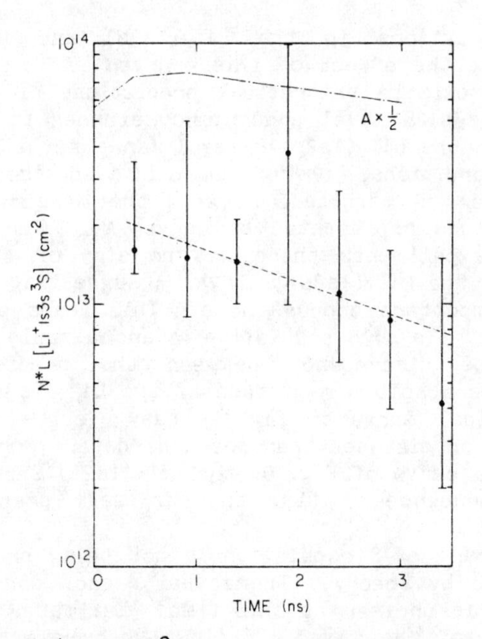

Fig. 12. N^*L [Li^+ $(1s3s)^3S$] as a function of time after the peak of the 1.06 μm pulse. Li density = 10^{17} cm^{-3}; distance from targets = 7 mm; energy incident on each of three targets = 2.5 J.

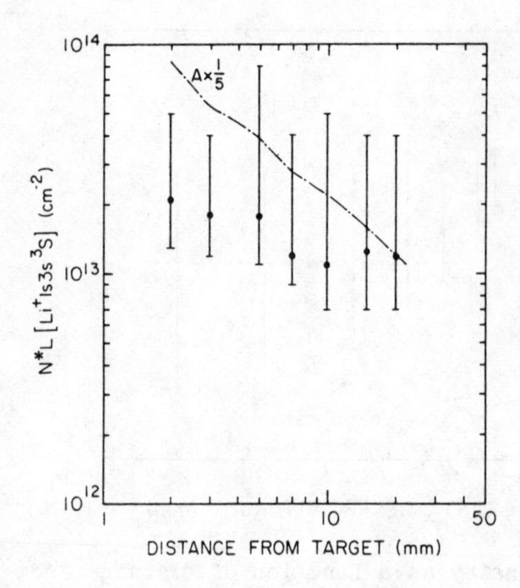

Fig. 13. N^*L [Li^+ (1s3s)^{3}S] as a function of distance from the three targets at 1.4 ns after the peak of the 1.06 μm pulse. Li density = 10^{17} cm^{-3}; energy per target = 2.3 J.

equilibrate the populations in the Li^+ (1s3s) and Li^+ (1s2p) levels.[12] To estimate the effect of this, curves A in Figs. 12 and 13 represent approximate theoretical predictions of the N^*L values for the Li^+ (1s3s)^{3}S level under circumstances in which it is in equilibrium with the Li^+ (1s2p)^{3}P level (as a result of ASE). Even under these conditions, there remains a discrepancy of approximately an order of magnitude between the measured values and the predicted values represented by curve A. Furthermore, in these particular experiments there was no sign of especially intense emission from the Li^+ (1s3s)^{3}S level - suggesting that ASE is probably not an important process here. This leads us to the conclusion that the Li^+ (1s3s)^{3}S population is anomalously low.

As well as the discrepancy between the predicted and measured values of the absolute magnitude of the Li^+ (1s3s)^{3}S population, the theoretical curve in Fig. 13 has a quite different functional dependence on distance than does the data. For comparison, in the analogous curve of Fig. 9, the Li^+ (1s2s)^{3}S population has a functional dependence on time that is well predicted by theory.

Clearly, the Li^+ (1s3s)^{3}S population is not being produced in the quantity predicted by theory. The actual reason for the low Li^+ (1s3s) population is unclear at this time. Estimates of rates for the de-excitation of the level Li^+ (1s3s)^{3}S by electrons, or photons from the laser-produced plasma, yield values that are too small to explain this result. Although anomalously fast deexcitation cannot be ruled out, the decay curve of Fig. 12, showing 2-3 ns decay time for the Li^+ (1s3s)^{3}S level makes this unlikely.

Similarly, anomalously small excitation rates for the Li^+ (1s3s) levels seem unlikely in the face of the observation that the Li^+ (1s2s) level and electron density measurements agree reasonably well with theory, and the well documented predictions of the shake-up ratio for Li^+ (1s3s) production. Further work is necessary in order to explain this disconcerting result.

Although laser action at 165.3 nm has not been demonstrated, the fundamental utility of this laser facility for creating large density-length products of excited levels has been verified in the form of the Li^+ $(1s2s)^3S$ measurements reported here. In the future it is expected that this facility will be used to investigate a variety of photoionization-pumped lasers in the VUV and XUV regions of the spectrum with emphasis on the laser classes produced by shake-up[6] and Auger[13] processes, as well as those that make use of metastable[14] and quasi-metastable[15] levels in atomic and ionic systems. It seems that these techniques should enable laser action to be demonstrated at a variety of wavelengths between 10 nm and 100 nm in the future.

ACKNOWLEDGEMENTS

The work described here was supported by the Army Research Office, the Air Force Office of Scientific Research, the Office of Naval Research, the Strategic Defense Initiative Organization, and the Lawrence Livermore National Laboratory.

REFERENCES

1. R. G. Caro, J. C. Wang, R. W. Falcone, J. F. Young, and S. E. Harris, Appl. Phys. Lett. 42, 9 (1983).
2. R. G. Caro, J. C. Wang, J. F. Young, and S. E. Harris, Phys. Rev. A 30, 1407 (1984).
3. J. C. Wang, R. G. Caro, and S. E. Harris, Phys. Rev. Lett. 51, 767 (1983).
4. W. T. Silfvast, O. R. Wood, II, and D. Y. Al-Salameh, "High-Gain VUV and Visible Inversions in He^+ and Ar^+ Produced by Two-Electron (Shakeup) Photoionization" (to be published).
5. P. J. K. Wisoff, R. G. Caro, and G. Mitchell, Opt. Comm. 54, 353 (1985) and "High Power Picosecond Blue Pulses from a Dye Laser Synchronously Pumped by a Q-Switched, Mode-Locked Nd:YAG Laser," Opt. Lett. (submitted for publication).
6. S. E. Harris and R. G. Caro, Opt. Lett. 11, 10 (1986).
7. F. P. Larkins, P. D. Adeney, and K. G. Dyall, J. Electron. Spectrosc. Relat. Phenom. 22, 141 (1981).
8. T. A. Ferrett, D. W. Lindle, P. A. Heimann, and D. A. Shirley, Lawrence Berkeley Laboratory, CA 94720 (personal communication).
9. P. Geard, Ph.D. Dissertation, University of Paris-Sud, Orsay, France (1984).

10. R. G. Caro and J. C. Wang, "Population Distribution in Li Vapor Excited by a Photoionization Electron Source, Phys. Rev. A, (to be published).
11. H. R. Griem, Spectral Line Broadening by Plasmas (Academic Press, New York, 1974).
12. R. W. Falcone, Physics Department, University of California, Berkeley, CA (private communication).
13. A. J. Mendelsohn and S. E. Harris, Opt. Lett. 10, 128 (1985).
14. S. E. Harris, Opt. Lett. 5, 1 (1980).
15. S. E. Harris, D. J. Walker, R. G. Caro, A. J. Mendelsohn, and R. D. Cowan, Opt. Lett. 9, 168 (1984).

MULTI-QUANTUM PROCESSES AT HIGH FIELD STRENGTHS

U. Johann, T. S. Luk, I. A. McIntyre, A. McPherson,
A. P. Schwarzenbach, K. Boyer, and C. K. Rhodes
Department of Physics, University of Illinois at Chicago
P. O. Box 4348, Chicago, Illinois 60680

ABSTRACT

Subpicosecond ultraviolet laser technology is enabling the exploration of nonlinear atomic interactions with electric field strengths in excess of an atomic unit. As this regime is approached, experiments studying multiple ionization, photoelectron energy spectra, and harmonically produced radiation all exhibit strong nonlinear coupling. Recent findings in connection with the ion spectra produced by such interactions are described, and it is shown that energy transfer from the radiation field to the atom depends importantly on the risetime of the field. The nonlinear atomic interaction is also discussed with respect to high laser intensities (> 10^{18} W/cm^2) in combination with high atomic density (> 10^{18} cm^{-3}), for which inter-atomic as well as intra-atomic coupling may play a significant role.

INTRODUCTION

A new experimental regime in which the optical field strength E is considerably greater than an atomic unit (e/a_0^2) is now becoming open for systematic study.[1] Indeed, electric field strengths on the order of 100 (e/a_0^2) should be attainable. This extraordinary range of electric fields has become accessible principally because of the availability of an ultraviolet laser technology capable of producing subpicosecond pulses[2,3] with energies approaching the joule level in low divergence beams at high repetition rates. Clearly, a technology of this genre will make possible the creation of physical conditions unachievable with any other known experimental means.

Figure (1) illustrates the parameters of the physical regime, in terms of pulse intensity (I) and pulse time scale (τ) that characterize our experimental studies. It is seen that ultraviolet laser technology will take us a considerable distance into the unexplored area, a zone associated with field strengths far greater than an atomic unit (e/a_0^2) and time scales that are advancing toward an atomic time τ_a. In addition, at the wavelength of 248 nm, intensities above ~ 5 x 10^{19} W/cm^2 will cause strongly relativistic motions to occur.

Recent experiments[4] have begun to explore the atomic response to intense 248 nm irradiation (~ 10^{16} W/cm^2) with subpicosecond pulses. Previous work had been conducted at a few specific wavelengths in the range between 193 nm and 10.6 µm with pulse lengths of $\geq$ 5 ps duration. Since the dynamics of electron ejection from an atom can[5-8] involve a time scale significantly shorter than ~ 5 ps, it is expected[9] that the nonlinear coupling with intense radiation will be significantly modified if sufficiently short pulses in the subpicosecond domain are used.

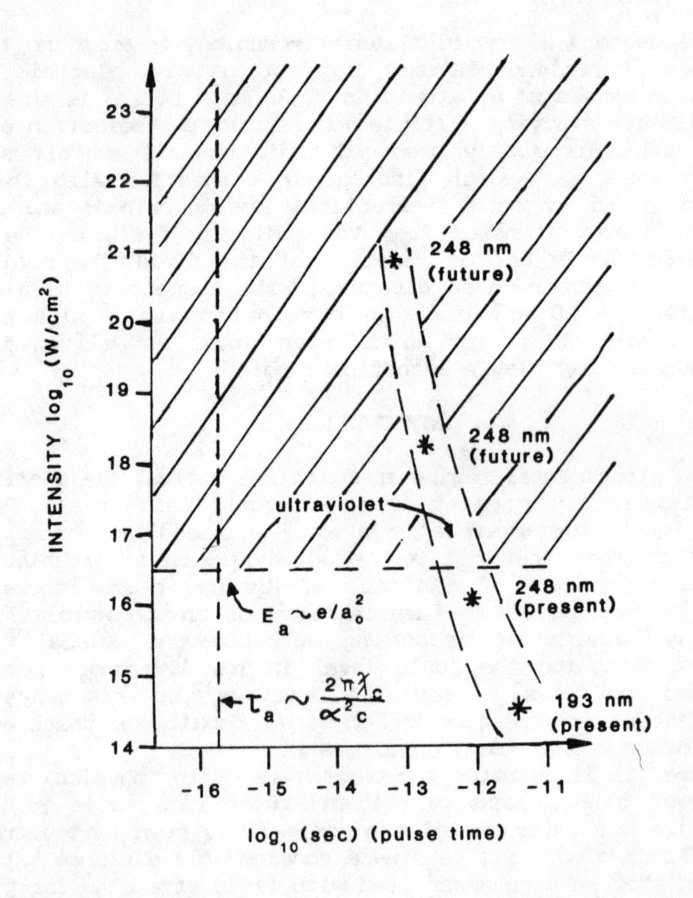

Fig. (1): Conditions of irradiation achievable with high brightness subpicosecond ultraviolet sources. The anticipated range of conditions that can be studied is shown as well as its relation to an atomic field strength (E_a) and an atomic time (τ_a).

EXPERIMENTAL DATA

A KrF^* (248 nm) laser[3] which produced pulses having an energy of 23 ± 2 mJ with a measured duration of 450 ± 150 fs was used for irradiation. This source had a divergence of approximately twice the diffraction limit and a spectral width of ~ 0.6 nm. This spectral breadth corresponds to a value of approximately four times the transform limit for the measured pulse width. The focussing system used in these studies was capable of producing a maximum intensity of ~ 10^{16} W/cm^2 in the experimental focal volume. The apparatus used for the measurement of the ion states[10] and the electron energy distributions[9] has been described previously.

The results of these subpicosecond studies (1) provide the first observation in ion spectra of the removal of an inner-electron in a direct multiquantum collision-free interaction, namely, an electron whose principal quantum number (n) is less than that characterizing the outer-most shell of the neutral atom, (2) demonstrate that the pulse width (τ) is an important parameter in the coupling, (3) reveal the characteristics of the spectra of the energetic electrons up to ~ 250 eV formed by the interaction, and (4) exhibit the production of coherent radiation at 22.6 nm,

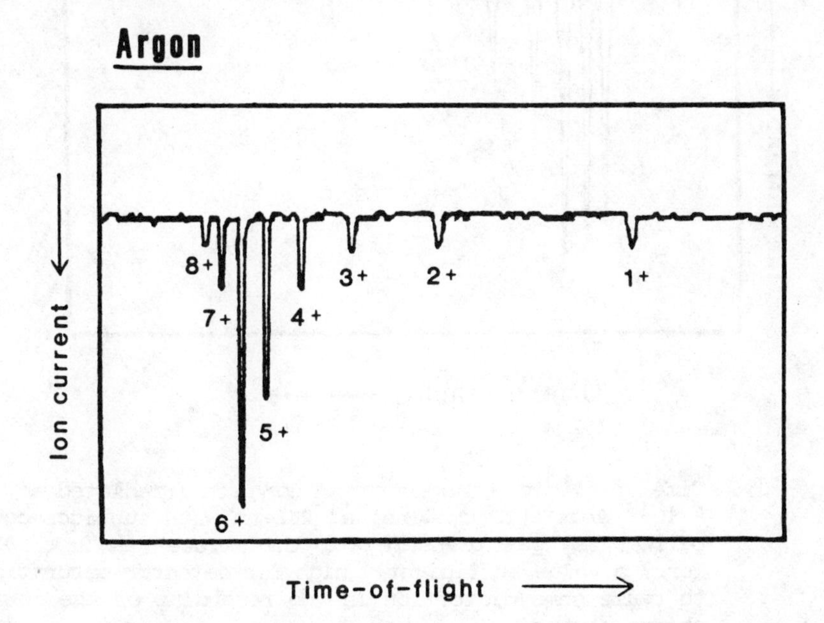

Fig. (2): Time-of-flight ion spectrum of argon irradiated at ~ 10^{16} W/cm^2 (20 cm lens) at 248 nm with subpicosecond pulses. The pressure for this trace was 2 x 10^{-5} Torr, a value sufficiently great for detector saturation to cause some distortion in the recording of the lower charge states.

the eleventh harmonic of 248 nm and the shortest wavelength produced by that means.

In comparison with the data obtained with the ~ 5 ps, (193 nm) source,[10] marked differences are seen in the ion charge state spectra[4] obtained with the subpicosecond 248 nm irradiation.

Figure (2) illustrates the results for argon at an intensity of ~ 10^{16} W/cm^2. In addition to the observation of Ar^{6+}, which was seen in the earlier studies[10] with 5 ps radiation, Ar^{7+} and Ar^{8+} are now visible, indicating the complete removal of both the $3p^6$ and $3s^2$ subshells.

A very similar result was also obtained for krypton, as shown in Fig.(3). Although only Kr^{6+} had been seen in the prior work,[10] Kr^{7+} and Kr^{8+} appeared clearly with the subpicosecond irradiation.

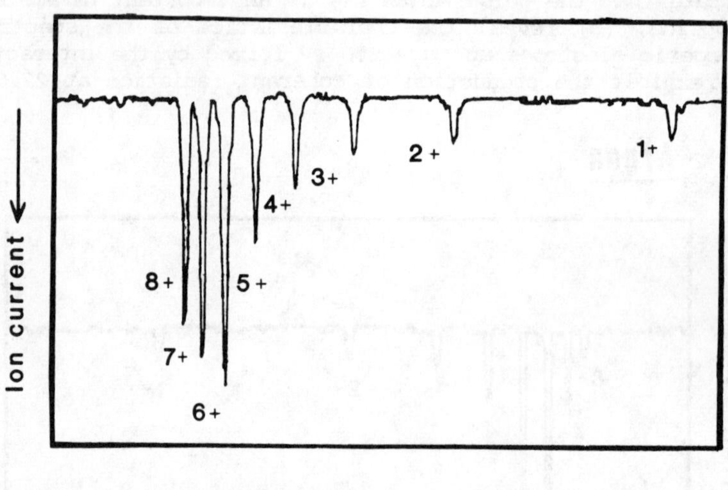

Fig. (3): Time-of-flight ion spectrum of krypton irradiated at ~ 10^{16} W/cm^2 (20 cm lens) at 248 nm with subpicosecond pulses. The gas pressure for this trace was 2 x 10^{-5} Torr, a value sufficiently high for detector saturation to cause some distortion in the recording of the lower charge states.

Considerably different behavior was also observed in xenon. The experiments[10] conducted at 193 nm with pulses having a duration of ~ 5 ps and an intensity of ~ 10^{16} W/cm^2, gave explicit evidence for the production of Xe^{8+}, a feature indicating complete removal of the outer-shell of the atom comprised of both the 5s and 5p subshells. A careful search failed to reveal the presence of Xe^{9+}, a species which would involve the additional ionization of an electron from the 4d atomic inner-shell. This charge state is now seen and Fig. (4) illustrates the detail of the Xe^{9+} signal which exhibits the expected isotopic signature,[10] an important feature confirming its identification, and indicates ionization of the 4d-shell and a minimum total energy transfer[11] of ~ 630 eV. In addition, the abundances of charge states with $q > 5$ are substantially augmented. Of course, since the pulse width at 248 nm is considerably shorter, the energy transfer <u>rate</u> has sharply increased, in this case by a factor of ~ 20. The xenon ion spectrum gives an average energy[11] transfer of ~ 130 eV, a value corresponding to an effective average cross section for the nonlinear interaction on the order of $\langle \sigma_{N\gamma} \rangle_{av} \cong 4 \times 10^{-21}$ cm^2. The observation of Xe^{9+} implies the presence of a peak energy transfer rate on the order of ~ 2×10^{-4} W/atom.

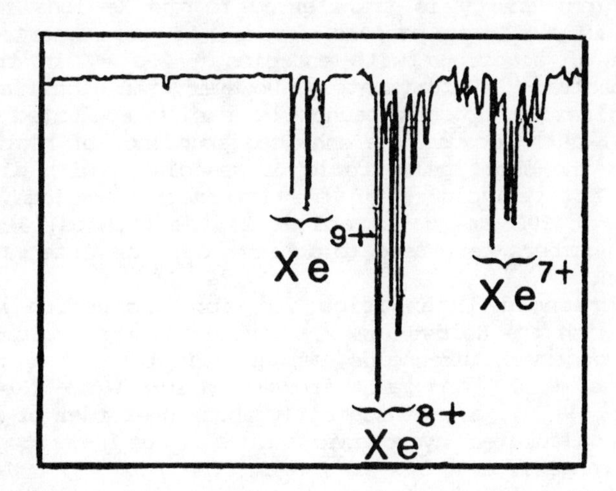

Fig. (4): Region of the ionic time-of-flight signal recorded at an intensity of ~ 10^{16} W/cm^2 in xenon with the 248 nm (450 ± 150 fs) laser pulse which specifically illustrates the presence of Xe^{9+} The peaks in the signals correspond to the distribution of known xenon isotopes.

At this stage of the investigations, it is difficult to determine fully which factors account mainly for the observed changes in the ion charge state distribution. An important role for the difference in wavelengths is believed to be unlikely for two reasons. First, the change in wavelength is relatively small. A comparison[10] of the 193 nm data with other studies[12,13] at 1.06 um and 532 nm does not reveal a wavelength dependence on ion production sufficiently strong, given the relative closeness of 193 nm and 248 nm, to cause an effect of this magnitude.

Moreover, the experimentally observed tendency is uniformly the <u>reverse</u> of that seen which, in specific comparisons[10] for xenon and krypton at ~ 10^{14} W/cm^2, showed that the ion production rates, particularly for the higher charge states, were reduced at longer wavelengths.

There are specific reasons, however, to expect an influence of the pulse risetime on the atomic response.[5-8,14] The sequential removal of outer electrons has been shown[5-8,10,14,15] to be the dominant mechanism of ionization for intensities $\leq 10^{15}$ W/cm^2 and pulse widths $\geq$ 5 ps. At a given intensity I_o, the sequential process involves independent acts of electron emission and yields a charge distribution that can be described by a rate equation analysis.[16] This model would actually associate shorter pulses having the same peak intensity with an unchanged or decreased abundance of the higher charge states unless a new physical process like a multiple ionization mechanism becomes important.[14] That sufficient energy is transferred to the Xe ions for double ionization in a direct process for $q > 5$ is supported by the observation of electrons with energies ~ 200 eV in the photoelectron spectra.[16] Unfortunately, however, the significance of a direct ionization process cannot be easily evaluated from our experimental data because the enhanced abundance of higher charge states with the short pulse could be associated with differences in the effective volumes of the interaction regions. In the experiments at 193 nm (5 ps) and at 248 nm (0.5 ps) lenses with different f-numbers were used to achieve the same intensity in the focal region.

The threshold intensities for ion production have been compared[16] with the Keldysh model[17] including the Coulomb correction. The observed thresholds, which correspond to a transition probability of ~ 10^{-3} and range from ~ 7 x 10^{11} W/cm^2 (Xe^+) to ~ 6 x 10^{15} W/cm^2 (Xe^{9+}), are consistently about one order of magnitude below the calculated hydrogenic values. For Ne, however, the thresholds range from ~ 8 x 10^{13} W/cm^2 (Ne^+) to ~ 7 x 10^{15} W/cm^2 (Ne^{4+}) and agree within a factor of two with the calculations.

In addition to the changes observed in the ion and electron spectra, the production of harmonic radiation is also strongly affected. Previously, with radiation at 248 nm in pulses of ~ 15 ps duration, 35.5 nm radiation was produced as the seventh harmonic.[18] In our experiment, the eleventh harmonic at 22.6 nm, the shortest wavelength available by such means was readily observed in neon, along with the lower harmonics at 27.6 nm, 35.5 nm, and 49.7 nm in neon and helium, as shown in Fig. (5). The harmonics were generated in a medium produced by a pulsed valve. It was also found, as illustrated in the inset of Fig. (5), that the signal strength fell rather slowly as the harmonic order increased. In our experiments with the subpicosecond pulses, the intensity of the ninth harmonic was less than a factor of thirty below that of the fifth, a finding that contrasts with the earlier work[18] which showed a decrement of several hundred between adjacent orders.

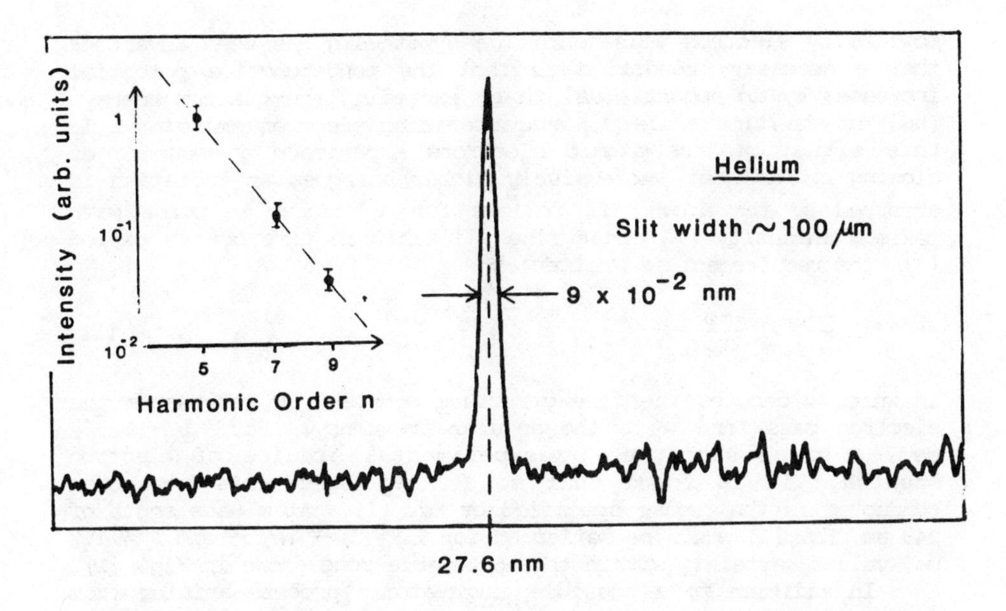

Fig. (5): Ninth harmonic (27.6 nm) signal produced by 248 nm in helium. The inset shows the relative strengths of the fifth, seventh, and ninth harmonics.

Therefore, all three classes of experiments, involving (a) the measurement of ion charge state distributions, (b) photoelectron energy spectra,[16] and (c) harmonically generated radiation, exhibit substantially stronger nonlinear coupling in the subpicosecond regime. A common physical basis for these phenomena is strongly indicated by the confluence of these findings.

DISCUSSION

It has been suggested[19-22] that ordered many-electron motions of outer-shell electrons could lead to enhanced rates of coupling from the radiation field to an atom. A mechanism which may enhance the probability for this kind of direct intra-atomic process in high intensity fields is suggested by studies[23-29] of photoelectron spectra generated by multiquantum ionization which have shown the importance of the ponderomotive potential in the suppression of photoelectron channels for intensities above certain specific values. The ponderomotive potential may, through the elevation of the effective ionization potential, tend to retard the total electron emission rate by reducing the number of channels available for the escape of electrons from the vicinity of the atomic potential.

Generally, it is required that energy be invested in electronic motion at a rate greater than that lost by damping of those motions by electron emission. The hypothesis has been advanced[9] that a necessary condition is that the ponderomotive potential increases by an amount equal to or exceeding the quantum energy ($\hbar\omega$) in the time scale τ_e characterizing electron emission. In this situation, the atomic electrons experience a sequence of closing channels at successively higher energies as radiation is absorbed by the atom. It follows for a triangular pulse with maximum intensity (I_o) that rises linearly in time over a period (τ), the requirement is written as

$$\frac{I_o}{\tau} \geq \frac{m\omega^3}{2\pi\alpha\tau_e} \tag{1}$$

in which α denotes the fine structure constant, m represents the electron mass, and ω is the angular frequency. For $\tau_e \sim 10^{-15}$ sec, a value supported by experimental studies of electron spectra,[9] it is found that so far no experiments have been conducted in the regime specified by Eq. (1). At a wavelength of 248 nm, Eq. (1) would be satisfied for $I_o \cong 10^{18}$ W/cm^2 and $\tau \cong 1.2$ ps, values certainly within the achievable zone shown in Fig. (1).

In addition to a possible intra-atomic process arising from driven many-electron motions, a corresponding inter-atomic coupling may occur when the intensity (I) and medium density (ρ) are sufficiently high. In the regime for which the electric field is greater than an atomic unit (e/a_o^2), nominally $I \gtrsim 10^{17}$ W/cm^2, the more loosely bound outer atomic/ionic electrons can be approximately modeled as free particles. Therefore, for those electrons, we can describe their motion as that of free electrons accelerating in intense coherent fields.[30,31] At $I \cong 10^{17}$ W/cm^2 with λ = 248 nm, the maximum excursion[32] of an electron is $x_o \cong$ 2.5 nm, a parameter that scales nonrelativistically as

$$x_o \cong 1.3 \times 10^{-20} \lambda^2 I^{1/2} \tag{2}$$

with x_o in (cm), λ in (nm), and I in (W/cm^2). With motions of this type, appreciable inter-atomic coupling could be expected for densities greater than a critical value $\rho_c \sim x_o^{-3}$, a relationship which yields a scaling of

$$\rho_c \cong 4.6 \times 10^{59} \lambda^{-6} I^{-3/2} \tag{3}$$

with ρ_c in (cm^{-3}), λ in (nm), and I in (W/cm^2).

Figure (6) illustrates the boundary separating the physical regimes in which inter-atomic and intra-atomic processes are expected to occur in the coupling. The Compton intensity (I_c) for 248 nm radiation is shown, the value for which the electrons acquire strongly relativistic velocities.[33] The absorption rate has a different density dependence for the intra-atomic and inter-atomic processes. The former depends linearly upon the density while the latter, on account of the collision induced nature of the interaction, would be expected to scale quadratically in the density. The quadratic density dependence is also a property of inverse bremsstrahlung, a process which has similarities to the inter-atomic mechanism under discussion, but differs in the intensity dependence.[34,35] In comparison to normal plasma processes, it should be noted that with 248 nm radiation at an intensity of ~ 10^{18} W/cm^2, $\rho_c \cong 1.8 \times 10^{18}$ cm^{-3}, a value nearly a full four orders of magnitude below the critical plasma density[32] $n_c \sim 1.6 \times 10^{22}$ cm^{-3} for that wavelength. Also, with the strong λ^{-6} scaling appearing in Eq. (3), if $\lambda \cong 100$ nm at $I \sim 10^{18}$ W/cm^2, the critical density is substantially increased to $\rho_c \sim 4.2 \times 10^{20}$ cm^{-3}, a density of approximately 16 amagat. Furthermore, if we assume an average ionization per atom of Z_{av}, the electron density $Z_{av}\rho_c$ will equal the critical density[32] n_c when,

$$\frac{\lambda^4 I^{3/2}}{Z_{av}} \simeq 4.6 \times 10^{32}. \tag{4}$$

We note that a problem rather similar to that discussed above has been examined by Pert and coworkers.[36]

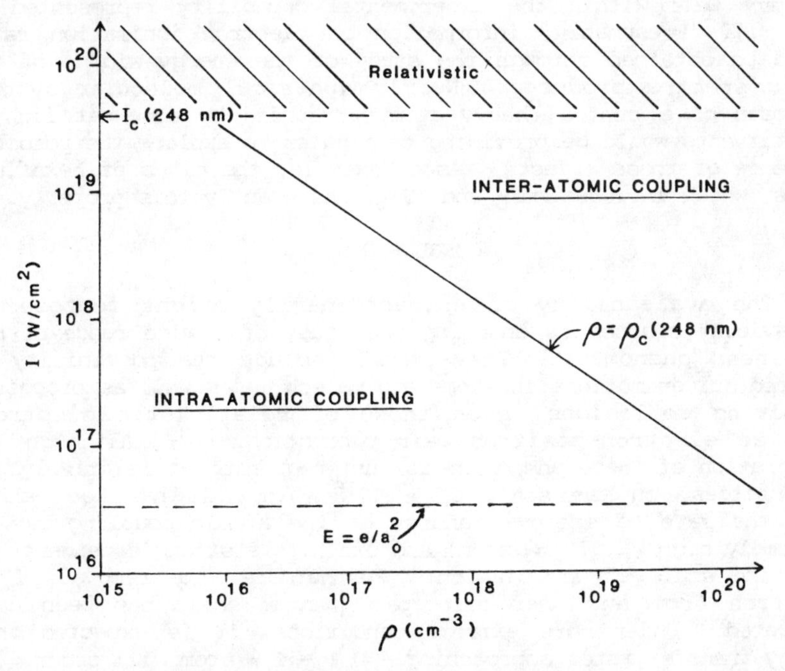

Fig. (6): Diagram showing the regions of intra-atomic and inter-atomic coupling as a function of laser intensity I and atomic density ρ. The Compton intensity I_c is indicated, above which strong relativistic motion occurs. $\rho = \rho_c$ corresponds to Eq. (3). The intensity corresponding to an electric field $E = e/a_o^2$ is also shown.

Finally, we speculate that such inter-atomic couplings could occur between neighboring atoms in molecules, independent of the medium density, provided that the radiative field is applied in a sufficiently short time. In order to estimate the conditions necessary for this phenomenon to occur, we examine the time τ for two atoms, with a reduced mass of M initially situated at a typical bond distance of ~ 2 a_o and ionized suddenly by the radiation to charge states Z_1 and Z_2, to undergo a coulomb explosion[37,38] and develop a separation of x_o, the parameter given in Eq. (2). For $x_o/a_o >> 1$, it can be shown that

$$\tau \simeq \frac{\lambda_c}{c} \frac{1}{\alpha^2} \sqrt{\frac{M}{m_e} \frac{1}{Z_1 Z_2}} \left[\frac{x_o}{a_o} + \ln \frac{2x_o}{a_o} \right] \tag{5}$$

in which α is the fine structure constant, m_e the electron mass, λ_c the electron Compton wavelength, and c the speed of light. At an intensity of ~ 2.5 x 10^{18} W/cm^2 for 248 nm, with M equal to twenty atomic units, Z_1 = 10, and Z_2 = 20, the time scale evaluates as $\tau \cong 100$ fs. These values, $I \cong 2.5 \times 10^{18}$ W/cm^2 and $\tau \cong 100$ fs, are well within the experimental capability represented in Fig. (1). Presumably, information on electron ionization rates could be obtained through the study of the energy widths of the ionic spectra produced. Nearly spherical molecular systems composed of a central heavy atom surrounded by somewhat lighter constituents would be promising candidates to explore the possible presence of these effects. As an example, the class of hexafluorides, which includes MoF_6 and UF_6, has exactly this property.

CONCLUSIONS

The availability of extraordinarily bright femtosecond ultraviolet sources is enabling the study of a wide range of new nonlinear phenomena. These will include the possibility of ordered driven motions in atoms and molecules as well as processes involving collisions, even those of relativistic electrons such as electron-positron pair production.[21,39] Although the exploration of these phenomena is just beginning at relatively low intensities, on the scale of ~ 10^{16} W/cm^2, it has been shown that the rate of energy transfer of the atomic coupling can be extremely high (~ 0.2 mW/atom) and exhibits a strong dependence on the time scale of irradiation. Furthermore, the removal of an electron from an inner principal quantum shell has been demonstrated. Under more extreme conditions, it is expected that energy transfer rates approaching ~ 0.1 – 1 W/atom will occur.

ACKNOWLEDGEMENTS

The authors wish to acknowledge the technical assistance of R. Slagle, J. Wright, T. Pack, and R. Bernico. This work was supported by the ONR, the AFOSR, the SDIO(ISTO), the DOE, the LLNL, the NSF, the DARPA, and the LANL.

REFERENCES

1. C. K. Rhodes, Science 229, 1345 (1985).
2. J. H. Glownia, G. Arjavalingham, P. P. Sorokin, and J. E. Rothenburg, Opt. Lett. 11, 79 (1986).
3. A. P. Schwarzenbach, T. S. Luk, I. A. McIntyre, U. Johann, A. McPherson, K. Boyer, and C. K. Rhodes, "Subpicosecond KrF* Excimer Laser Source," submitted to Optics Letters.
4. U. Johann, T. S. Luk, I. A. McIntyre, A. P. Schwarzenbach, K. Boyer, and C. K. Rhodes, "Subpicosecond Studies of Collision-Free Multiple Ionization of Atoms at 248 nm," submitted to Phys. Rev. Lett.
5. M. Crance and M. Aymar, J. Phys. B13, L421 (1980).
6. Y. Gontier and M. Trahin, J. Phys. B13, 433 (1980).
7. P. Lambropoulos, Phys. Rev. Lett. 55, 2141 (1985).
8. G. Wendin, L. Jönsson, and A. L'Huillier, Phys. Rev. Lett. 56, 1241 (1986).
9. U. Johann, T. S. Luk, H. Egger, and C. K. Rhodes, "Rare Gas Electron Energy Spectra Produced by Collision-Free Multiquantum Processes," Phys. Rev. A (in press).
10. T. S. Luk, U. Johann, H. Egger, H. Pummer; and C. K. Rhodes, Phys. Rev. A32, 214 (1985).
11. T. A. Carlson, C. W. Nestor, Jr., N. Wasserman, and J. C. McDowell, Atomic Data 2, 63 (1970).
12. A. L'Huillier, L. A. Lompré, G. Mainfray, and C. Manus, Phys. Rev. Lett. 48, 1814 (1982); ibid., Phys. Rev. A27, 2503 (1983).
13. A. L'Huillier, L. A. Lompré, G. Mainfray, and C. Manus, J. Phys. B16, 1363 (1983).
14. A. L'Huillier, L. A. Lompré, G. Mainfray, and C. Manus, J. Physique 44, 1247 (1983).
15. P. Agostini and G. Petite, Phys. Rev. A32, 3800 (1985).
16. U. Johann, T. S. Luk, I. A. McIntyre, A. McPherson, A. P. Schwarzenbach, K. Boyer, and C. K. Rhodes, "Multiphoton Ionization in Intense Ultraviolet Laser Fields," this volume.
17. L. V. Keldysh, Sov. Phys. -JETP 20, 1307 (1965).
18. J. Bokor, P. H. Bucksbaum, and R. R. Freeman, Opt. Lett. 8, 217 (1983).
19. A. Szöke, J. Phys. B18, L427 (1985).
20. C. K. Rhodes in Multiphoton Processes, P. Lambropoulos and S. J. Smith, editors (Springer-Verlag, Berlin, 1984) p. 31.
21. K. Boyer and C. K. Rhodes, Phys. Rev. Lett. 54, 1490 (1985).
22. A. Szöke and C. K. Rhodes, Phys. Rev. Lett. 56, 720 (1986).
23. M. Hollis, Opt. Commun. 25, 395 (1978).
24. P. Agostini, F. Fabre, G. Mainfray, G. Petite, N. K. Rahman, Phys. Rev. Lett. 42, 1127 (1979).
25. B. W. Boreham and B. Luther-Davis, J. Appl. Phys. 50, 2533 (1979).
26. P. Kruit, J. Kimman, and M. J. van der Wiel, J. Phys. B14, L597 (1981).

References, (cont.)

27. P. Kruit, J. Kimman, H. G. Muller, and M. J. van der Wiel, Phys. Rev. A28, 248 (1983).
28. K. G. H. Baldwin and B. W. Boreham, J. Appl. Phys. 52, 2627 (1981).
29. L. A. Lompré, A. L'Huillier, G. Mainfray, and C. Manus, J. Opt. Soc. Am. B2, 1906 (1985).
30. M. J. Feldman and R. Y. Chiao, Phys. Rev. A4, 352 (1971).
31. E. S. Sarachik and G. T. Schappert, Phys. Rev. D1, 2738 (1970).
32. T. P. Hughes, Plasmas and Laser Light (John Wiley and Sons, New York, 1975).
33. F. V. Bunkin et A. M. Prokhorov in Polarisation, Matiére et Rayonnement, édité par La Société Française de Physique (Presses Universitaires de France, Paris, 1969) p. 157.
34. P. Mora, Phys. Fluids 25, 1051 (1982).
35. P. D. Gupta, R. Popil, R. Fedosejevs, A. A. Offenberger, D. Salzmann, and C. E. Capjack, Appl. Phys. Lett. 48, 103 (1986).
36. G. J. Pert, J. Phys. B11, 1105 (1978); R. J. Dewhurst, G. J. Pert, and S. A. Ramsden, J. Phys. B7, 2281 (1974).
37. T. A. Carlson and M. O. Krause, J. Chem. Phys. 56, 3206 (1972).
38. W. Ebernardt, J. Stöhr, J. Feldhaus, E. W. Plummer, and F. Sette, Phys. Rev. Lett. 51, 2370 (1983).
39. J. W. Shearer, J. Garrison, J. Wong, and J. E. Swain, Phys. Rev. A8, 1582 (1973).

THEORY OF MULTIPHOTON IONIZATION OF ATOMS

Abraham Szöke
Lawrence Livermore National Laboratory
University of California
Livermore, CA 94550

ABSTRACT

A non-perturbative approach to the theory of multiphoton ionization is reviewed. Adiabatic Floquet theory is its first approximation. It explains qualitatively the energy and angular distribution of photoelectrons. In many-electron atoms it predicts collective and inner shell excitation.

INTRODUCTION

Picosecond visible and ultraviolet laser pulses of high intensity do violent things to atoms. In the last few years some spectacular results were obtained: above-threshold ionization, multiple ionization, striking angular distributions of photoelectrons, short wavelength lasers. The theoreticians not only did not provide guidance to the experimentalists, but there is a profound lack of agreement even on the approaches to be used to understand the phenomena.

In this paper I will review the particular approach I have been pursuing.[1-4] I hope to clarify some of the concepts that are incontroversial (or at least should be so) and point out where are the main unsolved problems (at least in my opinion).

Multiphoton ionization of atoms can be formulated as a scattering problem. On the incoming side there is an atom, and a large number of (more or less coherent) photons from the laser. On the outgoing side, most of the incoming photons go unhindered, some get scattered, some get absorbed, and the atom may get excited, singly or multiply ionized or all of the above. This is many-body scattering theory, so one cannot make too much progress. The fist simplification is to neglect all scattered photons, and spontaneously emitted ones. As the external field is strong, it can be represented by a classical field. Also, we are only interested in ionization by pulsed fields, whose spectrum is narrow compared to their frequency. These assumptions translate the problem to a solution of a time dependent Schrödinger equation

$$i\hbar \frac{\partial \Psi}{\partial t} = H\Psi \qquad (1)$$

0094-243X/86/1470169-5$3.00

where the Hamiltonian is

$$H = \sum_{i}^{Z} \frac{1}{2m}(\bar{p}_i + \frac{e}{c}\bar{A}(\bar{r}_i, t))^2 + \sum_{i \neq j}^{Z} \frac{e^2}{r_{ij}} - \sum_{i} Z \frac{e^2}{r_i} \tag{2}$$

with obvious notation [$e > 0$, $\bar{p}_i = - i\hbar \bar{\nabla}_i$, no electronic spin, and no relativity]. The external vector potential is an almost periodic function, with slowly varying amplitude and frequency (SVEA).

$$\bar{A}(\bar{r}, t) = \bar{A}_o(\tau_1) \cos [\int^t \omega(\tau_1') dt' - \bar{K}.\bar{r}]. \tag{3}$$

where we assumed a plane wave, and wrote τ_1 for the (slow) timescale of the pulse. It is natural to expand the solution of Eq. 1 in multiple timescales.[2,5] Defining the dimensionless fast time (actually it is the optical phase)

$$\tau_o = \int^t \omega(\tau_1') dt' \approx \omega t \tag{4}$$

and formally defining the slow time (also dimensionless) as

$$\tau_1 = \epsilon \tau_o \quad , \quad \epsilon = \text{const} << 1 \tag{5}$$

a hierarchy of equations is obtained for $\Psi \equiv \Psi(\bar{r}_i, \tau_o, \tau_1)$

$$\Psi = \Psi^{(o)} + \epsilon\, \Psi^{(1)} + \epsilon^2\, \Psi^{(2)} + \dots , \tag{6}$$

The first two are,

$$i\hbar\, \omega(\tau_1) \frac{\partial \Psi^{(o)}}{\partial \tau_o} - H\, \Psi^{(o)} = 0 \tag{7}$$

$$i\hbar\omega(\tau_1) \frac{\partial \Psi^{(1)}}{\partial \tau_o} - H\Psi^{(1)} = i\hbar\omega \frac{\partial \Psi^{(o)}}{\partial \tau_1} \tag{8}$$

The first one, Eq. (7), produces the wave functions and states of "Floquet theory": the Hamiltonian is exactly 2π periodic in τ_o, the states contain $A_o(\tau_1)$ and $\omega(\tau_1)$ as parameters. This provides a natural (if not always convenient) basis set for the solution. The slow time dependence of Ψ^o is obtained from the second equation by eliminating secular terms from it. [As shown by J. Garrison recently, one obtains as a bonus Berry's phase.[6]] The result is particularly clear when the pulse turns on adiabatically. The wave function is a "succession" of Floquet states,

$$\Psi_j^{(o)} = \exp [- i \int \Omega_j\, d\tau_o]\, \phi_j^{(o)}(A_o, \tau_o)$$

$$\phi_j^{(o)}(A_o, \tau_o + 2\pi) = \phi_j^{(o)}(A_o, \tau_o) \quad , \tag{9}$$

where state j correlates adiabatically to the (j'th) state, the unperturbed atom started from the beginning of the optical pulse. This should be properly called "adiabatic" Floquet theory. Extensive calculations were done on many systems of interest by Reinhardt, Chu and coworkers.[7]

I will now review informally the lessons and limitations of this theory. First, Eq. 7 has no "real" bound states, every state decays, i.e., ionizes (however slowly).[8] If outgoing wave boundary conditions are imposed on the wave function, the imaginary part of the eigenvalue, the so called quasi energy ($\Omega\hbar\omega$ in Eq. 9) gives the total ionization rate.[7] Second, the theory is <u>not</u> perturbative in the electromagnetic field strength, though its limitations will be discussed below. Third, it can be shown by model calculations that appreciable above threshold ionization is obtained.[2] Fourth, the absence of ponderomotive shifts in the observed electron energies and the sudden disappearance of low energy electrons due to "closure of channels" can be explicitly traced in the calculations,[2] in accordance with conservation energy arguments.[1]

The two main limitations to the validity of adiabataic Floquet theory are resonances and non-adiabaticity. As a blanket statement, when the first order theory breaks down most probably quantum chaos is obtained.[9] Non-linear resonances occur, when a multiple of the incident frequency approaches an energy level difference of the adiabatic Floquet states. These states, as a function of the field strength, can have avoided crossings. There are two complicating factors: the first one, mentioned above, is that all Floquet states belong to the continuum, and the second one that the quasi energies should properly be restricted to a finite range, $0 \leq E_F < \hbar\omega$.[7] Even disregarding the complications, at these crossings the adiabatic states "exchange" the character of their wave functions, so if the passage is adiabatic, there is really a "transition" and vice versa. The simplest way to estimate the conditions for adiabaticity is by the Landau-Zener formula,[10]

$$W = \frac{2\pi \mid V_{nm} \mid^2}{\hbar} \left| \frac{\partial}{\partial\tau} (E_m - E_n) \right|^{-1}_{E_m = E_n}$$

where E_m, E_n are the quasi energies of the diabatic levels, and $2 \mid V_{nm} \mid$ is the minimum energy separation at the avoided crossing. When $W \gtrsim 1$ the passage is adiabatic and vice versa. Thus it can be seen that both the strength of the laser field (through V_{nm}, and the dependence of E_n, E_m on it) and the shortness of the pulse is important. If many level crossings occur, the actual behavior of the atom may become quite uncalculable, and unpredictable.

Several aspects of multiphoton ionization are correctly given by adiabatic Floquet theory. The electron spectrum, as measured outside the laser field, consists of peaks corresponding to the absorption of a fixed (N) number of photons. These states correlate adiabatically to states close to the atom (inside the field!) that have their translational energy diminished by the ponderomotive potential, and "dressed" by a variable number of photons. In a linearly polarized field this number has a strong maximum when the electron's translational motion is parallel to the field's polarization. This implies that if an electron is emitted with a number of virtual photons attached to it, its energy spectrum (as measured outside the laser field) shows "above threshold ionization." Also, when the number of virtual photons becomes sizable (at low laser frequencies) we expect a strongly peaked angular distribution parallel to the laser's polarization.

C. Cerjan of LLNL has done computor solutions of simple model systems,[11] (without the adiabatic approximation) and the author attempted to get a "finite element" solution of the same model in the adiabatic Floquet approximation. These should be very useful to assess the importance (or unimportance) of detailed level structures of atoms in multiphoton ionization, and the presence or absence of non-linear resonances, non-adiabaticity and possible chaos.

Many electron atoms are more interesting but even more difficult to treat. Multiple scale expansion can be used to simplify time dependent Hartree Fock (TDHF) theory in almost periodic fields. The first approximation is then analogous to adiabatic Floquet theory and can be viewed as its extension to many electron systems.[3] The solution is again a succession of adiabatic states. Thus it is interesting to investigate their nature.

TDHF theory restricts the (properly antisymmetric) wave function to a single configuration. Denoting it by $|g\rangle^n$, (e.g., $|5p\rangle^6$ for the ground state of xenon) we can see that it includes both single-electron excitations, $|g\rangle^{n-1}|e\rangle$. (e.g., $|5p\rangle^5|6s\rangle$) and what we call collective excitations, $(\alpha|g\rangle + \beta|e\rangle)^n$, (e.g., $(1/\sqrt{2}\,|5p\rangle + 1/\sqrt{2}\,|6s\rangle)^6$). In a pure collective excitation, each electron is "dressed" coherently by the same number of photons. Ionization occurs when some electron's wave function has a component from the continuum. As a general rule, collective excitations occur when the atom is driven off resonance.

Collective excitations, if they don't decay, (see below) allow to put a large amount of energy into an outer shell of a many electron atom. This in turn can excite the inner shell of the same atom, possibly producing short wavelength lasers. Recently the rate for this process was estimated assuming collective excitation.[4,12]

The first hard question is whether, how strongly, and under what circucmstances are collective excitations excited. All the caveats raised for the single electron theory apply here too: there are problems with resonances, and non-adiabaticity. There are even serious questions as how to observe such a state if it is excited. The second hard question is the decay of these states. There are two main routes for their decay: autoionization, and the conversion of the internal energy into non-collective motion. In nuclear physics this is analogous to the decay of a giant dipole resonance into single nuclear emission or into a compound nuclear state (the latter is also called spreading).

The excitation question is within TDHF theory, it is being investigated through numerical solution (without an adiabatic approximation) by K. Kulander at LLNL.[13] The second problem, an even more difficult one may be addressed using diagrammatic perturbation with the adiabatic Floquet - TDHF states as a starting point.[14]

In summary, a non-perturbative theory of atomic multiphoton processes starts with adiabatic Floquet theory as its first approximation. Even at that stage, it predicts from first principles above threshold ionization, the closing of channels, collective excitation, and the transfer of energy to inner shells. The possible (and very probable) breakdown of the theory is due to resonances, non-adiabaticity and the decay of collective excitations due to electron-electron interaction (the same as collisions, the same as correlations). If these processes are dominant, we may end up with statistical concepts, similar to RRKM theory and quasi-continua in multiphoton dissociation of large molecules.

REFERENCES

1. A. Szöke, J. Phys. B. 18, L427 (1985).
2. A. Szöke, to be published.
3. A. Szöke, UCRL-93156 (1985).
4. A. Szöke, C. K. Rhodes, Phys. Rev. Lett. 56, 720 (1986).
5. J. Kervorkian, J. D. Cole, Perturbation Methods in Applied Mathematics, Springer, 1981.
6. J. Garrison, Adiabatic Perturbation Theory and Berry's Phase, to be published.
7. S-I. Chu, Adv. Mol. Phys. Vol. 21, 1985.
8. Ya. B. Zeldovich, UN 110, 139 (1973) [Sov. Phys. Uspekhi 16, 427 (1973)].
9. A. Casati, ed. Chaotic Behavior in Quantum Systems, NATO ASI Series B, Vol. 120, Plenum, 1985.
10. D. R. Bates, Quantum Theory, Vol. I, p. 293.
11. C. Cerjan, private communication.
12. A. L'Huillier, L. Jönsson, and A. Wendin, to be published.
13. K. Kulander, private communication.
14. M. Baranger, I. Zahed, Phys. Rev. C 29, 1005, 1010 (1984).

Electron angular distributions in above threshold ionization of xenon

M. Bashkansky[a], *P. H. Bucksbaum, R. R. Freeman, T. J. McIlrath*[b]

AT&T Bell Laboratories
Murray Hill, New Jersey 07974

L. F. Dimauro

Department of Physics and Astronomy, Louisiana State University
Baton Rouge, LA 70803

J. Custer

Department of Materials Science, Cornell University
Ithaca, NY 14853

ABSTRACT

Angular distributions of the photoelectrons produced in above threshold ionization of xenon by 100 psec 1.06 μm laser pulses have been measured. We find that the electron angular distributions, which are peaked along the polarization direction for linearly polarized light, differ for electrons emitted with different kinetic energies, corresponding to the absorption of different numbers of photons. Furthermore, for a given electron energy, the angular distribution is strongly dependent on the laser intensity. We propose that ponderomotive potential scattering is mainly responsible for the intensity dependence of the angular distributions.

The angular distributions of photoelectrons in multiphoton ionization (MPI) provide invaluable information about the nature of many-photon transitions. In non-resonant MPI, where there are no intermediate state resonances which dominate the cross section, the ionization probability is so small that extremely high laser intensities are required to achieve observable ionization rates. A number of new phenomena[1] may accompany the ionization process at these high photon densities, which are evidence of the highly nonlinear nature of matter-light interactions. The most dramatic of these is above threshold ionization (ATI), where electrons are emitted at a series of energies above the minimum for MPI, separated by the photon energy hν. Another, somewhat more subtle phenomenon, is the ponderomotive potential effect due to the coherent scatter-

[a] Physics Department, Columbia University
[b] On leave from IPST, University of Maryland

ing of laser photons by the charged particles in the light field. This ponderomotive potential effect tends to repel electrons from the region of tight focus. In this paper we will describe our observations of ATI electrons in xenon, which undergo ponderomotive scattering in the laser focus following ionization. The effect of this scattering is to alter the angular distributions in an intensity dependent way.

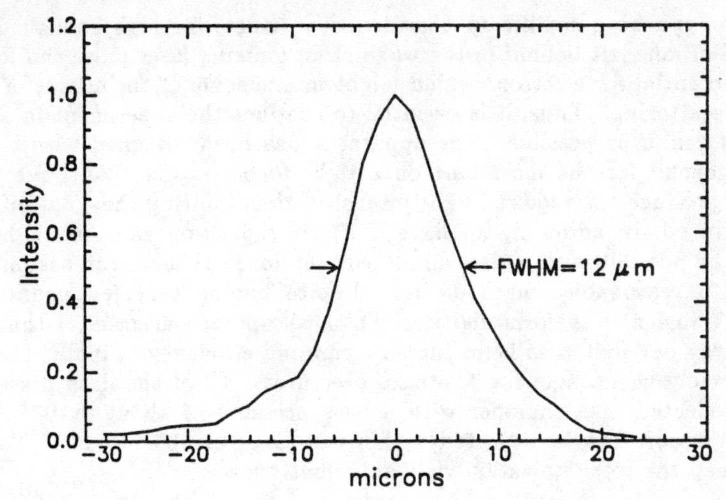

Figure 1 The beam profile at focus measured with a 12x magnification and a CCD diode array with 25 μm elements. A 15 cm (f/10) achromat lens was used to focus the 1064 nm beam.

To understand fully the angular distributions it is imperative to know the interaction region well. With this in mind a part of the laser beam was split off with an optically flat glass plate and focused with a 15cm (f/10) lens, similar to that used in the experiment. The focus was magnified by 12 and imaged onto a CCD array enabling us to monitor the beam profile at focus in a real time. The laser beam approximated a Gaussian profile in the focus (**figure 1**) with a full width at half maximum of 12 μm.

In our experiments we used transform-limited 100 picosecond duration mode-locked Nd:YAG pulses (1064 nm). These pulses were amplified to 50 mj, which enabled us to reach intensities of $\approx 10^{14}$W/cm^2 in the focal region. The light intensity was controlled by means of $\lambda/2$ plate and polarizer upstream, while the final polarization was adjusted by an additional $\lambda/2$ plate just outside the chamber. Energy fluctuations of better than 5% were maintained during a run.

The vacuum chamber houses the interaction region, the electron spectrometer, and the electron detector. While the base pressure in the chamber is maintained at 6x10^{-9} Torr, xenon density is varied from 2x10^{-5} Torr to 4x10^{-8} Torr by changing the rate of flow through a leak valve. The time-of-flight spectrometer consists of a 40cm long, magnetically shielded, tubular, copper Faraday cage. Electrons are detected by microchannel plates which subtend a 3 degree angle at the source. Signals are fed through an impedance matched conical anode to the transient waveform digitizer with an overall time resolution of 2 nanoseconds. Stray fields limit the overall resolution to approximately 0.03 eV at an energy of 1 eV. By the nature of the time-of-flight spectrometer the resolution deteriorates for the higher energy electrons. A time-of-flight spectrum is recorded for every laser shot, and single electron pulses are software discriminated and binned.

Space charge can be a significant consideration[2] since, for high enough densities, a cloud of ions left behind in the wake of an ionizing laser pulse can form a scattering potential for electrons which might mimic some of the effects of ponderomotive scattering. Thus, it is essential to conduct the experiment in as low a density of xenon as possible. The apparatus has been designed with this in mind. We depend for the most part on a tight focus, rather than large pulse energies, to produce the required light intensity, thus limiting the total number of ions produced. In addition, we have a 10 Hz repetition rate, much higher than would be possible with a glass amplified system, so that we can accumulate statistics in a reasonable length of time despite having very few counts per pulse. The transient waveform digitizer, which allows measuring more than one electron energy per shot, also helps increase running efficiency. Finally, the system is extremely clean, allowing low base pressures. All of the data presented here were collected in a chamber with a base pressure of about 6×10^{-9} Torr. The xenon pressure can be adjusted within the range of 10^{-8} to 10^{-5} Torr as needed to keep the total ionization yield sufficiently low.

As an additional guard against space charge induced scattering, we performed a series of auxiliary experiments to study its effects, and also did some computer modeling to attempt to understand it as completely as possible. We found that space charge induced effects manifest themselves in an asymmetric broadening and shift of the peaks in the electron energy spectrum, as well as a change in angular distributions, as low energy electrons produced late in the pulse experience electrostatic forces while leaving the region of the laser focus. Electron total yields are not greatly affected until much higher density. There is yet another effect, namely pulse pile-up, which can affect the number of detected electrons. Since we count single electron signals any two electrons overlapping in time will be counted as one. The width of the electron spectral features in the time domain depends inversely on the energy of the electrons. Consequently, one expects the lower energy electron peaks to dominate in the pulse pile-up governed data. We observe pulse pile-up to occur at about the same pressures as the space charge induced effects. For our apparatus, we found that all of these effects are avoided by adjusting the xenon pressure to maintain 2 or fewer detected electrons per laser shot, when the detection axis is along the polarization axis.

Previous studies of angular distributions for nonresonant MPI and ATI have concentrated on the use of electron distributions as a tool for understanding the atomic physics involved in the process of ionization[3]. Here we report a new effect: the tendency of the angular distributions to become isotropic in the plane perpendicular to **k**, the direction of propagation of the laser beam, when the intensity is raised. We shall presently argue that this phenomenon is independent of the atomic physics of ionization, and is rather a final state effect.

Figure 2 shows a typical 10,000 shot electron spectrum for MPI in xenon taken with 1064nm linearly polarized light. The vertical scale is normalized to number of counts per shot, per mTorr of xenon pressure, per meV energy bin. The laser polarization is along the detection axis. The principal feature is a series of ATI electron peaks.

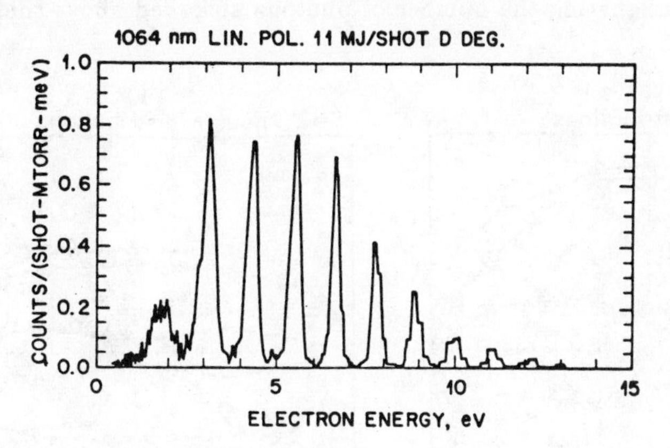

Figure 2 A typical 10,000 shot electron spectrum of ATI in xenon. The vertical scale is normalized to number of counts per shot, per mTorr of xenon pressure, per meV energy bin. The laser polarization is along the detection axis.

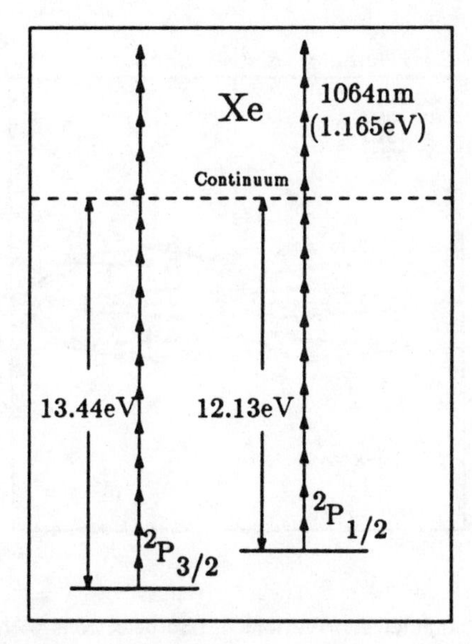

Figure 3 The ionization energy diagram for xenon showing multiphoton ionization (MPI) and above threshold ionization (ATI).

Xenon has two ionization potentials corresponding to the $^2P_{3/2}$ and $^2P_{1/2}$ fine structure levels, respectively, of the $5s^25p^5$ ground configuration of Xe^+ **(figure 3)** . For 1064nm light 11 photons are required to reach the lowest Xe^+ state $^2P_{3/2}$ and 12 photons to reach the higher $^2P_{1/2}$ state. One expects to see a series of closely spaced ($\approx$140 meV) pairs of peaks separated by the photon energy of 1.165 eV. However we find that the peaks corresponding to threshold ionization are absent from the spectrum. The peaks are labeled by the parameter S designating the number of photons absorbed above that necessary

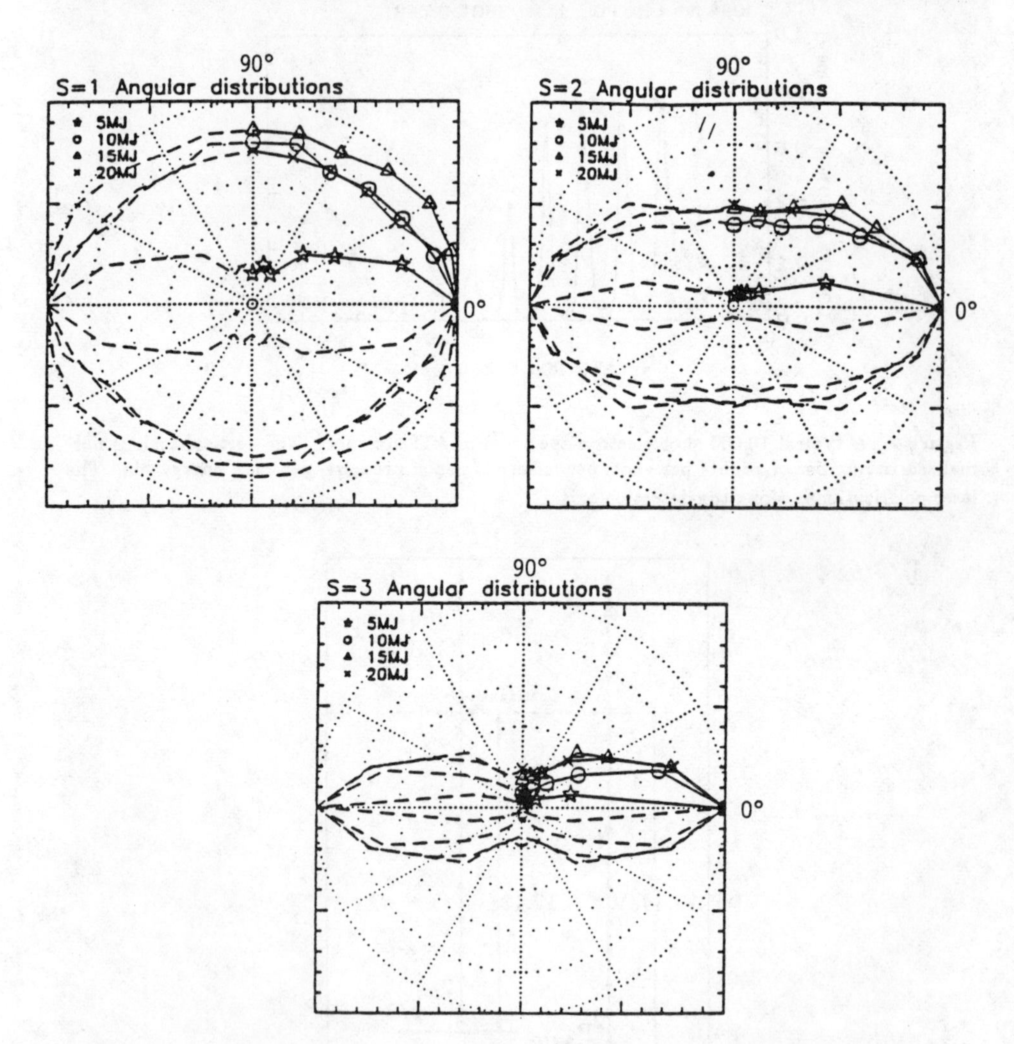

Figure 4 Polar graphs of angular distributions of photoelectrons with 1064 nm, linearly polarized light. 0 degrees is electron detection along E. 90 degrees is detection normal to E. Data was taken from 0 to 90 degrees and curves were continued to 360 degrees to aid in viewing.

for threshold ionization. The first observable peak (S = 1) corresponds to 12 and 13 photon absorption to the two fine structure ion states. This relative suppression of the threshold (S = 0) electrons in xenon has been reported previously[1]. The ponderomotive effect may be responsible for their absence. We will return to this point later, but now we will discuss the intensity dependence of angular distributions, shown in the **figure 4** .

To obtain one angular distribution the laser intensity is kept constant over the time of measurement, and the angle between light polarization and detector is changed in intervals of 15 degrees. The area under the peaks is integrated and plotted against angle in polar coordinates in **figure 4**. Measurements are performed from 0 to 90 degrees.

As the data in **figure 4** show, the angular distributions change significantly with laser intensity, each ATI feature becoming more isotropic as the intensity increases. This effect is stronger for the lower energy peaks, so that at each laser intensity, higher order ATI spectral features are more peaked in the light polarization direction than the lower order features. The angular distributions for the highest order features change very little over the range of laser intensities used here.

We believe that these strong changes in the electron angular distributions reflect a final state effect. After the electrons are produced, they can continue to interact with the radiation field gradient, or with electrostatic space charge fields of nearby ions. Field gradient, or ponderomotive scattering, dominates in regimes of high laser intensity; space charge dominates for high ion density.

Ponderomotive forces, which have been discussed extensively in the literature, are essentially derived from the oscillatory kinetic energy possessed by a charged particle, or a polarizible neutral particle, in the presence of an intense radiation field[4]. The ponderomotive effect on a charged particle, simply stated, is to add a positive effective potential energy, which really is just the oscillatory kinetic energy, whose magnitude is

$$\frac{q^2 E^2}{4m\omega^2} \qquad (1)$$

Here E is the amplitude of the laser electric field, and ω is its angular frequency. When a particle of charge q and mass m enters a region where the intensity is rapidly changing in space, as is the case in a tightly focused laser, it experiences a force proportional to the gradient of this potential. It can also be shown[5] that there is a net shift in the ionization energy of a neutral atom equal to equation 1.

It is not hard to see how this effect together with the Ponderomotive force affects the angular distributions in a way which qualitatively agrees with the results of our experiments:

> First assume some initial angular distribution for electrons emitted from the atom into an ATI channel. Our own data taken at low laser intensities show that these distributions are sharply peaked along the polariza-

tion direction, with an opening angle (not deconvolved) ranging from 15 degrees for the lowest ATI peak (spectroscopic designation S1) to 7 degrees for the higher peaks.

Next assume a profile of the laser beam at the focus to be approximately Gaussian as determined by direct measurement.

As a result of the extremely high intensity dependence of the ionization rate, the atoms tend to be ionized within a narrow range of intensities, so consider the electrons to be ionized in equal intensity rings.

Finally account for the temporal dependence by assuming a Gaussian time pulse with FWHM equal to 100 picoseconds.

Under these constraints, each electron follows a trajectory determined by its initial direction and speed, and by the gradient of its ponderomotive potential.

As an example, consider the behavior of the S1 peak as the light intensity is increased. It is convenient to think of the laser intensity in the units of the Ponderomotive potential, where 1 eV $\approx 10^{-13}$ I (W/cm^2) for 1064 nm light. The kinetic energy of an S1 electron is its low field ionization energy -- about 1.8 volts -- minus the ponderomotive potential energy shift, U_P, of the ionization potential. If ionization takes place at $U_P = 1$ eV, then the translational kinetic energy of the electron after leaving the ion is only .8 eV. The electron will regain its deficient kinetic energy upon leaving the region of intense focus, but in so doing, scatters from the gradient of the ponderomotive potential. The Ponderomotive potential for our model laser is axially symmetric and repulsive, and will tend to scatter the electrons in the radial direction from the origin. The higher energy electrons are born with much higher velocities in the polarization direction. Consequently, they will scatter through a smaller angle as they leave the focus.

The idea of simple Ponderomotive Scattering appears to predict qualitatively the experimental results. We are in a process of modelling electron trajectories in a rather complete computer simulation.

We wish to acknowledge useful conversations with M. Mittleman, A. Szoke, L. Armstrong, L. Pan and C. Greene, and technical advice and assistance by D. Schumacher. One of us (TJM) acknowledges the partial support of NSF grant CPE 81-19250.

REFERENCES

1. P. Agostini, F.Fabre, G.Mainfray, G.Petite and N.K. Rahman, Phys. Rev. Lett. *42*, 1127 (1979); P. Kruit, J.Kimman and M.J. van der Wiel, J. Phys. B *14*, L597 (1981).

2. M. Crance, J. Phys. B *19*, (1986) in press.

3. S. N. Dixit and P. Lambropoulos, J. Phys. B *16*, 1205 (1983); Phys. Rev. A *27*, 861 (1983); Adila Dodhy, R. N. Compton and J. A. D. Stockdale, Phys. Rev. Lett. *54*, 422 (1985); D. Feldman, G. Otto, D. Petring and K. H. Welge, J. Phys. B, in print (1986); H. J. Humpert, H. Schwier, R. Hippler and H. O. Lutz, Phys. Rev. A *32*, 3787 (1985); P. Lambropoulos, Adv. At. & Mol. Phys. *12*, 87 (1976); Appl. Opt. *19*, 3926 (1980); G. Petite, F. Fabre, P. Agostini, M. Crance and M. Aymar, Phys. Rev. A *29*, 2677 (1984).

4. T.W.B. Kibble, Phys. Rev. *150*, 1060 (1966); J.H. Eberly, Progress in Optics, 7, 359 (1969); M. H. Mittleman, *Introduction to the theory of laser-atom interactions* (Plenum, N.Y.) 1982.

5. P. Avan, C. Cohen-Tannoudji, J. Dupont-Roc and. C. Fabre, J. de Physique *37*, 993 (1976); L. Pan, L. Armstrong, Private communication.

MULTIELECTRON MULTIPHOTON IONIZATION

Anne L'Huillier,
Service de Physique des Atomes et des Surfaces
Centre d'études nucléaires de Saclay
91191 Gif sur Yvette, FRANCE

Lars Jönsson and Göran Wendin
Institute of Theoretical Physics
Chalmers University of Technology
S-412 96 Göteborg, SWEDEN

ABSTRACT

The experimental results obtained for Xenon at 532 nm are analysed in some details. Multiple ionization is interpretated as a stepwise process. The role of the spatial and temporal distribution of the laser pulse is pointed out. The influence of screening effects in multiphoton ionization is studied within the framework of many-body perturbation theory and the random phase approximation. The theory is applied to a calculation of the 2-photon ionization cross-section of Helium. Finally, the role of screening effects in the multi-ionization experiments is estimated.

INTRODUCTION

The production of multiply charged ions through multiphoton absorption has now been investigated[1-5] in a number of atoms ranging from Helium[2,6] to Uranium[3] and at different wavelengths from 193 nm[3,4] to 10 μm[5]. Some of the most spectacular results obtained so far are, for example, the production of He^{2+} or Xe^{6+} at 1064 nm, the removal of the entire n=5 shell of Xe and the formation of U^{10+} at 193 nm[3,4]. Finally, triple ionization of Xe has been observed at 10 μm[5]. To give an idea, the formation of Xe^{6+} represents an absorption of at least 250 eV, i.e. 210 photons at 1064 nm[2].

Although the amount of experimental results is growing rapidly, the understanding of the phenomena involved in the experiments is still far from being complete. One of the basic questions of course is the mechanism responsible for multiple ionization. Are all the electrons removed together, in a direct process ? or are they removed one at a time, in a sequential process ? Could inner shell electrons be involved in the process ?

0094-243X/86/1470182-11$3.00

Very different theoretical approaches have been proposed in order to explain the experimental data. First, statistical methods have been developed[7-9]. Aberg et al.[7] analyse the experimental data at 532 and 1064 nm within the framework of information theory. Geltman[9] uses a very simplified (independent-electron) description of a many-electron atom in a presence of a laser field. Both theories interpret multiple ionization as a direct process. On the contrary, the statistical description of Crance[8] is based upon the assumption that multiple ionization is essentially a stepwise process. Rhodes, Szöke and coworkers[4,11-12] suggest that, in very strong fields, the outermost shell of a many electron atom is collectively excited and ionized and could eventually transfer its energy to inner shells. Lambropoulos[12] points out the importance of the temporal distribution of the laser pulse. At least for the first multiple ionization processes, the atom is stripped in a sequential way, during the rise of the laser pulse.

Finally, the role of many-electron screening effects has been studied by Wendin et al.[13,14], within the framework of diagrammatic many-body perturbation theory. We have shown that the multiphoton ionization of an atom with several electrons on the outer shell is reduced due to screening of the field by the electrons. Moreover, during the stripping of an atom through stepwise ionization, the screening is progressively reduced, leading to enhanced yields of highly charged ions compared with singly charged ions.

The first section of this paper is devoted to a detailed analysis of the experimental result obtained in Xenon at 532 nm[1,15]. This result combines the main aspects of multi-electron multiphoton ionization phenomena, e.g, direct and stepwise double ionization, multiple ionization up to Xe^{5+}. Moreover, it has been obtained at not too high intensity, 10^{12}-10^{13} $W.cm^{-2}$. A great deal of information can be deduced in a simple way, by studying the the numbers of ions created as a function of the laser intensity.

In the second part of this paper, the theoretical formalism used for describing the role of many-electron effects in multiphoton ionization is briefly outlined[13-15]. It is applied to a calculation of the 2-photon ionization probability of Helium, using the local density random phase approximation. Finally, we shall use the results to interpret some aspects of the experimental data.

MULTIPLE IONIZATION OF XENON AT 532 NM

The experimental arrangment used in the experiments has been described elsewhere[1,2]. Briefly, a 50 ps Nd-YAG laser is focused into a chamber into which a rare gas has been released at a static pressure of 5×10^{-5} torr (no collisions). The ions created in the interaction region are extracted by a 1 kV.cm^{-1} electric field, then separated in charge and detected in a 20 cm-long time-of-flight spectrometer.

Figure 1a shows on a log-log plot the numbers of Xe ions created (up to Xe^{5+}) as a function of the laser intensity. The ionization energies and the numbers of ions required for the different processes are indicated in figure 1b. The Xe^{+} ions curve is typical of a 6-photon non-resonant ionization process, a straight line whose slope is equal to 6±0.5, followed by a saturation at $I_s = 8\times10^{11}$ W.cm^{-2}. This saturation effect plays a very important role in multi-electron multiphoton ionization and deserves some explanations. We assume that the neutral atoms are

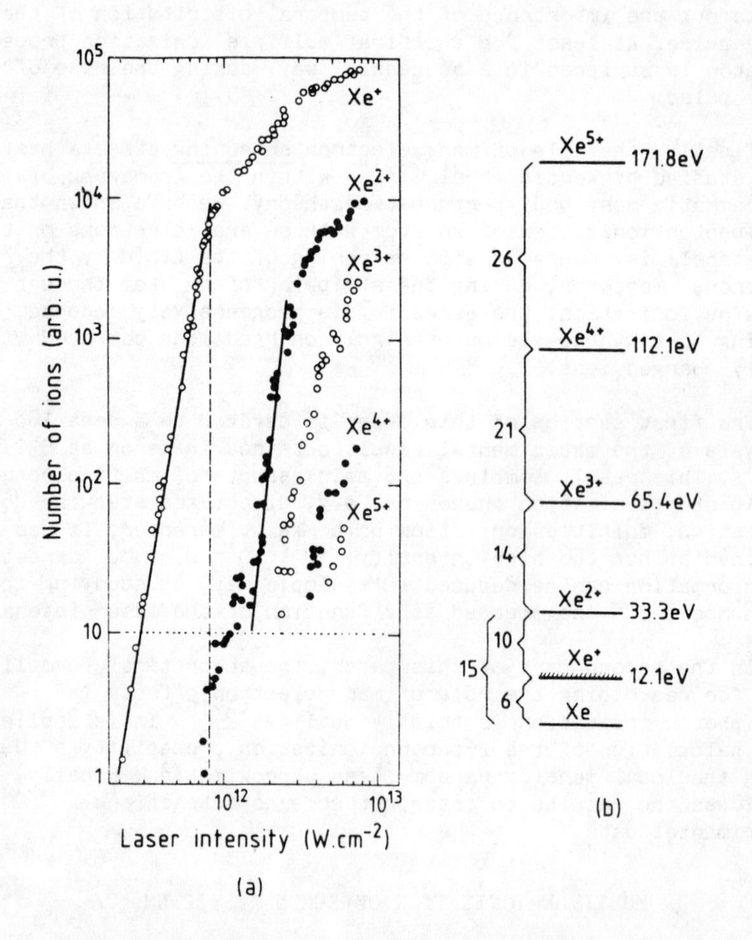

Fig.1a. Plot of the number of Xe ions as a function of the laser intensity at 532 nm.
1b. Schematic representation of the ionization energies and numbers of photons involved.

depleted by only a single N-photon ionization process, characterized by a generalized cross-section σ_N. The number of ions N_i can be written as

$$N_i = N_o \int \left(1-\exp(-\sigma_N I^N \tau_N f(\vec{r})^N)\right) d\vec{r} \qquad (1)$$

I is the laser intensity at the focus, $f(\vec{r})$ the spatial laser disribution (=1 at the focus); N_o is the density of atoms and τ_N the Nth order pulse duration ($= \int g(t)^N dt$, where g(t) is the temporal distribution of the laser pulse). As long as the expression in the exponential is small, the number of ions can be simply written as

$$N_i = N_o \; \sigma_N I^N \tau_N \; V_N \qquad (2)$$

with $V_N = \int f(\vec{r})^N d\vec{r}$. The saturation intensity is the intensity from which this approximation is not valid any more. It means that the ionization probability at the focus becomes equal to unity. However, outside the focal point ($|f(\vec{r})|<1$), the ionization probability may not yet be equal to unity. The increase in the number of ions beyond I_s is due to the extension of the volume where the ionization takes place. An ion signal detected at an intensity $I>I_s$ consists in : (i) ions created at the focus, before the laser pulse has reached its maximum, simply because the ionization probability becomes equal to 1 before the peak value of the pulse is reached ; (ii) ions created outside the focal region, at a lower intensity than that recorded. An ion detected at $I>I_s$ has always been created at a much lower intensity, either outside the region where the laser pulse is the most intense or during the rise of the laser pulse.

Let us consider now doubly charged ions (fig.1). At low intensities ($I<1.5\times10^{12}$ W.cm^{-2}), they are created through a direct process. They are first aligned on a straight line whose slope, measured at a shorter pulse duration 5 ps[16], is equal to 14.5±1. The onset of saturation occurs at the same laser intensity as for singly charged ions, which means that both single and double ionization processes pump on the same population of atoms. Doubly charged ions are then created through a direct 15-photon process from the neutral atoms. The sudden increase in the number of Xe^{2+} ions at $I= 1.5\times10^{12}$ W.cm^{-2} is due to the stepwise process $Xe \rightarrow Xe^+ \rightarrow Xe^{2+}$. In this intensity region, the first step is saturated, so that the intensity dependence of the number of ions reflects the second step of the process. i.e. a 10-photon ionization of the Xe^+ ion. The experimental points are set upon a straight line whose slope is equal to 11±1 ; a second saturation takes place at $I=2.5\times10^{12}$ W.cm^{-2}, when Xe^+ ions are depleted.

In summary, the relative importance of direct and stepwise processes depends on the laser intensity : at low intensity, doubly charged ions are created through a direct process ; at higher intensity, they are mostly created through a sequential process, via the singly charged ion.

Xe^{3+}, Xe^{4+} and Xe^{5+} ions appear at close laser intensities, in spite of very high non-linear orders for the last species (for example, 74 photons are required to create a Xe^{5+} ion). For clarity, we raise the experimental detection threshold (dotted line) by a factor of 10, thus eliminating consideration of the direct double ionization process. Xe^{k+} (k=2-5) ions are created in an intensity range for which processes leading to lower ionization stages are saturated. They are then created from $Xe^{(k-1)+}$ ions, the only ions whose population is not yet depleted in the interaction volume. An atom lying at the focus, exposed to a high laser intensity pulse (e.g. 10^{13} W.cm^{-2}), will be successively ionized as the laser pulse grows in time. First single ionization occurs at $I \leq 8\times10^{11}$ W.cm^{-2}, then double ionization at $I \leq 2.5\times10^{12}$ W.cm^{-2}, etc...At a time corresponding to the maximum of the laser pulse, the interaction volume consists in crowns of decreasing intensities, containing ions of decreasing charges. Xe^{k+} ($k<5$) ions detected at 10^{13} W.cm^{-2} come from the external region of the interaction volume.

In conclusion, from the experimental results presented in fig.1, it seems that multiply charged ions are created through a stepwise process, each step occuring successively as the laser intensity increases. The intensity dependences of the number of ions created reflect the last step of the multi-ionization process, because all the preceding steps are saturated. In this picture, the experimental curves describe single ionization processes on different atoms from Xe to Xe^{4+}. The main question raised by this interpretation is the following : the number of photons required to reach the different ionization thresholds increases rapidly from Xe (6) to Xe^{4+} (26). A 26-photon absorption process would be expected to take place at a much higher intensity than a 6-photon process. However, Xe^{+} and Xe^{5+} in fig.1 are detected at rather close laser intensities, separated by about a factor of 10.

MANY-ELECTRON EFFECTS IN MULTIPHOTON IONIZATION

This second section is devoted to a first investigation of the influence of many-electron effects in multiphoton ionization[13-15]. We shall only consider polarization (or screening) effects, which describe the response of an electronic shell to an external perturbation. The physical idea behind is very simple : an individual electron does not experience directly the external field, but an effective (screened) field, which is the sum of the external field and the fields induced by the perturbation of the electronic density.

The theoretical formalism is based upon diagrammatic many-body perturbation theory. Polarization effects are treated within the framework of the random phase approximation (RPA). This formalism has been described in details elsewhere[14]. In the present paper, we shall only review its essential features and apply it to a simple case. Although some strong field effects (dynamical Stark shift to first order in the intensity) can be introduced in the formalism without too much complications[14], we shall only consider here the weak field limit, i.e. the first non-vanishing order in the perturbation expansion for the electromagnetic field.

We use a one-electron basis set (with states $|i\rangle$, $|n\rangle$, $|\varepsilon\rangle$). $d_{\varepsilon i}=\langle\varepsilon|r|i\rangle$ is the dipole matrix element, $d_{\varepsilon i}(\omega)$ the effective dipole matrix element (ω is the photon energy). We assume that exchange effects are included in the one-electron basis. $d_{\varepsilon i}(\omega)$ is then solution of the simplified RPA integral equation[17]

$$d_{\varepsilon i}(\omega)=d_{\varepsilon i}-\sum_{n,j} C_{nj} \frac{2\omega_{nj}\langle\varepsilon j|1/r_{12}|in\rangle\, d_{nj}(\omega)}{\omega_{nj}^2-\omega^2} \tag{3}$$

$1/r_{12}$ is the Coulomb interaction.
$\omega_{nj}=\varepsilon_n-\varepsilon_j$ (ε_n, ε_j are one-electron energy eigenvalues).
C_{nj} is a spin/angular coefficient defined by

$$C_{nj}=\frac{(N_j-\delta_{ij})(2\ell_n+1)}{3}\begin{pmatrix}\ell_j & 1 & \ell_n\\ 0 & 0 & 0\end{pmatrix}^2 \tag{4}$$

$N_j=2(2l_j+1)$; $\delta_{ij}=1$ if i,j belong to the same shell, 0 otherwise. We can also define a local effective field $r(\omega)$ and an inverse dielectric function $\varepsilon^{-1}(r,\omega)$, which characterize the response of the shell to the external field.

$$d_{\varepsilon i}(\omega)=\langle\varepsilon|r(\omega)|i\rangle=\langle\varepsilon|r\varepsilon^{-1}(r,\omega)|i\rangle \tag{5}$$

$\varepsilon^{-1}(r,\omega)$ is solution of the integral equation

$$\varepsilon^{-1}(r,\omega)=1-\sum_{n,j} C_{nj} \frac{2\omega_{nj}\langle j|1/r_{12}|n\rangle\, d_{nj}(\omega)/r}{\omega_{nj}^2-\omega^2} \tag{6}$$

This theoretical treatment, or similar ones, has been extensively used for calculating one-photon cross-sections[18-21]. The random phase approximation allows to describe important collective effects in e.g. Xe or Ba. The purpose of the present work is to generalize this approximation to N-photon ionization.

We define a N-photon matrix element $d^N_{\varepsilon i}$

$$d^N_{\varepsilon i}=\sum_{k,m,n} \frac{d_{\varepsilon k}\ldots d_{mn}\, d_{ni}}{(\omega_{ki}-(N-1)\omega)\ldots(\omega_{mi}-2\omega)(\omega_{ni}-\omega)} \tag{7}$$

A straightforward generalization of the RPA to N-photon ionization consists in replacing the dipole matrix elements in eq.7 by the screened matrix elements. The effective N-photon matrix element is then defined as

$$d^N_{\varepsilon i}(\omega) = \sum_{k,m,n} \frac{d_{\varepsilon k}(\omega)\ldots d_{mn}(\omega) d_{ni}(\omega)}{(\omega_{ki}-(N-1)\omega)\ldots(\omega_{mi}-2\omega)(\omega_{ni}-\omega)} \tag{8}$$

In this approximation, we only consider the screening effects at frequency ω, i.e. the induced field which oscillates at the same frequency ω as the external field (linear response). In order to describe more completely screening effects in multiphoton ionization, one has to include screening at higher harmonics 2ω, 3ω,.. $N\omega$, i.e. induced fields oscillating at different harmonics (non linear response)[14].

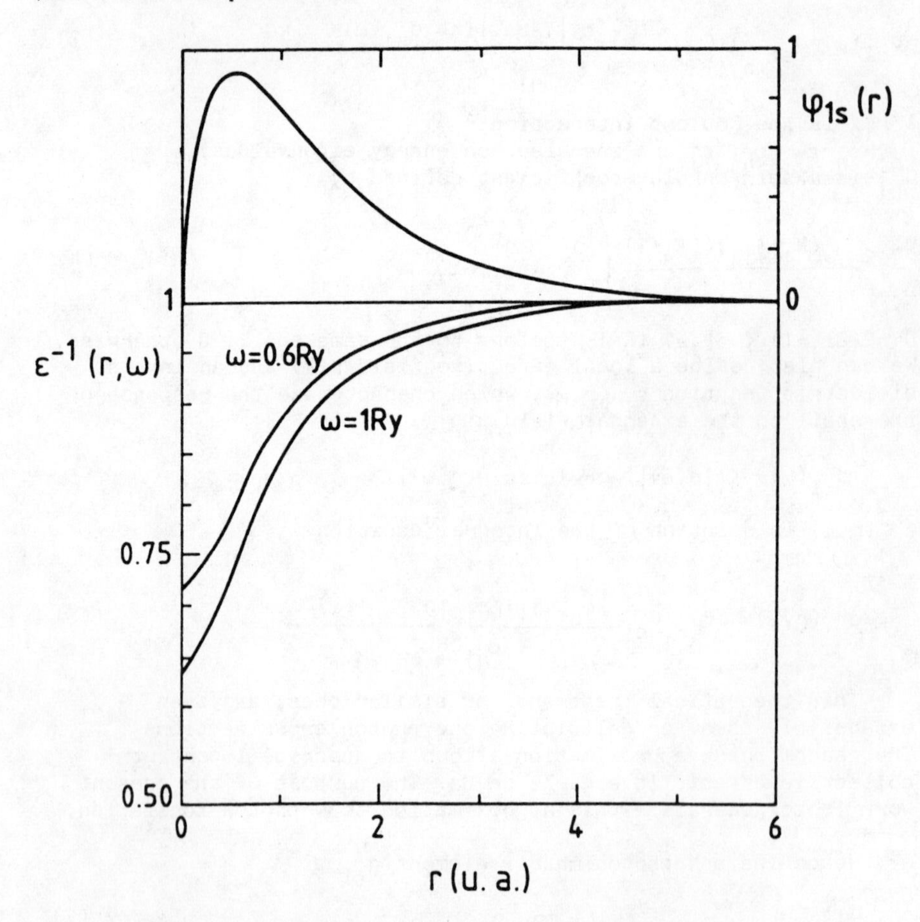

Fig.2. Plot of the inverse dielectric function as a function of r at ω=0.6 and ω=1 Ry.

We have applied this formalism to the simplest case, the calculation of the 2-photon ionization cross-section of Helium[14]. The 2-photon ionization rate is written as

$$p=2\pi\alpha c/a_0(I/I_0)^2(C_s|d^2_{\varepsilon s1s}(\omega)|^2+C_d|d^2_{\varepsilon d1s}(\omega)|^2) \qquad (9)$$

α is the fine-structure constant, c the light velocity and a_0 the Bohr radius. I_0 is a normalization factor[14] equal to 14.038 10^{16} $W.cm^{-2}$. C_s, C_d are spin/angular coefficients[14]. p is expressed in $W^{-2}.cm^4.s^{-1}$ unit. We use a local density approximation (LDA) one-electron basis set. The exchange/correlation potential is local, proportional to $n(r)^{1/3}$ (n(r) being the electronic density). The details of the numerical methods have been described elsewhere[14]. Briefly, $\varepsilon^{-1}(r,\omega)$ is calculated by solving eq.(7) using the Fredholm approximation[17]. The 2-photon matrix elements $d^2_{\varepsilon i}$ or $d^2_{\varepsilon i}(\omega)$ are calculated by performing explicitly the summation over the intermediate states.

Figure 2 shows the variations of the inverse dielectric function $\varepsilon^{-1}(r,\omega)$ as a function of r for two typical photon energies, 0.6 and 1 Ry, below the ionization threshold (1.81 Ry). The 1s wavefunction $\phi_{1s}(r)$ is plotted on the same scale at the top of the figure. $\varepsilon^{-1}(r,\omega)$ increases with r from 0.7 to 1. In a simple way, one can imagine a 1s electron to be dynamically dipole screened from the external field by the other 1s electron : the screening is maximum near the nucleus and is progressively reduced as r increases (there is no screening at large distances). At the maximum of the 1s wavefunction ($r \approx 0.5$) the local field is reduced by a factor 1/4 compared to the external field.

In figure 3, the variation of the 2-photon ionization cross-section p/I^2 is plotted as a function of the photon energy. The solid line is the RPA calculation whereas the dashed line corresponds to the one-electron approximation. The unusual shape of this 2-photon ionization spectrum, i.e. absence of resonances in the vicinity of bound excited states and very low ionization threshold (1.13 Ry) is due to the local density approximation used for calculating one-electron energies and wavefunctions. (The reason is that, in this approximation, there is imperfect cancellation between the self-interaction from the direct term and the self-interaction from the exchange term). The spectrum presented in fig.3 must be interpretated as follows :

(i) From 0.57 to 0.90 Ry, the curve obtained (solid line) is an average representation of the 2-photon absorption spectrum.

(ii) From 0.90 to 1.81 Ry, the curve is a 2-photon ionization spectrum. The vicinity of the LDA ionization threshold (1.13 Ry) is avoided for computational reasons. However, it is artificial and the curve can be extrapolated, as shown by the dashed-dot line. in the region of the 1snp resonances, the curve is an average representation of the 2-photon ionization spectrum.

(iii) Finally, above 1.81 Ry, the curve represents a 2-photon A.T.I (above-threshold-ionization) process, since the continuum is reached after the absorption of only one photon.

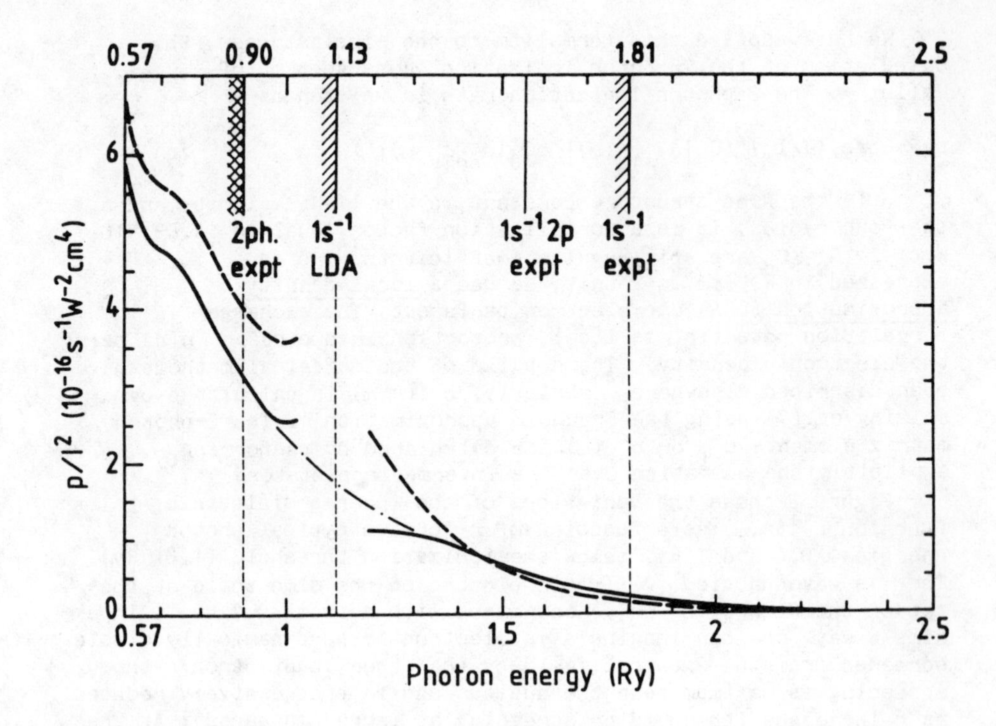

Fig.3. Plot of the 2-photon ionization cross-section of He as a function of the photon energy. (---) one-electron approximation. (——) RPA calculation.

An essential result shown in fig.3 is that screening effects are important, even for Helium. Their influence depend on the photon energy. By comparison to the one-electron approximation (dashed line), the screened 2-photon ionization cross-section (solid line) is <u>lowered</u> at low energies, up to a factor of 1.4 at 1 Ry. , then enhanced from 1.45 Ry. This result is in good agreement with other calculations performed by Victor and Ritchie, within the framework of the time dependent Hartree-Fock theory[22,23]. Let us emphasize that such effects will be much more important in heavier atoms, e.g. Xe or Ba.

Finally, we have estimated the influence of screening effects in multi-ionization multiphoton experiments, and we shall concentrate on the result obtained in Xe at 532 nm. The N-photon ionization rate can be written as

$$W_N = \sigma_N I^N \tag{10}$$

with, omitting the proportionality factor,

$$\sigma_N = |d^N_{\varepsilon i}(\omega)|^2 \qquad (11)$$

We write the screened dipole matrix element (eq(5)) as $d_{\varepsilon i}(\omega) = \langle\varepsilon^{-1}(r,\omega)\rangle\, d_{\varepsilon i}$. The N-photon ionization rate can then be approximated by

$$W_N = \sigma_N^{ind}\, I^N_{eff} \qquad (12)$$

in which $\sigma_N^{ind} = |d^N_{\varepsilon i}|^2$ is the independent-electron N-photon ionization cross-section, $I_{eff} = \langle\varepsilon^{-1}\rangle I$ is an effective intensity which represents the intensity experienced locally by each electron. We obtain the same expression (12) as for an independent electron system if the external laser intensity is replaced by the effective (local) intensity.

We have estimated the effective intensity for Xe at 532 nm by doing a local density RPA calculation. For a Xe 4d-electron, $I_{eff}/I=100$. This will obviously make extremely difficult a direct multiphoton ionization in an inner shell. For a Xe 5p-electron, $I_{eff}/I=3$. Screening effects therefore lower the ionization probability of an atom with several electrons in the outermost shell. This is quite general as long as the photon energy is low compared to typical excitation/ionization energies of the outer shell, as is usually the case in multiphoton ionization experiments.

However, during the stripping of an atom through stepwise ionization, the screening is progressively reduced from Xe to Xe^{4+} as the number of outer electrons decreases. A Xe^{5+} 5p-electron directly experiences the external intensity ($I_{eff}=I$). Let us imagine that the experimental curves presented in fig.1 are drawn as a function of the effective intensity. The curves would be shifted towards lower intensities ($I_{eff} \leq I$), by a factor depending on the ionic state (=3 for Xe^+, 1 for Xe^{6+}, if detected). The intensity range within with these multiple ionization processes appear would then be extended by a factor of approximately 3. This might, at least partly, help in understanding why multiple ionization processes in Xe are detected at close laser intensities[1,2,12,24].

In conclusion, at moderately high laser intensities, multielectron multiphoton ionization may be described as a succession of single ionization processes taking place during the rise of the laser pulse. It is essential to consider both the temporal and spatial laser distributions to understand the experimental data. The decrease of screening effects with the number of outer electrons during the stripping enhances the production of highly charged ions compared to singly charged ions. (The effective intensity increases much more quickly than the external intensity). This investigation of the influence of many-electron effects in multiphoton ionization, applied to a simple calculation of the 2-photon ionization cross-section of He shows that independent-electron approximations are not appropriate for describing multiphoton ionization of many-electron atoms. Preliminary results for 2-photon single ionization of the 5p-shell in Xe suggest that 5p-screening has to be included even in a qualitative description of the magnitude and frequency dependence of the 2-photon cross-section.

REFERENCES

1. A.L´Huillier, L.A.Lompré,G.Mainfray and C.Manus, Phys.Rev.A 27, 2503 (1983).
2. A.L´Huillier, L.A.Lompré, G.Mainfray and C.Manus, J.Phys.B 16, 1363 (1983).
3. T.S.Luk, U.Johann, H.Egger, H.Pummer and C.K.Rhodes, Phys.Rev.A 32, 214 (1985).
4. C.K.Rhodes, Science, 229, 1345 (1985).
5. S.L.Chin, F.Yergeau and P.Lavigne, J.Phys.B 18, L213 (1985).
6. L.A.Lompré, A.L´Huillier,G.Mainfray and C.Manus, Phys.Lett. 112A, 319 (1985).
7. T.Åberg, A.Blomberg, J.Tulkki and O.Goscinski, Phys.Rev.Lett. 52, 214 (1984).
8. M.Crance, J.Phys.B 17, 3503,4333,L435 (1984); 18, L155 (1985).
9. S.Geltman, Phys.Rev.Lett. 54, 1909 (1985).
10. K.Boyer and C.K.Rhodes, Phys.Rev.Lett. 54, 1490 (1985).
11. A.Szöke and C.K.Rhodes, to be published; A.Szöke, to be published.
12. P.Lambropoulos, Phys.Rev.Lett., 55, 2141 (1985).
13. G.Wendin, L.Jönsson and A.L´Huillier, Phys.Rev.Lett., in press.
14. A.L´Huillier, L.Jönsson and G.Wendin, Phys.Rev.A, in press.
15. A.L´Huillier, thèse de doctorat d´état, Paris, 1986.
16. A.L´Huillier, L.A.Lompré, G.Mainfray, and C.Manus, J.Physique, 44, 1247 (1983).
17. G.Wendin, in "New trends in Atomic Physics", p555, Les Houches, 1982, Session XXXVIII, eds. G.Grynberg and R.Stora, Elsenier Science Publishers B.V. (1984).
18. M.Ya.Amusia, Adv.At.Mol.Phys., 17,1 (1981).
19. G.Wendin, Phys.Rev.Lett., 53, 724 (1984).
21. A.Zangwill and P.Soven, Phys.Rev.A, 21, 1561 (1980).
22. G.A.Victor, Proc.Phys.Soc. 91, 825 (1967).
23. B.Ritchie, Phys.Rev.A, 16, 2080 (1977).
24. A.L´Huillier and M.Trahin, Phys.Lett.A, 112A, 377 (1985)

MECHANISMS OF MULTIPHOTON EXCITATION AND IONIZATION OF MULTI-ELECTRON ATOMS*

P. Lambropoulos and X. Tang
University of Southern California
Los Angeles, CA 90089-0484, U.S.A.
and
University and Research Center of Crete
Iraklion, Crete, 711 10 Greece

ABSTRACT

An analysis of multiphoton generalized cross sections shows that multiple electron ejection can be explained by a combination of sequential and direct processes. Events during the rise of a high power pulse are shown to play a decisive role and for pulses as short as 0.5 psec the sequential processes are found to be dominating. Results from a calculation of multiphoton double electron ejection in C are briefly discussed.

The transition probability per unit time for K-photon ionization to lowest non-vanishing order of perturbation theory[1] is given by

$$W_K = \hat{\sigma}_K F^K \tag{1}$$

where $\hat{\sigma}_K$ is the generalized cross section (in units of $cm^{2K} sec^{K-1}$) and F the photon flux per cm^2 per sec. As long as there are no resonances with intermediate atomic states[2], this expression has been found to be valid for intensities as high as $10^{15}\,W/cm^2$ and orders of K=25 and perhaps higher. As K becomes much larger tunneling is expected to become dominant. If there are intermediate resonances,, a single transition probability per unit time (rate) is not expected to adequately describe the process. A more detailed calculation in terms of a time-dependent set of equations becomes necessary in that case an example

*Work supported by the National Science Foundation, Grant Number PHY-8306263.

of which is discussed later on. It is often possible, however, to employ the above equation even in the presence of resonances if we are willing to settle for a picture of the average behavior. A modified value of $\hat{\sigma}_K$ dependent on F may then be necessary.

With the above caveats in mind, we explore the theoretical description of multiple ionization in terms of Eq(1). A general expression for $\hat{\sigma}_K$ is

$$\hat{\sigma}_K \cong \left| \sum_{a_{K-1}} \ldots \sum_{a_1} \frac{\langle f|r|a_{K-1}\rangle \ldots \langle a_2|r|a_1\rangle \langle a_1|r|g\rangle}{(\Delta E)_{K-1} \ldots\ldots (\Delta E)_2 \; (\Delta E)_1} \right|^2 \tag{2}$$

where $\langle b|r|a\rangle$ are dipole matrix elements between atomic states while ΔE are energy differences, the first one being $(E_{a_1} - E_g - \hbar\omega)$ etc. The multiple summations are over complete sets of states and ω is the frequency of the incident radiation. It is the general structure of $\hat{\sigma}_K$ and not the details of its calculation that interest us in this paper. A quantity that will also prove useful is obtained by taking the Kth root of $\hat{\sigma}_K$, namely

$$\Lambda_K = (\hat{\sigma}_K)^{1/K} \tag{3}$$

If we take a particular atom and examine how $\hat{\sigma}_K$ and Λ_K decrease with increasing K, we find that Λ_K decreases much more slowly. Moreover the rate of its decrease becomes rather small after $K \cong 10$, as we have shown elsewhere[3]. The meaning of this behavior will become clearer below.

Let us examine now the scaling of $\hat{\sigma}_K$ with Z, first in a hydrogen-like atom. There is more than one way to scale. We can change Z assuming that $\hbar\omega$ changes appropriately so as to leave K exactly the same. Each matrix element scales as Z^{-1} while each E scales as Z^2. As a result

$$\hat{\sigma}_K \sim \frac{1}{Z^{6K-4}} \tag{4}$$

while

$$\Lambda_K \sim \frac{1}{Z^{6-\frac{4}{K}}} \xrightarrow[K>>1]{} \frac{1}{Z^6} \tag{5}$$

Given a certain order K, the generalized cross section $\hat{\sigma}_K$ and consequently Λ_K will exhibit resonances with intermediate states as ω varies from the lowest to the highest value that span the range of K-photon ionization. If the atom is not hydrogenic and if double-electron and more complex excitations are possible, we can not expect the above simple scaling. Another way of looking at scaling in an average sense is to note that $\hat{\sigma}_K$ has the dimensions $(\text{length})^{2K}(\text{time})^{K-1} = L^{2K}T^{K-1}$, irrespective of how it was derived and what its detailed expression may be. Identifying L with an atomic radius and T with $\hbar$ divided by the ionization potential, I, for a hydrogenic atom, we obtain

$$\hat{\sigma}_K \sim \frac{1}{Z^{4K-2}} \tag{6}$$

and

$$\Lambda_K \xrightarrow[K>>1]{} \frac{1}{Z^4} \tag{7}$$

Without elaborating this issue any further here, we will use as an average working rule Eqs(6) and (7) keeping in mind that for atoms other than hydrogen-like, some effective Z ($Z_{eff.}$) must be used.

It was mentioned earlier that Λ_K decreases more slowly as K becomes larger than 10 or so. The physical reason for this behavior can be understood as follows. With ω being the photon frequency and I the ionization potential, a rough way to understand 2-photon ionization is to write it as $(\sigma_0 F)\tau_1(\sigma_1 F)$ where σ_0 is some cross section for the absorption of one photon by the initial state, τ_1 the lifetime of the intermediate state and σ_1 some cross section for the absorption of one photon by the intermediate state. If there is a real intermediate state, then τ_1 is its lifetime. If the process is non-resonant, τ_1 can roughly speaking be taken as the inverse of the detuning from the nearest state. On the basis of the uncertainty principle, one can argue that the lifetime of a virtual state is the inverse detuning. Assume now that K is large so that $\hbar\omega << I$. Then the detuning is of order I and a K-photon process can be thought of as a sequence

$$(\sigma_0 F)\tau_1(\sigma_1 F)\tau_2(\sigma_2 F)\ldots\ldots\tau_{K-1}(\sigma_{K-1}F)$$

Qualitatively speaking one might argue that for large K all σ_i should be of the order of the atomic size and all τ_i of the order of I. Then the limiting value of $\sigma_i\tau_i$ should be Λ_K for large K. In fact that is what it turns out to be. For hydrogen, Λ_K tends to something like 10^{-31} $cm^2 sec$ which is roughly of order unity in units of Bohr radii squared and 1/I. The intensity at which $\hat{\sigma}_K F^K$ varies very slowly corresponds to one photon per atomic area per lifetime of the virtual state. Needless to add that this argument can not be carried to the extreme of $K \to \infty$ because as K becomes large the multiphoton picture breaks down and tunneling takes over.

Let us consider now a laser pulse which we take to have the form

$$F(t) = F_0 \frac{e^2 t^2}{4\tau^2} e^{-t/\tau} \tag{8}$$

The total energy $\int_0^\infty F(t)dt$ is F_0 3.69τ while the full width at half maximum is about 3.4τ. Thus the parameter τ will be taken to be about $\tau_L/3.5$ where τ_L is the experimental laser width. The peak intensity is F_0. As the pulse rises, ionization will occur through sequential as well as direct processes illustrated in Fig.1. No matter how large F_0 is, the atom sees a low intensity at first and one of the questions is whether the atom or the first ion

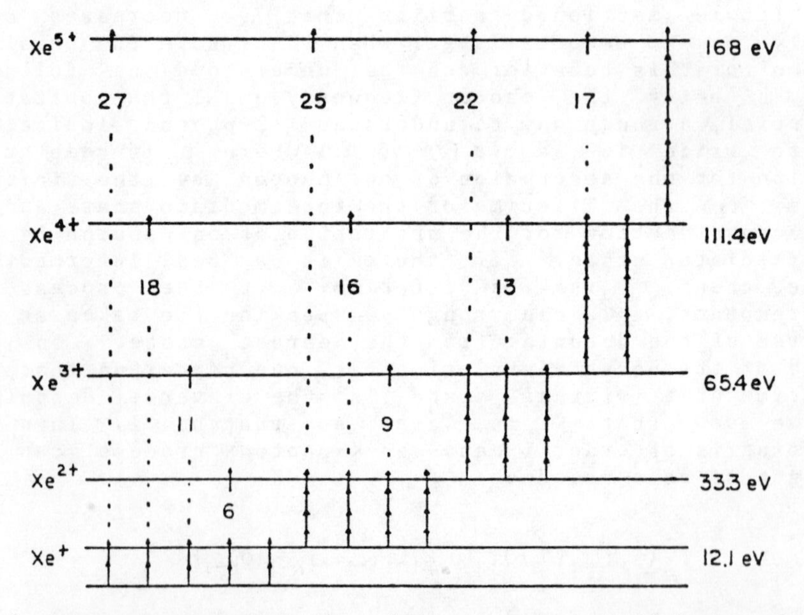

Fig.1. Schematic representation of direct and sequential channels for the creation of multiple ions of Xe up to Xe^{5+} with photons of energy 6.423 eV.

or the second ion etc. can survive long enough to be exposed to the peak intensity. A slightly different way of posing the same question is to inquire about the number of the various ionic species as a function of the peak power and the pulse duration.

Let N_0 be the initial number of neutral atoms in the interaction region, N(t) their number as a function of t during the pulse and $N_q(t)$ the number of the q^+ ionic species. If we define $n(t)=N(t)/N_0$ and $n_q(t)=N_q(t)/N_0$, then we have the number conservation equation

$$\sum_q n_q(t) = 1-n(t) \tag{9}$$

In principle, q runs from 1 to Z. For a given experiment, however, with specified pulse height, duration and photon frequency, q may be limited to a much smaller number. For example, in Xe up to 9^+ has been found in the experiments with UV frequencies and intensities up to 10^{16} W/cm^2.

The time-development of $n_q(t)$ is governed by a set of linear differential equations from which n(t) can be eliminated through Eq(9). These equations are based on the transition probability $\hat{\sigma}_K F^K(t)$. Thus, for example the rate of change of N(t) is

$$\frac{d}{dt} N(t) = -\sum_q \sigma_{K_q} F^{K_q}(t)\, N(t) \tag{10}$$

where K_q is the order of the multiphoton process in which q electrons are ejected directly without going through the sequence of ions of lower charge. For example K_1 is the order of the multiphoton process in which one electron is ejected from the neutral. Obviously K_q increases considerably as q increases. An example of the orders of ionization for Xe with q up to 5 is shown in Fig.1 with all the possible combinations of direct and sequential processes. We call direct any process in which q goes to q+2 or higher without going through q+1. In principle all of these processes contribute, although as we will see, the sequential are by far the dominant channels. Fig.1 does not show the important processes in which any of the ions may be left in an excited state thus making the next step much more probable. We return to this question later on. The functions $n_q(t)$ obey similar equations which are coupled. If the pulse were constant in time (which is not), the solutions could be expressed in closed form as linear combinations of exponentials. Otherwise, they must be solved numerically.

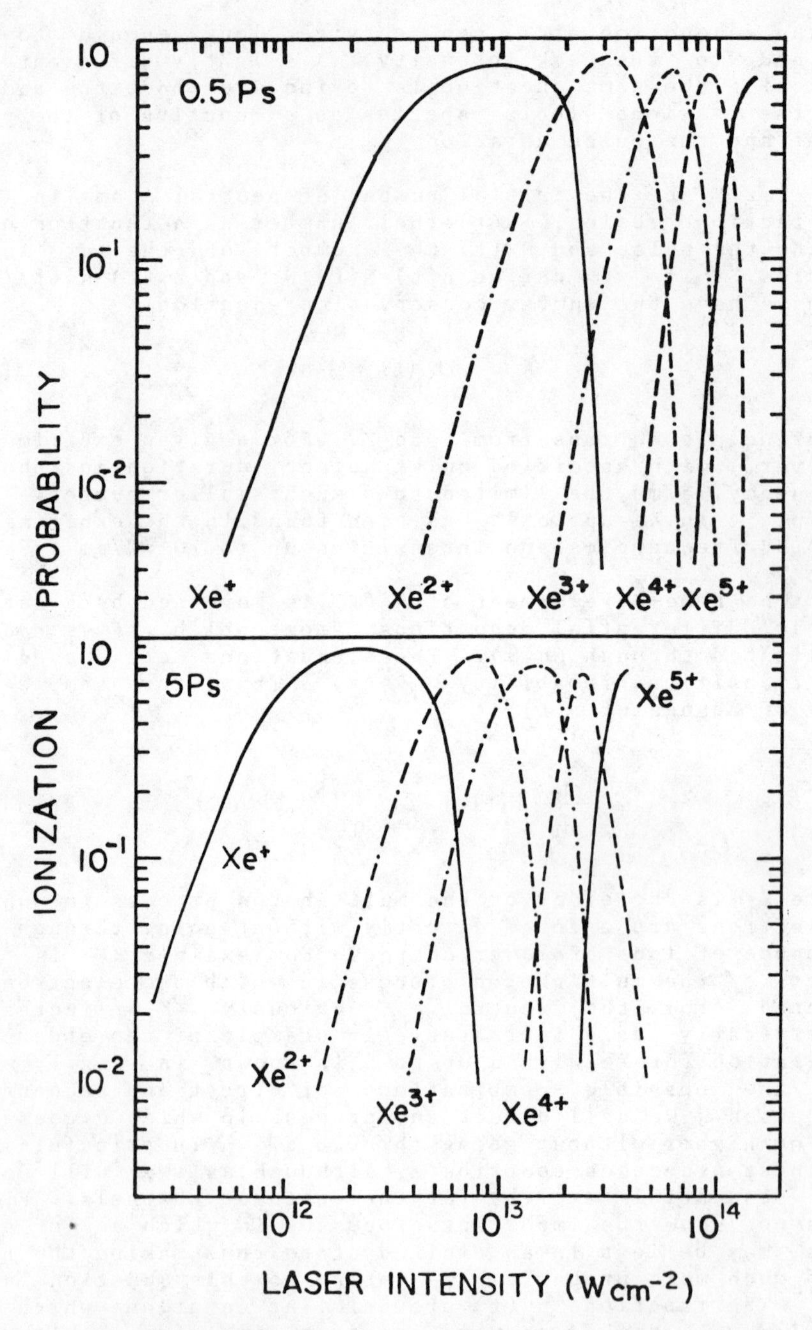

Fig. 2. Dependence of ionic species on laser peak intensity for two different pulse durations in the case of multiple ionization of Xe with photons of energy 6.423 eV.

We investigate now what these equations can tell us about multiple ionization. In Fig.2 we show the result of a sample calculation for, the production of up to Xe^{5+}. We have employed average values for $\hat{\sigma}_K$ appropriately scaled for the consecutive ionic species. The resulting curves of Fig.2 are typical and show that for pulses of peak heights in the range of 10^{13}-10^{14} W/cm^2 (depending on the atom, photon frequency and pulse duration) the atom is being stripped of its electrons one by one. Although the direct channels are included in the calculation, one finds that the dominant contribution is due to the sequential channels. We have explored the effect of pulse duration and shape. Some differences in detail exist. The main conclusion, however, remains the same. Stripping of the electrons is the dominant mechanism by which the ions are being created. It should be noted that as each electron is stripped away, the ion left behind is more tightly bound. Thus even though the intensity is rising, it does not necessarily exceed the perturbation theory limit. The order K keeps increasing and eventually tunneling would become the dominant mechanism. The stripping most probably stops when the value of the intensity (even at its peak) is too weak to cause singificant tunneling in an ion like Xe^{9+} .

Our calculations in this paper have ignored the spatial dependence of the laser intensity in the interaction region. This aspect is very important for a detailed interpretation of data, as is well known for more than 17 years now. The paper by L'Huillier in this volume is one more example of these effects. We have concentrated on the time aspect here because we wanted to investigate without any other complications what can be expected from a variation of the pulse duration. Moreover, the data with the UV source[4] have not included power dependences. In any case, we know what the space dependence will do. The populations of the various ionic species instead of bending downward, as they do in Fig.2, they will continue upward at a very low slope; owing to ionization from the lower intensity region of the interaction volume (see, for example, the paper by L'Huillier in this volume).

We should also note a further connection between our calculations and the Saclay work[5]. Beginning with the equations for $n_q(t)$ and employing values for cross sections derived from general considerations we obtain the evolution of $n_q(t)$. L'Huillier beginning with their data, employing similar equations and fitting to the data obtains certain values for $\hat{\sigma}_K$ which are in the general range we have obtained on theoretical grounds.

In summary, this part of our analysis has shown that although the peak intensity of a laser pulse may be extremely high, the neutral atom and several of its ionic species, never see the peak of the pulse. They have ionized long before that. By the time the peak is reached the ions still present have a much higher ionization potential. Thus even if the peak intensity is such that it corresponds to the non-perturbative regime of the atom it does not necessarily correspond to the non-perturbative regime of the ion present at the peak intensity. Recent experiments[4] are therefore readily understood in terms of this model without requiring any further hypotheses[6] about unusual forms of excitation. This, of course, does not mean that there are no loose ends in the understanding of the details; much remains to be done in that direction.

One of the rather fundamental questions related to multiple ionization is the role of double and possibly multiple electron excitations. Every time another ionization threshold is approached, a new manifold of multiple excitations must be crossed. We have addressed a part of that question by studying double ionization of carbon. Having chosen a photon energy ($\hbar\omega$=89710 cm^{-1}) that allows two-photon ionization of the atom, we have calculated double ionization through a succession of transitions between doubly excited states. Space does not permit elaboration on the details of this problem but two points should be made here. First, the sequential process is enormously enhanced by the fact that the ion can be left in excited states which make possible the ejection of a second electron by a lower order process. In general, scaling also favors sequential processes. Second, double-ionization multiphoton cross sections seem to be not too different from the corresponding single ionization multiphoton cross sections. The analysis of doubly excited states in a strong field reveals another very important feature that has wider implications for any sort of multiply excited state. If it takes more than one photon to reach a doubly excited state at a certain energy, then there is a number of channels of single and other excitations that lead to the same energy. The bigger the number of necessary photons, the larger the number of such channels. Moreover, the intensity necessary to excite such a state, causes significant transitions out of that same state by further photon absorptions. One must therefore think in terms of a state coupled to the field and study its behavior under those conditions before predicting the emission of any radiation from a purely atomic state of multiple excitation. In one of our calculations, for example, we have obtained transitions to the doubly excited state $3p^2$ ($^3P_0$) and from there to the 8p 13d($^3D_1$). One additional photon causes double ionization. We have found a rather large 4-photon double ionization cross section (10^{-110} $cm^8 sec^3$) which illustrates the fact that double ejection cross sections are not qualitatively different from those for single ejection. In this particular case, autoionization from the 8p 13d state tends to leave the ion in the excited 3d state from which it takes only one more photon to eject another electron. It is this type of calculation that requires a solution of the time-dependent problem in terms of the resolvent operator before an effective $\hat{\sigma}_K$ can be inferred. As we have shown in the context of Sr elsewhere[7], the branching ratio of excited to ground state ions depends quite strongly on the laser intensity. The main message of this work is that the richness of the phenomena appearing between thresholds at intensities above $10^{11} W/cm^2$ require painstaking analysis accounting for the relevant atomic structure. Details of this work will be published elsewhere.

REFERENCES

1. P. Lambropoulos, in Advances in Atomic and Molecular Physics (Academic, New York, 1976), Vol.12, p.87.

2. See, for example, M. Crance, in Multiphoton Ionization of Atoms, edited by S.L. Chin and P. Lambropoulos (Academic, New York, 1984), p.94.

3. P. Lambropoulos, Phys. Rev. Lett. 55, 2141(1985).

4. T.S. Luk, H. Pummer, K. Boyer, M. Shakidi, H. Egger, and C.K. Rhodes, Phys. Rev. Lett. 51, 110(1983).

5. A. L'Huillier, L.A. Lompre, G. Mainfray and C. Manus, Phys. Rev. Lett. 48, 1814(1982); also Phys. Rev. A27, 2503(1983).

6. K. Boyer and C.K. Rhodes, Phys. Rev. Lett. 54, 1490(1985).

7. X. Tang, Ph.D. thesis, University of Southern California 1985 (unpublished); and to be published.

MULTIPHOTON IONIZATION IN INTENSE ULTRAVIOLET LASER FIELDS

U. Johann, T. S. Luk, I. A. McIntyre, A. McPherson,
A. P. Schwarzenbach, K. Boyer, and C. K. Rhodes
Department of Physics, University of Illinois at Chicago
P. O. Box 4348, Chicago, Illinois 60680

ABSTRACT

The mechanism of collision-free multiphoton ionization of rare gases irradiated with ultraviolet radiation at an intensity of up to ~ 10^{16} W/cm^2 at 248 nm and a pulse length of ~ 0.5 ps and ~ 10^{15} W/cm^2 at 193 nm and ~ 5ps, respectively, has been studied by observing the electron energy spectra and ion charge state threshold intensities. The formation of multiply charged ions by a sequential process of ionization has been directly detected in the electron energy spectra in the form of a characteristic pattern of interwoven above threshold ionization (ATI) ladder line series. The threshold intensities for ion production have been compared with the Keldysh model and were found to be in good agreement for light ions (Ne) and consistently lower for heavier ions (Xe), scaling with the atomic number. These measurements, together with estimates of the ion population dynamics during the rise of the laser pulse can be reasonably understood on the basis of a single electron picture.

INTRODUCTION

A question of principal significance, especially in the context of coherent short wavelength generation, concerns the identification of the dominant laser-atom coupling mechanism under collision-free irradiation with an intense laser pulse.[1-6] Specifically, what are the important physical parameters governing the atomic response to the radiation field, and is the interaction adequately described by a single electron picture or one involving ordered many-electron motions?

The hypothesis has been advanced[7] that multiply excited states representing the coherent motion of outer-shell electrons may be an important agent for the population of inner-shell excited states and the subsequent reradiation of energy in the soft x-ray range. In order to study the conditions necessary for this desired atomic response to occur and the influence of possible loss mechanisms, ion charge state distributions and electron energy spectra produced by irradiation of rare gas atoms with ultraviolet radiation with different pulse lengths, wavelengths, and intensities have been investigated.

EXPERIMENTAL

For these measurements, two ultraviolet laser systems were available. One was a recently developed KrF* (248 nm) laser system[8] which produces pulses having a maximal energy of ~ 23 mJ with a pulse duration of ~ 0.5 ps. The second source used was an ArF* (193 nm) laser system[9] capable of producing a ~ 5 ps pulse with a maximal energy of ~ 40 mJ. The focussing systems used in these studies produced a maximum intensity of 10^{16} W/cm^2 at 248 nm and of 10^{15} W/wm^2 at 193 nm, respectively, in the experimental volume. The focussing lens was not corrected for spherical aberration and had, therefore, a focal diameter[5] of ~ 20 µm which limited the achievable laser intensity, but made the experiments relatively insensitive to the detailed spatial properties of the laser beam. The apparatus used for the measurement of the ion states[10] and the electron spectra have been described previously. Both are time-of-flight type spectrometers; for the electron measurements a magnetic mirror collimates the electrons while the ions are extracted by a static electric field.

Rare Gas Electron Energy Spectra With 193 nm Irradiation

Fig. 1 illustrates typical electron time-of-flight spectra produced at a peak laser intensity of ~ 5 x 10^{14} W/cm^2 at 193 nm for helium and at an intensity of ~ 2 x 10^{13} W/cm^2 for xenon. Also shown are the relevant energy level diagrams with the multiphoton transitions and the electron emission lines indicated. Similar recordings have been obtained for neon, argon, and krypton and their specific features have been discussed in Ref. 5.

In addition to the generally prominent lowest order ionization lines, corresponding in the two examples to the absorption of four photons by helium and two photons by xenon, processes which leave the singly charged ions in their respective ground states, several above threshold ionization (ATI) lines spaced by the photon energy (6.41 eV) are present. In xenon, the resolved line at ~ 5.7 eV indicates that the first ion is preferably produced in its lowest excited ns^2 np^5 $^2P^o_{1/2}$ state after absorption of three 193 nm photons.

The intensity of the observed ATI-lines drops more rapidly in a monotone manner with increasing order in the ultraviolet as compared to corresponding results obtained at visible and infrared wavelengths, in accordance with calculations performed by several authors.[11-14]

For the range of radiative intensities studied ($\leq 10^{15}$ W/cm^2), the distributions on the ATI-lines seem to be unaffected in all species examined by highly excited configurations of the neutral atoms above the ionization threshold or by excited ionic configurations, which represent possible competing decay channels, when energetically open. Furthermore, if the intensities of the ATI-lines are interpreted as a measure of the photon absorption rate for high order processes, the probability for sufficient energy transfer in order to allow direct double ioni-

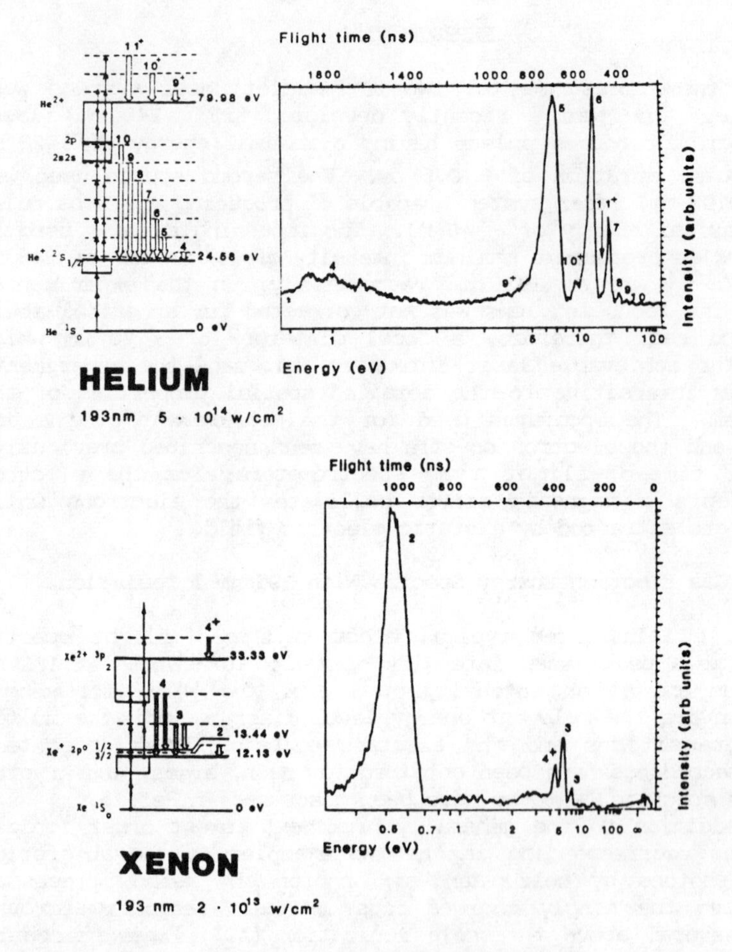

Fig. 1a,b: Electron time-of-flight spectra for He and Xe irradiated with a 193 nm, 5 ps laser pulse. The line identification is given in the schematic energy level diagrams showing the relevant energy level, multiphoton transitions, and electron emission lines. Excited neutral or ionic configurations are indicated by the boxes below the ionization limits.

zation or excitation is fairly low, and therefore, cannot account for the observed relative ion charge state abundance.[5,10] On the other hand, evidence for the sequential removal of outer-shell electrons is readily observed in these spectra in the form of additional sharp electron lines. For helium, the prominent ATI-ladder from the first ionization step to the He^{+} ground state after absorption of four or more photons, is accompanied by a much weaker ATI-ladder originating from the sequential removal of the second electron after absorption of 9, 10, or 11 photons. Similar

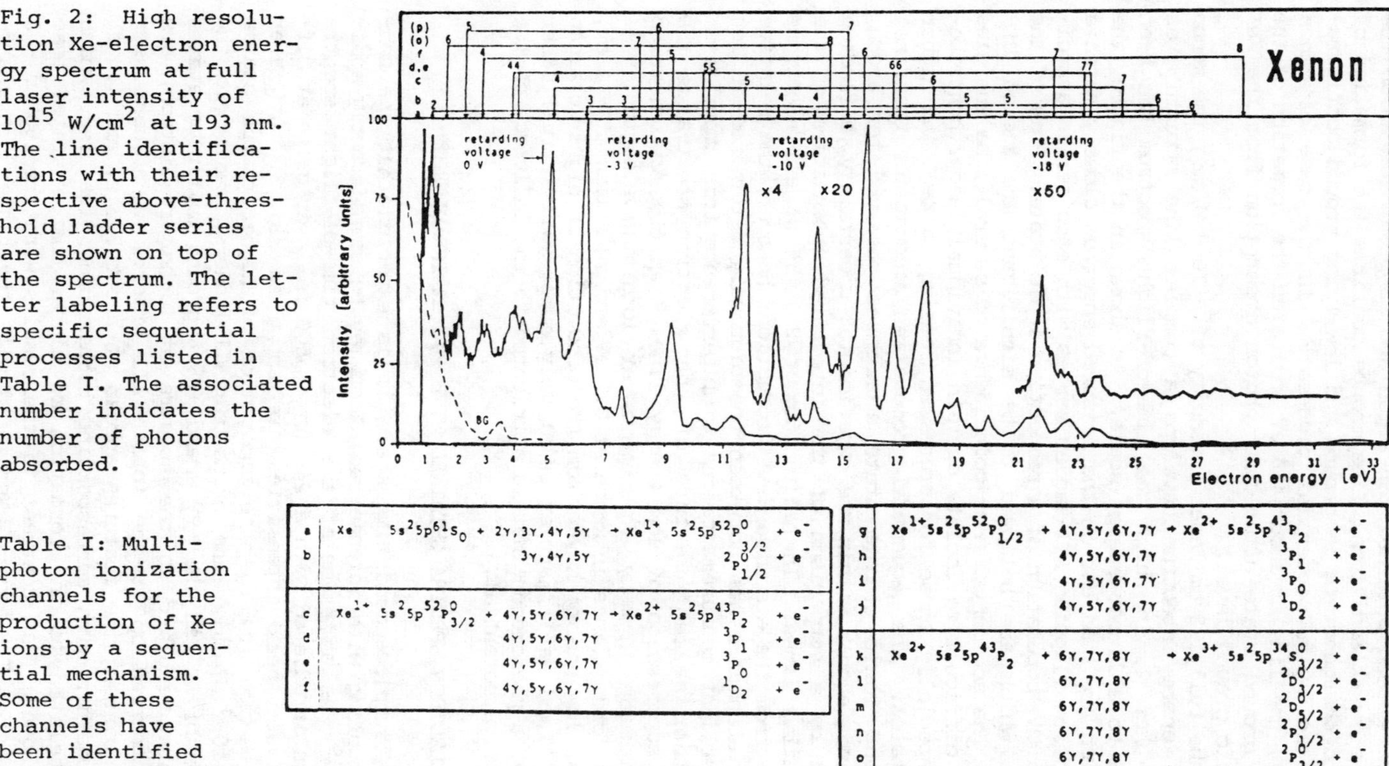

Fig. 2: High resolution Xe-electron energy spectrum at full laser intensity of 10^{15} W/cm^2 at 193 nm. The line identifications with their respective above-threshold ladder series are shown on top of the spectrum. The letter labeling refers to specific sequential processes listed in Table I. The associated number indicates the number of photons absorbed.

	Initial state	Photons		Final state	
a	$Xe\ 5s^2 5p^6\ ^1S_0$	+ 2γ,3γ,4γ,5γ	→	$Xe^{1+}\ 5s^2 5p^5\ ^2P^0_{3/2}$	+ e^-
b		3γ,4γ,5γ		$^2P^0_{1/2}$	+ e^-
c	$Xe^{1+}\ 5s^2 5p^5\ ^2P^0_{3/2}$	+ 4γ,5γ,6γ,7γ	→	$Xe^{2+}\ 5s^2 5p^4\ ^3P_2$	+ e^-
d		4γ,5γ,6γ,7γ		$^3P_1$	+ e^-
e		4γ,5γ,6γ,7γ		$^3P_0$	+ e^-
f		4γ,5γ,6γ,7γ		$^1D_2$	+ e^-
g	$Xe^{1+}\ 5s^2 5p^5\ ^2P^0_{1/2}$	+ 4γ,5γ,6γ,7γ	→	$Xe^{2+}\ 5s^2 5p^4\ ^3P_2$	+ e^-
h		4γ,5γ,6γ,7γ		$^3P_1$	+ e^-
i		4γ,5γ,6γ,7γ		$^3P_0$	+ e^-
j		4γ,5γ,6γ,7γ		$^1D_2$	+ e^-
k	$Xe^{2+}\ 5s^2 5p^4\ ^3P_2$	+ 6γ,7γ,8γ	→	$Xe^{3+}\ 5s^2 5p^3\ ^4S^0_{3/2}$	+ e^-
l		6γ,7γ,8γ		$^2D^0_{3/2}$	+ e^-
m		6γ,7γ,8γ		$^2D^0_{5/2}$	+ e^-
n		6γ,7γ,8γ		$^2P^0_{1/2}$	+ e^-
o		6γ,7γ,8γ		$^2P^0_{3/2}$	+ e^-
p	$Xe^{2+}\ 5s^2 5p^4\ ^1D_2$	+ 5γ,6γ,7γ	→	$Xe^{3+}\ 5s^2 5p^3\ ^4S^0_{3/2}$	+ e^-

Table I: Multiphoton ionization channels for the production of Xe ions by a sequential mechanism. Some of these channels have been identified with observed line series in Fig. 2.

results have been obtained for argon and krypton, for which the presence of weak additional line groups confirms the formation of the second charge state in ground and low-lying ground configuration states by a sequential process.[5] In the case of xenon (Fig. 1b), the line at about 4.5 eV indicates the formation of the Xe^{2+} $^3P_2$ ground state after four photon absorption from the Xe^+ $^2P^o_{3/2}$ ionic ground level.

At the full 193 nm laser intensity of ~ 10^{15} W/cm^2, the xenon electron energy spectrum develops a complex line structure as shown in Fig. 2, in which the time-of-flight spectrum has been converted to a linear energy scale. As shown in the figure, the lines which are manifest in the measured spectrum closely match a pattern of overlapping ATI-ladder series with each series associated with the population of a respective ionic state. Some of the processes which are believed to be significant are listed in Table I. The accidental near coincidence of several line energies makes it difficult to isolate certain individual ladder contributions, especially those for processes leading to Xe^{3+} and higher charge states. The sequential process can account for the first and second ionic charge state abundances observed for argon, krypton and xenon[10] within the experimental uncertainty, a finding indicated by a comparison of the integrated intensities of the corresponding lines. In helium, about 2 - 3% of all electrons originate from the stepwise He^{2+} production; however, the relative He^{2+} ion abundance has not been unambiguously determined in ion time-of- flight measurements due to interference from impurities.

Another salient feature of the electron spectra is the relative suppression of low energy lines,[5] an observation which has been made in many experiments at longer wavelengths.[15-17] The lowest order lines in xenon at 0.7 eV and in helium at ~ 1 eV, as well as low energy lines from the production of higher charged ions (Xe^{2+} and He^{2+}), appear relatively suppressed at full laser intensity, a phenomenon that is attributed to the influence of the ponderomotive potential.

ELECTRON SPECTRA OBTAINED WITH 248 nm IRRADIATION

On a subpicosecond time scale it is expected that the nonlinear coupling with intense radiation may be significantly modified.[18,19] As an example, the electron time-of-flight spectra from argon irradiated with a 248 nm, ~ 500 fs laser pulse at different laser peak intensities are shown in Fig. 3. At laser intensities too low for the production of higher charge states (e.g. ~ 10^{14} W/cm^2), the principal ATI-ladders leading to the production of Ar^+ in the $^2P^o_{1/2}$ and $^2P^o_{3/2}$ ground multiplet states (which are resolved in the spectra at higher resolution) are readily observed up to the fourth ATI order. At higher laser intensities, new line features appear at about 3.5 eV and 4.5 eV energy together with corresponding higher order features, which can be related to the formation of higher charge states (see below). However, in contrast to the results with the 193 nm ~ 5 ps pulse, the ATI-ladders associated with the higher ion charge states seem to be shifted in energy up to a value of about 1.5 eV, a finding which is confirmed in comparable spectra from helium, neon, and xenon at 248 nm. Again, similar to observations at 193 nm, the low energy and shifted five photon line for Ne at 2 eV and the three photon absorption line for krypton at 0.97 eV are suppressed relative to the next higher order lines. Another general feature of all electron spectra at 248 nm is the appearance of relatively high energy electrons with energies up to a maximum of about 200 eV for xenon and ~ 120 eV for argon. These were observed in the laser intensity range spanning from 10^{15} W/cm^2 to 10^{16} W/cm^2.

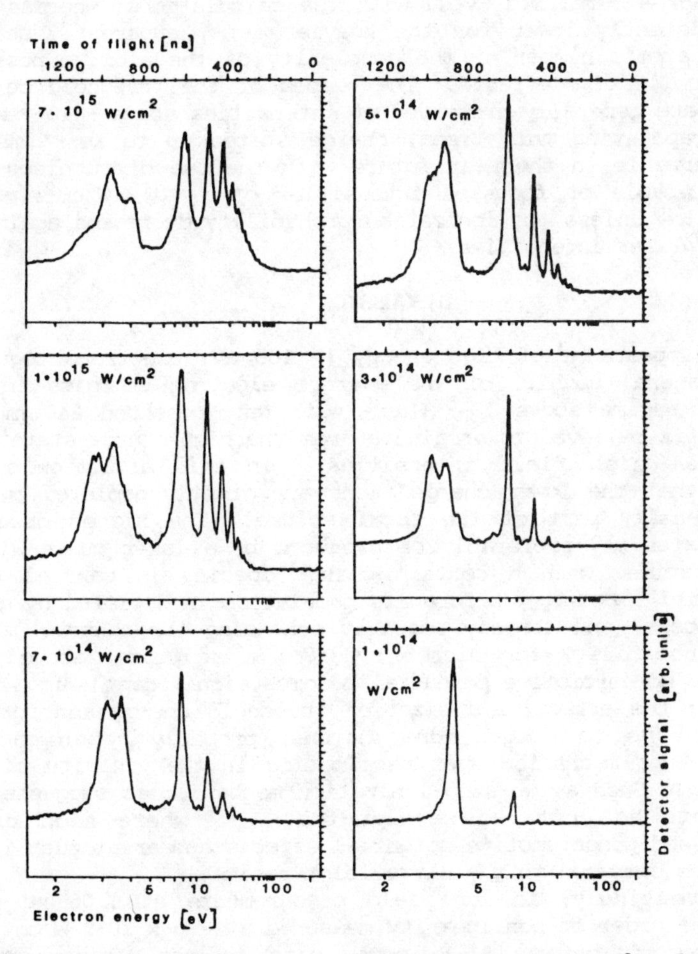

Fig. 3: Photoelectron time-of-flight spectra for Ar produced at different peak intensities in a 248 nm ~ 500 fs laser pulse.

IONIZATION THRESHOLD INTENSITIES AT 248 nm

The ion time-of-flight technique has been used to measure the threshold laser intensities for all charge states of the rare gases which appear below 10^{16} W/cm^2. The experimental values are plotted in Fig. 4 as functions of the ionization energies for each charge state. They are compared to calculated threshold laser intensities for a transition probability of about 10^{-3}, which corresponds to the detection limit of the apparatus, using Keldysh's formula[20] (including coulomb correction) in the tunneling approximation ($\gamma \ll 1$) and the Keldysh-Raizer formula[21] in multiphoton approximation ($\gamma \gg 1$). The observed thresholds for neon, which range from ~ 8 x 10^{13} W/cm^2 (Ne^+) to ~ 7 x 10^{15} W/cm^2 (Ne^{4+}), agree remarkably well with the calculations, whereas they are consistently lower for the heavier ions, apparently scaling with the atomic number or the complexity of the atom, a possible hint to collective effects. The shapes of the threshold curves, however, are generally preserved at intensities above ~ 10^{14} W/cm^2 and, extrapolating this trend, charge states up to Xe^{23+} may be well observable in the near future with the use of subpicosecond sources capable of focussed intensities of ~ 10^{18} W/cm^2. This would follow unless the ionization probability decreases again for ultrahigh laser intensities.[22]

DISCUSSION

The appearance of high energy (> 100 eV) electrons together with a general upshift of the average electron energies in the intensity regime above 10^{15} W/cm^2 with subpicosecond 248 nm irradiation is believed to originate from the high charge state production at high field intensities. In this situation it is expected that the lower charged ions are greatly depleted in the high intensity part of the focal volume. The higher observed charge states may preferably be produced by a laser pulse with a short risetime, when a certain charge species is carried to a higher field intensity, before its population is depleted by a low order process, and thereby possibly enhancing the coherent transition probability for higher order. In addition, above 10^{15} W/cm^2, the ponderomotive potential becomes significant (Fig. 4) by increasing the effective ionization threshold energy and forcing the electrons to absorb more photons, probably by an inverse Bremsstrahlung mechanism when oscillating in the vicinity of the parent ion. Such a two-step ionization mechanism is suggested by experiments at infrared wavelengths,[23-25] where high order processes and ponderomotive potential effects appear at much lower intensities compared to the ultraviolet regime.

Interestingly, in ion yield measurements at 1.06 µm with helium, the order of nonlinearity measured at ~ 5 x 10^{14} W/cm^2 for the He^+ production was 21, closely matching the minimum number (22) of photons needed to overcome the low field ionization threshold. However, the quiver barrier at this intensity should have shifted the effective ionization threshold to about 75 eV,

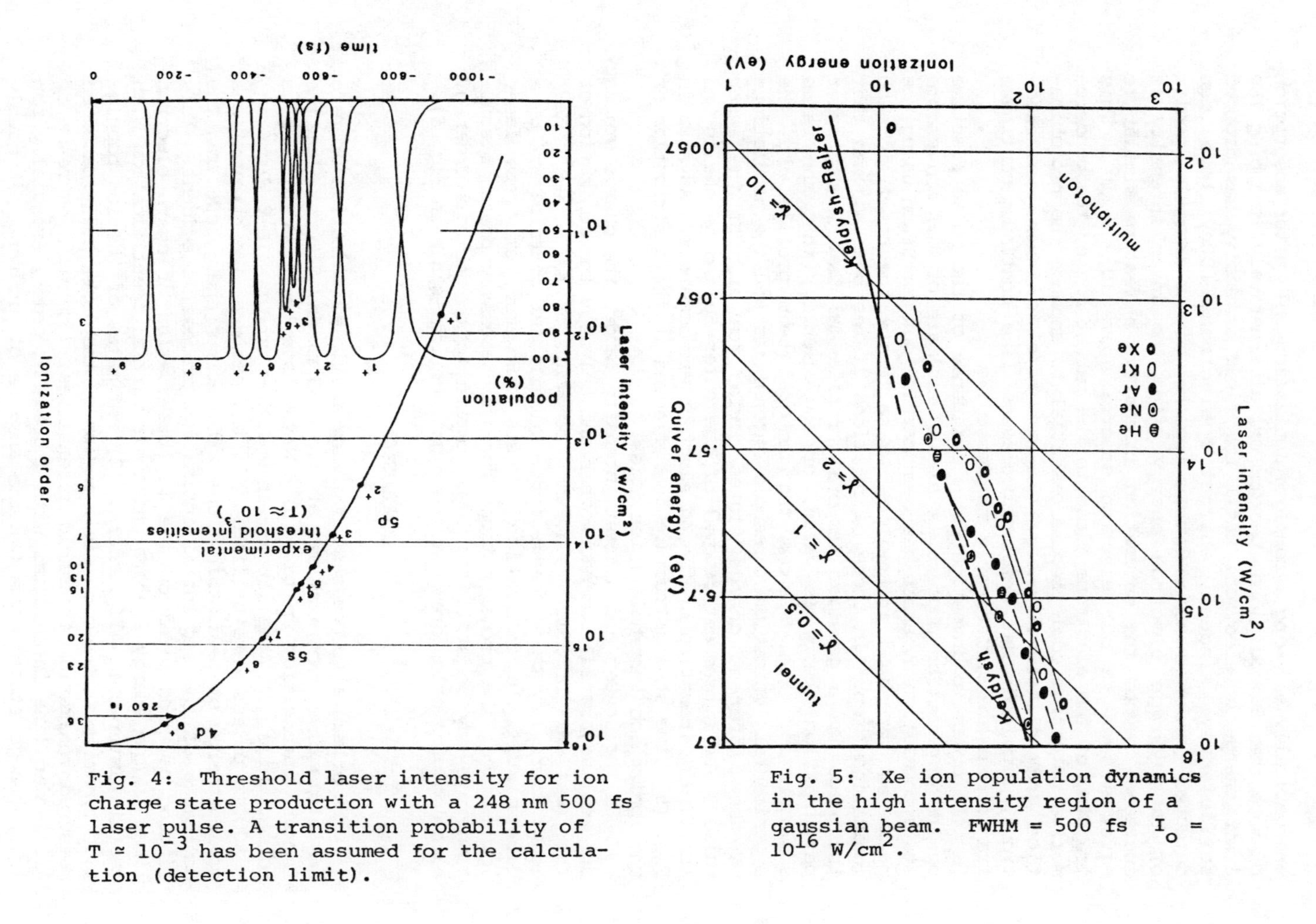

Fig. 4: Threshold laser intensity for ion charge state production with a 248 nm 500 fs laser pulse. A transition probability of $T \simeq 10^{-3}$ has been assumed for the calculation (detection limit).

Fig. 5: Xe ion population dynamics in the high intensity region of a gaussian beam. FWHM = 500 fs $I_o = 10^{16}$ W/cm^2.

energetically allowing a complete ionization only after absorption of more than 50 additional photons or, alternatively, trapping the lower energetic electrons until the field intensity has dropped sufficiently. Indeed, energetic electrons (80eV) have been detected[25] under these conditions together with a dramatic upshift of the average electron energy after the laser intensity is increased beyond the saturation intensity, a feature similar to effects seen in our electron spectra above 10^{15} W/cm^2. This trapping of the electrons close to the parent ion by the ponderomotive potential could become essential to enhance the recombination and collisional energy transfer rate to inner core electrons, provided that damping processes like continuum-continuum transitions are sufficiently suppressed.

In order to investigate the influence of the pulse risetime on the interaction process, a simple estimate of the ion charge state population dynamics has been carried out for xenon using the measured threshold intensities to calibrate the onset of each ionization step. In these calculations it has been assumed that the pulse has a gaussian temporal profile and that lowest order perturbation theory is valid, or equivalently, that the rates scale as I^N. Fig. 5 shows the temporal beam profile with the experimental ionization threshold intensities and the populations of the different charge states during the rise of the laser pulse. The subsequent removal of the outer 5p and 5s shells and the final liberation of the 4d electron are apparent. As seen in Fig. (5), the shell structure is reflected in the distribution of the ionization threshold intensities.

In the lowest and highest intensity range, the charge states are well separated from each other and only for the ionization of the last 5p-electrons (e.g. the formation of Xe^{4+}, Xe^{5+}, Xe^{6+} ions) there may be a chance for a significant contribution by a direct multiple ionization process. However, at these laser intensities, the cross-sections for the necessary photon absorption rate may be still too low (compare to the minimum order of each ion production drawn at the right vertical scale in Fig. 5).

CONCLUSION

Experimental processes involving multiple ionization of rare gas atoms have been examined at both 193 nm and 248 nm. For the former, intensities up to ~ 10^{15} W/cm^2 have been used in ~ 5 ps pulses while, for the latter, intensities up to ~ 10^{16} W/cm^2 in ~ 500 fs pulses were utilized. On the basis of the observations of ion charge state distributions and electron energy spectra, the dominance of the sequential ionization process is evident. This finding is predicated on the (1) observation of sharp electron lines in the energy spectra and their partial identification with certain interwoven ATI-ladders, (2) estimates of the ion population dynamics in the high field region of the laser pulse based on measured ionization threshold intensities, (3) rapid decrease of ATI-ladder line intensities with increasing order, especially for lower charge states, and (4) the absence of influence of doubly excited states or excited ionic configurations on the ATI-ladder line intensities. These results indicate, for the experimental regime studied, that the atomic behavior can be reasonably understood within the framework of a single electron picture. Finally, we note, however, that certain other important characteristics of the interaction, such as the energy transfer rate, have been seen to change sharply with the use of subpicosecond radiation as discussed in other work appearing in this volume.[18]

The intensity thresholds for ion production exhibit shell effects and are in good agreement with calculations based on the Keldysh model for neon. The heavier atoms, however, deviate in a consistent manner with increasing atomic number and exhibit ionization at intensities less than that represented by the Keldysh picture.

ACKNOWLEDGEMENTS

The authors wish to acknowledge the technical assistance of R. Slagle, J. Wright, T. Pack, and R. Bernico. This work was supported by the ONR, the AFOSR, the SDIO(ISTO), the DOE, the LLNL, the NSF, the DARPA, and the LANL.

REFERENCES

1. C. K. Rhodes in Multiphoton Processes, P. Lambropoulos and S. J. Smith editors (Springer-Verlag, Berlin, 1984) p. 31; L. A. Lompré and G. Mainfray, ibid., p. 23; D. Feldman, H. -J. Krautwald, and K. H. Welge, ibid., p. 223; W. B. Delone, V. V. Swan, and B. A. Zon, ibid., p. 235.
2. P. Lambropoulos, Phys. Rev. Lett. 55, 2141 (1985).
3. P. Agostini and G. Petite, J. Phys. B 18, L287 (1985).
4. P. Agostini and G. Petite, Phys. Rev. A 32, 3800 (1985).
5. U. Johann, T. S. Luk, H. Egger, and C. K. Rhodes, "Rare Gas Electron Energy Spectra Produced By Collision-Free Multiquantum Processes," Phys. Rev. A (in press).
6. G. Wendin, L. Jönsson, and A. L'Huillier, Phys. Rev. Lett. 56, 1241 (1986).
7. A. Szöke and C. K. Rhodes, Phys. Rev. Lett. 56, 720 (1986).
8. A. P. Schwarzenbach, T. S. Luk, I. A. McIntyre, U. Johann, A. McPherson, K. Boyer, and C. K. Rhodes, "Sub-picosecond KrF* Excimer Laser Source," submitted to Opt. Lett. (1986).
9. H. Egger, T. S. Luk, K. Boyer, D. R. Muller, H. Pummer, T. Srinivasan, and C. K. Rhodes, Appl. Phys. Lett. 41, 1032 (1982).
10. T. S. Luk, U. Johann, H. Egger, H. Pummer, and C. K. Rhodes, Phys. Rev. A32, 214 (1985).
11. Y. Gontier and M. Trahin, J. Phys. B13, 4383 (1980).
12. M. Crance and M. Aymar, J. Phys. B13, L421 (1980).
13. S. I. Chu and J. Cooper, Phys. Rev. A32, 2769 (1985).
14. W. B. Delone et. al., J. Phys. B16, 2369 (1983).

References, (cont.)

15. P. Kruit, J. Kimman, H. G. Müller, and M. J. van der Wiel, Phys. Rev. A28, 248 (1983).
16. L. A. Lompré, A. L'Huillier, G. Mainfray, and C. Manus, J. Opt. Soc. Am. B2, 1906 (1985).
17. T. J. McIlrath, M. Barhkansky, P. Bucksbaum, and R. R. Freeman, "Suppression of Multiphoton Ionization with Circularly Polarized Light," AIP Conference Proceedings, this volume.
18. U. Johann, T. S. Luk, I. A. McIntyre, A. McPherson, A. P. Schwarzenbach, K. Boyer, and C. K. Rhodes, "Multiquantum Processes at High Field Strengths," AIP Conference Proceedings, this volume.
19. A. L'Huillier, L. A. Lompré, G. Mainfray, and C. Manus, J. Physique 44, 1247 (1983).
20. L. V. Keldysh, Sov. Phys. -JETP 20, 4 (1965).
21. Yu. P. Raizer, Sov. Phys. -USP. 8, 650 (1966).
22. G. J. Pert, J. Phys. B8, L173 (1975).
23. K. G. H. Baldwin and B. W. Boreham, J. Appl. Phys. 52, 2627 (1981).
24. P. Kruit, H. G. Müller, J. Kimman, and M. J. van der Wiel, J. Phys. B16, 2359 (1983).
25. A. L'Huillier, L. A. Lompré, G. Mainfray, and C. Manus, J. Phys. B16, 1363 (1983).

LASER SPECTROSCOPY OF THE 109.1 nm TRANSITION IN NEUTRAL Cs

D. P. Dimiduk, K. D. Pedrotti,* J. F. Young, and S. E. Harris
Edward L. Ginzton Laboratory, Stanford University
Stanford, CA 94305

ABSTRACT

Absorption of tunable coherent VUV radiation is used to measure fine-structure splitting and oscillator strength, and to estimate hyperfine splitting of the Cs 109.1 nm transition, which originates on a quasi-metastable level.

INTRODUCTION

Certain core-excited quartet levels in alkali-like atoms and ions, termed quasi-metastable, have slow autoionizing rates and comparable (relatively fast for quartets) VUV radiative rates.[1] This circumstance, desirable for laser transitions, is due to angular momentum and parity selection rules on these quartets and the doublets to which they may couple.

The 109.1 nm transition in Cs (Fig. 1) is between Cs $(5p^5 5d6s)^4P^o_{5/2}$ and Cs $(5p^6 5d)^2D$, and is a prototype of a class of transitions which originate on quasi-metastable levels. It has been observed in emission from a pulsed hollow-cathode discharge.[2] We here report an experiment using the same discharge to populate the lower level of this transition; laser-generated tunable VUV radiation probes the discharge to make absorption measurements at near Doppler-limited resolution.

EXPERIMENTAL

The tunable VUV radiation for the absorption experiment was generated using two-photon resonant four-wave sum-frequency mixing (4WSFM) using a process similar to that of Jamroz, et al.[3] We chose the particular scheme shown in Fig. 2, which permits generation using only 532 nm pumped dye lasers.

A single Q-switched, frequency-doubled Nd:YAG laser was used to pump two dye laser systems. The first was operated with an intracavity etalon to narrow the linewidth, tuned to 542.2 nm, then frequency-doubled to make the two-photon resonant wave. The second dye laser was tuned around a small region near 558 nm, thus tuning the generated VUV. The beams were combined and focused into the Zn cell using a f = 50 cm lens. The overall experimental diagram is shown in Fig. 3.

---

*Now with Rockwell International, 1049 Camino Dos Rios, Thousand Oaks, CA 91360.

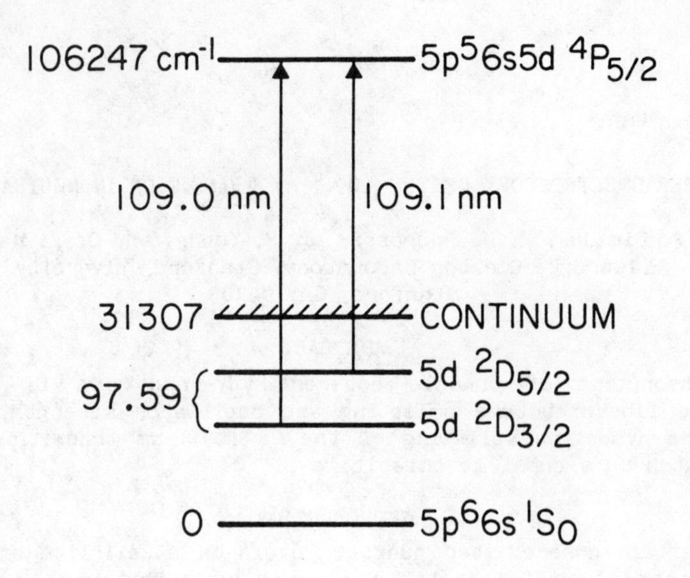

Fig. 1. Selected Cs I energy levels showing the transitions observed.

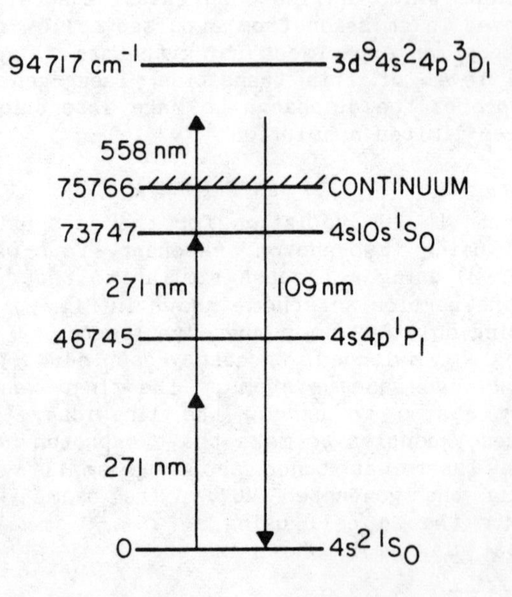

Fig. 2. Energy levels for resonant VUV generation in Zn vapor.

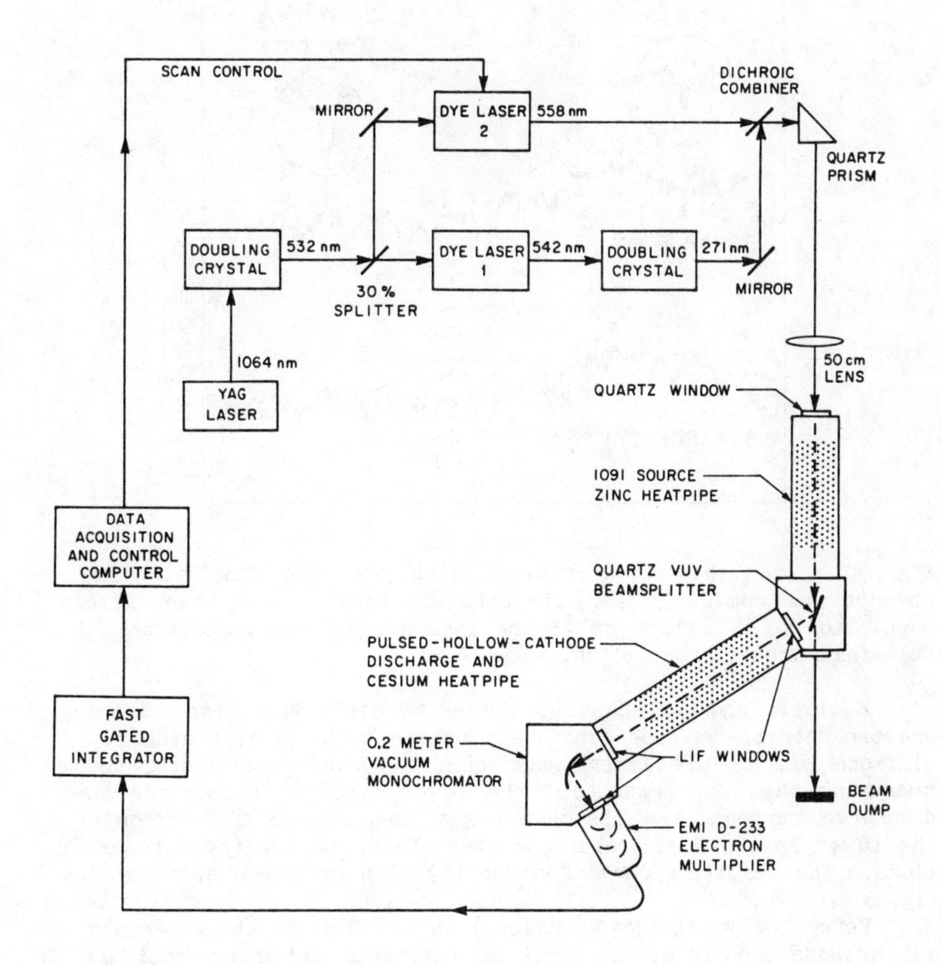

Fig. 3. Schematic of experimental apparatus.

The scanned VUV was absorbed by the excited Cs $(5p^6 5d)^2D$ atoms in the discharge. An absorption trace showing both components of the transition is shown in Fig. 4. An absolute calibration of the 109.1 nm wavelength was accomplished by comparing the wavelengths of the dye lasers to emission lines of Kr and Hg. The result for the stronger $^2D_{5/2}$ absorption is 109.111 ± 0.0035 nm.

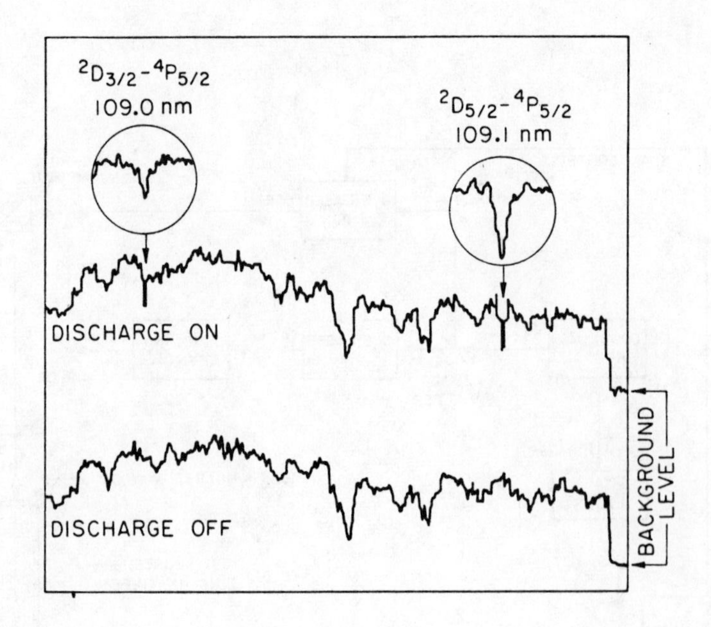

Fig. 4. VUV absorption scan showing the two fine structure absorptions from Cs $(5d)^2D$. The data was taken with a lower level population of 3×10^{13} cm^{-3}; the insets show the absorptions at full instrumental resolution.

Multiple shorter scans were made to study each absorption in greater detail. We also separately measured the excited atom populations via visible laser absorption on 5d-nf transitions, thus measuring the NL product of the lower levels. By varying the discharge current, the absorption was measured as a function of the lower level NL , yielding curves-of-growth for these transitions. The resulting curve for the 109.1 nm component is shown in Fig. 5.

Reference 4 provides additional information on the experimental methods used in making these measurements and on the analysis.

RESULTS AND DISCUSSION

Careful study of both the visible and the VUV absorption data yielded the oscillator strengths of both components of this transition. The f values are determined by the frequency integrated absorption at small NL , which is insensitive to linewidth or hyperfine-splitting effects. At intermediate and large NL , the curves-of-growth deviated from that expected for a single Doppler + Lorentz broadened line. We believe the additional absorption is due to hyperfine splitting of the upper level. Using this assumption, additional computer modeling of the absorption profile yielded an experimental estimate of the hyperfine splitting constant for the upper state. The results are shown in Fig. 5 and Table 1.

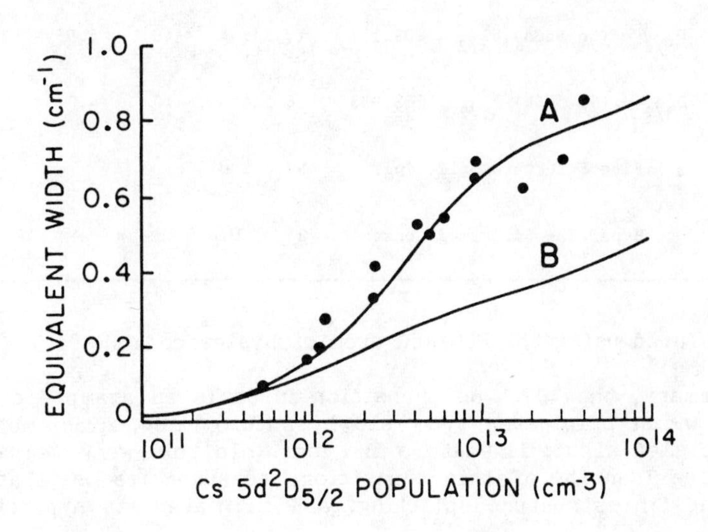

Fig. 5. Equivalent width vs. lower level population for the 109.1 nm absorption, assuming a 30 cm column length. The points are experimental data. The solid lines are calculated values from computer-modeled curves-of-growth assuming Doppler, Stark and collision broadening of 0.14, 1.3×10^{-3}, and 1.5×10^{-3} cm^{-1}, respectively. Curve A also includes hyperfine splitting with a constant of $A = 0.025$ cm^{-1}.

The measured oscillator strengths agreed well with our calculations using the code RCN/RCG.[5] We also approximately calculated the hyperfine splitting constant for the transition. The nuclear interaction was determined using the known value of the Cs ground state splitting constant, and RCN calculated the value of $|\psi_{6s}(0)|^2$ for both the ground state and the quasi-metastable configurations. The resulting theoretical estimate,[6] $A = 0.021$ cm^{-1}, agreed well with our experimental estimate of $A = 0.025$ cm^{-1}.

The hyperfine splitting reduces the gain cross section for lasers operating on the 109.1 nm transition. If hyperfine structure is neglected, the calculated cross section is 4.0×10^{-14}, based on our measured oscillator strength and assuming a Doppler width of 0.15 cm^{-1}. If, however, we assume that the upper level's population is distributed among its hyperfine components according to their degeneracies and sum the contributions of the various Doppler-broadened components, the peak value of the effective cross section is reduced to 1.2×10^{-14} cm^2.

Table 1. Summary of results from absorption measurements on the 109 nm transitions.

Transition	λ [nm]	$f_{exp.}$	$f_{calc.}$ [(a)]
$Cs(5p^6 5d)^2D_{5/2}$ - $(5p^5 5d6s)^4P^o_{5/2}$	109.111	$(7.2 \pm 4) \times 10^{-3}$	6.95×10^{-3}
$Cs(5p^6 5d)^2D_{3/2}$ - $(5p^5 5d6s)^4P^o_{5/2}$	108.998	$(6.5 \pm 2) \times 10^{-4}$	4.8×10^{-4}
Fine Structure Splitting:		97.3 ± 0.3 cm^{-1}	
Hyperfine Splitting Constant:		a = 0.025 cm^{-1}	

[(a)]Calculated using the RCN/RCG atomic physics code (Ref. 5).

In summary, the 109.1 nm transition in Cs is an example of a transition which originates from slowly autoionizing, quasi-metastable levels of alkali-like atoms and ions. In this work we have confirmed the identity of this transition, measured its oscillator strength and fine-structure splitting, and estimated its hyperfine splitting.

ACKNOWLEDGEMENTS

The work described here was supported by the Army Research Office and the Air Force Office of Scientific Research. Captain Dimiduk, USAF, acknowledges support of the Air Force Institute of Technology.

REFERENCES

1. S. E. Harris, D. J. Walker, R. G. Caro, A. J. Mendelsohn, and R. D. Cowan, Opt. Lett. 9, 168 (1984).
2. D. E. Holmgren, D. J. Walker, and S. E. Harris, in Laser Techniques in the Extreme Ultraviolet, S. E. Harris and T. B. Lucatorto, eds. (AIP, New York, 1984), p. 496.
3. W. Jamroz, P. E. LaRocque, and B. P. Stoicheff, Opt. Lett. 7, 617 (1982).
4. K. D. Pedrotti, D. P. Dimiduk, J. F. Young, and S. E. Harris, "Identification and Oscillator Strength Measurement of the 1091. nm Transition in Neutral Cs," Opt. Lett. (June 1986) (to be published).
5. R. D. Cowan, Mail Stop B212, Los Alamos National Laboratory, Los Alamos, New Mexico 87545.
6. I. I. Sobel'man, An Introduction to the Theory of Atomic Spectra (Oxford, Pergamon Press, 1972), pp. 205-216.

OPTICAL CHARACTERISTICS OF SUPER-RADIANT HARMONIC RADIATION FROM A FEL AND OPTICAL KLYSTRON - 3D MODEL

A. Gover, A. Luccio, A. Friedman, A.M. Fauchet
Brookhaven National Laboratory, Upton, NY 11973

ABSTRACT

In free-electron sources of coherent light, bunching of the electrons allows the super-radiant production of coherent harmonic radiation, up to substantially high harmonic numbers. This paper presents a fully three dimensional model formulation for evaluating coherent harmonic radiation from a free-electron laser or optical klystron. The radiated electromagnetic field is expanded in terms of a set of orthonogonal free-space modes. Expansion coefficients can be evaluated once a generalized form of the FEL pendulum equation has been solved numerically. Preliminary numerical computation of the electron bunching are given. The super-radiant radiation is characterized in terms of spectral brightness parameters and the enhancement in this factor for super-radiant emission as compared to conventional harmonic wiggler radiation is estimated.

INTRODUCTION

Frequency multiplication from a free-electron laser (FEL), in an oscillator configuration or using an external laser, might allow to obtain coherent radiation in a region of the spectrum (λ < 1000 Å) where the construction of an FEL oscillator seems beyond the capability of presently operating accelerators and storage rings. This paper presents a 3D model formulation of super-radiant harmonic radiation from a FEL that lends itself well to numerical computation and parameters evaluation. We also present some preliminary results from the computer model calculation of the electron modulation (bunching) process using a new FEL pendulum equation model with slowly varying parameters. The formalism can easily be extended to other types of free-electron devices.

The basic system is described in Fig. 1. A strong coherent electromagnetic wave (laser beam) at frequency ω_m modulates the <u>energy</u> of a relativistic electron beam inside a wiggler (or an undulator). After drift in a uniform (Fig. 1b) or a dispersive section (Fig. 1a), the electron beam <u>density</u> is modulated at all harmonic frequencies $\omega_n = n\omega_m$, n integer, of the e.m. wave. If allowed to radiate, such a charge density produces coherent radiation at frequencies $\omega_n = n\omega_m$.

The physical mechanism for super-radiant harmonic generation in an FEL is similar to the mechanism responsible for frequency multiplication in a microwave klystron. The UV free-electron frequency multipliers, commonly considered, use either an FEL (single wiggler) or

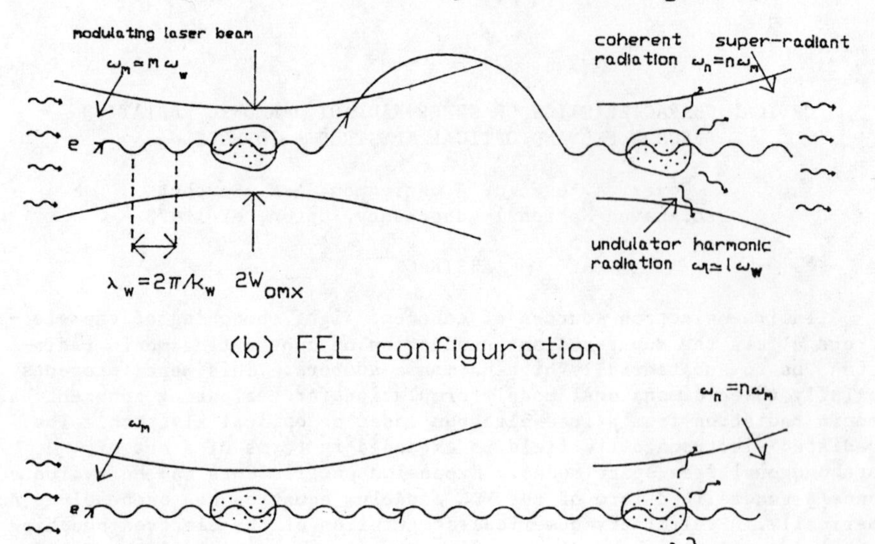

Fig. 1. Schemes for super-radiant (bunched beam) coherent harmonic generator.

a "transverse optical klystron" (TOK) configuration (two wigglers separated by a magnetic dispersive drift space) to couple the electron beam to the electromagnetic wave.[1,2] These two configurations are shown in Figs. 1b and 1a respectively. The modulating electromagnetic field can either be an external laser or the intra-cavity field of an FEL or TOK oscillator.

Part I of this paper presents the main steps of the calculation of the radiated harmonic electromagnetic field. As is common in FEL analysis the single electron equations of motion are solved to yield the evolution of the energy $\gamma_j(t)$ and position $\vec{r}_j(t)$ of electron j under the combined action of the two external fields: the undulator magnetic field and the modulating electromagnetic field. The radiated field is then expanded in terms of a complete set of orthonogonal free-space modes. The expansion coefficients of the modes are given simply and explicitely in terms of the single electron trajectories $\vec{r}_j(t)$. They can therefore be readily computed numerically, or analytically in some simple cases. In Part II we evaluate the optical parameters which characterize the radiation and particularly the spectral brightness (brilliance) of the source.

PART I: MODELLING THE ELECTRON BEAM MODULATION AND RADIATION

The following calculation is restricted to a single linear, and uniform undulator but the equations can easily be modified to apply to a more general magnetic structure. The magnetic field is taken to be:

$$\vec{B} = B_o \sin k_w z \, \hat{e}_y , \qquad 0 \leq z \leq L \tag{1}$$

where $k_w = 2\pi/\lambda_w$, and L is the undulator length. The modulating electromagnetic wave is assumed to be a gaussian mode, linearly polarized along x, with a cylindrical focus:[3]

$$\vec{E}_m(x,y,z,t) = \hat{e}_x E_o \frac{w_{xm}(0) w_{ym}(0)}{w_{xm}(z) w_{ym}(z)}$$
$$\mathrm{Re}\left(\exp - \left[\frac{x^2}{w_{xm}^2(z)} - \frac{y^2}{w_{ym}^2(z)} + i\Phi(x,y,z) + i\frac{\omega_m}{c} z - i\omega_m t\right]\right) \tag{2}$$

$$\Phi(x,y,z,\omega_m) = \frac{1}{4}\frac{\omega_m}{c}\left(\frac{x^2}{R_{xm}(z)} + \frac{y^2}{R_{ym}(z)}\right)$$
$$- \frac{1}{2}\left(\arctan\left(\frac{z-L/2}{z_{Rxm}}\right) + \arctan\left(\frac{z-L/2}{z_{Rym}}\right)\right)$$

$$\left(z_R(\omega_m)\right)_{x,y} = \pi \frac{\left(w_{om}\right)^2_{x,y}}{\lambda_m}$$

In Equation (2), $w_{xm}(z)$ and $w_{ym}(z)$ are the spot sizes, z_{Rxm} and z_{Rym} the Rayleigh lengths and $R_{xm}(z)$, $R_{ym}(z)$ the radii of curvature of the mode with the usual definitions for those quantities.
The trajectory and velocity of electron j in the magnetic field alone can be written

$$x_j(t) = x_{jo} + c\beta_{oj}\phi_{xjo}(t-t_{jo}) - \frac{K}{\gamma k_w} \sin(k_w z_j(t))$$

$$y_j(t) = y_{jo} + c\beta_{oj}\phi_{yjo}(t-t_{jo}) \tag{3}$$

$$\vec{\beta}_{j\perp}(t) = \hat{e}_x \frac{K}{\gamma} \cos(k_w z_j(t)) + c\beta_{oj}\phi_{xjo}\hat{e}_x + c\beta_{oj}\phi_{yjo}\hat{e}_y$$

where t_{jo} is the entrance time of electron j into the wiggler; x_{jo}, y_{jo} the initial transverse coordinates, and ϕ_{xjo}, ϕ_{yjo} the initial off axis entrance angles. The average electron trajectories of (3) are straight lines as a result of neglecting betatron oscillation along the wiggler, assuming it is short enough. Inclusion of betatron oscillation in the trajectory expressions is straightforward.[4]

The modulating electromagnetic field modifies $\underline{r}_j(t)$ and $\underline{\beta}_j(t)$. For modelling the modulation process, in the most general case one has to solve the Lorentz force equation under the combined action of the magnetic and electromagnetic fields.

To describe the radiation emission process we expand the radiated electromagnetic field $\vec{E}(\vec{r},\omega)$ in terms of a complete set of power orthonormalized free space radiation modes

$$\vec{E}(\vec{r},\omega) = \sum_q c_q(\omega)\, \vec{E}_q(\vec{r}) \tag{4}$$

It can be shown[5] that:

$$c_q(\omega) = -\frac{1}{4P_q}\, \Delta W_q(\omega) \tag{5}$$

where P_q is the normalization power of mode q and

$$\Delta W_q = \sum_{j=1}^{N_T} -e \int_{t_{oj}}^{t_j(L)} \vec{E}_q(\vec{r}_j(t))\, e^{i\omega t} \cdot \vec{v}_j(t)\, dt \tag{6}$$

can be interpreted as the complex sign-reversed work performed by mode q on the N_T electrons in the e-beam pulse.

The choice of the orthogonal modes set is a matter of ecconomizing the computation process. A successful choice of a set is one which permits to include only a small number of modes in the infinite sum (4) and still decribe adequately the radiation field. For an electron beam with a Gaussian density profile an Hermite-Gaussian orthongonal set[3] is expected to be a good choice.

$$\vec{E}_q(r,t) = \hat{e}_x\, \vec{\mathscr{E}}_q(r)\, e^{-i\omega t + i\frac{\omega}{c}z + i\phi_q(\vec{r})} \tag{7}$$

$$\mathscr{E}_q(r) = E_o \frac{w_x(o)w_y(o)}{w_x(z)w_y(z)}\, H_{q_x}\left(\sqrt{2}\,\frac{x}{w_x(z)}\right) H_{q_y}\left(\sqrt{2}\,\frac{y}{w_y(z)}\right) e^{-\frac{x^2}{w_x^2}-\frac{y^2}{w_y^2}} \tag{8}$$

For spontaneous emission and super-radiant emission computation it is enough to calculate (6) to first order in the radiation field. The transverse coordinates $\vec{v}(t)$ can be taken to be the same as in the static wiggler (Eq. 3). However the longitudinal modulation parameters $\gamma_j(t)$, $p_{zj}(t)$, $z_j(t)$ must be calculated in a nonlinear model which includes the combined effects of the magnetostatic wiggler and the modulating laser. Writing the force equation in terms of the relativistic proper time $d\tau = dt/\gamma_j(t)$ the force equations reduce into pendulum equations with slowly varying amplitude and phase (due to the diffraction of the modulating laser).[5] Assuming the modulating laser frequency is nearly an m multiple of the fundamental wiggler undulator frequency - $\omega_m \simeq m\omega_w = 2m\gamma^2 ck_w/(1+K^2/2)$, the modulation of the beam at frequency ω_m takes place via the m harmonic of the wiggler. The modulation equations are then:

$$\frac{d\gamma_j}{d\tau} = \frac{1}{2} (-)^{\frac{n-1}{2}} \frac{e\mathscr{E}_m(x_j(t),y_j(t))}{mc} K \left[J_{\frac{n-1}{2}}(u) - J_{\frac{n+1}{2}}(u) \right]$$

$$\sin \left[\psi_{mj} + \Phi_{mj}(x_j y_j z_j \omega_m) \right]$$

$$\frac{d\psi_{mj}}{d\tau} = \frac{k_m + mk_w}{m} p_{zj} - \omega_m \gamma_j \tag{9}$$

$$\left[J_{\frac{n-1}{2}}(u) - J_{\frac{n+1}{2}}(u) \right] \omega_m p_{zj} - \left[(k_m+k_w)\, J_{\frac{n-1}{2}}(u) - (k_m-k_w)\, J_{\frac{n+1}{2}}(u) \right]$$

$$mc^2 \gamma_j = \text{const}_j$$

where

$$u \equiv \frac{\omega_m}{8k_w} \frac{K^2}{c\beta_z^2 \gamma_j} \simeq \frac{m}{4} \frac{K^2}{1+K^2/2}$$

$x_j(t)$ and $y_j(t)$ are obtained from Equations (3) and the m harmonic phase is: $\psi_{mj} = (k_m+mk_w)\, z_j(t) - \omega_m t$. The third equation in (9) defines an harmonic interaction constant of the motion analogous to the one identified by Ginzburg[6] for the limt $K << 1$. The initial conditions for the set of equations (9) are:

$$\psi_j(\tau = 0) = \psi_{jo} = - \omega t_{jo}$$

$$z_j(\tau = 0) = 0$$

$$\gamma_j(\tau = 0) = \gamma_{jo}$$

At this stage of the 3-D model code development only the modulation equations (9) were reduced into a numerical computation program. Figures 2-4 display some results of the numerical computations. Figures 2 and 3 depict phase and energy bunching effects of 10 sample electrons per optical period in the fundamental harmonic frequency ($m = 1$). All the electrons, starting at equi-spaced initial phases ψ_{jo}, propagate along the axis of the modulating Gaussian mode (2). Figure 4 depicts the phase space trajectory of an electron trapped in the ponderomotive field of the modulating laser operating in the third harmonic ($m = 3$). A distinct 3-D effect is noted in Fig. 4: due to the diffraction related phase delay of the Gaussian wave (2) at $z > 0$, the center of synchrotron oscillations of the electrons drifts backward in phase, and the phase space trajectories do not overlap!

One possible use of the trajectories data produced by the numerical solution of the modulation equations (9), is to calculate average bunching parameter like the average phase $\langle\psi_m(z)\rangle_j$, average density harmonic modulation $\eta_n(z) = \langle e^{in\psi_m(z)}\rangle_j$, and average energy harmonic modu-

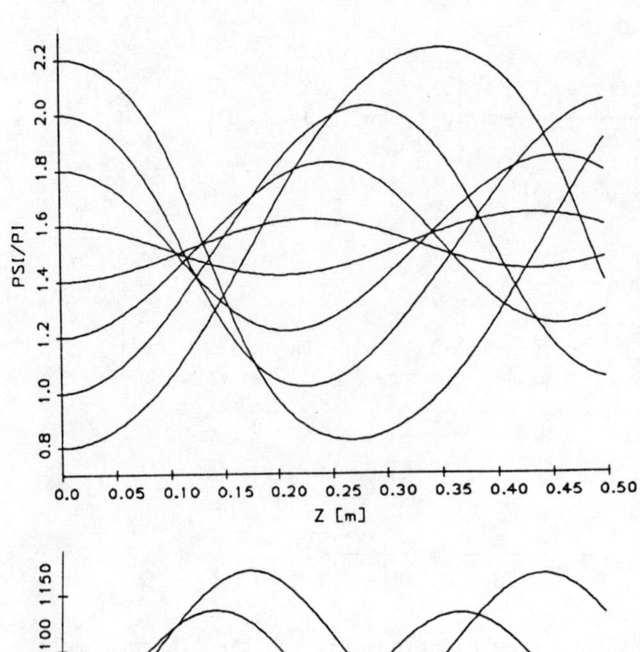

Fig. 2 Optical bunching (m = 1).

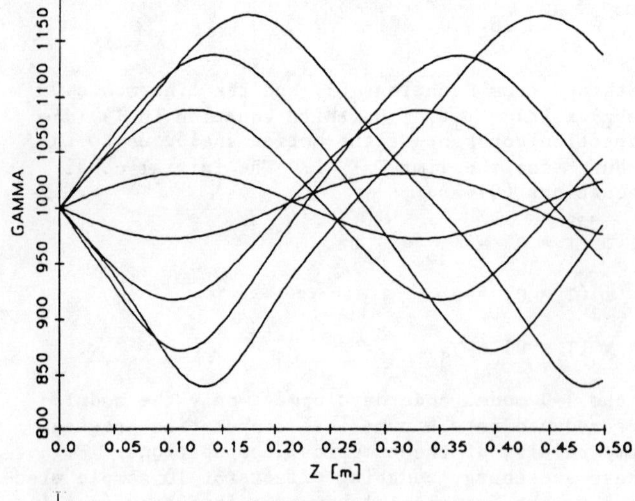

Fig. 3 Energy modulation (m = 1).

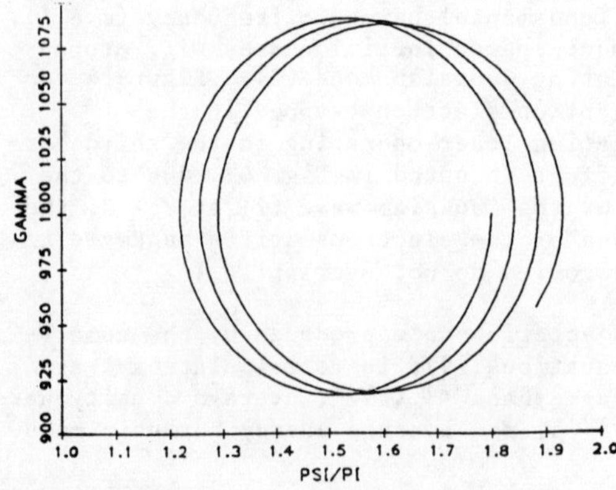

Fig. 4 Trapped electron phase-space trajectories (m = 1).

lation $\langle \gamma_j(z) e^{in\psi_m(z)} \rangle_j$. The averaging process is defined by

$$\langle\ (\ )\ \rangle_j = \frac{1}{2\pi} \int_o^{2\pi} d\psi_{mj} \int_o^{\infty} d\Delta\gamma_{oj} f(\Delta\gamma_{oj}) \iiiint_{-\infty}^{\infty} dx_{oj} dy_{oj} d\phi_{xoj} d\phi_{yoj} f(x_{oj}, y_{oj}, \phi_{xoj}, \phi_{yoj})\ (\ ) \tag{10}$$

where $f(x_{oj}, y_{oj}, \phi_{xoj}, \phi_{yoj})$ is the initial electron distribution in position and angle and $f(\Delta\gamma_{oj})$ is the initial energy distribution. In a storage ring all of these functions can be taken as Gaussian distributions.

The average bunching parameters can be used for computing the radiation field at harmonic n of the modulating laser frequency ($\omega = n\omega_m \backsim nm\omega_w$), within a simplified fluid charge radiation model.[1] This may be a reasonable model for a structure made of two distinct modulation and radiation sections (Fig. 1a), provided that the turbulant charge flow in the e-beam (due to the finite emittance) does not invalidate the fluid model. In the more general case, and particularly in a uniform FEL configuration (Fig. 1b), one must use the more exact radiation model, given by Eqs. (4-6), which takes into account the individual complex energy contribution of each electron in the ensemble distribution.

PART II: THE EMITTED RADIATION CHARACTERISTICS

For complete definition of the emitted radiation we need to compute explicitly the amplitudes of the mode expansion (4). Substituting the general mode expression (7) in Eq. 6, we can rewrite the energy emission into mode q at wiggler harmonic frequency ℓ as

$$\Delta\hat{W}_q^{(\ell)} = \sum_{j=1}^{N_T} \Delta\hat{W}_{jq}^{(\ell)}$$

$$\Delta\hat{W}_{jq}^{(\ell)} = \int_{t_{oj}}^{t_j(L)} b_j(t)\, e^{-i\psi_{\ell j}(t)} dt \tag{11}$$

$$b_j(t) = -\frac{ecK}{2} \frac{1}{\gamma j} \left[J_{\frac{\ell-1}{2}}(u) - J_{\frac{\ell+1}{2}}(u) \right] \mathscr{E}_q^{\omega *}(x_j(t), y_j(t), z_j(t))$$

$$\psi_{\ell j} = \left(\frac{\omega}{c} + \ell k_w\right)\bar{z}_j(t) + \phi_q^{\omega}(x_j(t), y_j(t), \bar{z}_j(t)) - \omega t$$

A significant saving in computation is obtained by recognizing the symmetry of the modulation problem under translation $t \rightarrow t + T_m$, where $T_m = 2\pi/\omega_m$ is the modulating laser period. The summation over all the electrons in the electron beam pulse may be then split into an infinite summation over all optical periods and an internal summation over N_{br} electrons in each period (bunch) indexed by subscript number $r = t/T_m$.

This enables writing

$$\Delta\hat{W}_q^{(\ell)} = N_T F_p(\omega) F_b(\omega)$$

$$F_p(\omega) = \frac{1}{N_T} \sum_{r=-\infty}^{\infty} e^{-ir\omega T_m} N_{br} \qquad (12)$$

$$F_b(\omega) = \langle \Delta\hat{W}_{jq}^{(\ell)} \rangle_j$$

where $N_T = \sum_{r=1}^{\infty} N_{br}$ is the total number of electrons in the e-beam pulse, and the averaging takes place over the electrons in <u>only one period</u> T_m of the modulating laser (10).

The interference factor $F_p(\omega)$ reflects the fact that each electron emits coherently during its entire transit time through the wiggler. Consequently $F_p(\omega)$ is a very narrow line function, Fourier transform limited by the e-beam pulse duration, and it determins the spectral width of the emitted radiation. The "form factor" $F_b(\omega)$ is a wide spectrum function and must be computed numerically on the basis of the data computed from the modulation equations solution (9,11). The interference factor $F_p(\omega)$ may be computed analytically. It is shown in Fig. 5 for an example of a rectangular pulse, for which $F_p(\omega)$ is the well known multi-beam interference function $|F_p(\omega)| = \sin(\pi N_p \omega/\omega_m)/N_p \sin(\pi\omega/\omega_m)$, where $N_p = T/T_m$ is the number of optical periods in the electron beam pulse duration T.

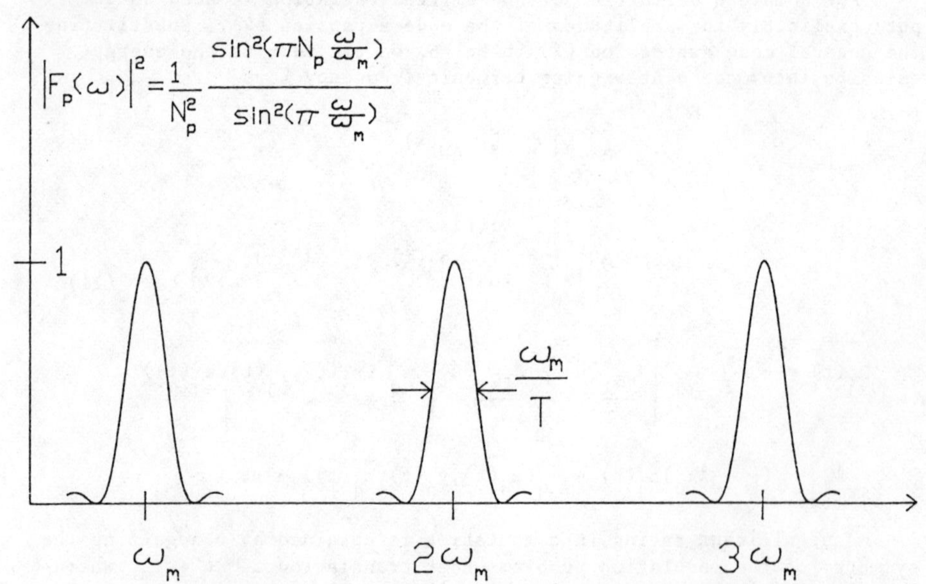

Fig. 5. The super-radiance interference function for a retangular current pulse of duration $T = Np2\pi/W_m$.

The optical characteristics of a radiation source at any practical distance from the source may be most fully described in terms of its spectral brightness function $d^3P/dA\, d\Omega d\omega$. We can write the brightness in terms of the Wigner distribution function of the radiation field[7]

$$\frac{d^3P}{dA\, d\Omega d\omega} = \frac{1}{\pi}\frac{1}{\lambda^2}(\varepsilon_o/\mu_o)^{\frac{1}{2}}\, W_E(\vec{r}\,,\vec{k}\,,\omega,t) =$$

$$= \frac{1}{\pi}\frac{1}{\gamma^2}(\varepsilon_o/\mu_o)^{\frac{1}{2}}\sum_{qq'} W_{c_q c_{q'}}(\omega, t-z/c)W_{\mathscr{E}_q\mathscr{E}_{q'}}(\vec{r}\,,\vec{k}\,) \qquad (13)$$

Here we used the relation $E_q = \sum_q c_q(\omega)\mathscr{E}_q(\vec{r})\exp[i(\omega/c)z]$ and the Wigner distribution definition:

$$W_{fg}(\omega,t) \equiv \int_{-\infty}^{\infty} d\tau < f(t+\tau/2)f^*(t-\tau/2)> e^{i\omega\tau}$$

$$= \frac{1}{2\pi}\int_{-\infty}^{\infty} d\Omega < F(\omega+\Omega/2)F^*(\omega-\Omega/2)e^{-i\Omega t}$$

where F,G are the Fourier transforms of functions f,g.

To obtain more explicit expressions we now note that the transverse beam profile of the DC current (and consequently of the optical frequency current) is approximately Gaussian (at least in a storage ring FEL device). Consequently it would be a reasonable assumption that predominately a single mode - the fundamental Gaussian mode-is excited with an optical beam waist equal to the e-beam radius r_b. For this approximation to hold we must also neglect diffraction of the mode, so that its Rayleigh length is sufficiently longer than the radiation section in the wiggler:

$$r_b > \left(\frac{\lambda_n L}{4\pi}\right)^{\frac{1}{2}}$$

When only a single mode has significant amplitude in the emitted radiation, we may set

$$W_{\mathscr{E}_q\mathscr{E}_q}(\vec{r}_\perp = 0, \vec{k}_\perp = 0) = 8\,(\mu_o/\varepsilon_o)^{\frac{1}{2}}P_q \qquad (14)$$

Combining (13), (14) and (10) we obtain an approximate expression for the spectral brightness of the super-radiant FEL emission on axis, assuming a rectangular current pulse $I(t) = I_o\text{rect}[(t-t_o)/T]$:

$$\frac{d^3P(\omega,t)}{dA\, d\Omega d\omega} \simeq \frac{1}{2\pi}\frac{N_P^2}{P_q\lambda^2 T}\frac{\sin^2(\pi N_P\omega/\omega_m)}{\sin^2(\pi\omega/\omega_m)}F_b(\omega)\text{rect}\left[(t - \frac{z}{c} - t_o)/T\right] \qquad (15)$$

The spectral width of the emitted radiation is dominated by the narrow multibeam interference function $\sin^2(\pi N_P\omega/\omega_m)/\sin^2(\pi\omega/\omega_m)$ which is Fourier tranform limitted by the electron beam pulse duration (it

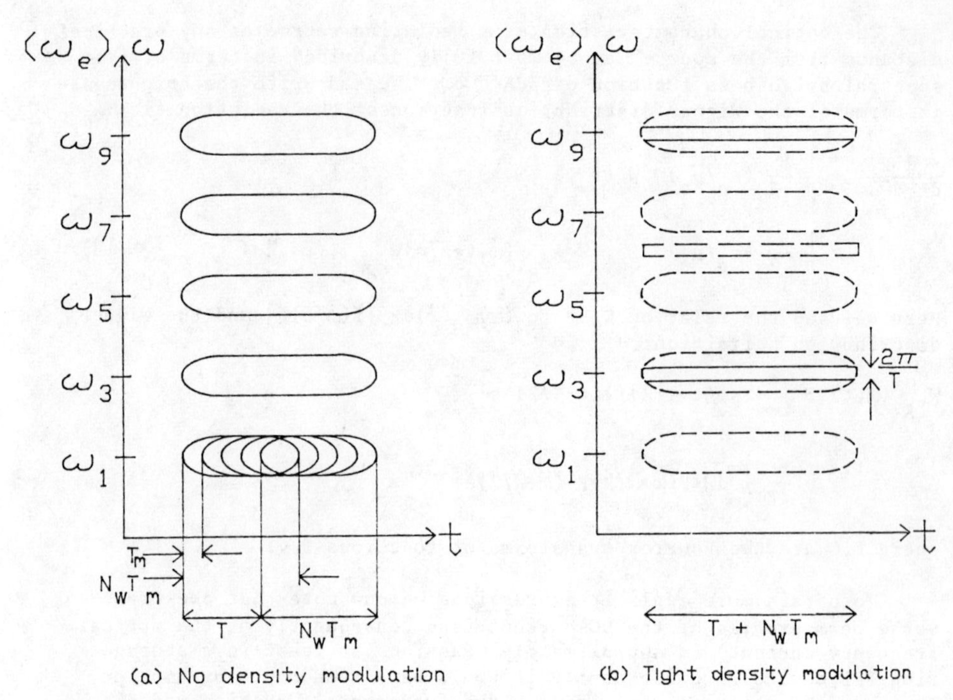

Fig. 6. Time-frequency phase-space picture.

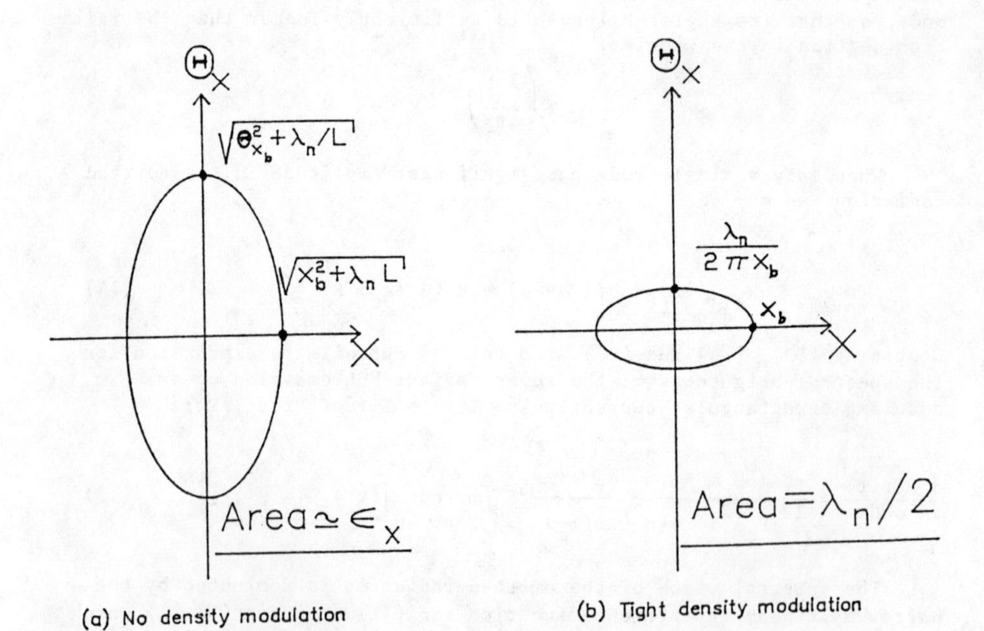

Fig. 7. Space-angle phase space picture.

was implicitely assumed that the beam pulse duration is much shorter than the modulating laser coherence time).

The brightness enhancement feature of super radiant emission is well illustrated in phase space diagrams. Figure 6 and Figure 7 describe respectively in time-frequency phase space and space-angle phase space the domains of radiation emission of both unmodulated and tightly modulated (bunched) electron beams. A substantial contraction is obtained in time-frequency phase space area - by a factor $(\omega/N_W):(2\pi/T)$. The contraction in space-angle phase space area is substantial only for the regime $\varepsilon > \lambda/2$ (where ε is the e-beam emittance defined as the beam phase space area). In this limit the conventional spontaneous emission radiation phase space domain is dominated by the e-beam emittance, and the contraction factor with super-radiant emission is thus $\varepsilon_x\varepsilon_y:(\lambda/2)^2$. The total phase space contraction then turns out to be:

$$\text{Reduction Factor} = 4\varepsilon_x\varepsilon_y cT/(\lambda_n^3 N_W) \tag{16}$$

This will also be approximately the enhancement factor in the radiation spectral brightness if we assume that the spontaneous emission power is emitted, in the tight modulation limit almost entirely into the contracted phase space domain.

Note, that the total phase space area contraction of the super radiant FEL is even bigger than indicated by (16), because some emission bands vanish. For example in Fig. 6 we note that wiggler harmonics 1, 5 and 7 vanish in the tight modulation limit. The reason for that is that the interference function $F_P(\omega)$ vanishes due to desctructive intereference of all the electrons in the pulse. Also note that all the even harmonics of the laser (see n = 2) vanish because there is no even harmonic emission of the wiggler radiation (see n = 2 ℓ = 6), and consequently the form factor $F_b(\omega)$ nulls.

REFERENCES

(1) G. Vignola, R.R. Freeman, B.M. Kincaid, C. Pellegrini, A. Luccio, J. Murphy, J. Galayda and A. Van-Steenbergen, Nucl. Inst. and Meth. in Phys. Res. A 239, 43 (1985).
(2) B. Girard, Y. Lapierre, J.M. Ortega, C. Bazin, M. Billardon, P. Elleaume, M. Bergher, M. Velghe, Y. Petroff, Phys. Lett. 53, 2405 (1984).
(3) A. Yariv. Introduction to Quantum Electronics, Wiley (1975).
(4) A. Gover, H. Freund, F.L. Granatstein, J.H. McAdoo, C.M. Tang, In Infrared and Millimeter Waves, Vol. 11, p. 291 (K. Button ed.) Accademic 1984.
(5) A. Gover, A. Friedman, unpublished.
(6) N.S. Ginzburg and M.D. Tokman, Sov. Phys. Tech. Phys. 29 (6), 604 (1984).
(7) M.J. Bastiaans, J. Opt. Soc. Am. 69, 1710 (1979).
(8) S. Krinsky, M.L. Perlman, R.E. Watson, Handbook on Synchrotron Radiation Vol. 1, p. 65, ed. E.E. Koch, North Holand (1983).

STIMULATED RAMAN SCATTERING EXPERIMENTS OF HYDROGEN MOLECULES IN THE VUV REGION AROUND 126nm

Yoichi Uehara, Kou Kurosawa and Wataru Sasaki
Department of Electronics, University of Osaka Prefecture,
Mozu-Umemachi, Sakai, Osaka 591, Japan

Etsuo Fujiwara, Yosiaki Kato and Chiyoe Yamanaka
Institute of Laser Engineering, Osaka University,
Yamada-oka, Suita, Osaka 565, Japan

Masanobu Yamanaka
Course of Electromagnetic Energy Engineering,
Osaka University, Yamada-oka, Suita, Osaka 565, Japan

ABSTRACT

High power tunable argon excimer laser was used for a pumping source of stimulated Raman scattering in molecular hydrogen. Intense Stokes radiations up to second order were observed. The threshold power of the stimulated scattering and the conversion efficiency of the first Stokes radiation were measured. An effect of the dispersion of the refractive index of H_2 on stimulated Raman scattering and feasibility of generation of the anti-Stokes radiation are discussed.

INTRODUCTION

Recently we reported a frequency-tunable argon excimer laser,[1] which was the strongest light source around the wavelength of 126nm. However the tuning range was not so wide and in between 124.5nm and 127.5nm. For applications such as spectroscopic use it would be desired to extend the obtainable spectral range.

Stimulated Raman scattering is considered to be a powerful way for obtaining new lines by the frequency conversion of the laser radiation. Several researchers have reported high efficiency energy conversion from the laser radiation to the Stokes or the anti-Stokes radiations, in visible and ultraviolet spectral region.[2] To our knowledge, however, any experiments of stimulated Raman scattering have not been carried out in the VUV region around 126nm yet. So we are studing feasibility of the frequency conversion of the argon excimer laser radiation by stimulated Raman scattering. Molecular hydrogen is one of the most suitable Raman media because it has a large cross section of stimulated Raman scattering and no absorption line in the spectral region around 126nm.

In this paper, we observed stimulated Raman scattering in H_2 and its characteristic is reported.

EXPERIMENTALS

Figure 1 shows the experimental setup for stimulated Raman experiments. The details of the argon excimer laser apparatus will be described elsewhere in this proceeding. Basically, the argon gas filled in the anode pipe was excited by electron-beam accelerated by 700kV. Optical elements for the laser cavity were set at both ends of the anode pipe.

The output mirror was an uncoated MgF_2 wedged plate. When we used an uncoated quartz plate as the reflector at the other end of the cavity, we obtained broadband oscillation at 125.8nm with the spectral width of 1nm. For wavelength selection, a MgF_2 prism backed with a quartz plate was used as shown in Fig.1. In this case, the laser radiation was tuned from 124.5nm to 127.5nm with the spectral width of 0.3nm. So we call this case the narrowband oscillation.

As shown in Fig.1, the laser radiation, which was transmitted to the vacuum vessel through a MgF_2 window, was focused by a 50cm focal length MgF_2 lens into a 100cm-long Raman cell in which high purity H_2 gas was filled.

A part of the output beam, which contained both the laser and the Raman radiations, was reflected by the MgF_2 beam splitter and injected into a Seya-Namioka type 50cm focal length spectrometer. In order to monitor the power of the laser and the Raman radiations a photodiode(Hamamatu photonics R1193U-04), whose sensitivity is calibrated before the experiment, was set at a place close to the exit slit . Otherwise the spectra were recorded on Kodak SWR films. The waveform of the output beam, which came through the beam splitter in Fig.1, was monitored using a photodiode (Hamamatu photonics R1328U-04).

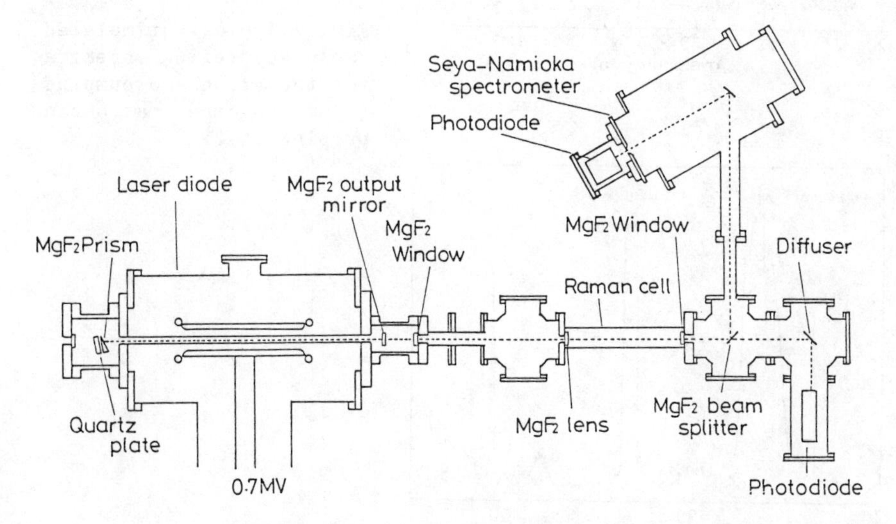

Fig. 1 Schematic drawing of experimental setup.

RESULTS

Broadband pumping

First Stokes line was observed in the pressure range between 0.25atm and 8atm, while second Stokes line was observed only at 4atm. Figure 2(a) shows a typical Raman spectrum, which was obtained at the H_2 pressure of 4atm by broadband pumping with the peak power of 1.1MW. The threshold power of the stimulated Stokes Raman scattering was measured at the H_2 pressure of 4atm to be 450kW. Conversion efficiency was determined by dividing the peak power of the first Stokes radiation by the peak power of the input laser. In Fig.3(a) is plotted the conversion efficiency against the H_2 pressure. It is found that the efficiency is as high as about 40% at the pressure of 1atm and it is almost independent of the pressure in the range above 1atm.

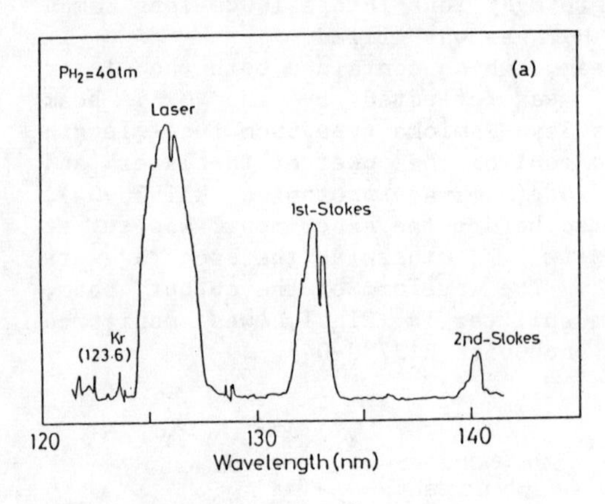

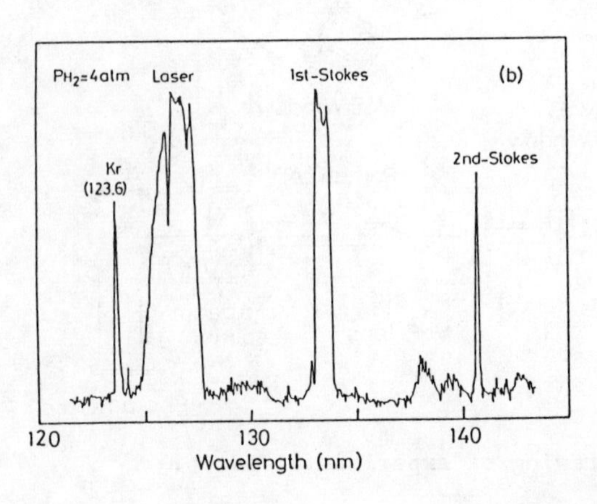

Fig. 2 Typical stimulated Raman scattering spectra for the broadband pumping (a) and the narrowband pumping (b).

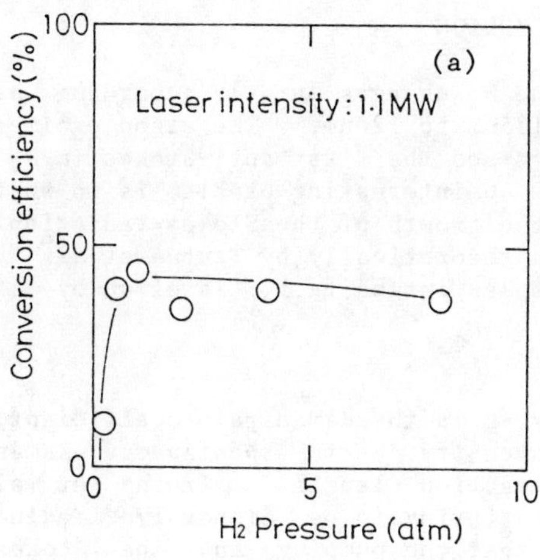

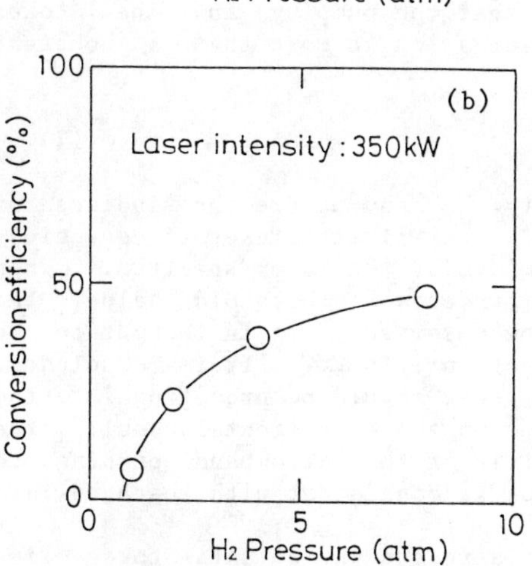

Fig. 3 Pressure dependence of the conversion efficiency of the first Stkoes radiation for the broadband pumping (a) and the narrowband pumping (b).

Narrowband pumping

In the case of using narrowband oscillation as the pumping source, the first Stokes Raman line was observed at a pressure ranging from 0.25atm to 8atm, and furthermore the second Stokes Raman line was obtained at a H_2 pressure above 2atm. The threshold power of the stimulated Stokes Raman scattering was measured at the H_2 pressure of 4atm to be 170kW. This is much lower than the value of 450kW of the threshold power in the broadband pumping. Figures 2(b) and 3(b) illustrate a typical Raman spectrum obtained at the H_2 pressure of 4atm with the pumping of 350kW and the conversion efficiency of the first Stokes Raman line. The efficiency increases to about 40% with increasing the H_2 pressure. It should be noted here that any anti-Stokes Raman lines were not observed at all as well as for the case of the broadband pumping.

DISCUSSION

The refractive index of the H_2 changes largely according as the wavelength decreases from 135nm to 120nm.[3] The argon excimer laser, the first Stokes in H_2 and the first anti-Stokes in H_2 are in this spectral region. An interesting problem is in what manner the dispersion affects the growth of the Stokes radiation. This problem has been discussed theoretically by Trutna et al.[4]

The growth of the first Stokes intensity I_s is given by

$$I_s = I_{so} \exp(g I_l z) \tag{1}$$

where I_l is the laser intensity, g is the Raman gain coefficient, I_{so} is the initial Stokes intensity due to spontaneous Raman scattering and z is an interaction length. Trutna et al. analyzed stimulated Raman scattering in a dispersive medium theoretically and concluded that the pumping and the Stokes radiations cannot interact over a length more than a coherent length, l_{coh}, defined by

$$l_{coh} = \frac{c\pi}{2(n_l - n_s)\Delta\omega_L} \tag{2}$$

where c is the light velocity, n_l and n_s are the indices of refraction at the laser and the first Stokes frequencies, respectively, $\Delta\omega_L$ is the bandwidth of the laser spectrum. When the value of $gI_l z$ in eq. (1) exceeds a threshold value, the stimulated scattering should be observed. z is thought to be l_{coh}, which is inversely propotional to $\Delta\omega_L$. It is concluded, therefore, that the threshold power should be propotional to the bandwidth. This is reflected on the experimental result; the ratio of the threshold-intensitiy of the narrowband pumping to the broadband pumping , 170/450, is consistent with the bandwidth ratio of them, 0.3/1.0.

We are also interested in a reason why no anti-Stokes line was observed. The process of the generation of the anti-Stokes line is similar to well-known CARS process[5] with the difference that the Stokes radiation is not externally supplied but is I_s given by eq.(1). When the laser radiation I_l and the Stokes radiation I_a interact collinearly over a length of l, the intensity of the generated anti-Stokes radiation I_a is given by[6]

$$I_a = \left(\frac{8\pi^2\omega_a c^2}{\hbar\omega_s^4}\right)\cdot\left[\frac{N}{\Gamma}\left(\frac{d\sigma}{d\Omega}\right)\right]^2 I_l^2 \cdot I_s \cdot l^2 \cdot \left(\frac{\sin(\Delta k l/2)}{\Delta k l/2}\right)^2 \tag{3}$$

where ω_a is the frequency of the anti-Stokes line, N is the density of H_2, Γ is the damping constant $(d\sigma/d\Omega)$ is the scattering cross section of spontaneous Raman scattering, Δk is a phase mismatch defined as $\Delta k = k_a + k_s - 2k_l$. I_s is given by eq.(1) and g is expressed by

$$g = 2\left(\frac{8\pi^2 c^2}{\hbar\omega_s^3}\right)\cdot\frac{N}{\Gamma}\left(\frac{d\sigma}{d\Omega}\right) \tag{4}$$

where $\hbar$ is the Planck's constant, ω_s is the frequency of the Stokes line.

Using these, relations, eq.(3) is transformed as follows,

$$I_a = \ln^2(\frac{I_s}{I_{so}})\cdot(\frac{1}{2})^2\cdot\frac{1}{z^2}\cdot l^2\cdot(\frac{\sin(\Delta kl/2)}{\Delta kl/2})^2\cdot I_s \qquad (5)$$

where ω_a is approximated with ω_s. z is estimated by l_{coh} and l is estimated by using the well known relation $\pi=\Delta k\cdot l$. Therefore following relation is obtained;

$$I_a = \ln^2(\frac{I_s}{I_{so}})\cdot(\frac{1}{l_{coh}\Delta k})^2\cdot I_s$$

$$= \alpha I_s \qquad (6)$$

The intensity ratio α (and also the power ratio) of the anti-Stokes and the Stokes radiations is calculated to be $2x10^{-3}$ and $2x10^{-2}$ for the narrowband pumping and for the broadband pumping, respectively. As mentioned above, the Stokes lines were observed with the peak power of a few hundreds kilowatts. The peak power of anti-Stokes line is estimated to be at the most a few kilowatts. We are not able to detect such a weak signal less than a few kilowatts. This is considered to be the reason why anti-Stokes radiations were not detected in this experiment.

CONCLUSION

The stimulated Raman scattering in H_2 was investigated in the pressure range from 0.25atm to 8atm with the tunable argon excimer laser. The stokes radiations up to second order were observed. The conversion efficiency of the first Stokes radiation was above 40%. No anti-Stokes line was observed. The effect of the dispersion on stimulated Stokes Raman scattering and the feasibility of the generation of the anti-Stokes radiation were discussed. It is also found that the power of the anti-Stokes radiation was estimated to be from a few hundreds W to a few kW.

ACKNOWLEGEMENT

The authors are indebted to Prof. J. Fujita of Nagoya University for his generous support. This research was partly supported by a grant-in-aid for a special research project on nuclear fusion from the Ministry of Education, Science and Culture of Japan.

REFERENCES

1 Y.Uehara, W.Sasaki, S.Kasai, S.Saito, E.Fujiwara, Y.Kato, C.Yamanaka, M.Yamanaka, K.Tsuchida, and J.Fuita, Opt.Lett. 10, 478(1985).

2 D.W.Traior, H.A.Hyman, and R.M.Heinrichs, IEEE J. Quantum Electron. QE-18, 1929(1982).
3 G.A.Victor and A.Dalgarno, J.Chem.Phys. 50, 2535(1968).
4 W.R.Trutna,Jr.,Y.K.Park, and R.L.Byer, IEEE J.Quantum Electron. QE-15, 648(1979).
5 J.W.Nibler and G.V.Knighten, in Raman spectroscopy of Gases and Liquids, ed. by A.Weber (Springer, Berlin, 1979) p.261.
6 W.Kaiser and M.Maier, in laser handbook ed. by F.T.Arecchi and E.O.Schulz-DuBois, (North-Holland, Amsterdam, 1972) vol.1, p.1092.

VARIED-SPACE GRAZING INCIDENCE GRATINGS IN HIGH RESOLUTION SCANNING SPECTROMETERS

Michael C. Hettrick and James H. Underwood

Lawrence Berkeley Laboratory
Center for X-ray Optics
Berkeley, California 94720

ABSTRACT

We discuss the dominant geometrical aberrations of a grazing incidence reflection grating and new techniques which can be used to reduce or eliminate them. Convergent beam geometries and the aberration correction possible with varied groove spacings are each found to improve the spectral resolution and speed of grazing incidence gratings. In combination, these two techniques can result in a high resolution ($\lambda / \Delta\lambda > 10^4$) monochromator or scanning spectrometer with a simple rotational motion for scanning wavelength or selecting the spectral band.

INTRODUCTION

The increased demand for intense, coherent sources of soft x-ray and extreme ultraviolet radiation[1] motivates the development of new spectroscopic instruments operating in the grazing incidence regime λ = 10-1000 Å. X-ray holography[2,3], photoelectron spectroscopy[4] and grating microscopy[2,5] all require a pre-monochromator with spectral resolving power $\lambda / \Delta\lambda = 10^3 - 10^4$ and higher. For example, at a wavelength of 30 Å, a coherence length of 60 microns converts to a resolving power of approximately 2×10^4.

Existing spectroscopic instruments are not capable of delivering such high performance without severely compromising other requirements. For example a conventional Rowland circle grating, constrained to the geometry of equidistant grooves, results in considerable astigmatism, an oblique spectrum, and complicated motions for scanning wavelength[6]. In applications where only low or modest spectral resolution is required, $\lambda/\Delta\lambda < 10^3$, the spherical grating surface can be replaced by a toroid[7], resulting in a near removal of astigmatism. Visible holography can also be used to alter the focal surface of spectra and to improve the imaging[8]. However, each of these techniques is severely limited in its potential for dramatic aberration correction at short wavelengths.

During the past several years, a new technique in grating fabrication has emerged as a demonstrated tool for aberration correction. Ruling engines outfitted with state-of-the-art interferometric readout systems and computer control have made it possible to vary the spacings of grooves in a continuous manner across the grating ruled width[9,10]. This technological advance has been successfully exploited in the field of soft x-ray and extreme ultraviolet spectroscopy, providing erect focal surfaces for imaging of spectra[9,11] at high resolution. Curved grooves have also been recently demonstrated with a mechanical ruling engine[10]. With these new degrees of freedom it is now possible to first specify the desired performance, and then to deduce the mechanical ruling corrections necessary to yield these characteristics. This is a reversal of the situation confronted by grating scientists since the time of Rowland.

In addition to new groove patterns, the use of unconventional beam geometries has allowed the use of simple optical surfaces. A plane grating surface placed in convergent incident light has been shown to provide imaging free of astigmatism[12]. Such a geometry is particularly convenient in space-borne instruments, due to the presence of a large telescope for collection of the incident light. This not only increases the effective speed of the optical system, but also employs a simple optical surface which can be made to the accuracy required for high resolution.

The purpose of this paper is to apply the principles of varied spacing and simple (plane or spherical) grating surfaces to the task of designing grazing incidence laboratory monochromators. We indicate the design options available today and in the near future for the construction of high spectral resolution monochromators and scanning spectrometers. We emphasize the potential of an advanced varied space grating to maintain high resolution ($\lambda/\Delta\lambda > 10^4$) over a wide wavelength region through a simple rotational scanning motion, particularly convenient in the environment of an ultra-high vacuum.

THE LIGHT-PATH FUNCTION

In the short wavelength domain, below approximately 1000 Å, the physical diffraction-limited resolution of most optics is insignificant and the main task is the minimization of its geometrical aberrations. The analytical formalism which is most instructive for the purpose of understanding the geometrical aberrations of a diffraction grating is based on Fermat's principle. It states that a light ray will trace a path through an optical system so as to minimize variations in its effective path-length. The effective path-length, F, equals the physical length traversed, L, minus the phase shift of grating groove N:

$$F(w,\ell) = L(w,\ell) - m\lambda N(w,\ell) \tag{1}$$

where L and N are each functions of the position (w,ℓ) at which the light ray strikes the optical aperture. If a normal incidence focal plane is placed a distance r' from the grating center, then Fermat's principle can be used to find the focal position (x,y):

$$x = r' \, dF/dw \, / \cos\beta, \quad y = r' \, dF/d\ell \tag{2}$$

where β is the angle made with the grating surface normal by the ray as it is diffracted, and x is in the direction of spectral dispersion. Equation

2 indicates the ray positions at which maximum constructive interference is achieved for light diffracted from the immediate vicinity of grating coordinate pair (w,ℓ). Given a finite grating size, x and y will drift over a range of values, resulting in an image whose size represents the total geometrical aberration of the optic.

When the grating sizes w and ℓ are small in comparison to the object distance r, it is useful to expand the light-path function as a power series in these grating coordinates:

$$F(w,\ell) = \Sigma F_{ij}(w,\ell)\, w^i \ell^j \quad (3)$$

$$\text{where} \quad F_{ij}(w,\ell) = L_{ij}(w,\ell) - m\lambda N_{ij}(w,\ell). \quad (4)$$

In the case of a spherical surface with radius R the path-length coefficients, L_{ij}, are well known[13]:

$L_{00} = r + r'$ = length of the principal ray;

$L_{10} = \sin\beta - \sin\alpha$;

$L_{20} = \frac{1}{2}(\cos^2\alpha/r - \cos\alpha/R) + \frac{1}{2}(\cos^2\beta/r' - \cos\beta/R)$;

$L_{02} = \frac{1}{2}(1/r - \cos\alpha/R) + \frac{1}{2}(1/r' - \cos\beta/R)$;

$L_{30} = -\frac{1}{2}\sin\alpha/r\,(\cos^2\alpha/r - \cos\alpha/R) + \frac{1}{2}\sin\beta/r'\,(\cos^2\beta/r' - \cos\beta/R)$;

$L_{12} = -\frac{1}{2}\sin\alpha/r\,(1/r - \cos\alpha/R) + \frac{1}{2}\sin\beta/r'\,(1/r' - \cos\beta/R)$;

... and higher-order terms. (5)

The coefficients associated with the interference term (N_{ij}) are dependent on the distance between adjacent grooves (σ) and on the groove pattern (linear, circular, elliptical, etc). Following Harada & Kita[9] we expand the groove density as a power series in w along the mid-plane of the grating (ℓ=0):

$$1/\sigma = 1/\sigma_0 + a_1 w + a_2 w^2 + a_3 w^3 + \ldots \quad (6)$$

where the coefficients a_n provide the effect of varied spacing. Thus, for the otherwise classical case of straight grooves formed at the intersection of the grating spherical surface and parallel planes oriented normal to this surface at the grating pole (w=0), we have:

$$N_{10} = 1/\sigma_0\,; \quad N_{20} = \tfrac{1}{2} a_1\,; \quad N_{30} = 1/3\, a_2\,; \ldots \quad (7)$$

It should be noted that only the interference term N_{10} is present for conventional equally spaced grooves, and varied spacing itself effects the aberration terms dependent only on w. Aberration correction of terms involving the groove length ℓ (dominant F_{02} and F_{12}), requires groove curvature. In the case of a spherical surface, Harada & Kita have shown that a conventional rectilinear ruling motion can supply a small amount of groove curvature by tilting the ruling planes relative to the grating normal at its center[9]. The groove curvature radius obtained with this technique is approximately R/tanθ as viewed from the grating normal, where θ is the ruling plane tilt which can be as large as 30°. An even more direct means of obtaining groove curvature has been developed by Hirst, being a "circular ruling engine"[10]. This makes it possible to rule concentric grooves with a maximum radius of curvature of approximately 500 mm, independent of the grating surface shape. These radii are typically much shorter than obtained with the technique of Harada, and thus the two methods complement each other. If the distance from the grating center to the groove rotation axis is D_o, then one can derive the aberration correction coefficients N_{ij} by the substitution $w = D_o - \sqrt{(D_o - w)^2 + \ell^2}$ in eq'n 6. In particular, this results in non-zero values for N_{02} (astigmatism correction) and N_{12} (astigmatic coma correction). However, in the discussion which follows, we consider only varied-space straight grooves.

LINE PROFILES

Ideally, the path-lengths (eq'n 5) and their interference shifts (eq'n 7) would cancel each other, resulting in individual terms F_{ij}=0. Such a system images a point source without aberration, and is referred to as stigmatic. Of course, this is difficult to realize in any practical optical system, and one must consider the effect of various aberrations. Apart from the term F_{10}, which via the grating equation is zero, each aberration will produce a point-spread function of the image at the focal plane. For example, in the direction of dispersion,

$$dI/dx = c\ x^{(2-i)/(i-1)} \tag{8}$$

where I is the intensity and c is a normalization factor. In Fig. 1 we sketch the image profiles for non-zero values of F_{20} (i=2), F_{30} (i=3) and F_{40} (i=4). These images are distinguishing in shape, and are usually referred to as "de-focusing" (i=2), "coma" (i=3) and "spherical aberration" (i=4) by grating designers. Similar aberrations are in general present in the perpendicular direction, y, along the height of the image, the first of these being "astigmatism" (j=2).

CLASSICAL SCANNING ABERRATIONS

In the classical case of a spherical grating with equidistant straight grooves, both de-focusing and coma are absent along the Rowland circle. The image shape is dominated in the dispersion direction by spherical aberration and in the image height direction by considerable astigmatism. The situation worsens quickly when one deviates from the Rowland circle at grazing incidence. For example, given an immovable source and exit slit, a convenient rotation of the grating about its pole is commonly used to scan wavelength, such as used in a conventional toroidal grating monochromator[7]. However given the classical ruling constraint a_n=0, eq'ns 5 and 7 reveal there is only enough freedom to independently choose r and r', and thus to remove the de-focusing

Fig. 1. Line profiles of ray aberrations at a focal plane. The vertical direction is proportional to the intensity of photons diffracted per unit displacement in the dispersion direction (horizontal). De-focusing (top) yields a uniform intensity whereas coma (middle) and spherical aberration (bottom) are peaked at the ray position from the grating center. Extremum image size is defined by the sharp edges corresponding to rays diffracted from the grating width edges. The dominant aberration in magnitude of extremum width, if left uncorrected, would be de-focusing.

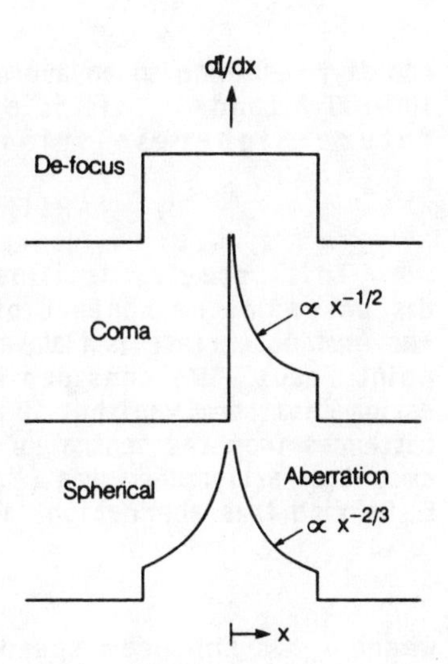

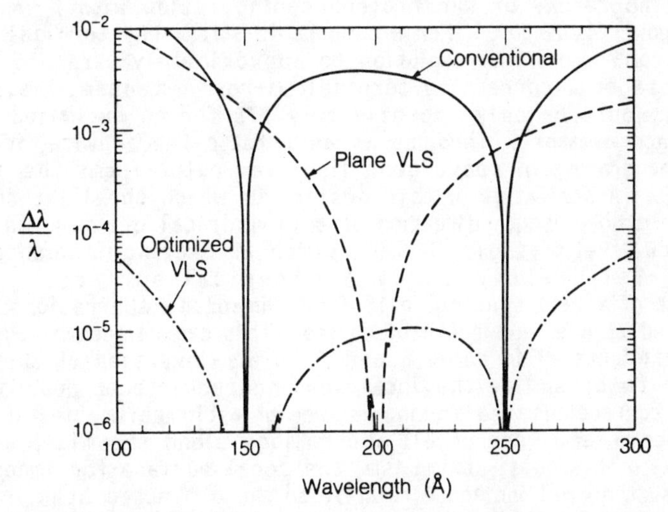

Fig. 2. Fractional resolution versus wavelength, scanned with a a simple rotation of the grating about its central groove, using fixed entrance and exit slits. Only the de-focusing aberration is plotted here (F_{20} in text). A spherical or toroidal grating with equally spaced grooves (solid) is compared to a varied-space plane grating (dashed) and an improved version where the radius of grating curvature is large, and the de-focusing term is stationary at 150 Å.

aberration (F_{20}) at two wavelengths. In Fig. 2 we show the spectral resolution due to the de-focusing term as a function of wavelength for a 1500 g/mm grating with 20 meter radius of curvature, a focal length r' of approximately 2.5 meters, a graze angle of approximately 9^{0}, a numerical aperture of 0.02 and an angular deviation of 162^{0}. The optimum wavelengths are 150 Å and 250 Å, away from which the aberration grows rapidly, resulting in an average spectral resolution of only 300 over the 100-300 Å band. This is a factor of 10-100 smaller than desired for future high-resolution studies.

VARIED-SPACE PLANE GRATING

This geometry is illustrated in Fig. 3, and has been previously discussed in the context of extreme ultraviolet spectroscopy[12,14,15]. The grating surface is flat, and operates in a light beam converging to a point focus. We consider the use of straight grooves for which the astigmatism term vanishes if the virtual source and real focus are equal distances from the central groove of the grating. In such a mounting, the amount of astigmatic coma (F_{12}) is very small. Setting $r = -r'$ in eq'n 5, we find this aberration limits the spectral resolving power to[14,15]:

$$\lambda / \Delta\lambda = 8 f_y^2, \tag{9}$$

where f_y is the beam speed along the grooves. In contrast to most gratings, this result is independent of the graze angle. For example, in a typical laboratory or synchrotron configuration with $f_y = 50$, the resolving power is 20,000. For a comparable soft x-ray toroidal grating, astigmatic coma limits the resolution to approximately $\lambda/\Delta\lambda = 11$; even with the use of a correcting toroidal mirror in tandem, the spectral resolution would be only approximately 275 for an optimized toroidal grating monochromator[7]. Thus, as an astigmatic-free device, it appears that a plane grating in convergent light far outperforms the available alternatives. A derivative of this design, in which the light converges only in the groove length direction of a cylindrical grating, would have the same low level of sagittal coma, which at grazing incidence can be approximately recovered by a spherical surface.

By use of varied spacing, all of the in-plane aberrations F_{n0} can be eliminated at one wavelength of choice. This can be seen by adjustment of the coefficients a_n in eq'ns 6 and 7, or by inversion of the grating equation[14] to determine the local spacing required at coordinate w. With these corrections, the grating is free of astigmatism, nearly free of astigmatic coma, and free of all aberrations along the mid-plane (ℓ=0). In addition to this quasi-stigmatism, the focal surface for imaging of a spectrum is at normal incidence relative to the diffracted beam, resulting from the fact that both the chosen wavelength and the zero order image are at equal distances from the grating center.

We now consider adopting this geometry for a monochromator, and scan wavelength by simply rotating the grating about its central groove, keeping the source and exit slit fixed. In Fig. 2 we show the increase in the de-focusing aberration (F_{20}) away from a corrected wavelength of 200 Å. Over a scanning region of approximately 30 Å, the spectral resolving power is kept better than 10,000. This insensitivity to small rotations of the grating has been previously noted in the context of a misalignment aberration[15]:

$$\Delta\lambda / \lambda = \tau_y (\cos \beta_* / \cos \alpha_* - 1)/f_x \tag{10}$$

Fig. 3. Geometry of a varied space plane grating in convergent light. One wavelength of choice in addition to zero order· is stigmatic across the ruled width. The residual aberration at this wavelength is sagittal coma, also called "astigmatic" coma, which depends on the square of the sagittal focal speed f_y. Lower panel shows the grating plane, containing straight grooves.

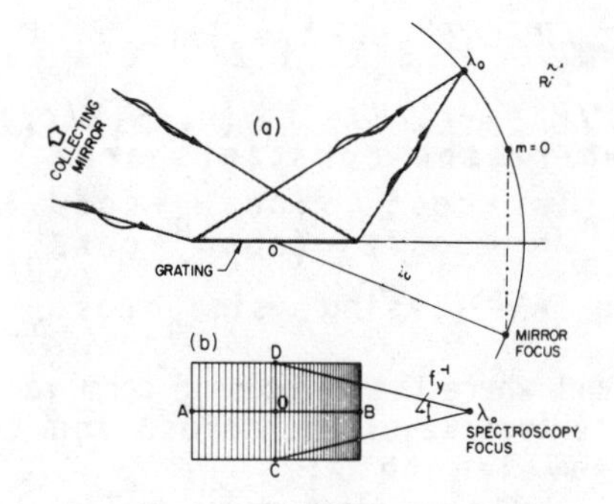

where f_x is the focal speed of the grating across its ruled width, τ_y is the rotation angle for selection of wavelength, and λ_* is the quasi-stigmatic wavelength.

MODIFICATIONS TO THE BASIC DESIGN

This performance can be significantly enhanced by a slight relaxation of two assumptions which have been made in the design of space instruments based on this varied-space grating[12,14,15]:

a) The flat grating surface provides a stigmatic zero order image in reflection, which facilitates alignment and wavelength calibration of a spectrum. However, by curvature of the grating surface the defocusing aberration of a second wavelength (i.e. in addition to λ_*) can be removed. This is particularly useful in extending the usable scanning range of a monochromator. Due to the use of converging incident light, the required deviation from flatness is extremely small, typically corresponding to hundreds of meters in radius. Optimally, this radius could be obtained by use of a cylindrically bent grating substrate, for which eq'n 9 still holds and astigmatic coma thereby remains small. However, with such large radii, the sagittal focusing is in practice negligible, allowing the use of a spherical grating in most circumstances.

b) The use of converging incident light onto the grating requires a fore-optic to refocus the entrance slit or source. In space astronomy, this is provided automatically by a large aperture telescope required in any event to collect the starlight. Such telescopes typically provide a single focus, which leads to the requirement of r = -r' in order to remove astigmatism. However, by using a Kirkpatrick-Baez (K-B) mirror system[16] in a laboratory instrument, the focal distances in the two orthogonal directions can be made unequal. This allows the virtual distance r to take on two values: r_t in the plane of reflection and r_s along the grooves. By setting $-r_s$ equal to the grating focal distance r', astigmatism will still be removed and eq'n 9 still valid. This leaves a free parameter, r_t, which can be used for further aberration correction. We find the most powerful such adjustment to be a removal of the derivative $dF_{20}/d\tau_y$ at one of the two wavelengths where F_{20} has also been set equal to zero[17]. In combination with a finite radius of curvature, this leads to the following focusing condition:

$$r_t/R = (t_2 s_1 - t_1 s_2)/(t_2 u_1 - t_1 u_2) \quad (11)$$

$$r'/R = -r_s/R = t_1/(u_1 - s_1 R/r_t) \quad (12)$$

where the constants are:

$$s_n = -\cos^2\alpha_n\ (\cos\alpha_n + \cos\beta_n) + 2\cos\alpha_n\ \sin\alpha_n\ (\sin\beta_n - \sin\alpha_n)$$

$$t_n = -\cos^2\beta_n\ (\cos\alpha_n + \cos\beta_n) - 2\cos\beta_n\ \sin\beta_n\ (\sin\beta_n - \sin\alpha_n)$$

$$u_n = -2(1 - \sin\alpha_n\ \sin\beta_n + \cos\alpha_n\ \cos\beta_n) \quad (13)$$

and where the defocusing term vanishes at wavelengths λ_n (n=1,2). The derivative of the defocusing term will also vanish at correction wavelength λ_1.

The dot-dash curve in Fig. 2 shows the result of applying these fine aberration corrections assuming the same basic parameters as before (1500 g/mm, 162° angular deviation, r'=2.5 m, 9° graze angle and a numerical aperture of 0.02). The de-focusing vanishes at 150 Å and 250 Å, rising to only approximately 10^{-5} between these wavelengths. Higher order scanning aberrations (e.g. F_{30}, F_{40}) must also be considered, but these can be made zero at one wavelength and decrease as the square and higher powers of the numerical aperture.

In Fig. 4 we illustrate one possible optical configuration consistent with the optimized design discussed above. As the grating accepts a large numerical aperture (.02), the entrance slit would be fed by a single mirror which de-magnifies the source size (e.g. synchrotron radiation) in the dispersion direction, thus providing high spectral resolution. A system of two bent glass or bent metal mirrors[18,19] in an orthogonal (K-B) arrangement provides a convergent beam to the grating. The bent mirror approach results in cylindrical optical surfaces for which the imaging in each direction is de-coupled, and thus absent of the mixed aberrations which dominate a single aspherical mirror. The mirror M1 refocuses the entrance slit to an optimum point V behind the grating, whereas M2 images the original source onto the exit slit in the direction perpendicular to dispersion. Over selected bandpasses, the K-B re-focusing system can be replaced by a single spherical mirror operated off-axis to form the two desired foci. Using the numerical ray trace program SHADOW (developed by F. Cerrina), we have verified the expected quasi-stigmatic imaging property of an entire spectroscopic system of this type. With a 100 micron wide entrance slit, a spectral resolution of 12,500 was obtained for a plane varied-space grating at a wavelength of approximately 200 Å. The point source response represented a resolution of 30,000 limited by astigmatic coma, and an image height of 30 microns. The variation in groove spacing required is only 20%, easily accommodated with existing numerically controlled ruling engines[9,10]. For the widest bandpass during scanning, a grating curvature radius of 500 meters would be used.

Advantages of the proposed monochromator design are as follows:

1) No astigmatism;
2) Small astigmatic coma;
3) Simple rotational motion for selecting wavelength;
4) Negligible scanning aberrations over a wide spectral band;
5) Straight groove pattern on a plane or spherical grating surface;
6) Erect focal surface for use as a spectrometer;

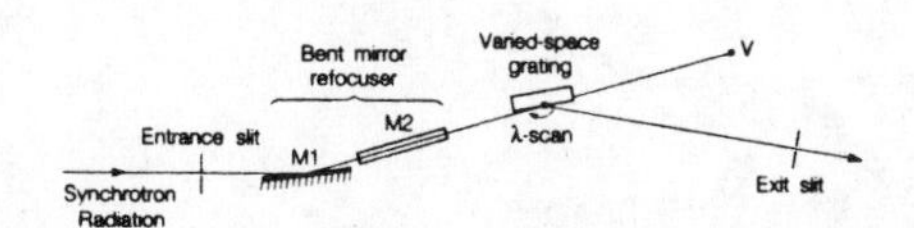

Fig. 4. High resolution grazing incidence monochromator. An approximately planar grating in convergent light can be rotated to select wavelength with minimum aberration. The re-focuser can be replaced by a normal incidence spherical multilayer placed offaxis.

We thank F. Cerrina for use of the ray trace program SHADOW, and modifications executed to permit varied-space mechanical rulings. We are not unaware of other related work being done in this area, including the use of a holographically recorded grating to construct a monochromator with similar properties to those discussed above[20,21]. This work was supported by the Office of Basic Energy Sciences, U.S. Department of Energy, under Contract No. DE-AC03-76SF00098.

REFERENCES

1. D. Attwood, K. Halbach and K.-J. Kim, Science 228, 1265(1985).
2. M. Howells and J. Kirz, AIP Conf. Proc. No. 118 on Free Electron Generation of Extreme Ultraviolet Coherent Radiation, ed. by J.M.J. Madey and C. Pellegrini, (Amer. Inst. Phys., New York, 1983), p. 85.
3. P.L. Csonka, J. Appl. Phys. 52, 2692(1981).
4. M.O. Krause, in Synchrotron Radiation Research, ed. by H. Winnick, (Plenum Press, 1980), p. 101.
5. M.C. Hettrick, Aplanatic Grazing Incidence Diffraction Grating, submitted to Appl. Opt. (1986).
6. S.L. Hulbert, J.P. Stott, F.C. Brown and N.C. Lien, Nucl. Instrum. Meth. 208, 43(1983).
7. C.T. Chen, E.W. Plummer and M.R. Howells, Nucl. Instrum. Meth. 222, 103(1984).
8. R.J. Fonck, A.T. Ramsey, and R.V. Yelle, Appl. Opt. 21, 2115(1982).
9. T. Harada and T. Kita, Appl. Opt. 19, 3987(1980).
10. G. E. Hirst, Proc. Soc. Photo-Opt. Instrum. Eng. 560, xxx(1985).
11. M.C. Hettrick, S. Kahn, W.R. Craig and R. Falcone, Stigmatic Grazing Incidence Laboratory Plasma Spectrometer with Variable Line-Space Grating, in preparation (1986).
12. M.C. Hettrick, S. Bowyer, R. Malina, C. Martin and S. Mrowka, Appl. Opt. 24, 1737(1985).
13. T. Namioka, J. Opt. Soc. Am. 49, 446(1959).
14. M.C. Hettrick and S. Bowyer, Appl. Opt. 22, 3921(1983).
15. M.C. Hettrick, Appl. Opt. 23, 3221(1984).
16. P. Kirkpatrick and A.V. Baez, J. Opt. Soc. Am. 38, 766(1948).
17. Aberration correction of a rotational mounting has been discussed for the case of normal incidence by R. Iwanaga and T. Oshio, J. Opt. Soc. Am. 69, 1538(1979).
18. J.H. Underwood and D. Turner, Proc. Soc. Photo-Opt. Instrum. Eng. 106, 125(1977).
19. Acton Research Corporation, Acton, MA 01720.
20. M. Pouey, Proc. Soc. Photo-Opt. Instrum. Eng. 597, xxx(1985).
21. B. Lai, F. Cerrina and M. Pouey, Extreme UV Circle Spectrometer for Line Profile Measurements, this volume.

MEASUREMENT OF OSCILLATOR STRENGTHS OF Hg BY FOUR-WAVE MIXING

A. V. Smith and W. J. Alford
Sandia National Laboratories, Albuquerque, NM 87185

ABSTRACT

We have developed techniques for accurate determination of oscillator strengths based on four-wave mixing. Using both refractive index matching and the frequency dependence of the nonlinear susceptibility, we have obtained about 50 oscillator strengths important for four-wave mixing in Hg.

INTRODUCTION

Atomic mercury vapor is an attractive candidate as a medium for the generation of light in the 120-140 nm range by nonlinear sum-frequency mixing. It has a two-photon resonance at a convenient wavelength (the 6S ground state to 7 singlet S transition lies in the tuning range of a frequency doubled DCM dye laser), and its nonlinear susceptibility is expected to be relatively high as well. The utilization of Hg has been hampered, however, by the lack of knowledge of its dipole transition matrix elements. Theoretical results are unreliable because of the large number of electrons (80) involved, and experimental values are scarce and not very accurate. This has made it difficult to predict and optimize mixing efficiency.

We have applied new techniques based on sum frequency mixing to measure those matrix elements involved in mixing through the 7S two-photon state--namely, the 6S-nP and 7S-nP matrix elements. For the measurement of the 6S-nP matrix elements, we rely on the requirement of refractive index matching in sum and difference frequency mixing to obtain f values with typical accuracies of 10%. The 7S-nP matrix elements were derived by requiring that the calculated nonlinear susceptibility pass through the nulls which we located experimentally. Using these results, one can predict refractive index mismatch, mixing efficiency, and the effects of efficiency limiting processes such as Stark shifts and reverse Raman conversion.

6S-nP f VALUE MEASUREMENT

The oscillator strengths of the 6 (singlet) S ground state to n (singlet and triplet) P (n = 6 - 13) transitions were measured by taking advantage of the requirement of refractive index matching in nonlinear frequency conversion.[1] Effective conversion with unfocused input light beams occurs only when the propagation, or k,

0094-243X/86/1470246-7$3.00

vector of the generated frequency matches the sum of the k vectors of the input beams. Since the k vector of the output beam is strongly dispersed near the 6S-nP transitions, there is a frequency near each transition where index matching is achieved, and hence where the intensity of the output light peaks. In practice, two or three lasers generate the input light. The first two photons reach the 7S level while the energy of the third photon is scanned to locate the index matching frequency. If the k vectors of the input beams are known, this determines the refractive index at the index match point. We use two methods to systematically vary the sum of the input k vectors to map complete refractive index dispersion curves near the nP levels. The first involves two input lasers. One is tuned to the 6S-7S two-photon resonance, and the second is scanned. The input k vector is varied by changing the small (0-2 degree) crossing angle of the beams. The second method involves three input lasers. The frequencies of the first two always sum to the 6S-7S transition frequency, but one is set to various small (10-300 wavenumbers) detunings from the 6S-6 (triplet) P transition, thus varying the input k vector sum.

The results of these measurements are dispersion curves such as that shown in Figure 1. Because extensive portions of the curves are measured rather than a single point, the deduced f value for each transition is almost independent of the f values of the other transitions.[2] This is particularly important in Hg because the continuum contribution to the refractive index is substantial. Accuracies of 10% have been achieved. This high accuracy is due largely to the fact that our measurements are based on frequency measurements alone for the three laser method and on frequency, crossing angle, and Hg density only for the crossed beam method.

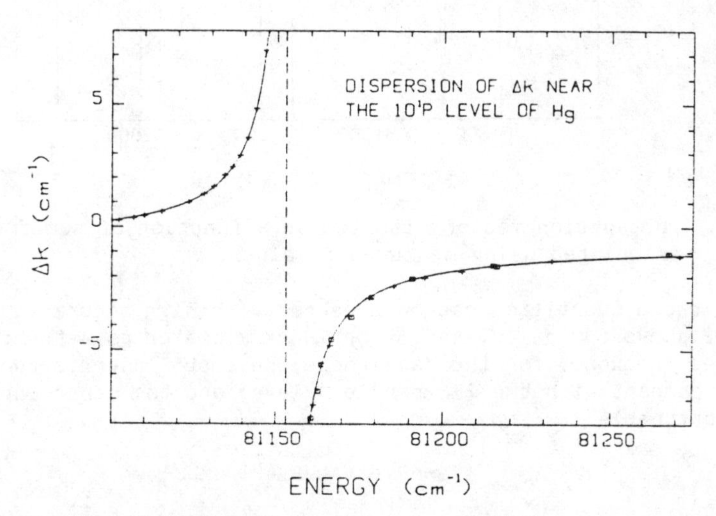

Fig. 1. Propagation vector mismatch as a function of sum frequency for a number density of 2.41 x 10(16)/cc. One laser is two-photon resonant with 7S.

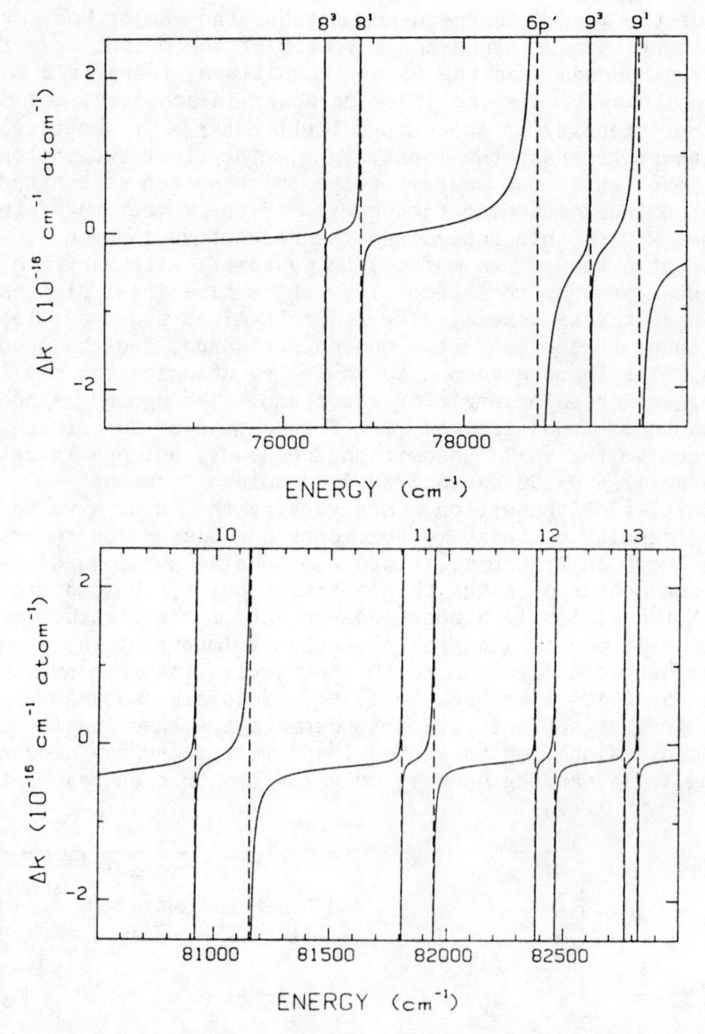

Fig. 2. Propagation vector mismatch as a function of sum frequency calculated using measured f values.

Each of these quantities can be measured with high accuracy. Figure 2 shows our calculated Δk or index mismatch as a function of generated frequency for the case where one input laser is two-photon resonant with the 7S (singlet) level and the other input laser is tunable.

7S-nP f VALUES

The expression for the nonlinear susceptibility responsible for sum and difference frequency mixing consists of a sum of matrix element products divided by resonance denominators. This leads to interference nulls between resonances. For the measurement of 7S (singlet and triplet) to nP (singlet and triplet) f values, we located as many interference nulls or deep minima in the nonlinear susceptibility as possible. An example of a null is shown in Figure 3. From both the 7 singlet and 7 triplet S states, we located 12 nulls. We then adjust the set of $\langle 7S|d|nP\rangle$ matrix elements to cause the calculated nonlinear susceptibility to pass through the nulls.[3] Since there are 18 f values to fit in each case, this is insufficient data to make a unique determination of f values. Fortunately, there is supplementary data available from other sources which allows us to achieve adequate results. This data consists of lifetimes for the 7S states and a few f values measured by Mosburg and Wilke[4] in an Hg discharge. By imposing in addition the constraint that the f values sum to unity, we arrived at the values listed in Table I. Uncertainties in the resulting f values are expected to be approximately 50%.

This method of measurement has provided us not only with the magnitudes of the f values but also with the signs of the matrix element products $\langle 6S|d|nP\rangle\langle 7S|d|nP\rangle$ which occur in the calculation of the nonlinear susceptibility. It is also very sensitive, as can be seen from the size of the smaller f values we measured. Also, like the 6S-nP measurements, it is based on measurements of frequencies only. Our best estimate of the frequency dependence of the susceptibility when the 7 (singlet) S is two-photon resonant is shown in Figure 4 along with the measured null positions. The major sources of uncertainty in these results are the large error limits of Mosburg and Wilke and the lack of knowledge of the 7S to continuum matrix elements.

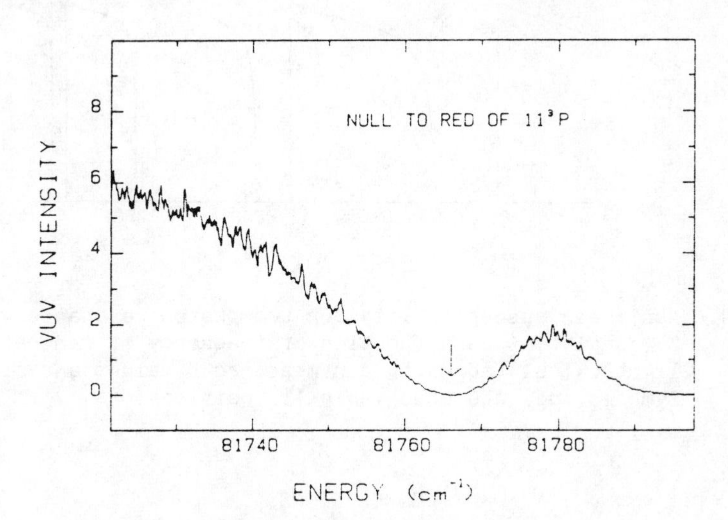

Fig. 3. Example of an observed null in the nonlinear susceptibility near the 11P for two-photon resonance with the 7 (singlet) S. Null occurs at position indicated by arrow. Zero at right edge is due to index mismatch.

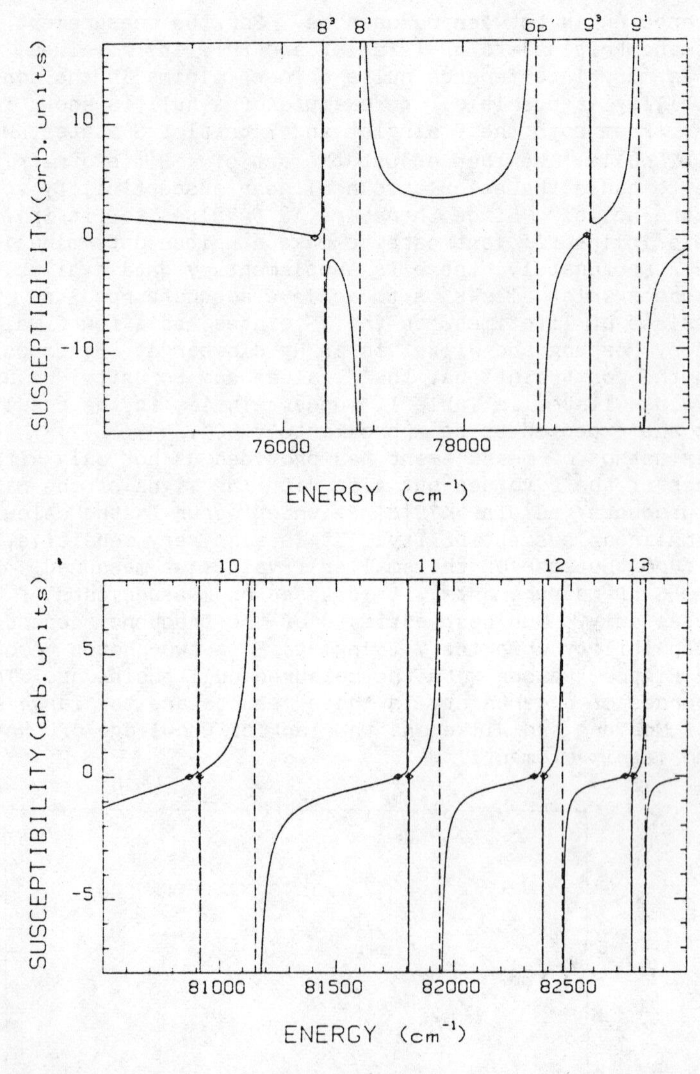

Fig. 4. Nonlinear susceptibility for two-photon resonance with 7 (singlet) S as a function of frequency of generated light calculated using our measured f values and signs. Symbols indicate observed null positions.

CONCLUSIONS

We have measured those dipole transition matrix elements of importance in nonlinear frequency mixing in Hg through the 7S intermediate levels. These results allow calculations of refractive index mismatch and nonlinear susceptibilities with very

Table I - Measured Hg f Values

	6^1S	7^1S	7^3S		6^1S	7^1S	7^3S
6^1P	1.15	.14	.0051	6^3P	.024	.0029	.11
7^1P	.020	1.19	.030	7^3P	$<10^{-5}$	.072	.52
8^1P	.0105	.076	4.1×10^{-4}	8^3P	.0014	.0011	.011
6p'	.15	.011	6.0×10^{-4}				
9^1P	.070	.012	8.3×10^{-4}	9^3P	.0055	2.0×10^{-5}	5.0×10^{-4}
10^1P	.0155	.0034	1.9×10^{-4}	10^3P	.0018	8.7×10^{-8}	1.1×10^{-4}
11^1P	.0050	.0015	7.4×10^{-5}	11^3P	.0011	5.0×10^{-7}	2.6×10^{-5}
12^1P	.0023	6.3×10^{-4}	2.2×10^{-5}	12^3P	.0008	1.8×10^{-6}	1.3×10^{-5}
13^1P	.0011	2.4×10^{-4}	1.1×10^{-5}	13^3P	.0005	3.2×10^{-6}	6.8×10^{-6}

much greater accuracy than previously was possible. Although we used three input lasers, it is expected that in most applications, two input lasers will be used. Our results indicate that in this case, index matching can still be achieved over 93% of the 120-140 nm range by adding the proper proportions of Xe or Kr.

Although we measured f values from both the 7 singlet and triplet S states, the susceptibility for the singlet intermediate is considerably larger except for very small ranges near the triplet P levels. Because of that and the necessity of three input lasers for mixing through the triplet (selection rules require unequal frequencies for 6S-7 triplet S transition), it is expected that mixing through the triplet will have very limited application.

The techniques developed in this work have proven to give highly accurate f values for transitions from the ground state. Because a significant portion of the refractive index dispersion curve is measured, the resulting f values are largely independent of the f values of other transitions including continuum contributions. The f values from the excited states are not as accurate because they depend on supplemental data with large uncertainty. The technique is very sensitive, however, and yields signs as well as magnitudes of matrix elements.

ACKNOWLEDGMENT

This work was performed at Sandia National Laboratories, Albuquerque, New Mexico, supported by the U.S. Department of Energy under contract number DE-AC04-76DP00789 for the Office of Basic Energy Sciences.

1. A. V. Smith and W. J. Alford, to appear in Phys. Rev. A, May 1986, and references therein.
2. S. D. Kramer, C. H. Chen, and M. G. Payne, Proceedings of the Second Topical Meeting on Laser Techniques in the Extreme Ultraviolet, ed. S. E. Harris and T. B. Lucatorto (AIP, New York, 1984).
   H. Puell and C. R. Vidal, Opt. Commun. 19, 279 (1976).
   J. J. Wynne and R. Beigang, Phys. Rev. A, 23, 2736 (1981).
   R. Mahon and F. S. Tomkins, IEEE J. Quantum Electron. QE-18, 913 (1982).
3. W. J. Alford and A. V. Smith, to be submitted to JOSA B,
   A. Snitzer, K. Hollenberg and W. Behmenburg, Laser Techniques in the Extreme Ultraviolet, ed. S. E. Harris and T. B. Lucatorto, AIP (1984), p. 361.
4. E. R. Mosburg and M. D. Wilke, J. Quant. Spectrosc. Radiat. Transfer 19, 69 (1978).

X-RAY BRAGG OPTICS

V.V. Aristov, A.A. Snigirev, Yu.A. Basov, A.Yu. Nikulin

Institute of Problems of Microelectronics Technology and Superpure Materials, USSR Academy of Sciences, 142432 Chernogolovka, USSR

ABSTRACT

X-ray diffraction optics is being in progress towards applications of conventional methods adopted from visible light optics. When evaluating prospects of its employment it should be kept in mind that the index of reflection for X-rays differs from unit by the magnitude of an order $\Delta n = 10^{-3} - 10^{-5}$, so that thickness t of the diffraction element is to amount to $(\Delta n)^{-1}$ of wavelengths. It means that any X-ray optic element is three-dimensional that is why potentialities of "plane" X-ray optics are limited. In the past years a number of papers have proposed Bragg diffraction phenomena to be employed for developing X-ray optic systems[1-4]. The idea to use Bragg diffraction integrally with diffraction on an artificially fabricated Bragg-grating-based structure enables one to get over a number of difficulties and shortcomings inherent in "plane" X-ray optics. Below are given experimental results of the studies and the analysis of some distinguishing features of Bragg-Fresnel optics (BFO).

1. BRAGG DIFFRACTION ON A MONOBLOCK CRYSTALLINE LATTICE (KINEMATIC APPROXIMATION)

Bragg diffraction is commonly considered as diffraction of X-ray wave $\vec{K}_0$ on a single crystal. In this case an angular spectrum of a plane wave diffracted by a crystal is determined by interplanar spacing d, reciprocal grating vector $\vec{H}(|\vec{H}| = 1/d)$ and Laue interferometric function L, which depends on the crystal sizes a, b and t in directions x, y and z, respectively.

$$L(H) = \mathrm{sinc}(\pi \Delta H_x a)\,\mathrm{sinc}(\pi \Delta H_y b)\,\mathrm{sinc}(\pi \Delta H_z t) \qquad (1)$$

here $\mathrm{sinc}X=\frac{\sin X}{X}$. The wave angular spectrum in the direction close to $\vec{K}_H$ ($\vec{K}_H=\vec{K}_0+\vec{H}$) is to be found using the Ewald construction, where the node of a reciprocal grating L(H) in size is intersected with a sphere of $|\vec{K}|=1/\lambda$ radius.

Assume that two or more crystals take part in scattering simultaneously. In this case the waves diffracted by each of the crystals interfere so that the Laue function and spectrum $\Delta\vec{K}$ change. Find the spectrum view at the plane wave scattering on N equal crystals spaced at regular intervals with period C. Let $N \gg 1$, $b \gg t$, $b \gg a$. In this case:

$$L(H)=\delta(\Delta H_y)\,\delta\left(\Delta H_x - m\frac{1}{c}\right)\mathrm{sinc}(\pi\Delta H_x a)\,\mathrm{sinc}(\pi\Delta H_z t) \quad (2)$$

Here m is integer. As is seen from (2), the spectrum is descrete and can vary with the experimental geometry.

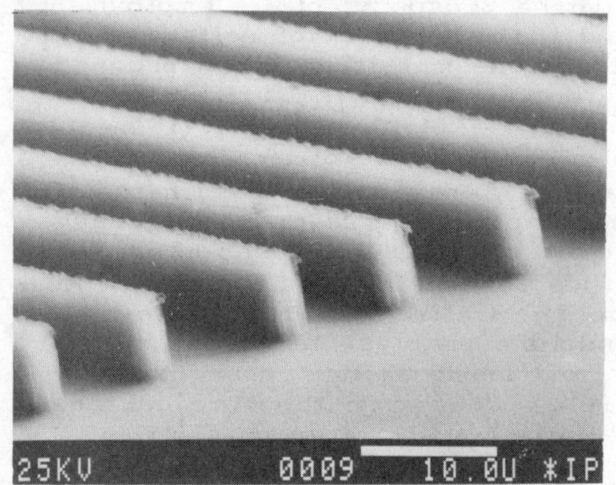

Fig. 1. Periodic structure on the surface of a perfect single crystal.

Fig. 1 displays the photo of a grating consisting of equal coherently arranged crystals, the Laue function of which is given by formula (2). The grating is fabricated on the surface of a monoblock perfect Si crystal by microlithography. Fig. 2 presents wave spectra scattered by the grating during symmetric Bragg diffraction of a plane wave on (III) planes (λ=1.54Å). The changes observed in the spectrum pattern are described within the framework of the kinematic approximation of the diffraction theory. It should be noted that there is the identity between the spectra of waves scattered in the direction close to $\vec{K}_H$ and $\vec{K}_0$ (in the absence of Bragg reflection). The advantage of use of Bragg scattering lies in the fact that varying the diffraction conditions (and thus the experimental geometry) may affect the spectrum pattern (e.g. to reduce relative intensity of zero order). Let us consider characteristic properties of imaging in the Fresnel zone during Bragg diffraction.

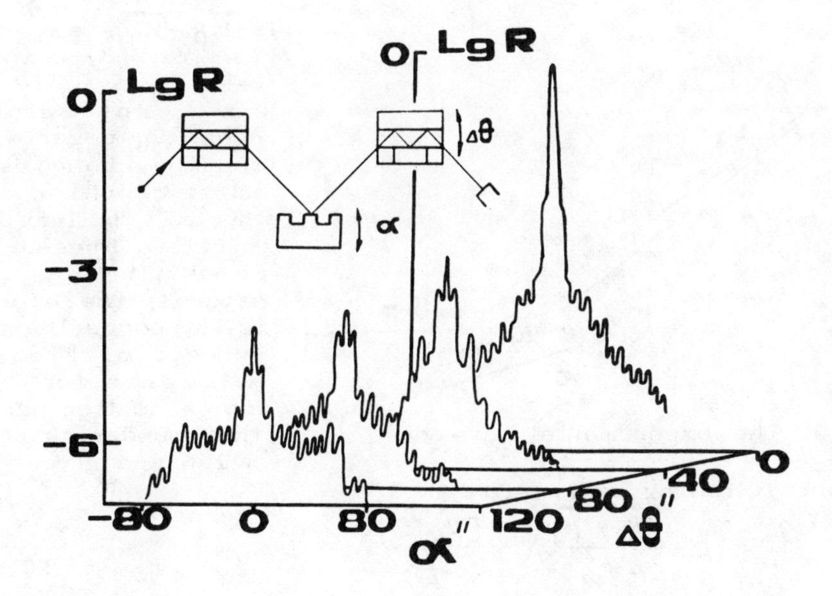

Fig. 2. Diffraction image of the X-ray wave scattering on a periodic structure.

2. IMAGING DURING BRAGG DIFFRACTION (KINEMATIC APPROXIMATION)

The classical theory considers X-ray Bragg diffraction in the framework of the approximation of Fraunhofer diffraction. It gives an idea of Bragg diffraction as diffraction in a strictly preset direction. Solution of the problem posed in the section calls for developing a diffraction theory allowing for X-ray wave scattering at small source-to-crystal (L_1) and crystal-to-observation plane (L_2) distances. Such a theory has been developed in [5,6]. It is based on evaluating (as well as the Fresnel diffraction theory) wave phases φ to P at scattering of an incident wave from $S(L_1)$ by different crystal points r (Fig. 3). At $|r| \ll |L_1|$ and $|r| \ll |L_2|$

$$\varphi(r)=2\pi\left[\frac{(L_1+L_2)}{\lambda}+(\vec{K}_0-\vec{K}_H)\vec{r}+\frac{[\vec{K}_0\times\vec{r}]^2}{2L_1}\lambda+\frac{[\vec{K}_H\times\vec{r}]^2}{2L_2}\lambda\right] \quad (3)$$

At $(\vec{K}_0-\vec{K}_H)\vec{r}=n$, where n is integer, i.e., when at point r=0 the Bragg condition is fulfilled the change in phase $\varphi(r)$ is determined but by the last-mentioned terms in(3).

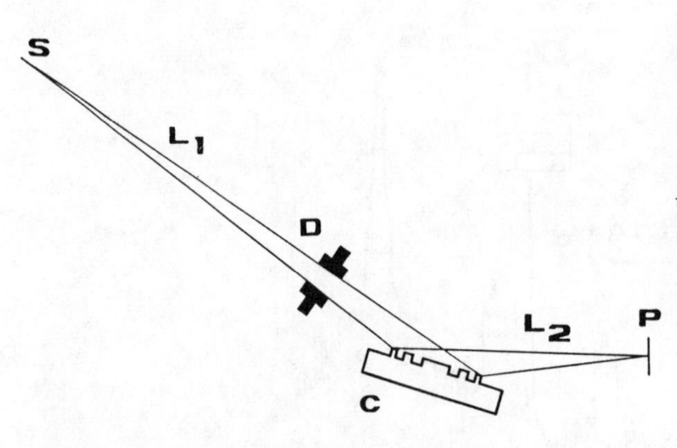

Fig.3. The experimental lay-out.

If $\overrightarrow{K_0} = \overrightarrow{K_H}$, expression (3) transforms to a well-known expression in optics, describing the change in the wave phase, which is utilized to obtain Fresnel zones. It is natural, therefore, to introduce the concept of Fresnel zones for Bragg scattering, the boundaries of which are given by the following expression:

$$(m-1) \leq \frac{[\overrightarrow{K_0} \times \overrightarrow{\tau}]^2}{L_1}\lambda + \frac{[\overrightarrow{K_H} \times \overrightarrow{\tau}]^2}{L_2}\lambda \leq m \qquad (4)$$

Here, m is integer. Diffraction on such a structure, as well as in optics, can result in focusing of a plane or spherical wave in direction $\overrightarrow{K_H}$. Fig.4 presents a photo of a linear zone structure fabricated on the surface of a Si single crystal by microlithography. The last zone is 0.5 μm in size, the relief depth is 2.5 μm. It follows from (4) that the vertical shape of the zone structure is dictated by the distance equality between L_1 and L_2. At $L_1 > L_2$ the zones rotate in the direction of the observation point. Yet, the vertical zone structure permits achieving a good focusing at $L_1 \gg L_2$ as well. Fig.3 displays a scheme of the Bragg diffraction exper-

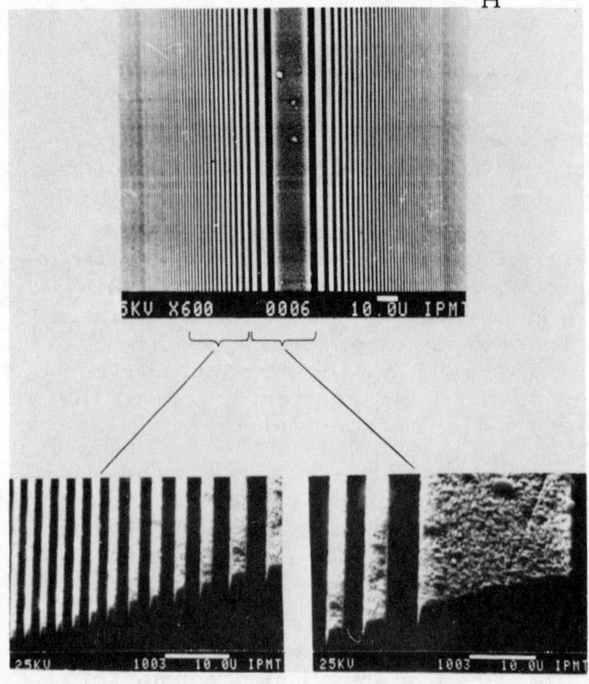

Fig.4. A Fresnel zone plate.

iment on (III) radiation planes CuK_α : $L_1 = 1$m, $L_2 = 0.04$m. The X-ray topograph (Fig.5) demonstrates a reduced image of the tube focus. The presence of the background is likely to be due to divergence in shape of zone plates from the optimal one. The experiment reveals the fact that as in the case of plane Fresnel optics, one is able to vary distance L_1 over rather a wide range and to achieve focusing with different magnification indices.

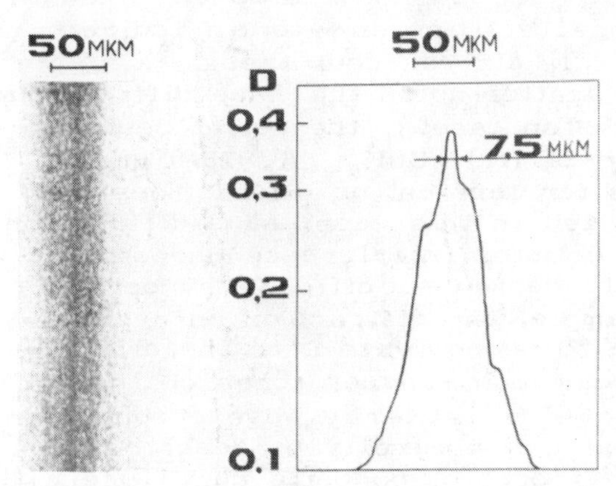

Fig.5. The X-ray tube focus image.

3. SOME CHARACTERISTIC PROPERTIES OF BRAGG FRESNEL OPTICS (BFO) DICTATED BY MULTIPLE (DYNAMIC) SCATTERING

It is a matter of general experience that the phenomenon of Bragg diffraction becomes considerably complicated at multiple scattering. At present the analysis of distinguishing features of dynamic Bragg Fresnel optics without any systematic studies involves difficulties. Some conclusions, however, can be drawn on the base of qualitative arguments.

a) In the "reflection case" the radiation penetration depth in crystal during Bragg diffraction is defined by extinction depth Λ_e, which is*, for instance, for silicon (III) CuK_α equal to 1.5 μm. Then it is hoped that X-ray Bragg optics of a high (close to unit) diffraction efficiency with spatial resolution up to 0.1–0.2 μm can be developed (we believe that a structure with a width-to-height ratio of an order 10 can be readily fabricated). Extinction length for Ge is three times less which permits increasing resolution to 300 Å–500 Å. The inclusion of dynamic scattering during Bragg diffraction in the "reflection case" may bring about variations in ratios of spatial frequency amplitudes compared to the amplitudes computed

* In the case of soft X-ray radiation of multilayer interference mirrors we can, evidently, restrict ourselves to kinematic approximation.

on the Fourier-analysis base. It should also be taken into account thatfalse reflections due to an indirect excitation mechanism are likely to occur there.

b) In the case of scattering in the Laue diffraction geometry ("the transmission case"), the relief depth is to be much greater (for Si(III), CuK_α Λ_e=18.5 μm).It means that possibilities for fabricating diffractions with high resolution are limited in this case. Laue diffraction, however, gives rise to complex interference phenomena which can be employed to design specific X-ray optic elements. Thus, for example, the diffraction on a plane-parallel plate is known to be equivalent to the diffraction on a slit with marking out Fresnel zones on the crystal across its thickness [5,6]. X-ray wave propagation in a thick ($t>\Lambda_e$) crystal can nominally be described by propagation of two wave fronts of opposite curvature with wavelength $2\Lambda_e$ travelling along atomic planes [1]. For symmetric Laue diffraction the first Fresnel zone size across the crystal thickness is $t_1=[\lambda(L_1+L_2)]^{\frac{1}{2}}/\sin\theta$ (θ is the Bragg angle). Setting t_1 equal to the size of the first zone for Bloch waves $t_B=(2\Lambda_e t)^{\frac{1}{2}}$, we obtain the crystal thickness

$$t_s=\frac{(L_1+L_2)\lambda}{2\Lambda_e \sin^2\theta}$$

, at which wave divergence in vacuum is compensated for one of the Bloch waves (the wave front curvature is opposite to that in vacuum) that is focusing occurs there. For another wave the imaginary focus shifts aside. Changes in crystal distances and thickness cause these waves to interfere forming a complicated diffraction pattern [2]. For a wedge-shaped crystal focusing is observed for fixed thickness t_s. Unfortunately, at other thicknesses focusing is absent since the extinction length is an incident angle function so that the divergence compensation occurs for the first zone only. This experimental results shows that in the Laue diffraction case the Fresnel zone structure should be fabricated with regard to dynamic scattering. In a specific case of an incident plane wave upon the crystal surface and L=0 discussed first in [1], the focusing Fresnel zone structure is given by the Bessel function.

CONCLUSIONS

The experimental results demonstrate potentialities of X-ray optics of a new type, that is, of Bragg Fresnel optics (BFO). The BFO elements are found to be durable: phase focusing elements can be fabricated. Focusing and diffraction efficiency of BFO elements can be easily operated. Resolution of BFO as well as other types of X-ray elements is governed by fabrication precision of its structure.

REFERENCES

1. V.L.Indenbom, V.I.Kaganer. Metallofizika. 1, 17(1979).
2. V.V.Aristov, V.I.Polovinkina, A.M.Afanas'ev, V.G.Kohn. Acta Cryst. A36, 1002 (1980).
3. F.N.Chukhovskii, P.V.Petrashen'. DAN 228, 1087(1976).
4. V.V.Aristov, A.Yu.Nikulin, A.A.Snigirev. XII Hungarian Diffraction Conference, Sopron (1985).
5. E.V.Shulakov, V.V.Aristov. IIIVsesoyuznoye soveshaniye "Coherent Irradiation Interaction with Matter", Uzhgorod, 39 (1985).
6. E.V.Shulakov, V.V.Aristov. Acta Cryst. A38, 454 (1982).

EXTREME ULTRAVIOLET MULTILAYER REFLECTORS

Marion L. Scott, Paul N. Arendt, and Bernard J. Cameron
Materials Science and Technology Division

Brian E. Newnam
Chemistry Division
Los Alamos National Laboratory
Los Alamos, New Mexico 87545

David Windt and Webster Cash
University of Colorado
Boulder, Colorado 80309

ABSTRACT

We have investigated the design, fabrication, and reflectance measurements of a multilayer silver/silicon reflector for use at 58.4 nm. Our results indicate that reflectors in the extreme ultraviolet do not perform as well as predicted due to the presence of surface oxides and other surface contamination layers. In addition, we have found that the correct optical constants for silver have now been published. We find also that these multilayer coatings can be utilized as reflective polarizers in the EUV with an extinction ratio of 75:1 and a throughput of 28% for the s-polarized component of the beam.

INTRODUCTION

Multilayer dielectric reflectors and simple metal coatings do not, in general, provide adequate reflectance for laser cavities or other optical requirements in the extreme ultraviolet spectrum. Various multilayer combinations of metals, semiconductors, and dielectrics are required at different wavelengths within the EUV spectrum to achieve the desired reflectance performance. In this work, we have investigated the properties of a multilayer coating of silver and silicon at 58.4 nm. One of our principal objectives has been to determine what factors limit the performance of these multilayers when utilized in the EUV.

DESIGN

The design of multilayer reflectors requires knowledge of both the refactive indexes and the extinction coefficients of the materials involved in the coating at the design wavelength. If these optical constants are not accurately known, the resulting performance of the coating design cannot be predicted. We began our design effort by ascertaining the optical constants of silver and silicon from the literature. Initially, we had no way of inferring whether these constants were the correct ones. Our initial four layer design for this coating is shown in Fig. 1. The reflectance versus thickness and angle of incidence predicted for this design are shown in Figs. 2 and 3, respectively. The predicted normal incidence reflectance from these figures is 38%.

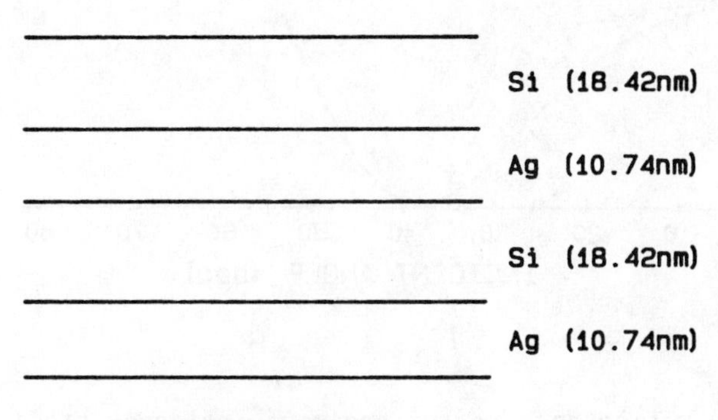

Fig. 1. A four layer reflective coating designed for use at 58.4 nm using silver and silicon on a silicon substrate.

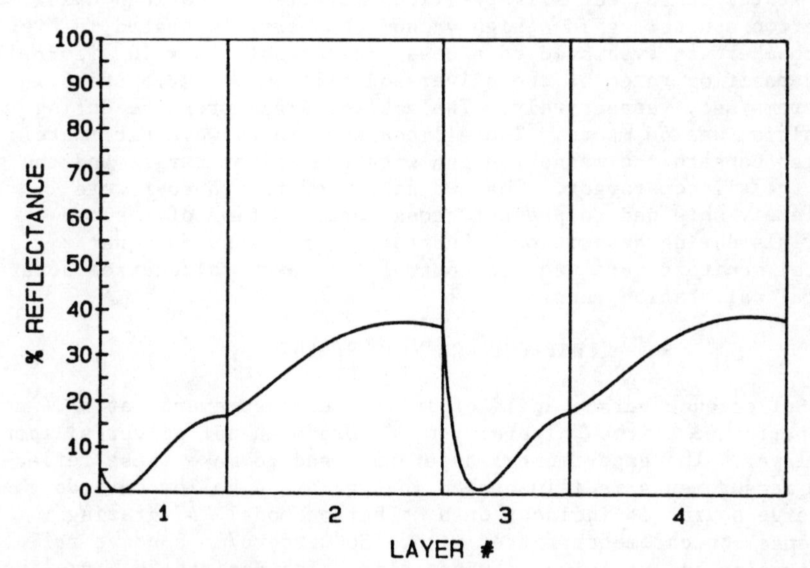

Fig. 2. The reflectance versus thickness for the four layer silver/ silicon multilayer coating.

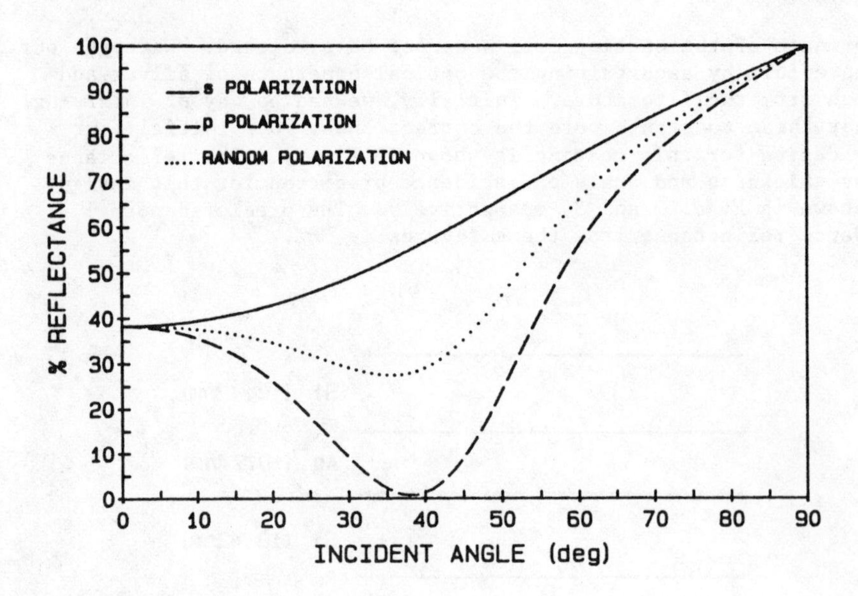

Fig. 3. The reflectance versus angle of incidence predicted for the four layer silver/silicon multilayer coating.

FABRICATION

We fabricated our silver/silicon multilayer coatings using DC magnetron sputtering in a high vacuum chamber illustrated in Fig. 4. The chamber was evacuated to a base pressure of 1.4×10^{-6} Torr. The deposition rates of the silver and silicon were 3.6 and 6.1 Angstroms/sec, respectively. The ambient Argon pressure during deposition was 40 mtorr. The silicon substrates were alternately rotated beneath the magnetron gun with the silver target and the gun with the silicon target. The two halves of the chamber were completely shielded to prevent cross contamination of the two materials during deposition. Shutters on the guns and quartz crystal monitors were used to control the layer thicknesses after several calibration runs.

REFLECTANCE MEASUREMENTS

Reflectance versus angle of incidence measurements at 58.4 nm were performed at the University of Colorado on our silver/silicon multilayer. The experimental apparatus used to make these reflectance measurements is illustrated in Fig. 5. A hollow cathode gas-discharge source is incident on a McPherson Model 247 grazing incidence monochrometer fitted with a 300 groove/mm concave reflection grating and scanning entrance slit. The desired 58.4 nm line of Helium is transmitted by the monochrometer to the sample with a resolution of .2 nm and angular divergence of 7 arcmin. The results

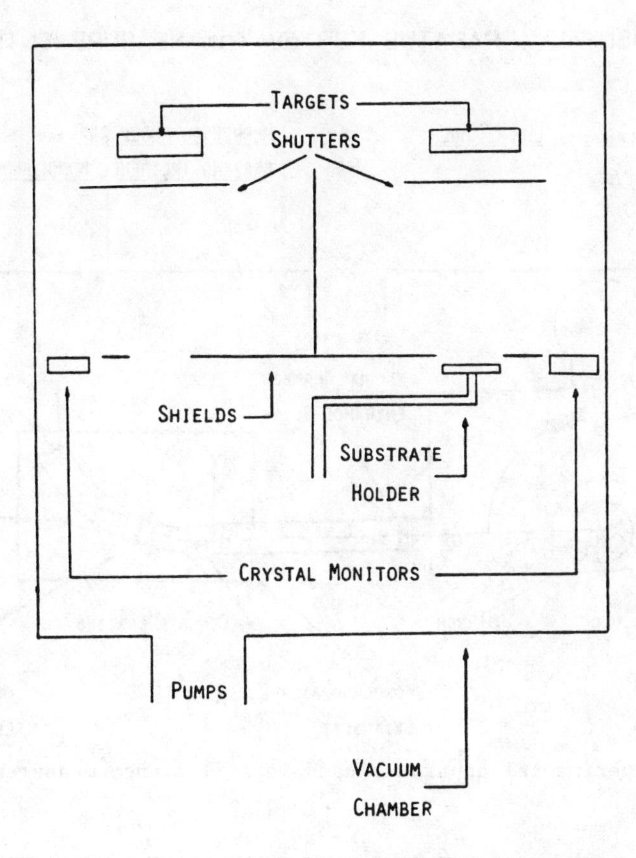

Fig. 4. Schematic diagram of the high vacuum chamber utilizing for DC magnetron sputtering the silver/silicon multilayer coating.

of these measurements are compared to our original design calculation in Fig. 6.

NEW MODEL REQUIRED

The significantly lower reflectance values measured for this coating required that we remodel the coating with surface contamination layers of silicon oxide and carbon (illustrated in Fig. 7) as well as reviewing the optical constants of the materials. We found optical constants for silver in reference 1 that better described our data when used in conjunction with the new model. An excellent fit to the measured reflectance versus angle of incidence data was obtained with the new model as indicated in Fig. 8. This fit was obtained by varying only the thicknesses of the two surface contamination layer thicknesses. The optical constants for silicon dioxide and carbon were obtained from references 1 and 2 respectively. The values obtained for the thicknesses of these two layers were 1.2 nm

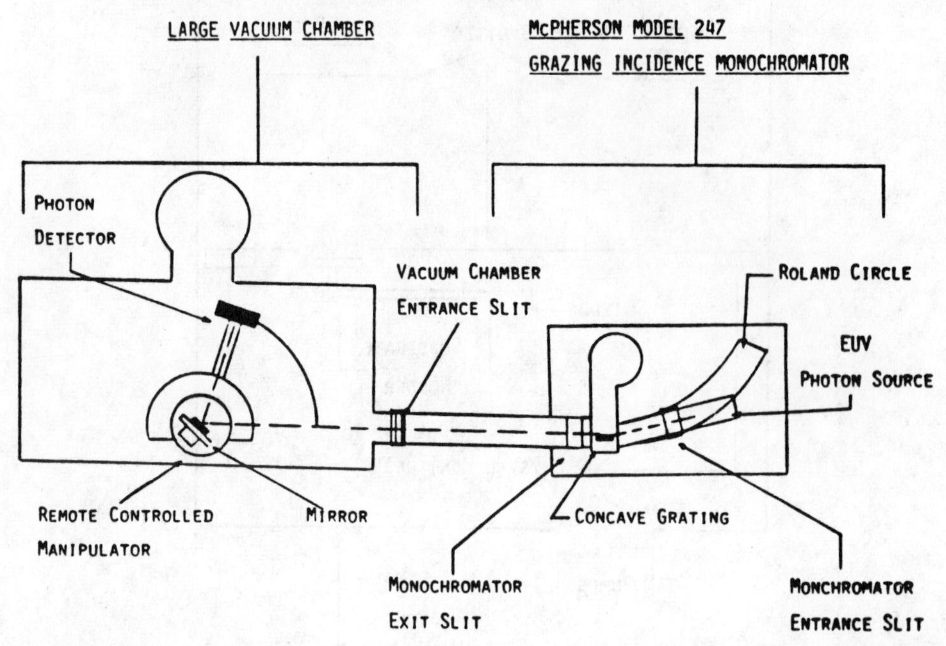

Fig. 5. Experimental apparatus used in reflectance measurements at 58.4 nm.

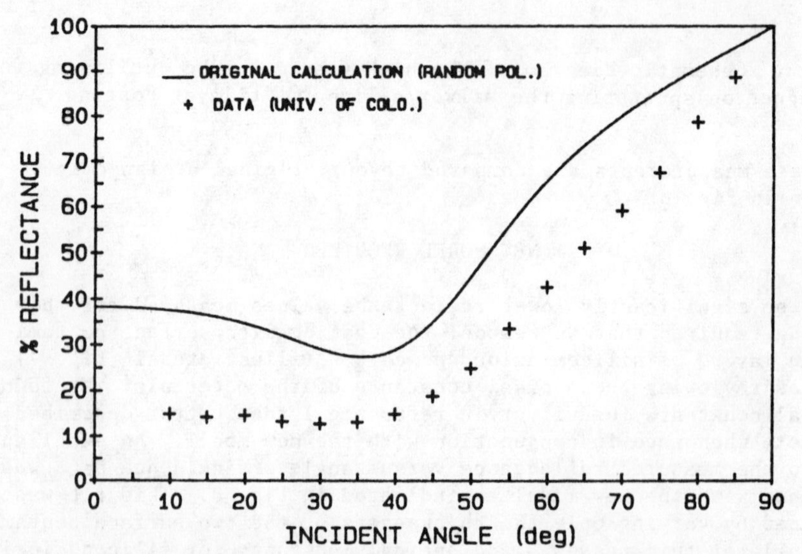

Fig. 6. A comparison of our original design calculation of the reflectance versus angle of incidence with the measured values.

for the silicon oxide and 1.1 nm for the carbon. These results are in agreement with previous surface science experience in terms of the type and thickness of surface contamination on silicon.

We wondered why such thin surface layers should have such a profound effect on the reflectance of our multilayer coating. The answer is a combination of two factors. The first factor is that the optical constants of these materials indicate that the layers

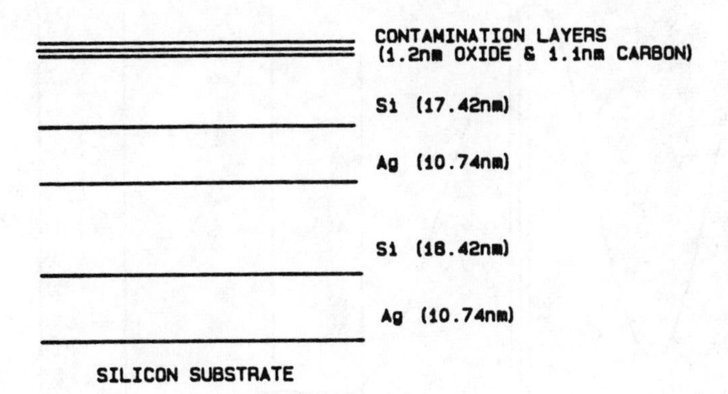

Fig. 7. New model of our silver/silicon multilayer coating incorporating surface contamination layers of silicon oxide and carbon.

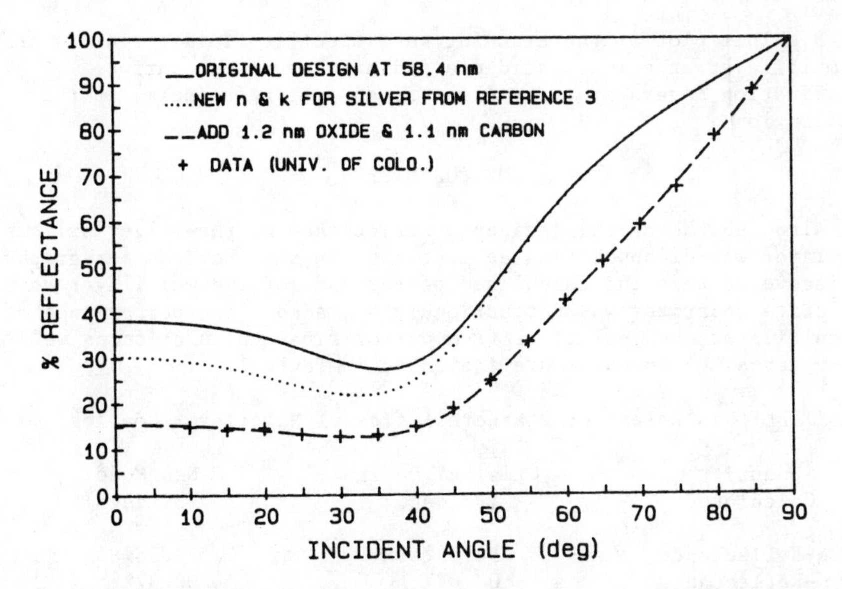

Fig. 8. Comparison of calculations from new model and measured reflectance versus angle of incidence data at 58.4 nm.

are highly absorbing at this wavelength (58.4 nm). The second factor is that these surface layers occur very near the anti-node of the standing wave electric field set up by the incident and reflected EUV beams as shown in Fig. 9. This combination causes the surface contamination layers to exhibit a maximum absorption of the beam, and therefore, a maximum reduction in the reflectance of the coating.

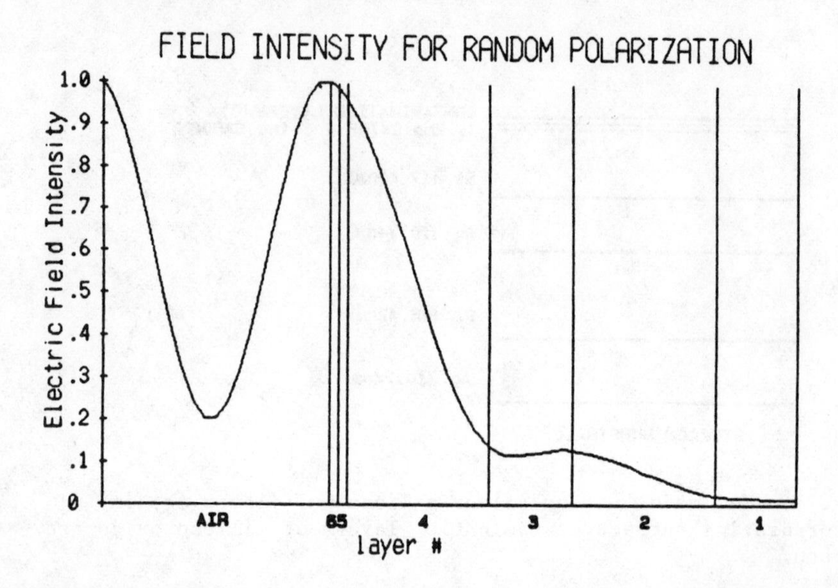

Fig. 9. This plot of the standing wave electric field intensity in the multilayer at normal incidence indicates that the surface contamination layers are very close to an anti-node (peak) in this distribution.

EUV POLARIZER

Although the normal incidence reflectance of the silver/silicon multilayer was disappointing as mentioned in the previous paragraphs, we discovered that the calculated performance of the multilayer as a reflective polarizer was not seriously degraded. The performance of the multilayer as a polarizer from our original calculation as well as the actual performance are indicated in Table I.

TABLE I. Polarizer Characteristics of Multilayer Coating

Quantity Calculated	Original Design (38 deg.)	New Model (39 deg.)
s-Reflectance	54.92%	27.58%
p-Reflectance	00.64%	00.37%
Rs / Rp	86:1	75:1

Although the s-polarized reflectance of the coating decreased by a factor of two from the original calculation, the extinction ratio remained about the same because the p-polarized reflectance also decreased by approximately a factor of two.

CONCLUSIONS

We have found optical constants that agree well with the reflectance measurements of our multilayer coating of silver/silicon, including those of the surface contamination layers. Surface contamination layers are very important in the EUV and must be modeled to achieve satisfactory agreement between theory and experiment. We also conclude that multilayer coatings can be utilized as reflective polarizers in the EUV.

REFERENCES

1. E. D. Palik, Ed., Handbook of Optical Constants of Solids (Academic Press, Inc., Orlando, 1985).

2. H. J. Hagemann, W. Gudat, and C. Kunz, DESY SR-74/7 (1974).

OVERTONE PRODUCTION OF SOFT X-RAYS WITH FREE-ELECTRON LASERS*

M. J. Schmitt, C. J. Elliott, K. Lee, B. D. McVey
Los Alamos National Laboratory, Los Alamos, NM 87545

ABSTRACT

Two one-dimensional free-electron laser codes have recently been written that include harmonic generation. A comparison of the results of these codes show that a self-consistent treatment of the harmonic interaction is not required in the presence of a strong fundamental field. Use of these codes to predict the effects of emittance on harmonic production have been conducted. The effects of wiggler-field amplitude fluctuations and odd-harmonic wiggler-field components on the harmonic-radiation production are also discussed.

INTRODUCTION

The harmonics generated in free-electron laser (FEL) oscillator experiments are produced primarily through coherent-spontaneous radiation of the fundamentally-bunched electron beam. This mechanism dominates over the much weaker harmonic linear-gain mechanism which cannot overcome cavity mirror losses. The amplitude of each harmonic is determined by its Fourier component of the transverse current, scaled by the harmonic coupling coefficient.

This paper explores the effects of emittance, wiggler-field amplitude fluctuations and odd-harmonic wiggler-field components on harmonic production. A comparison between the one-dimensional codes is discussed in Section I. In Section II the effects of emittance on the harmonic amplitudes are calculated assuming constant gain in the fundamental. Small amplitude fluctuations in the wiggler magnetic field are modeled in Section III. Modifications of the harmonic coupling coefficients resulting from a small third-harmonic wiggler-field component are given in Section IV along with their consequences on harmonic production.

---

* Work performed under the auspices of, and supported by, the Division of Advanced Energy Projects of the U. S. Department of Energy Office of Basic Energy Sciences.

0094-243X/86/1470268-7$3.00

I. HARMONIC FEL CODES

The equations describing the interaction of odd-harmonic frequencies with an electron beam in a FEL are well documented[1,2]. These equations have been incorporated into a one-dimensional (1D) code written by one of us (C. J. Elliott) to model both the radiation produced by the electron beam and the effect of the harmonic fields back on the electrons. This self-consistent code integrates the phase-averaged field equations given by

$$\frac{\partial \mathcal{E}_f}{\partial x} = -2\pi\rho K_f(\xi) \left\langle \frac{e^{-i(f\psi_j+\varphi_{of})}}{\gamma_j} \right\rangle_{electrons} \tag{1}$$

and the energy equation for each electron that states

$$\frac{d\gamma_j}{dt} = \sum_{f=1}^{\infty} \frac{e\mathcal{E}_f}{2m_e c\gamma_j} K_f(\xi)\cos(f\psi_j + \varphi_{0f}) \tag{2}$$

where j denotes the jth electron and f is the harmonic number. This code has been compared to another code (HFELP) originally written by B. D. McVey and later modified by M. J. Schmitt to include the harmonic radiation produced by a fundamentally bunched electron beam. The effects of the harmonic radiation fields on the electrons are not included in HFELP so that the electrons are only affected by the fundamental electromagnetic field that is assumed to be much larger than the harmonic fields.

To analyze the coherent-spontaneous emission of a FEL operating at the fundamental we assume the oscillator to have reached saturation so that a large fundamental wave exists. A single-pass calculation is then performed assuming

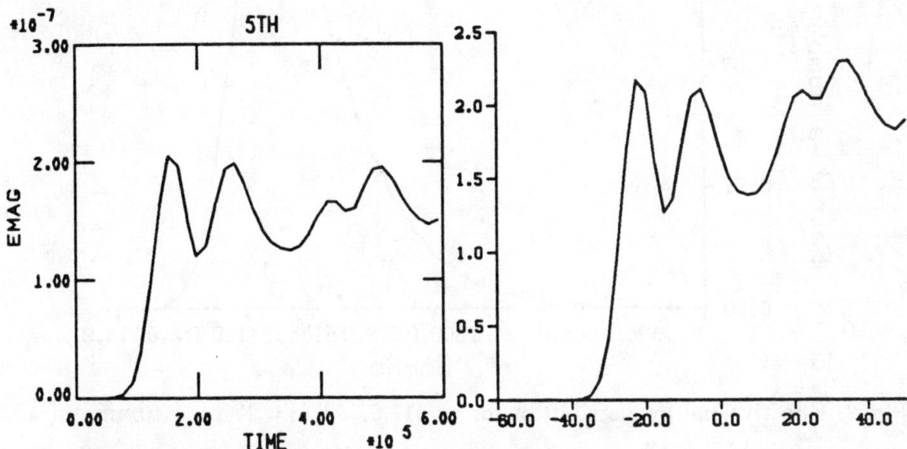

Fig. 1. Harmonic output from the self-consistent code (a) and from HFELP (b).

the saturated fundamental signal exists at the entrance to the wiggler. This calculation is valid assuming the mirror losses at the harmonics prevent the harmonic radiation from circulating in the cavity. Gain at the fundamental is adjusted through an effective fill factor to match that of the 3D calculation[3]. The results of both codes, for input parameters resembling those of the Los Alamos oscillator experiment[4], are shown in Figures 1a,b for the fifth harmonic.

Results at the first few odd harmonics observed thus far all yield identical results. From this comparison it is seen that for small harmonic powers ($P_f/P_1 \ll 1$) a self-consistent analysis is not required to obtain the correct harmonic output levels. Thus, calculation of the harmonic radiation can be done economically.

II. EMITTANCE EFFECTS

The growth of electron beam emittance has deleterious effects on FEL gain and efficiency. To evaluate how emittance affects harmonic production, simulations were performed with the 1D code HFELP. Owing to the one-dimensional nature of the code, electron-beam emittance was modeled as an effective electron energy spread. The energy spread represents the effective axial energy of the electrons. A typical distribution for the electron beam used in the simulations for the proposed Los Alamos XUV FEL[5] is shown in Figure 2 for the case $\epsilon_n = 40\pi$mm·mrad.

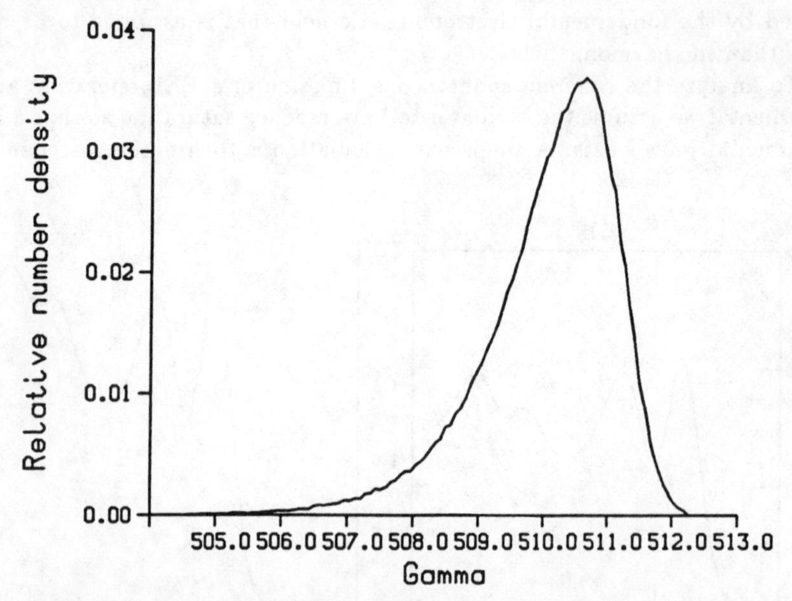

Fig. 2. Effective energy distribution for γ=511.5, $\Delta\gamma/\gamma$=.2%, ϵ_n=40πmm·mrad.

The parameters used in the XUV simulations were; γ=511.5, $\Delta\gamma/\gamma$=.2%, λ_s=500Å, I=150Amps, B_w=.75T, L_w=800cm, λ_w=1.6cm and the emittance was varied from 20π to 40π. The harmonic electric-field amplitudes at the end of the wiggler for these conditions are plotted in Figure 3 for the first three odd-harmonics. All harmonics show a significant reduction in field amplitude. The third-harmonic amplitude shows the greatest effect with a reduction of ten in field or two orders-of-magnitude in power. These results clearly show that the minimization of emittance is critical for maximization of harmonic output.

III. WIGGLER FIELD ERRORS

The magnetic field that results after the construction of a wiggler magnet is not a pure sinusoid. Differences in the magnetization of each magnet and machining tolerances guarantee that the resultant field will have variations in both amplitude and wavelength. In this section we consider the effects of magnetic amplitude fluctuations on harmonic production. Although we are not treating magnetic wavenumber fluctuations directly, their treatment can be handled using this same formalism. Since this is a 1D treatment it is tacitly assumed that any random walk of the electron beam is compensated at the appropriate intervals to keep the electron and optical beams concentric.

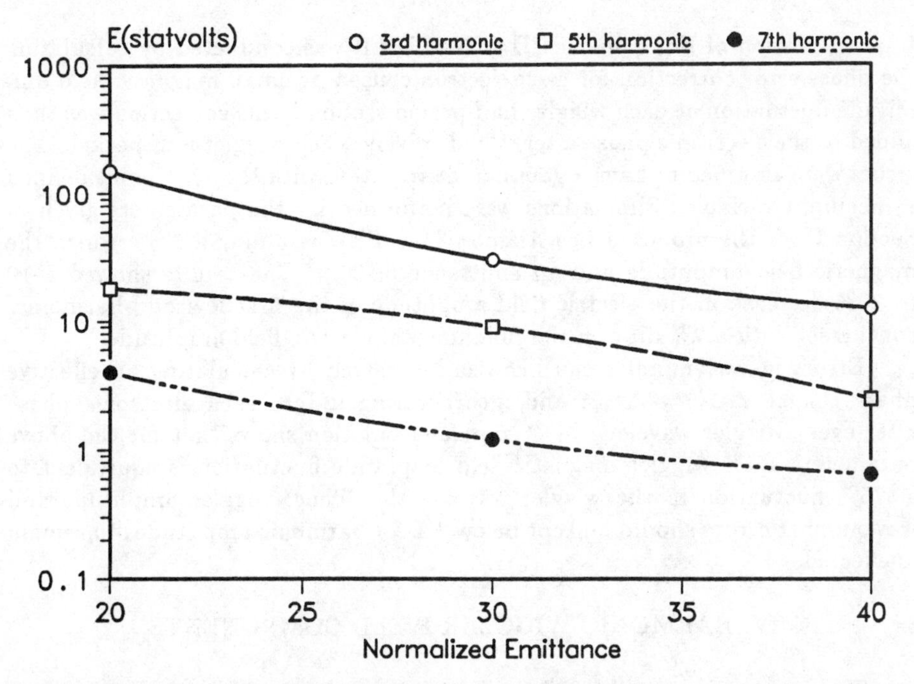

Fig. 3. Harmonic electric-field amplitudes as a function of normalized emittance (30=30πmm·mrad) for the proposed Los Alamos XUV FEL.

To model the variation of the amplitude of the wiggler field in the phase-averaged equations one can calculate an effective variation in the electron phase ψ due to the change in the magnetic-field amplitude. The electron phase in the wiggler and optical fields is given by

$$\psi = (k_w + k_s)\bar{z} - \omega t \tag{3}$$

where $\bar{z}$ is determined by the average electron velocity in the axial direction. Assuming a perturbation in the wiggler field amplitude given by ΔB_w, a resultant perturbation in the axial position will result that will in turn produce a variation in the electron phase given by

$$\Delta\psi = (k_w + k_s)\Delta\bar{z} \tag{4}$$

and upon substitution for the appropriate value of $\Delta\bar{z}$ gives

$$\Delta\psi_j = -\frac{\pi}{2}\frac{(1 + k_s/k_w)}{\gamma_j^2}a_w\Delta a_w \tag{5}$$

where a_w is the dimensionless magnetic vector potential defined

$$a_w = \frac{eB_w}{m_e c^2 k_w} \tag{6}$$

Implementation of Eq(5) in the 1D code HFELP was conducted by calculating the phase error correction for each electron caused by small magnetic-field amplitude fluctuation at each wiggler half-period section. This correction was then added to the electron's phase after it had traverse each wiggler half-period. The errors were assumed to have a gaussian distribution with the $1/e^2$ point defined as an input variable. Simulations were performed for the parameters given in Section II for the proposed Los Alamos XUV FEL assuming a 1% error in the magnetic field amplitude and an emittance of 20π. The results showed a 10 to 20% decrease in the electric field amplitude of the first few odd-harmonics concurrent with a 7% drop in the fundamental electric-field amplitude.

Errors in wavenumber can likewise be treated by calculating an effective phase change $\Delta\psi_\lambda = \Delta k_w \bar{z}$ and incorporating it into each electron's phase after every wiggler wavelength. A simple calculation shows that for the above parameters a 1% wiggler magnetic-field amplitude fluctuation is equivalent to a .75% fluctuation in the wiggler wavelength. Thus, wiggler amplitude and wavenumber errors should be kept below 1% for harmonic amplitudes to remain uneffected.

IV. HARMONIC WIGGLER FIELD COMPONENTS

To increase the strength of the FEL interaction, experimenters strive to achieve normalized magnetic vector potentials of order unity or greater. The construction of such wigglers results in highly peaked magnetic field intensities near the pole faces – significantly deviating from the analytically assumed sinusoid. A Fourier decomposition of these fields yield third-harmonic field components approaching 20% or more that of the fundamental[6]. The effects of these harmonic wiggler fields *cannot* be included by linear superposition. We have calculated how these components change the coupling coefficients for the odd-harmonic radiation and examine below how a small third-harmonic wiggler field modifies the fundamental and third-harmonic radiation.

To include the effect of a third-harmonic wiggler field we first define the transverse velocity produced by a sum of harmonic wiggler fields as

$$\beta_\perp^2 = \left(\frac{eA_w}{m_e c^2 \gamma}\right)^2 = \frac{e^2}{m_e^2 c^4 \gamma^2} \left\{\sum_{m=1}^{\infty} A_{wm} \cos(mk_0 z)\right\}^2 \tag{7}$$

which can be used to derive a new resonance condition in the high-γ limit given by

$$k_0 = \frac{k_s}{2\gamma_0^2} \left\{1 + \sum_{m=1}^{\infty} \frac{a_{wm}^2}{2}\right\} \tag{8}$$

These equations can then be used to derive new coupling coefficients for use in Eqs(1),(2). Assuming only a third-harmonic wiggler field with $a_{w3}/a_{w1} \ll 1$, the new coupling terms for the fundamental and third-harmonic are

$$\begin{aligned} \mathcal{K}_1(\xi) = {} & J_0\left(\xi_1 \frac{a_{w3}}{a_{w1}}\right) J_0(\xi_3) \\ & \cdot \left[a_{w1}\left\{J_0(\alpha_1) - J_1(\alpha_1)\right\} + a_{w3}\left\{J_1(\alpha_1) + J_2(\alpha_1)\right\}\right] \end{aligned} \tag{9}$$

and

$$\begin{aligned} \mathcal{K}_3(\xi) = {} & J_0\left(3\xi_1 \frac{a_{w3}}{a_{w1}}\right) J_0(3\xi_3) \\ & \cdot \left[a_{w1}\left\{J_2(\alpha_3) - J_1(\alpha_3)\right\} + a_{w3}\left\{J_0(\alpha_3) - J_3(\alpha_3)\right\}\right] \end{aligned} \tag{10}$$

where

$$\alpha_n = n\xi_1 \left(1 + 2\frac{a_{w3}}{a_{w1}}\right) \tag{11}$$

and

$$\xi_m = \frac{a_{wm}^2}{4m} \left(1 + \sum_{n=1,3} \frac{a_{wn}^2}{2}\right)^{-1} \tag{12}$$

A comparison of these new coefficients (assuming a 10% third-harmonic contribution) with the old[1] coefficients for parameters consistent with those of the Los Alamos oscillator experiment show that while the fundamental term is unaffected, the third-harmonic term has changed by 33%. Note that we have assumed the magnitude of a_{w3} opposite that of a_{w1} such that the resultant magnetic field is more peaked than a sinusoid. For this configuration the third-harmonic power is enhanced by 75%. This example dramatically shows how the harmonics can be affected by harmonic wiggler components. A more thorough analysis of this effect is presented elsewhere[7].

REFERENCES

1. W.B. Colson, IEEE JQE **QE-17**, 1417 (1981).
2. J.M.J. Madey and R.C. Taber, **Physics of Quantum Electronics**, vol. 7, Jacobs *et al.*,Eds., (Addison-Wesley, Reading, MA, 1980), ch. 30.
3. B.D. McVey, in *Proceedings of the Seventh International Conference on Free-Electron Lasers*, held in Tahoe City, CA, 1985, in publication.
4. B.E. Newnam *et al.*, IEEE JQE **QE-21**, 867 (1985).
5. J.C. Goldstein *et al.*, in *Proceedings of the International Conference on Insertion Devices for Synchrotron Sources*, held in Stanford, CA, on Oct. 27-30, 1985, to be published.
6. K. Halbach, Journal de Physique, Colloque C1, supplément au n°2, **Tome 44**, février 1983.
7. M.J. Schmitt and C.J. Elliott, to be published.

GAIN PHYSICS OF RF-LINAC-DRIVEN XUV FREE-ELECTRON LASERS*

John C. Goldstein, Brian D. McVey, and Brian E. Newnam
University of California, Los Alamos National Laboratory
Los Alamos, New Mexico 87545 USA

ABSTRACT

In an rf-linac-driven XUV free-electron laser oscillator, the gain depends on the details of the shape of the electron beam's phase-space distribution, particularly the distribution of electrons in the transverse (to the direction of propagation) position and velocity coordinates. This strong dependence occurs because the gain in this device is inhomogeneously broadened. Our previous theoretical studies have assumed that the transverse phase space distribution is a product of uncorrelated Gaussian functions. In the present work, we present the results of a theoretical study of the gain for non-Gaussian phase-space distributions. Such distributions arise either from a better representation of the electron beam from an rf-linac or from an emittance filter applied to the beam after the linac.

INTRODUCTION

In several previous theoretical studies[1,2,3] we have investigated the properties of a free-electron laser (FEL), operating in the extreme ultraviolet (XUV, 10-100 nm) portion of the optical spectrum, which is driven by an electron beam produced by a radio-frequency linear accelerator. The primary problem to be overcome by any FEL oscillator operating in this spectral region is the necessity of achieving large small-signal gain values (at least several hundred percent) in order to exceed the threshold of laser oscillation in an optical resonator with relatively poor reflectors. Mirror reflectances of 50%-80% are expected to become available in the XUV spectral range.[4-6] Extrapolations from presently available accelerator technology suggest that both rf-linacs and electron storage rings can generate pulses of high-energy electrons having sufficiently large peak currents to drive XUV FELs. However, one expects that electron pulses from an

*Work performed under the auspices of, and supported by, the Division of Advanced Energy Projects of the U.S. Department of Energy, Office of Basic Energy Sciences.

rf-linac will have a larger emittance (i.e., occupy a larger volume in phase-space) than those from a storage ring in which synchrotron radiation effects reduce the emittance. A direct consequence is that the gain of an rf-linac-driven XUV FEL will be severely inhomogeneously broadened. Nonetheless, our previous theoretical studies [1,2,3] indicated that sufficient gain will be available to allow for laser oscillation and that an rf-linac-driven XUV FEL would exceed by 4 or 5 orders of magnitude the light intensity in this spectral range available from present or planned synchrotron sources. Other points related to the design and operation of an rf-linac-driven XUV FEL are discussed in the paper by Newnam et al.[7] in these proceedings.

This paper presents the results of a theoretical investigation of the dependence of the magnitude of the small signal gain of an rf-linac-driven XUV FEL upon the shape of the distribution that leads to the inhomogeneous broadening. It is analogous to changing the distribution of atomic velocities in a low-pressure Doppler-broadened gas laser from a Maxwellian to other distribution functions. In a gas, the Maxwellian is the unique physically important distribution; however, there can be many different distributions in the case of an electron beam. The "correct" distribution function of electrons' transverse motion depends upon details of the design and operation of a linear accelerator and is not known a priori. In an FEL, the magnitude of the transverse emittance determines the amount of inhomogeneous broadening and, therefore, the accompanying reduction of the small-signal gain, much like the temperature determines the amount of Doppler broadening in a low-pressure gas laser. However, the physics of an rf-accelerator allows the possibility of generating electron beams with the same value of the transverse emittance but having different underlying distributions of transverse particle motion. The dependence of the gain upon different distributions, each one having the same value of the emittance, is the subject of this study.

Our previous studies[2,3] assumed that the transverse phase-space distribution of the electron beam is a product of uncorrelated Gaussian functions. The inhomogeneous broadening reduces the gain from very large values (thousands) for a perfect (zero emittance) electron beam to about ten for a beam with a finite emittance. We would like to assure ourselves that the gain is not reduced further through a dependence on the shape of the phase-space distribution. Also, it is important to quantitatively understand the reduction of small-signal gain due to the inhomogeneous broadening in order to place realistic limits on other parameters (such as the peak current, intrinsic energy spread, and undulator length) in order to assure sufficiently large gain for the laser to exceed threshold with the available mirrors.

TRANSVERSE EMITTANCE

Let $\hat{z}$ be the axis of the optical resonator and the undulator magnet. The transverse coordinates of an electron are $x, x\prime, y$, and $y\prime$ where $x\prime = dx/dz \simeq \beta_x$ and $y\prime = dy/dz \simeq \beta_y$. We are considering the case of highly relativistic electrons having $\gamma_o \sim 400$, where $\gamma_o mc^2$ is the total energy of an electron, for which the transverse velocities $\beta_x c$ and $\beta_y c$ are very small compared to the axial velocity (i.e., $\beta_x, \beta_y \ll 1$). The distribution of electrons in the beam is described by a function $P(x, x\prime, y, y\prime)$, which is normalized to unity, such that the current of electrons with coordinates x to x + dx, x$\prime$ to x$\prime$ + dx$\prime$, etc. is given by

$$dI(x, x\prime, y, y\prime) = I_o P(x, x\prime, y, y\prime)\ d\tau \ . \tag{1}$$

Here $d\tau = dx\ dx\prime\ dy\ dy\prime$ and I_o is the total current. The mean value of any function $f(x, x\prime, y, y\prime)$ of the transverse variables is given by $\langle f \rangle = \int d\tau P f$. We think of electron trajectories in which the transverse coordinates are functions of the axial coordinate z.

We shall use the definition of Fraser et al.[8] for the transverse emittance:

$$\epsilon_x = 4\pi \left[\langle x^2 \rangle \langle x\prime^2 \rangle - \langle x\ x\prime \rangle^2 \right]^{1/2} , \tag{2}$$

$$\epsilon_y = 4\pi \left[\langle y^2 \rangle \langle y\prime^2 \rangle - \langle y\ y\prime \rangle^2 \right]^{1/2} , \tag{3}$$

and

$$\epsilon = \sqrt{\epsilon_x \epsilon_y} \ . \tag{4}$$

In the following analysis, we will assume that $\epsilon_x = \epsilon_y = \epsilon$ and then calculate small-signal gain values for an XUV FEL driven by electron beams having the same total current I_o and emittance ϵ but having different phase-space distribution functions P.

We restrict our attention to a particular class of distributions P. Let us define B by

$$B = (x/\bar{x})^2 + (x\prime/\bar{x}\prime)^2 + (y/\bar{y})^2 + (y\prime/\bar{y}\prime)^2 , \tag{5}$$

where $\bar{x}, \bar{x}\prime, \bar{y}$, and $\bar{y}\prime$ are fixed constants. We consider only distributions P of the form

$$P = P(B) \ . \tag{6}$$

We assume that our FEL has an untapered parabolic-pole-face[9] undulator magnet. Such an undulator magnet, with equal focusing in x and y, induces electrons to make harmonic oscillations in the transverse variables. The wavenumber of these betatron oscillations is:

$$k_\beta = \frac{a_w k_w}{2\gamma_o} , \tag{7}$$

where $k_w = 2\pi/\lambda_w, \lambda_w$ is the wavelength of the constant-period undulator magnet, and

$$a_w = \frac{|e|B_w}{mc^2 k_w} \tag{8}$$

where B_w is the peak on-axis value of the magnetic field of the undulator.

Since the transverse motion of an electron is simple harmonic with wavenumber k_β (neglecting the wiggle motion itself which is the source of the basic FEL mechanism), one can show that if the following matching conditions are satisfied, the form of the function P(B) is the same anywhere inside the undulator; that is, for any two values of z inside the undulator, say z_1 and z_2, $P\left(B(z_1)\right) = P\left(B(z_2)\right)$. Again, this neglects high-frequency variations due to the wiggle motion itself. The matching conditions connect the four constant parameters of P:

$$\bar{x'} = k_\beta \bar{x} , \tag{9}$$

and

$$\bar{y'} = k_\beta \bar{y} . \tag{10}$$

Since, under these conditions, the fundamental form of P is invariant anywhere inside the undulator, so is any mean value calculated from P such as the electron beam dimensions $\langle x^2\rangle$, $\langle y^2\rangle$, and $\langle r^2\rangle = \langle x^2\rangle + \langle y^2\rangle$.

FOUR SPECIFIC TRANSVERSE PHASE-SPACE DISTRIBUTIONS

We will consider the following four distribution functions which are functions of the transverse coordinates through the variable B which is defined by Eq. (5):

Case 1: Gaussian

$$P_1 d\tau = N exp(-B) d\tau \tag{11}$$

Case 2: "Almost Gaussian"

$$P_2 d\tau = 1.67097 N exp(-B^{1.5}) d\tau \tag{12}$$

Case 3: "Almost Uniform"

$$P_3 d\tau = 2.25305 N exp(-B^4) d\tau \tag{13}$$

Case 4: Uniform

$$P_4 d\tau = 2N d\tau \tag{14}$$

for all values of $x, x\prime, y,$ and $y\prime$ that lie within the four-dimensional ellipsoidal surface defined by B = 1.

The factor N is a normalization factor:

$$N = \left(\pi^2 \bar{x}\ \bar{x}\prime\ \bar{y}\ \bar{y}\prime\right)^{-1} \quad . \tag{15}$$

We exhibit explicitly for these four distributions expressions for the emittance. Let Z ($\bar{Z}$) refer to any of the four quantities x, $x\prime$, y, $y\prime$ ($\bar{x}$, $\bar{x}\prime$, $\bar{y}$, $\bar{y}\prime$). 4

Case 1: Gaussian

$$\langle Z^2 \rangle = 0.5 \bar{Z}^2 \tag{16}$$

$$\epsilon_x = 2\pi\ \bar{x}\ \bar{x}\prime; \quad \epsilon_y = 2\pi\ \bar{y}\ \bar{y}\prime \tag{17}$$

Case 2: Almost Gaussian

$$\langle Z^2 \rangle = 0.278 \bar{Z}^2 \tag{18}$$

$$\epsilon_x = 1.1122\pi\ \bar{x}\ \bar{x}\prime; \quad \epsilon_y = 1.1122\pi\ \bar{y}\ \bar{y}\prime \tag{19}$$

Case 3: "Almost Uniform"

$$\langle Z^2 \rangle = 0.17175 \bar{Z}^2 \tag{20}$$

$$\epsilon_x = 0.687\pi\ \bar{x}\ \bar{x}\prime; \quad \epsilon_y = 0.687\pi\ \bar{y}\ \bar{y}\prime \tag{21}$$

Case 4: Uniform

$$\langle Z^2 \rangle = (1/6)\ \bar{Z}^2 \tag{22}$$

$$\epsilon_x = (2/3)\pi\ \bar{x}\ \bar{x}\prime; \quad \epsilon_y = (2/3)\pi\ \bar{y}\ \bar{y}\prime \tag{23}$$

In the numerical examples below, we will always assume that $\epsilon_x = \epsilon_y = \epsilon$. Hence, specifying ϵ, together with the matching conditions Eqs. (9) and (10), is sufficient to fully determine the four parameters of each distribution function.

EFFECTIVE ENERGY DISTRIBUTION FUNCTION

The effects upon FEL performance of a beam with a finite transverse emittance can be understood in terms of the shape of the associated effective energy distribution.[2,3] If one has a monoenergetic beam in which each electron has energy $\gamma_o mc^2$ but also has a nonzero transverse motion, then to the FEL the electron behaves as if it had an energy $\gamma_{eff} mc^2$ where γ_{eff} is given by

$$\gamma_{eff} = \gamma_o \left[\frac{1 + 0.5\, a_w^2}{1 + 0.5\, a_w^2 + \gamma_o^2 \left[\beta_x^2 + \beta_y^2 + (k_\beta x)^2 + (k_\beta y)^2 \right]} \right]^{1/2} . \quad (24)$$

Here, the electron's transverse coordinates and velocities are x, y and $\beta_x c, \beta_y c$, and k_β and a_w are given by Eqs. (7) and (8). Physically, the FEL resonance is a relation between the undulator magnet properties, the optical wavelength, and the electron's axial velocity. In a beam with a finite emittance, an electron's axial velocity is reduced if it has a finite transverse motion. This reduction is expressed by using γ_{eff} instead of γ_o in the FEL equations of motion.

Note that given a distribution of transverse coordinates P, we can uniquely obtain an associated distribution of $\gamma_{eff}, f(\gamma_{eff})$. Also, our selection of a class of distributions P which can be matched to the undulator means that $f(\gamma_{eff})$ is the same anywhere within the undulator, as is P. (This, of course, is true for untapered undulator magnets whose characteristics do not depend on the longitudinal position z of an electron, and also holds only if the FEL interaction converts only a small fraction of the electron's initial energy $\gamma_o mc^2$ into light.) Our previous studies[2,3] have shown that, for cases of interest for an XUV FEL, the significance of $f(\gamma_{eff})$ is that the gain is approximately proportional to $df/d\gamma_{eff}$: that is, maximum gain occurs where the slope is largest (and positive), and, therefore, distributions f that have large maximum slopes are desirable in that they lead to large values of the gain.

Another consequence of considering distributions P that can be matched to the undulator is that, if the beam is not matched, then the distribution $f(\gamma_{eff})$ would be different at different axial positions z along the undulator. Hence, conceptually, the conditions of peak gain would vary along the undulator in this nonstationary situation, and it would be much harder to understand the processes that lead to the gain.

The specific distributions of effective energy $f(\gamma_{eff})$ that correspond to the four transverse phase-space distributions P above are shown in Figs. 1-4. The undulator and electron beam properties used to derive these curves are summarized in Table I. Note that we are considering the electron beam to have zero intrinsic energy spread, that is, to be monoenergetic. From the discussion above, it is clear that one would expect Case 4, the uniform distribution, to yield the highest gain since $f(\gamma_{eff})$ for that case, Fig. 4, has a prominent sharp edge where the slope is very large.

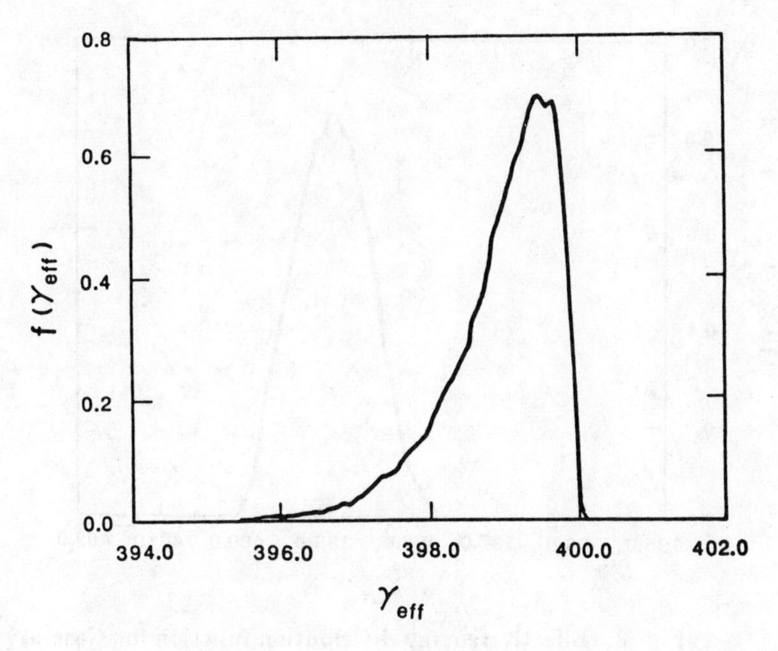

Fig. 1. Effective energy distribution function for Case 1.

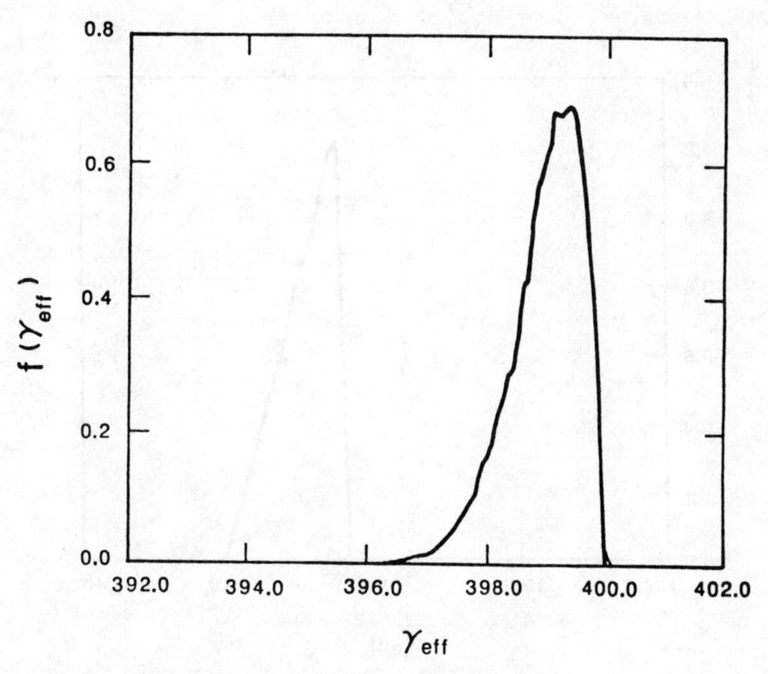

Fig. 2. Effective energy distribution function for Case 2.

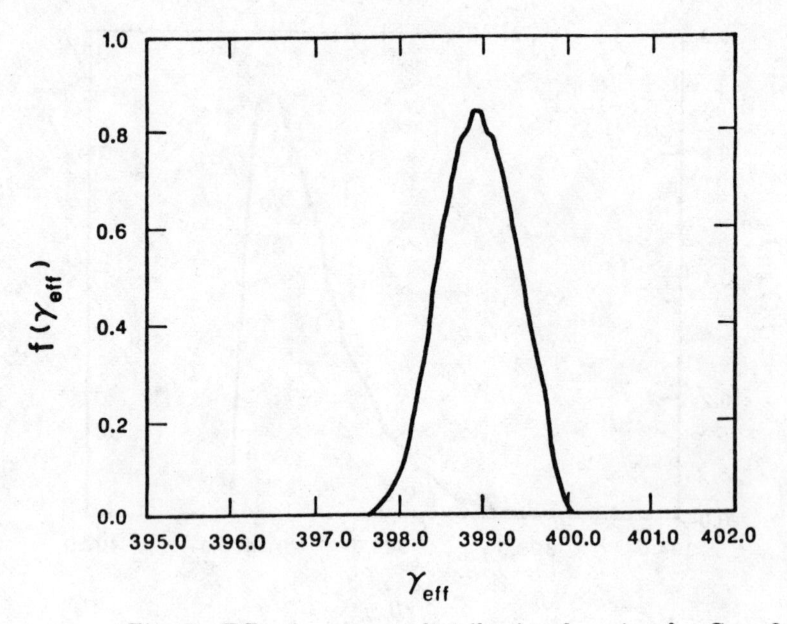

Fig. 3: Effective energy distribution function for Case 3.

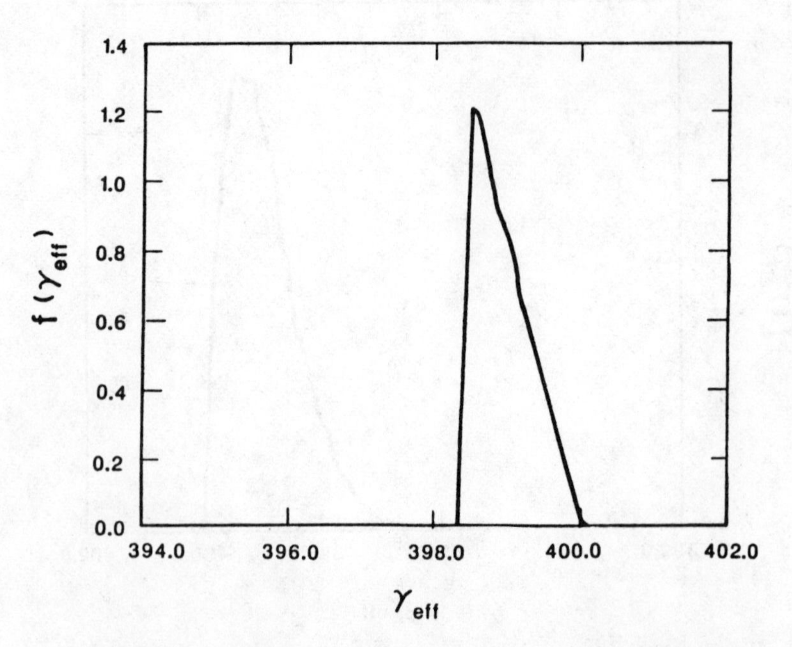

Fig. 4. Effective energy distribution function for Case 4.

TABLE I
PARAMETER VALUES FOR GAIN CALCULATIONS

Optical Parameters	
a) wavelength	variable, near 82 nm
b) Rayleigh range	variable, 50-1200 cm
Untapered Undulator Parameters	
a) wavelength, λw	1.6 cm
b) peak magnetic field, Bw	0.75 T
c) a_w	1.12
d) full gap	0.4 cm
e) length	6-m or 12-m
f) parabolic-pole-face with equal focusing in x and y	
g) magnet material	$SmCo_5$
Electron Beam Parameters	
a) peak current	100 A
b) E/mc^2	$\gamma_o = 400$
c) ϵ	3.0630×10^{-5} cm rad
d) $\epsilon_n = \gamma_o\ \epsilon$	39π mm-mrad
e) $\lambda_\beta = 2\ \pi/k_\beta$	1.143×10^3 cm
f) fractional intrinsic energy spread $\Delta\gamma/\gamma_o$	0

RESULTS OF NUMERICAL SMALL-SIGNAL GAIN CALCULATIONS

We emphasize that the ideas associated with $f(\gamma_{eff})$ are helpful in physically understanding the gain mechanism. However, we have used the 3-D FEL simulation code FELEX[10] to numerically.compute the results presented below. In this code, the 3-D particle motion and optical diffraction are handled numerically. FELEX does not directly use the distribution function $f(\gamma_{eff})$ in computing the optical gain. All of the results presented below are single-pass gains, not multiple-pass calculations in which the optical mode shape and the gain are calculated iteratively until a self-consistent solution for both is reached.[3] We assume that the light at the entrance to the undulator is in the lowest-order Gaussian mode which is specified by a Rayleigh range and an optical wavelength.[11] The amplitude of the optical field is fixed by specification of an initial light intensity. For each value of the Rayleigh range, the optical wavelength is varied to obtain the maximum gain. These are single wavefront calculations which neglect the pulsed nature of the electron and optical beams.

Figure 5 shows the results of numerically calculating the small-signal single-pass power gain versus the Rayleigh range of the incident light. The other system parameters are specified in Table I, and these results are for a 12-m undulator. The four curves correspond to the four distributions listed above. Note that the dotted curve, which is for Case 4, the uniform distribution, has been reduced in ordinate by a factor of ten. Hence, this case, which gives a maximum gain of about 130 for a Rayleigh range of 600 cm, yields the highest gain of the four cases, all of which have the same normalized emittance ($\gamma_o \epsilon = \epsilon_n$) of $39\pi \times 10^{-4}$ cm rad. Note that the highest gain exceeds the lowest gain by almost a factor of ten. The different dependences on Rayleigh range for the four cases is of interest: if one knew that the electron beam was Gaussian, one would need an optical resonator that produced a Rayleigh range of 300 cm, whereas 600 cm is needed if the beam is correctly described by distribution P_4.

Figure 6 shows results for identical conditions but using a 6-m undulator. The magnitude of the gain is lower than for the 12-m undulator, but Case 1, the Gaussian, shows the highest gain. Note that the optimum Rayleigh range for each case has also changed. For a simple two-mirror optical resonator with mirrors of fixed radii of curvature, the Rayleigh range can be changed by changing the length of the resonator.[11] However, in an FEL oscillator, the length of the optical resonator is not arbitrary but is closely tied to the time interval between successive electron pulses from the accelerator.[12] Hence, different phase-space distributions can substantially impact the design of the optimum FEL optical resonator.

Figure 7 shows the effect of relaxing one assumption made up to this point: the gain for three different distributions is plotted versus intrinsic fractional energy spread ($\Delta\gamma/\gamma_o$). The transverse emittance is the same as before, but one sees that for large energy spreads $\sim 0.4\%$, all cases look similar (and have substantially depressed gains). In this regime, the physics of the inhomogeneous broadening is totally dominated by the real energy spread of the electron beam, not the emittance. The distribution $f(\gamma_{eff})$ would be almost identical with $f(\gamma)$, the distribution of actual electron energies which in these calculations is taken to be Gaussian in shape about $\gamma_o = 400$. The full width at e^{-1} points is the abscissa of Fig. 7. Evidently, the intrinsic energy spread must be held to a few tenths of a percent if this laser is to exceed threshold.

Figure 8 shows the effect of changing the current upon the magnitude of the small-signal gain for two of the phase-space distributions. The upper pair of curves, one solid and one dotted which intersect at a current of 100 A and a gain of 130, are for the uniform distribution, Case 4. The lower pair of curves, which intersect at a current of 100 A and a gain of 32, are for the Gaussian distribution, Case 1. The solid lines show the effect of changing the current while the emittance is held constant. The dotted lines show the effect of keeping

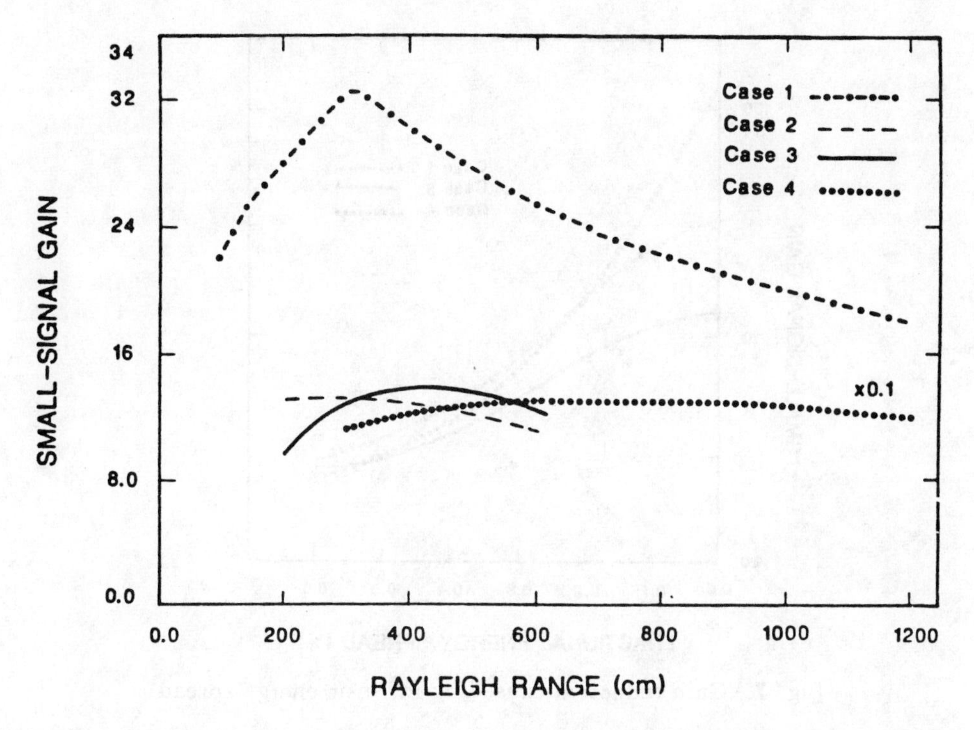

Fig. 5. Small-signal power gain vs. Rayleigh range for 12-m undulator.

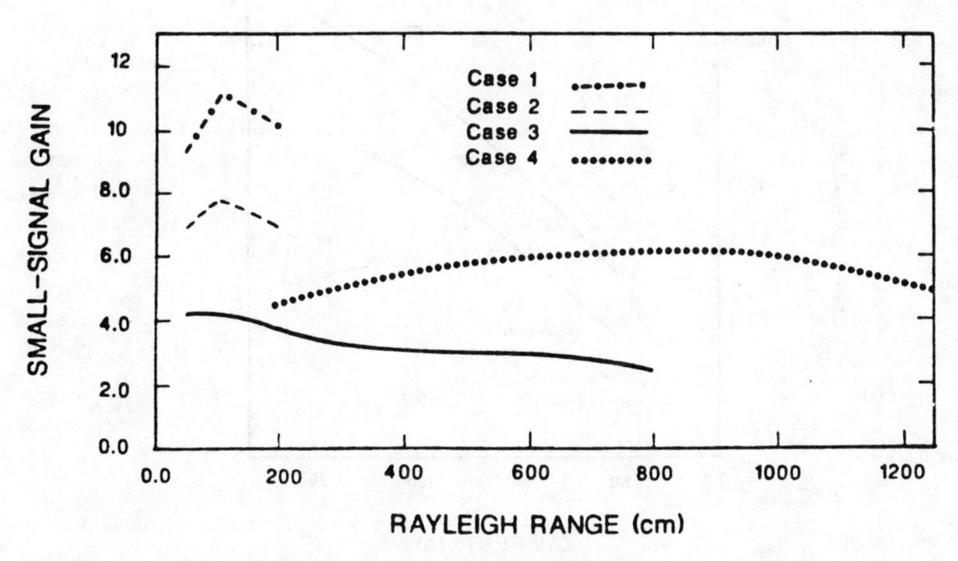

Fig. 6. Small-signal power gain vs. Rayleigh range for 6-m undulator.

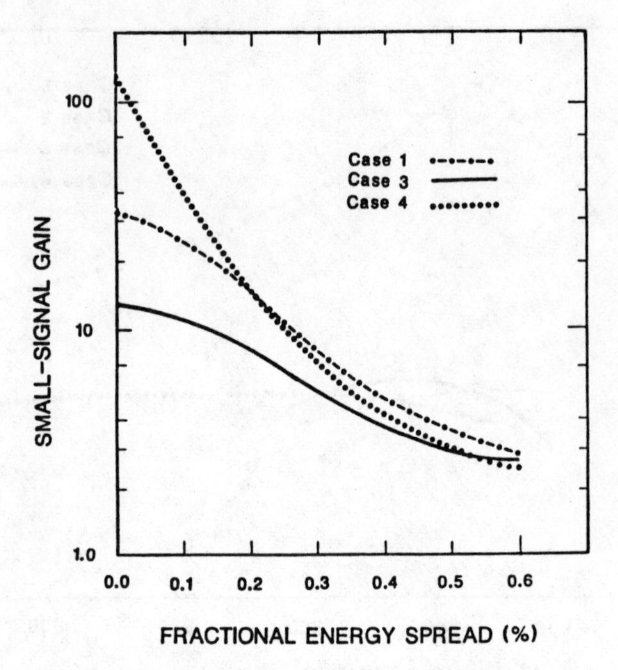

Fig. 7. Gain reduction caused by intrinsic energy spread.

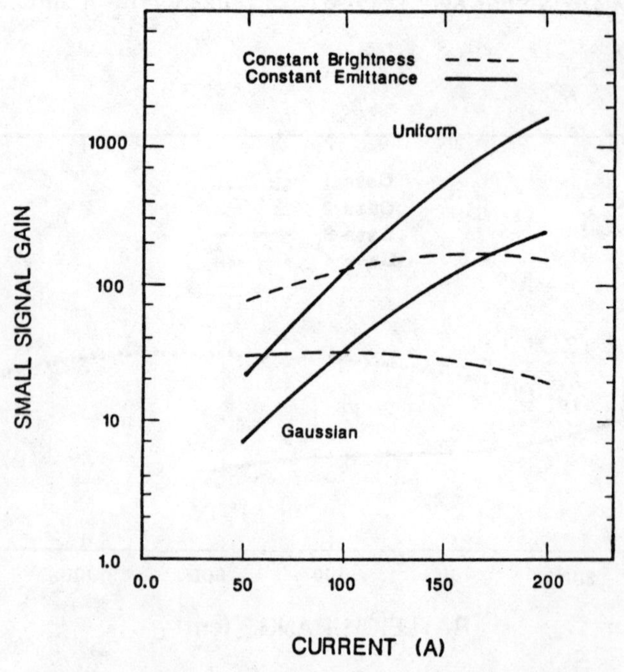

Fig. 8. Variation of gain with current for 12-m undulator.

the brightness constant: the brightness is proportional to the current divided by the square of the emittance. Hence, for the dotted curves the emittance changes like the square root of the change in current.

We emphasize that the most important parameter in determining the magnitude of the gain is the value of the emittance (whatever the underlying distribution). Looking at the upper pair of curves for the uniform distribution, we see that gain drops by almost a factor of 10 for a $\simeq$ 40% increase in the emittance at 200 A current.

DISCUSSION

We have found that, for the range of parameters of interest for an rf-linac-driven XUV FEL, the particular distribution that yields the highest gain depends upon the length of the undulator magnet used. The physical insight provided by the shape of the effective energy distribution $f(\gamma_{eff})$ must be supplemented by considerations of the transverse spatial overlap between the electron and light beams. We have not seen strong optical guiding effects[13] in any of these calculations; rather, the optical mode is distorted from that of free-space propagation. Hence, we use optical beam size variations—in the qualitative arguments below—based upon free-space propagation mode sizes. In free space, the mean squared optical mode radius $\langle w^2 \rangle$ averaged over the undulator length for a mode focused at the middle of the undulator can be written as

$$\langle w^2 \rangle = \left(\frac{\lambda}{\pi}\right)\left[Z_R + (L_w/2)^2/(3Z_R)\right] \tag{25}$$

where L_w is the undulator length, Z_R is the Rayleigh range, and λ is the optical wavelength. For a fixed L_w, the minimum $\langle w^2 \rangle$ occurs for $Z_R = L_w/(2\sqrt{3})$. In that case

$$\langle w^2 \rangle_{min} = \lambda \; L_w/\pi\sqrt{3} \; . \tag{26}$$

The results of Figs. 5 and 6 show that maximum gain for Case 1, the Gaussian phase-space distribution, is achieved under approximately this condition. On the other hand, maximum gain for Case 4, the uniform distribution, is attained for large values of $\langle w^2 \rangle$ where Eq. (25) reduces approximately to

$$\langle w^2 \rangle \simeq \frac{\lambda}{\pi} Z_R \tag{27}$$

i.e., the Rayleigh range is large and $\langle w^2 \rangle$ is approximately independent of the undulator length. We note that the physical-space current densities corresponding to Case 1 and Case 4, obtained from $J(x,y) = I_o \int \; dx\prime \; dy\prime \; P$, are

$$J_1 = (2k_\beta \; I_o/\epsilon) \; exp\left(-\left[(x/\bar{x})^2 + (y/\bar{y})^2\right]\right) \; , \tag{28}$$

and

$$J_4 = (4k_\beta\ I_o/3\epsilon)\left[1 - (x/\bar{x})^2 - (y/\bar{y})^2\right] \ , \tag{29}$$

so that the Gaussian case has a higher current density on axis than the uniform case by 50% for the same emittance:

$$J_4(0,0)/J_1(0,0) = 0.67 \ . \tag{30}$$

Hence, for small optical spot sizes $\langle w^2 \rangle$, the Gaussian has the higher gain. The electrons that contribute to the sharp edge of Fig. 4, $f(\gamma_{eff})$ for the uniform distribution, have a large transverse energy, since $\gamma_{eff} \approx \gamma_o$ − Const. × (transverse energy). Those electrons clearly lie near the surface of the ellipse B = 1 and are in some sense on the "outside" of the real-space current density J_4. In order for those electrons to be effective in generating gain, there must be light at large radius, and so maximum gain in this case occurs for large $\langle w^2 \rangle$. The fact that Case 4 gives the highest gain for a long undulator, but not for the shorter undulator, may be connected with the finite "delay length" needed before exponential gain becomes apparent; Case 1 may never be in the exponential gain regime. Note that the variation with undulator length of the gain is also different for the four cases.

The dependence of the magnitude of the gain upon the Rayleigh range of the incident light, the length of the undulator, and the size of the electron beam for a Gaussian electron beam shape has been discussed by Colson and Elleaume[14] for low-gain conditions with no inhomogeneous broadening or betatron motion, and fixed lowest-order Gaussian optical mode shape. As pointed out in Ref. 14, the small-signal power gain curves of Figs. 5 and 6 roll off at large Rayleigh range because the energy extracted from the electron beam, with a fixed on-axis light intensity at the undulator's entrance, becomes smaller relative to the total power in the light beam due to the increased optical beam transverse dimension with increasing Rayleigh range. The curves roll off at small Rayleigh range due to this effect as well as the rapid phase variation of the light ($\phi\ \alpha\ tan^{-1}\ (z/Z_R)$) which shifts the light out of resonance with the electron beam. We note that arbitrary values of Z_R are not allowed since the optical beam must be smaller than the undulator gap. In our case, this means that $60 \leq Z_R \leq 1500$ for the 6-m case and $240 \leq Z_R \leq 1500$ for the 12-m case to avoid severe vignetting at the ends of the undulator.

SUMMARY AND CONCLUSIONS

The dependence of the small-signal gain of an rf-linac-driven XUV FEL upon the shape of the electrons' transverse phase-space distribution function has been studied. Single-pass small-signal gain values were calculated using the three-dimensional FEL simulation code FELEX.[10] A particular class of distribution functions was considered. Electron beams characterized by a distribution belonging to this class can always be matched to the undulator magnet so that the phase-space distribution, as well as averages computed from it, are invariant with respect to axial position along the magnet's length.

We have found that, for a given value of the emittance, different transverse phase-space distributions can yield substantially different (by factors of three to ten) small-signal gain values. The dependence of the magnitude of the gain upon the Rayleigh range of the incident light is different for different phase-space distributions, as shown in Figs. 5 and 6. The dependence of the results upon undulator length, as shown in these two figures, leads to the conclusion that the transverse spatial overlap between the optical and electron beams is an important factor, along with the shape of the effective energy distribution function $f(\gamma_{eff})$, which strongly affects the magnitude of the small signal gain.

Figure 7 shows that intrinsic energy spreads larger than a few tenths of a percent drastically reduce the gain for the cases studied. Recent numerical simulations of the energy spread in a 200-500 Mev linac using the Los Alamos code PARMELA[7] show that, by proper phasing of the rf-fields in accelerator cavities, the energy spread can be held to less than 0.1%.

Figure 8 reminds us that, for zero intrinsic energy spread, the dominant property that determines the small-signal gain is the magnitude of the emittance. Figure 8 suggests that the variation of the gain with emittance is different for the four different cases, but we have not yet made such calculations. Also to be done in the future is a study of how the saturated gain varies among these four different transverse phase-space distributions, as well as calculations that extend the results of this paper to other optical wavelengths, i.e., 50 nm and 12 nm. Multiple-pass, self-consistent oscillator solutions[3] should be calculated to obtain information about the optical quality of the light beam, and to obtain more accurate values of the small-signal gain particularly in high-gain cases. Finally, we remark that gain calculations using numerically calculated transverse phase-space distributions generated by PARMELA, or some other linear accelerator simulation code, may be done in the future.

ACKNOWLEDGMENT

One of the authors (J.C.G.) thanks Walter P. Lysenko for useful discussions on the numerical generation of arbitrary phase-space distributions.

REFERENCES

1. B. E. Newnam, J. C. Goldstein, J. S. Fraser, and R. K. Cooper, in Free Electron Generation of Extreme Ultraviolet Coherent Radiation, AIP Conference Proceedings, No. 118, J. M. J. Madey and C. Pellegrini, Eds. (American Institute of Physics, N.Y., 1984), p. 190.
2. J. C. Goldstein, B. E. Newnam, R. K. Cooper, and J. C. Comly, Jr., in Laser Techniques in the Exteme Ultraviolet, AIP Conference Proceedings

No. 119, S. E. Harris and T. B. Lucatorto, Eds. (American Institute of Physics, N.Y., 1984), p. 293.

3. J. C. Goldstein, B. D. McVey, and B. E. Newnam, in Proceedings of the International Conference on Insertion Devices for Synchrotron Sources held at SLAC, Stanford, Calif., Oct. 27-30, 1985, to be publ., 1986.
4. T. W. Barbee, Jr., in Ref. 1, p. 53.
5. E. Spiller, in Ref. 2, p. 312.
6. B. E. Newnam, M. L. Scott, and P. N. Arendt, Paper TuE15, these proceedings; also B. E. Newnam in Laser Induced Damage in Optical Materials : 1985, NBS Spec. Publ., H. E. Bennett, A. H. Guenther, D. Milam, and B. E. Newnam Eds., to be publ., 1986.
7. B. E. Newnam, B. D. McVey, J. C. Goldstein, C. J. Elliott, M. J. Schmitt, K. Lee, T. S. Wang, B. E. Carlsten, J. S. Fraser, R. L. Sheffield, M. L. Scott, and P. N. Arendt, Paper MA3, these proceedings; also, B. E. Newnam, B. D. McVey, and J. C. Goldstein, Paper J3, presented at the Seventh International Free-Electron Laser Conf., Tahoe City, Calif. Sept. 8-13, 1985, to be publ. in Nucl. Inst. and Methods in Phys. Res., 1986.
8. J. S. Fraser, R. L. Sheffield, E. R. Gray, and G. W. Rodenz, IEEE Trans. on Nucl. Sci. NS – 32, 1791 (1985).
9. E. T. Scharlemann, J. Appl. Phys. 58, 2154 (1985).
10. B. D. McVey, Paper B13, presented at the Seventh International Free-Electron Laser Conf., op. cit.
11. H. Kogelnik and T. Li, Proc. IEEE 54, 1312 (1966).
12. W. B. Colson and A. Renieri, J. de Physique 44, Colloque C1, Supplement 2, 1 (1983).
13. D. A. G. Deacon, J. M. J. Madey, and J. LaSala, Paper TuE9, these proceedings.
14. W. B. Colson and P. Elleaume, Appl. Phys. B.29, 101 (1982).

PRELIMINARY RESULTS FROM A GAIN EXPERIMENT IN A NEON-LIKE COPPER PLASMA

R. C. Elton, W. A. Molander and T. N. Lee
Naval Research Laboratory, Washington, DC 20375

ABSTRACT

A series of experiments recently carried out at NRL were directed towards utilization of laser-heated ultra-thin foils for generating elongated plasmas of sufficient uniformity to produce significant amplification on 3p-3s transitions (mostly) in neon-like copper ions. The NRL Pharos III laser operating at 100-330 J in 2-6 ns pulses was used as a driver. Variations were made in the plasma length, the laser energy and pulse shape, and the copper thickness. Besides vuv grazing-incidence spectroscopy, x-ray crystal spectroscopy and pinhole photography were used for diagnostics. A novel slotted copper foil target design was also successfully tested. Various explanations for the lack of measureable gain so far are discussed. Also, a number of spectral lines from a selenium target exposure are included.

INTRODUCTION

Early conceptualization of x-ray lasers demonstrated[1] the feasibility of extrapolating isoelectronically 3p-3s transition lasers from the visible/near-uv into the vuv and possibly soft x-ray regions, using electron-collisional pumping in high temperature plasmas. Following several partially successful attempts at various laboratories, a dramatic demonstration was recently carried out at Lawrence Livermore Laboratory (LLNL) with neon-like selenium (Z=34) ions ceated by a line-focused laser[2]. While yttrium (Z=39) also showed lasing, germanium at lower Z (=32) failed to lase after several attempts. This surprising result, along with a number of others not predicted, have generated almost as much curiosity and interest as have the spectacular gains achieved on the J=2 lines.

With this background, we have attempted to produce gain in neon-like copper (Z=29) on similar transitions and with similar targets; but using a lower-power and longer-wavelength laser, which should produce plasma conditions suitable for similar gain coefficients as in the selenium experiments, according to a recent Z-scaling model[3]. The potential gain lines were measured in the vuv region, and other diagnostics included x-ray spectroscopy and pinhole photography. We also tested a novel slotted (thicker) copper foil design

We had an opportunity to irradiate a sample selenium target from LLNL. While no gain could be measured on one shot, the neon-like species were generated and a number of selenium lines were measured in the region of interest.

EXPERIMENT

The experimental arrangement, as shown in Fig. 1, consisted of a vacuum target chamber in which were placed formvar substrates of thickness 1500 Å and lengths varying from 4 to 24 mm, coated with copper to layer thicknesses of 375-1500 Å. Each target was checked for thickness, uniformity, and pinhole-free with a microdensitometer. Copper was chosen as a coating both for ease of manufacture and as a desireable target element for the available driver conditions, according to the modeling. After manufacture, the targets were stored in an argon atmosphere, both to inhibit oxidation of the copper surface and to prevent absorption of water vapor by the formvar. The targets were irradiated with a 1.05 μm Nd:glass laser pulse of nominally 2 ns fwhm and energy variable from 140-325 J, focused to a line of width of 100 μm and lengths of 6, 12, 18 and 30 mm. The elongated plasma formed was viewed axially for gain determination in the 200-300 Å region with a 1-meter grazing (88°) incidence vacuum spectrograph equipped with a 1200 grooves/mm grating. The data were recorded on Kodak type 101 film. This spectrograph was accurately aligned for the 10-20 mrad emission cone for amplified spontaneous emission; the alignment was verifiable on a shot-to-shot basis by point projection of a fine-mesh screen on the vuv film. A limiting aperture consisting of two glass plates forming a slit of width 600 μm parallel to the foil plane was provided on some shots at the end of the plasma to purposely control the amount of emission arriving outside this field of interest.

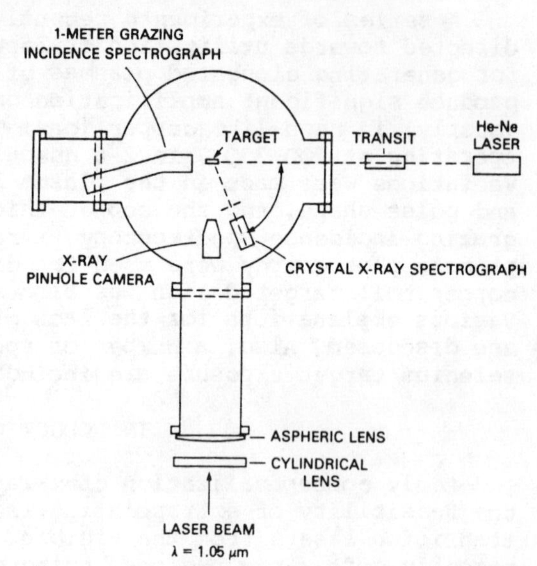

Fig. 1. Experimental setup

Included in the diagnostics was a flat crystal x-ray spectrograph covering the 8-17 Å wavelength range, equipped with a 100 μm wide slit mounted perpendicular to the crystal surface to provide spatial resolution registered along the spectral lines. The spectrograph was used to measure the ionic distribution both along the length of the plasma, and also perpendicular to the foil surface in an end view.

The plasma distribution along the length was also photographed in the x-ray (1-1.6 keV) region with a pinhole (35 μm diameter) camera mounted somewhat above the laser beam.

RESULTS: COPPER TARGETS

A total of 45 grazing-incidence spectrograms were recorded with accurate alignment along the plasma axis under varying conditions. A portion of a typical spectrum in the region of interest for 3p-3s lasing in Cu XX lines is shown in Fig.2, where the cutoff shorter than 170 Å is due to a 1200 Å thick aluminum filter used to reject higher order lines in the region of interest shown. About one-half of the data were obtained with a gaussian 2 ns fwhm driving laser pulse shape; while the latter half were obtained with a linearly-rising pulse of the same width but reaching peak in somewhat less than 1 ns (it was found from previous exploratory experiments with a 6-10 ns pulse that the shorter pulse was required for plasma axial uniformity). The following presentation of data is in order of increasing ion state beginning with sodium-like (even though lower states such as Mg- and Al-like were also observed).

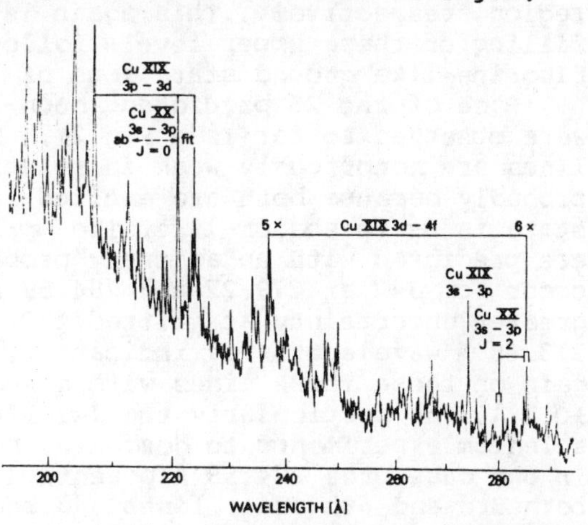

Fig. 2. Copper vuv spectrum

<u>Sodium-Like Copper</u>: The vuv spectra recorded time-integrated in the 30 - 300 Å wavelength region were dominated by Na-like Cu XIX lines[4], with the time-integrated intensity attributable to persistent recombination from long-lived neon-like ions in the ground state, followed by cascade through high ℓ states. This is consistent with the x-ray resonance spectra which showed a dominating (lingering) concentration neon-like lines, as expected with the high ionization potential. Particularly prominent in the Cu XIX spectra were the n=4 to n=3 transition lines in the 30-50 Å range, as well as the 5-4 lines in the

70-110 Å region. The 3-3 transitions at longer wavelengths in the 200-300 Å region were present but were relatively less intense, particularly for shots taken with the more rapidly-rising (linearly) pulse. It is quite possible that refraction[5] of the beam along the laser plasma axis (scaling as wavelength squared) served to decrease the relative emission of these 3-3 lines, and that an increased electron density gradient associated with shock waves and compression[6] in the plasma with the faster-rising pulses enhanced this effect.

Neon-Like Copper: From x-ray data[7], neon-like Cu XX lines are dominant, with fluorine- and sodium-like lines less intense. However, in the vuv spectral region, 4d-3p lines of Cu XX (wavelengths obtained from 4d-2p, 3d-2p, and 3d-3p[8]) are quite weak compared to lines between lowest excited levels such as 5-4 in Cu XIX described above, even accounting for different instrument sensitivity between the 30 and the 100 Å region, respectively. This again is consistent with cascade filling of these upper levels following recombination from fluorine-like ground state ions of less integrated abundance.

None of the 25 predicted[8] neon-like Cu XX 3-3 lines were observed so far (see Fig.2). Such 3-3 lines in neon-like lines are notoriously weak in relative spontaneous emission, probably because both are excited levels whereas the ground state is 3s in sodium-like. The 3p-3s lines of lasing interest are predicted with an accuracy probably as good as ±0.5 Å to occur for J=2 at 279.22 and 284.59 Å; and for J=0 with much greater uncertainty at "fitted" 221.38 Å and "ab"-initio 213.68 Å wavelengths as indicated in the figure. A search for gain on these three lines with a measurement capability of ±0.2 Å, and particularly the J=2 lines expected from the selenium experiments to dominate, proved generally negative. In one case, the 284.59 Å potential gain line overlapped with both 3rd and 6th order lines and showed some shot-to-shot variation, but without sufficient correlation with plasma length to claim an exponential dependence unique to ASE. These interpretations are preliminary and necessarily cautious.

Fluorine- to Nitrogen-Like Copper: For ions in species above neon-like Cu XX, the x-ray crystal data[7] indicate fluorine-like Cu XXI lines on most shots. In the vuv region, lines in the 65-120 Å wavelength range from 2p-2s transitions[9] indicate that some oxygen-like Cu XXII was also present on most shots; and that nitrogen-like Cu XXIII might have been present in a very low concentration on some shots. There was no evidence on any shots of carbon-like Cu XXIV ions. The presence of these higher stages of ionization may indicate excessive ionization beyond the desired neon-like species resulting in decreased gain, a situation created in our attempt to uniformly vaporize the foil with high irradiance; and indeed no stages higher than oxygen-like have been reported from the selenium experiments at LLNL. On the other hand, one attempt[10] to explain the low gain on the transition beginning with a J=0 level in selenium does model population in the oxygen-like species.

Oxygen and Carbon "Impurities": Spectral lines from lithium-like O VI, and particularly the 3d-2p transition, were present to varying degrees on most shots; and mostly came to be associated with an oxide layer formed on somewhat aged copper targets. Helium-like O VII (and perhaps hydrogenic O VIII) lines were observed at wavelengths shorter than 128 Å. Carbon lines were generally weak, including the hydrogenic C VI Balmer series lines at 182.2 Å and shorter wavelengths; the C VI Lyman series lines in the 30-40 Å wavelength region were also considerably weaker than Cu XIX lines in the vicinity. The exception was a series of six shots in which the formvar side of the target was irradiated, with the 1000 Å thick copper coating to the rear. Then these oxygen and carbon lines became as intense as the nearby copper lines. This indicates that the formvar was not strongly radiating when heated behind the copper layer; but on the other hand, the copper did reach high radiance when heated behind the formvar.

Another interesting result is that we detected no evidence of enhanced C VI 182.2 Å Balmer-α radiation (compared to other Balmer- and Lyman-series lines) that would be indicative of gain, such as reported recently in a quite similar experiment with selenium coatings, even when the plasma length reached 23.5 mm.

RESULT: A SELENIUM TARGET SHOT

We had an opportunity to irradiate a single LLNL-sample selenium/formvar target with 320 J at 1 μm wavelength in a 3 ns pulse. The laser was focused into a line of length 12 mm and width 100 μm for an overall irradiance of 2×10^{13} W/cm .

X-ray crystal spectrographic data (Fig. 3) obtained at 8 Å indicate that we were able to generate neon-like Se XXV ions on which vuv lasing was obtained[2] at LLNL. Fluorine-like Se XXVI lines are weak or absent.

Magnesium-like[12] Se XXIII and sodium-like[4] Se XXIV lines are identified in the vuv spectrum shown in Fig. 4. (For this particular case, the vuv spectrograph was 5-10° off axis). After eliminating lines from these two

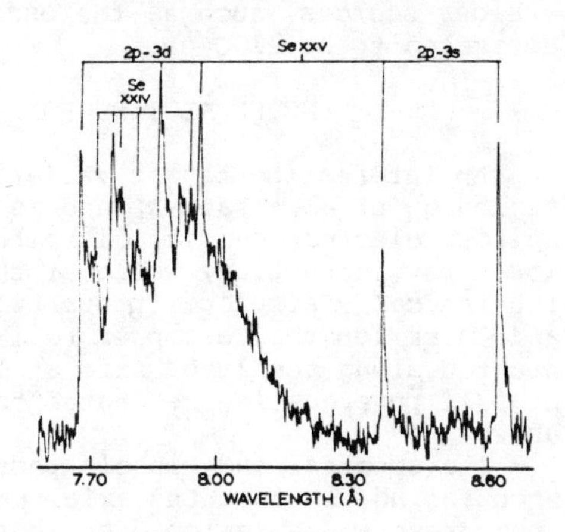

Fig. 3. Se x-ray spectrum

species and from C and O (O VI 2p-3d and C VI Balmer-α shown), we associate 11 other lines at: 181.48, 182.86, 184.79, 185.50, 189.24, 189.83, 195.15, 196.06, 202.84, 205.59, and 209.36 (±0.2) Å to selenium, perhaps from 3p-3s transitions. Of particular interest are lines near the J=0 182.4 (measured--183.17 calculated[13]) and the J=2 206.3 and 209.6 Å wavelengths where lasing has been reported. Lasing cannot be shown on a single shot such as this; however these results are most encouraging for achieving lasing with this particular driverr in future experiments.

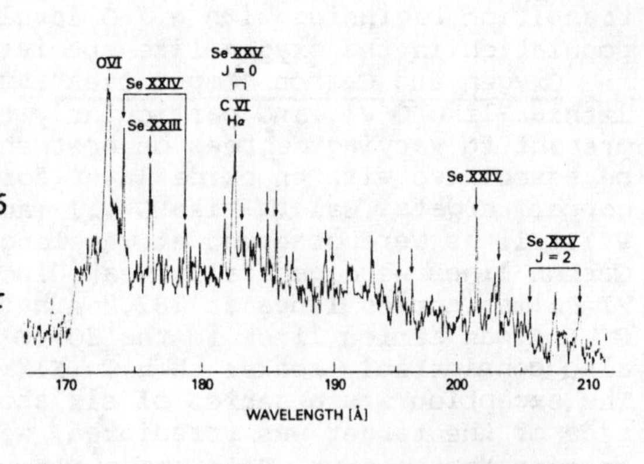

Fig. 4. Se vuv spectrum

TEMPERATURE MEASUREMENTS

The electron temperature of the copper plasma may be inferred[7] from line intensity ratios of neon-like Cu XX 2p ^1S- nd ^1P series lines, assuming local thermal equilibrium (LTE) among the nd levels. The average electron temperatures thus obtained with 375, 750, and 1500 Å thick copper layers are 300, 400 and 650 eV, respectively, i.e., there is a tendency towards a higher temperature for the thicker copper coatings. This kT is 1/3 to 1/6 the ionization probability for Cu XX. The uncertainty in the temperature measurement arises from various sources, such as the oscillator strengths used, and is estimated to be ±100 eV.

NOVEL SLOTTED-FOIL TARGET RESULTS

An interesting target variation has been suggested by Dahlburg, et al.[14] at NRL and is predicted to give a uniform electron density distribution across the plasma for low x-ray refraction, while at the same time being intrinsically simpler. In our two initial tests so far, at 5 and 10 mm lengths, a copper foil of 1.3 μm thickness was slotted along the laser axis at low power and then irradiated at full power as in the coated formvar foil targets described above.

The spectral and pinhole photographic results were most encouraging, in that the axial uniformity was comparable, the front/rear ion-emission distribution very symmetrical (no carbon and oxygen plasma components present) and the spectral species almost identical. The slotted foil therefore seems to live up to theoretical expectations, at least in regard to the uniformity of the plasma distribution obtained. Again, however, no gain lines were observed, at least on the two attempts to date.

SUMMARY

No clear evidence of gain at the predicted wavelengths on the 3p-3s neon-like Cu XX transitions has been measured in this first series of tests. This is as puzzling as was the negative result at LLNL on germanium; and so far numerical codes do not provide any physics reason for such low gain at decreasing Z except perhaps overionization in an electron-collisional pumping model. Certain effects such as refraction of the axial beam out of the field of view, along with the somewhat limited discrimination of time-integrated spectroscopy, may be rendering the overall 3p-3s emission below the observable level so far. Using the simple split-foil and obtaining spectral results so similar to those of the formvar-backed thin copper films is very encouraging for future experiments. Also, the measurment of neon-like Se XXV lines at irradiances 2-3 times less than that at LLNL and with laser operation at the 1.05 μm fundamental (instead of second harmonic) is most encouraging for extension of the selenium studies.

ACKNOWLEDGEMENTS

We gratefully acknowledge expert technical assistance of J.L. Ford in carrying out these experiments; as well as the cooperation of other NRL colleagues in the Laser Physics Branch, the Microelectronic Processing Facility, and the Laboratory for Computational Physics. Valuable discussions with Prof. H.R. Griem are also recalled with appreciation. A.K. Bhatia kindly provided oscillator strengths for the temperature estimates. This research was supported by DNA, ONR and SDIO.

REFERENCES

1. R.C. Elton, Appl. Optics 14, 97 (1975).
2. M.D. Rosen, et al., and D.L. Matthews, et al. Phys. Rev. Letters 54, 106 and 110 (1985).
3. U. Feldman and J.F. Seely, J. Appl. Phys. 56, 2475 (1984).
4. E. Ya. Kononov, A.N. Ryabtsev and S.S. Churilov, Physica Scripta 19, 328 (1979).
5. V.A. Chirkov, Sov. J. Quantum Electron. 14, 1497 (1984).
6. J.P. Dahlburg (private communication, 1986).
7. T.N. Lee, W.A. Molander, J.L. Ford and R.C. Elton, Rev. Sci. Instruments (to be published).

8. J.H. Scofield (private communication, 1985).
9. W.E. Behring, et al., J. Optical Soc. Am. B 2, 886 (1985).
10. J.P. Apruzese, J. Davis, M. Blaha, P.C. Kepple and V.L. Jacobs, Phys. Rev. Letters 55, 1877 (1985).
11. J.F. Seely, et al., Optics Comm. 54, 289 (1985).
12. B.C. Fawcett and R.W. Hayes, J. Opt. Soc. Am. 65, 623 (1975).
13. J.H. Scofield (to be published).
14. J.P. Dahlburg, J.H. Gardner, M.H. Emery and J.P. Boris (to be published).

TRANSVERSE CORRELATIONS IN START-UP OF A FREE ELECTRON LASER FROM NOISE*

L.H. Yu and S. Krinsky
National Synchrotron Light Source
Brookhaven National Laboratory
Upton, New York 11973

ABSTRACT

Linearized Vlasov-Maxwell equations are used to derive a partial differential equation determining the 3-dimensional slowly varying envelope function of the radiated electric field. The equation is solved analytically. From the correlation function $\langle E(z,\vec{r},t)\, E^*(z',\vec{r}\,',t)\rangle$ of the electric field averaged over the stochastic ensemble describing the initial shot noise in the beam, we compute the longitudinal and transverse correlation lengths $\sigma_\parallel$ and $\sigma_\perp$. The radiated power S per unit cross-sectional area of the electron beam is

$$S = \frac{\rho S_e}{9 n_o V_c} \exp(\sqrt{3}\, 4\pi N_w \rho),$$

where $V_c = (2\pi)^{3/2}\, \sigma_\parallel \sigma_\perp^2$ is the coherence volume, n_o the electron density, $S_e = (\gamma_o mc^2)\, n_o c$ the power per unit area in the electron beam, N_w the number of wiggler periods and ρ the Pierce parameter. The angular distribution of the radiation is characterized by the Gaussian factor $\exp(-\theta^2/2\sigma_\theta^2)$, where $2\pi\sigma_\theta\sigma_\perp = \lambda$ (radiated wavelength). Our analysis is applicable for wiggler length $L = N_w \lambda_w$ long enough for the exponential regime to be reached, but short enough so that $L\,\sigma_\theta \lesssim a$, the electron beam radius.

INTRODUCTION

There is great interest in using a free electron laser (FEL) operating in the high-gain regime for the generation of high intensity coherent radiation at wavelengths below 1000 Å. Amplification in a long wiggler magnet of the initial spontaneous radiation emitted by individual electrons has the attractive feature that the use of an optical resonator is avoided. The process of self-amplified spontane-

*This work has been performed under the auspices of the U.S. Department of Energy.

ous emission is still not well understood. Three key issues which require further elucidation to facilitate the design of a single pass FEL are the start-up[1-3] of the laser from the shot noise in the electron beam, the guiding[4,5] (or self-focusing) of the radiation by the electron beam, and the saturation of the exponential growth of the radiation field due to nonlinear effects.

In this paper we present the results of an analysis of the start-up of an FEL from shot-noise. Linearized Vlasov-Maxwell equations have been used to derive a partial differential equation determining the three-dimensional slowly varying envelope function of the emitted radiation, extending an earlier one-dimensional treatment[2,3] to include transverse variations. The problem with the one-dimensional model is that individual electrons are treated as two-dimensional charge sheets, hence the angular distribution of the radiation cannot be properly described, and the total radiated power cannot be correctly determined. In the three-dimensional calculation which we shall present, individual electrons are described as point charges (Fig. 1), allowing us to determine the angular distribution of the emitted radiation and the build-up of transverse correlations.

Initially each electron radiates independently of all others, and the angular distribution of the radiation is that of the spontaneous radiation from a point charge. As the electron beam proceeds down the wiggler magnet different electrons communicate via their emitted radiation and correlations build up. As the transverse correlation length increases, the angular distribution of the radiation narrows. The description of the development of transverse correlations and the narrowing of the angular distribution are the key subjects of this paper.

ENVELOPE EQUATION

Suppose a highly relativistic electron beam is moving in the positive z-direction through a periodic helical wiggler with vector potential $\vec{A}_w = A_w(\hat{e}_- e^{ik_w z} + c.c.)/\sqrt{2}$, where $\hat{e}_\pm = (\hat{e}_1 \pm i\hat{e}_2)/\sqrt{2}$ and $\hat{e}_1$ and $\hat{e}_2$ are orthogonal unit vectors transverse to $\hat{z}$. The transverse electron velocity is approximated by $\vec{v}_\perp \simeq -e\vec{A}_w/m\gamma$ and the longitudinal velocity by $v_\parallel \simeq c(1 - \frac{1+K^2}{2\gamma^2})$, where γ is the electron energy in units of its rest mass and $K = eA_w/mc$ is the wiggler strength parameter. The electron beam is assumed to be initially monoenergetic with all electrons having energy γ_o and longitudinal velocity $v_\parallel(\gamma_o) = v_o$. The spontaneous radiation emitted by the electrons in the forward direction is left circularly polarized with wave number k_o and frequency $\omega_o = k_o c$. The combined action of the static wiggler field and the radiation field produces a ponderomotive potential, which has the dependence $e^{ik_o z - i\omega_o t}\, e^{ik_w z}$. Because the electron beam moves with velocity v_o, the modulation should be of the form $e^{ik_r(z-v_o t)}$. To be in resonance, these two exponential expressions should be the same, hence we have

$$k_r = k_o + k_w . \tag{1}$$

and

$$k_r v_o = k_o c = \omega_o . \tag{2}$$

It follows that $k_o/k_w = v_o/(c-v_o) = 2\gamma^2/(1+K^2)$ and $k_r = \omega_o/v_o = \omega_w/(c-v_o)$, where $\omega_w = k_w c$.

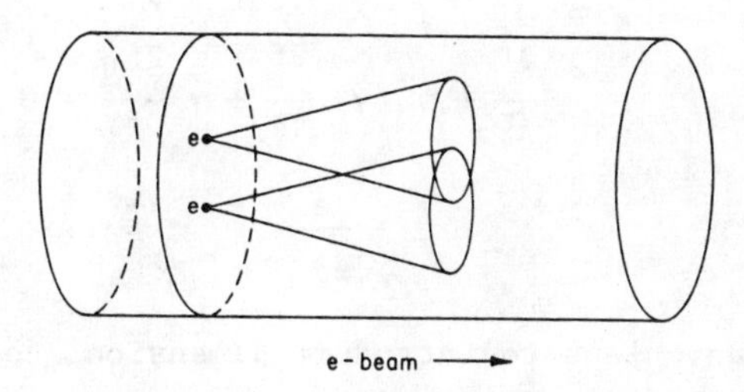

Fig. 1. In the three-dimensional calculation, individual electrons are described as point charges allowing the proper determination of angular distribution and transverse correlations of the emitted radiation.

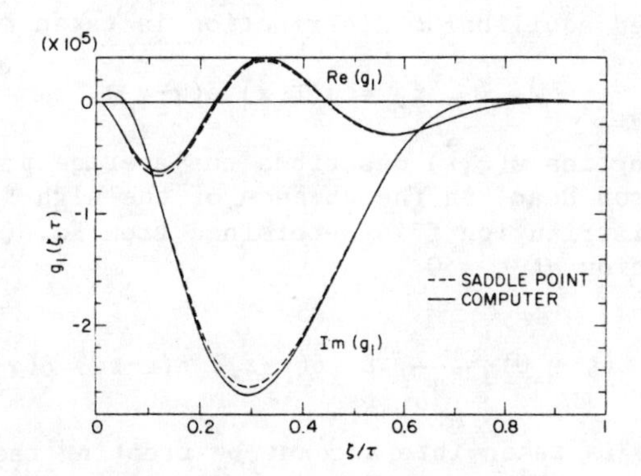

Fig. 2. Numerical evaluation of the one-dimensional Green's function $g_1(\zeta,\tau)$ introduced in Eq. (21). Note the saddle point approximation is very accurate. The case shown corresponds to $\rho = 3\mathrm{x}10^{-3}$, $N_W = 300$, $2\ \rho\tau = 3.6\ \pi$.

The radiated electric field $\vec{\varepsilon}$ satisfies the wave equation, in mks units,

$$(\nabla^2 - \frac{1}{c^2}\frac{\partial^2}{\partial t^2})\vec{\varepsilon} = \mu_o \frac{\partial \vec{j}}{\partial t} \tag{3}$$

The current density $\vec{j}$ is given by

$$\vec{j} = en_o \int \vec{v}_\perp \, f d\gamma \tag{4}$$

with n_o being the peak density of the electron beam and $n_o f(z,\vec{r},\gamma,t)$ $dzd^2rd\gamma$ being the number of electrons in element $dzd^2rd\gamma$. (Transverse coordinates denoted by $\vec{r}$.) Writing the distribution as $f = f_0 + f_1$, the linearized Vlasov equation is

$$\frac{\partial f}{\partial t} + v_\parallel(\gamma)\frac{\partial f}{\partial z} + \dot{\gamma}\frac{\partial f_0}{\partial \gamma} = 0 \tag{5}$$

where

$$\dot{\gamma} = \frac{e}{mc^2}\,\vec{v}_\perp \cdot \vec{\varepsilon}\,. \tag{6}$$

It is convenient to introduce dimensionless variables measuring spatial and temporal variations:

$$\tau = \omega_w t\ ,\ \zeta = k_r(z - v_o t),$$

$$\vec{x} = \sqrt{2k_o k_w}\,\vec{r}\ , \qquad \nabla_\perp^2 = \frac{\partial^2}{\partial x_1{}^2} + \frac{\partial^2}{\partial x_2{}^2}\ .$$

The unperturbed equilibrium distribution is taken to be

$$f_o = u(\zeta,\vec{x})\ \delta(\gamma-\gamma_o), \tag{7}$$

the smooth function $u(\zeta,\vec{x})$ describes the average properties of the initial electron beam, in the absence of the high-frequency shot noise. The distribution f is determined from Eq. (5) subject to the initial condition at t = 0,

$$f(t=0) = \frac{1}{n_o}\sum_i \delta(z-z_i)\ \delta(\vec{r}-\vec{r}_i)\ \delta(\gamma-\gamma_o)\ . \tag{8}$$

The shot noise is taken into account by treating the initial coordinates $z_i,\vec{r}_i$ of the i th electron as stochastic variables and determining physical quantities as averages over the ensemble of possible $z_i,\vec{r}_i$.

Introducing the slowly varying envelope function E by writing $\vec{\varepsilon} = \frac{1}{\sqrt{2}} E \exp(ik_o z - i\omega_o t)\, \hat{e}_+$, and using the paraxial approximation, the coupled Vlasov-Maxwell equations can be shown to take the form:

$$\left(\frac{\partial}{\partial\tau} + \frac{\partial}{\partial\zeta} - i\nabla_\perp^2\right) E = J \quad , \tag{9}$$

$$\frac{\partial^2 J}{\partial\tau^2} = \alpha \left(\frac{\partial}{\partial\tau} + \frac{\partial}{\partial\zeta} + i\right)(uE) \ , \tag{10}$$

$$J \equiv \frac{n_o \mu_o e^2 c^2 A_w}{2m\omega_w} e^{-i\zeta} \int \frac{d\gamma}{\gamma} f \ . \tag{11}$$

The constant α in Eq. (10) is related to the Pierce parameter ρ of Bonifacio et al.[6] by

$$\alpha = (2\rho)^3 = \frac{n_o \mu_o e^4 A_w^2}{2m^3 \gamma_o^3 \omega_w^2} . \tag{12}$$

Eqs. (9) and (10) immediately lead to the envelope equation:

$$\frac{\partial^2}{\partial\tau^2}\left(\frac{\partial}{\partial\tau} + \frac{\partial}{\partial\zeta} - i\nabla_\perp^2\right) E = \alpha \left(\frac{\partial}{\partial\tau} + \frac{\partial}{\partial\zeta} + i\right)(uE) \ . \tag{13}$$

GREEN'S FUNCTION

For an initially uniform electron beam, $u(\zeta,\vec{x}) = 1$, we introduce the Green's function $g(\zeta,\vec{x},\tau)$ via

$$\left[\frac{\partial^2}{\partial\tau^2}\left(\frac{\partial}{\partial\tau} + \frac{\partial}{\partial\zeta} - i\nabla_\perp^2\right) - \alpha\left(\frac{\partial}{\partial\tau} + \frac{\partial}{\partial\zeta} + i\right)\right] g = \delta(\zeta)\ \delta(\vec{x})\ \delta(\tau) \ . \tag{14}$$

Solving by Fourier-Lapace transform yields

$$g(\zeta,\vec{x},\tau) = \int_{-\infty+is}^{\infty+is} \frac{d\Omega}{2\pi i} \int \frac{dq_\parallel d^2q_\perp}{(2\pi)^3} \frac{\exp(iq_\parallel\zeta + i\vec{q}_\perp\cdot\vec{x} - i\Omega\tau)}{D(\Omega,q_\parallel,\vec{q}_\perp)} , \tag{15}$$

with

$$D(\Omega,q_\parallel,\vec{q}_\perp) = \Omega^3 - (q_\parallel + q_\perp^2)\,\Omega^2 + \alpha\Omega - \alpha(1 + q_\parallel) \ . \tag{16}$$

It follows that

$$g(\zeta,\vec{x},\tau) = \int dq_\parallel d^2q_\perp\ G_q(\tau) e^{iq_\parallel\zeta + i\vec{q}_\perp\cdot\vec{x}} \tag{17}$$

where

$$G_q(\tau) = \frac{-1}{(2\pi^3)}\left[\frac{e^{-i\Omega_1\tau}}{(\Omega_1-\Omega_2)(\Omega_1-\Omega_3)} + \frac{e^{-i\Omega_2\tau}}{(\Omega_2-\Omega_1)(\Omega_2-\Omega_3)} + \frac{e^{-i\Omega_3\tau}}{(\Omega_3-\Omega_1)(\Omega_3-\Omega_2)}\right] \quad (18)$$

and $\Omega_1,\Omega_2,\Omega_3$ are the three solutions of $D(\Omega,q_\parallel,\vec{q}_\perp) = 0$.

Another useful representation of the Green's function can be obtained from Eq. (15) by employing the identity

$$\frac{i}{D} = \int_0^{\infty+i\varepsilon} d\nu\, e^{i\nu D} \quad (19)$$

which allows the integrals over $q_\parallel$ and $\vec{q}_\perp$ to be performed, one obtains

$$g(\zeta,\vec{x},\tau) = \frac{i\theta(\zeta)\theta(\tau-\zeta)}{8\pi^2\zeta}\, e^{ix^2/4\zeta} \int_{-\infty+is}^{\infty+is} \frac{d\Omega}{\Omega^2} \exp\left[-i\Omega(\tau-\zeta) - \frac{i\alpha\zeta}{\Omega^2+\alpha} + \frac{i\alpha x^2}{4\Omega^2\zeta}\right] \quad (20)$$

We chose $s > \sqrt{\alpha}$ in Eq. (20), so the integral vanishes[7] for $\tau - \zeta < 0$.

In the regime of exponential growth, $\tau \to \infty$, τ/ζ finite, the integral in Eq. (18) is dominated by a saddle point[8] at $\Omega_o = [2\alpha\zeta/(\tau-\zeta)]^{1/3}\, e^{2\pi i/3}$. We consider $\rho \ll 1$ so $\alpha \ll |\Omega_o^2|$. Then we found the saddle point method is valid if $\rho\tau \gg 1$, and

$$g(\zeta,\vec{x},\tau) \simeq g_E(\zeta,\vec{x},\tau)\, g_1(\zeta,\tau)\, R(\zeta,\vec{x},\tau)\ , \quad (21)$$

where

$$g_E(\zeta,x,\tau) = -\frac{i\theta(\zeta)\theta(\tau-\zeta)}{4\pi\zeta}\, e^{ix^2/4\zeta}\ , \quad (22)$$

$$g_1(\zeta,\tau) = \frac{1}{4\pi\rho}\sqrt{\frac{\pi}{6\rho\zeta}}\, \exp\left\{-\frac{3}{2}\, e^{2\pi i/3}\, [2\zeta(\tau-\zeta)^2]^{1/3}\, 2\rho i - \frac{\pi}{4} i\right\}, \quad (23)$$

$$R(\zeta,\vec{x},\tau) = \exp\left\{\frac{1}{8}\, i\, e^{2\pi i/3}\, [2\zeta(\tau-\zeta)^2]^{1/3}\, 2\rho\, \frac{x^2}{\zeta^2}\right\} \quad (24)$$

Here, g_E is the Green's function (within the paraxial approximation) for the low density limit $\alpha = 0$; g_1 is the Green's function for the one-dimensional model[3] [see Fig. 2]; R describes the transverse fall-off of the radiation. From the derivation of these results, it can be seen that the transverse fall-off is a consequence of the term $\alpha\partial g/\partial\zeta$ in Eq. (14) and correspondingly of the term $\alpha q_\parallel$ in Eq. (16). This term is negligible in the one-dimensional treatment, but is of primary importance in three-dimensions, since zero detuning corresponds to $q_\parallel + q_\perp^2 = 0$ as shown later [see Eq. (36)], so $q_\parallel \approx -q_\perp^2$ and the term $\alpha q_\parallel$ determines the divergence angle of the radiation, and the transverse size of the radiation cone.

The maximum of $g_1(\zeta,\tau)$ is at $\zeta = \tau/3$. From Eq. (24), we find

$$\left|R\ (\zeta = \tau/3,\ \vec{x},\ \tau)\right|^2 = e^{-x^2/2\sigma_x^2} \tag{25}$$

with

$$\sigma_x^2 = \frac{\tau}{3\sqrt{3}\,\rho}\ . \tag{26}$$

FEL START-UP FROM SHOT NOISE

We wish to solve the envelope equation (13) subject to initial conditions specified at $t = 0$. In particular, we specify $E(\zeta,\vec{x},o) = E_o(\zeta,\vec{x})$, $J(\zeta,\vec{x},o) = J_o(\zeta,\vec{x})$ and $\dot{J}(\zeta,\vec{x},o) = \dot{J}_o(\zeta,\vec{x})$, where the dot denotes $\partial/\partial\tau$ and the current J was introduced in Eqs. (9-11). The envelope function is then determined by

$$E(\zeta,\vec{x},\tau) = \int d\zeta' d^2x'\ [E_o(\zeta',\vec{x}')\ \ddot{g}(\zeta-\zeta',\ \vec{x}-\vec{x}',\ \tau)$$
$$+ J_o(\zeta',\vec{x}')\ \dot{g}(\zeta-\zeta',\vec{x}-\vec{x}',\tau) + \dot{J}_o(\zeta',\vec{x}')\ g(\zeta-\zeta',\vec{x}-\vec{x}',\ \tau)] \tag{27}$$

where $g(\zeta,\vec{x},t)$ is the Green's function defined in Eq. (14). Here, E_o represents an initial electric field possibly due to an external laser; J_o describes the initial spatial bunching of the electron beam and $\dot{J}_o$ corresponds to an initial energy modulation of the electron beam.

We assume the absence of an external radiation field, $E_o = 0$, and describe the shot noise by

$$J_o = \frac{n_o \mu_o e^2 c^2 A_w}{2m\omega_w}\, e^{-i\zeta} \int \frac{d\gamma}{\gamma}\ f \tag{28}$$

$$f = \frac{1}{n_o} \sum_i \delta(z-z_i)\ \delta(\vec{r}-\vec{r}_i)\ \delta(\gamma-\gamma_o)\ , \tag{29}$$

where the coordinates z_i, $\vec{r}_i$ of the i th electron are treated as independent random variables. For the purposes of the present discussion we ignore the spread in energies of the electrons, hence $\dot{J}_o = 0$. Although $\langle E\rangle = 0$, averages of quantities quadratic in E do not vanish. In particular, the correlation function of the electric field at two different spatial points is found to be expressed in -

terms of the Fourier transform of the Green's function $G_q(\tau)$ [see Eqs. (17-18)] via

$$C(z,\vec{r},\tau) = \frac{1}{Z_o} \langle E^*(z,\vec{r},\tau)\ E(o,o,\tau)\rangle$$

$$= 4\pi^3 \alpha c(\gamma_o mc^2) \int dk_\parallel d^2k_\perp \left|\dot{G}(q_\parallel,\vec{q}_\perp,\tau)\right|^2 e^{ik_\parallel z + i\vec{k}_\perp\cdot\vec{r}} , \quad (30)$$

where $k_\parallel = k_r q_\parallel$, $\vec{k}_\perp = \sqrt{2k_o k_w}\,\vec{q}_\perp$, and $Z_o = \mu_o c$ is the impedance of free space. Denoting the radiated power per unit area by S and the radiated power per unit area, per unit solid angle, per unit frequency $dP/dAd\Omega d\omega$, we see that

$$S = C(o,o,\tau) = \int d\omega d\Omega \ \frac{dP}{dAd\Omega d\omega} \ . \quad (31)$$

Using $dk_\parallel d^2k_\perp = k^2 dk d\Omega$ with $k = \omega/c$, Eqs. (30) and (31) show

$$\frac{dP}{dAd\Omega d\omega} = 4\pi^3 \alpha m\gamma_o \omega^2 \left|\dot{G}(q_\parallel,q_\perp^2,\tau)\right|^2 . \quad (32)$$

In the high gain regime, keeping only the dominant term,

$$\dot{G}_q(\tau) \simeq \frac{i}{(2\pi)^3} \ \frac{\Omega_1 e^{-i\Omega_1\tau}}{(\Omega_1-\Omega_2)(\Omega_1-\Omega_3)} , \quad (33)$$

where $D(\Omega_1,q_\parallel,\vec{q}_\perp) = 0$ [Eq. (16)] and $\mathrm{Im}\Omega_1 > 0$. We consider $\rho \ll 1$ and $2\rho\tau > 1$. Introducing μ by

$$\Omega = 2\rho(1 + q_\parallel)^{1/3}\ \mu , \quad (34)$$

to good approximation μ is determined from

$$\mu^3 - \Delta\mu^2 - 1 = 0 , \quad (35)$$

$$\Delta = \frac{q_\parallel + q_\perp^2}{2\rho(1+q_\parallel)^{1/3}} \simeq \frac{\omega - \omega_1(\theta)}{2\rho\omega_1(\theta)} , \quad (36)$$

where[9] $\omega_1(\theta) = 2\gamma^2\omega_w/(1+K^2+\gamma^2\theta^2)$. Note that $\mathrm{Im}\mu_1$ is maximum at $\Delta = 0$, so $q_\parallel \approx -q_\perp^2$, and

$$\left|\dot{G}_q(\tau)\right|^2 \simeq \frac{1}{(2\pi)^6}\,\frac{1}{9(2\rho)^2} \exp\left[\sqrt{3}\ 2\rho\tau\left(1 - \frac{\Delta^2}{9} - \frac{q_\perp^2}{3}\right)\right] . \quad (37)$$

From Eq. (32) we find

$$\frac{dP}{dAd\Omega d\omega} = \frac{1}{9(2\pi)^3} \rho(\gamma_o mc^2)k_o^2 \, e^{\sqrt{3}\,2\rho\tau} \, e^{-(\omega-\omega_1(\theta))^2/2\sigma_\omega^2} \, e^{-\theta^2/2\sigma_\theta^2}, \tag{38}$$

with

$$2\sigma_\omega^2 = \frac{9(2\rho\omega_o)^2}{\sqrt{3}\,2\rho\tau} \tag{39}$$

$$2\sigma_\theta^2 = \frac{\sqrt{3}\,(1+K^2)}{\gamma^2 2\rho\tau} \tag{40}$$

Integrating Eq. (38) over frequency and solid angle we obtain the radiated power per unit cross-sectional area of the electron beam

$$S = \frac{\rho S_e}{9(2\pi)^{3/2}} \frac{k_o^2\sigma_\theta^2}{n_o\sigma_\parallel} e^{\sqrt{3}\,2\rho\tau} , \tag{41}$$

where

$$S_e = (\gamma_o mc^2)n_o c \tag{42}$$

is the power per unit area in the electron beam and

$$\sigma_\parallel = c/\sigma_\omega \tag{43}$$

is the longitudinal correlation length.

In a similar fashion, the correlation function of Eq. (30) can be evaluated, yielding

$$C(z,\vec{r},\tau) = \frac{\rho S_e e^{\sqrt{3}\,2\rho\tau}}{9n_o V_c w(z)} \exp\left(-\frac{z^2}{2\sigma_\parallel^2}\right) \exp\left(-\frac{r^2}{2\sigma_\perp^2 w(z)}\right) , \tag{44}$$

where the transverse correlation length for $z = 0$ is

$$\sigma_\perp = 1/k_o\sigma_\theta , \tag{45}$$

the coherence volume V_c is

$$V_c = (2\pi)^{3/2} \sigma_\parallel\sigma_\perp^2 \tag{46}$$

and

$$w(z) = 1 + \frac{i\sqrt{3}\,k_r z}{2\rho\tau} . \tag{47}$$

The total power radiated per unit area given in Eq. (41) can now be rewritten as

$$S = C(o,o,\tau) = \frac{1}{9}\,\rho S_e\, e^{\sqrt{3}\,2\rho\tau}\,\frac{1}{n_o V_c}\,, \tag{48}$$

and is seen to be inversely proportional to the number of electrons in a coherence volume.

In our analysis the transverse variation of the electron density has been neglected, i.e. u appearing in Eq. (13) has been taken to be unity. This is a reasonable approximation for times short enough that radiation emitted by the electrons in the bulk of the beam has not reached the edge, i.e. (ct) $\sigma_\theta \lesssim a$, where a is the electron beam radius. Once the radiation reaches the edge of the beam, the transverse fall-off of u becomes important and may lead to self-focusing or guiding.[4,5]

One-dimensional analysis gives the total radiated power[2,3]

$$P_{1D} = \frac{1}{9}\,\rho P_e\, e^{\sqrt{3}\,2\rho\tau}\,\frac{1}{N_c} \tag{49}$$

where N_c is the number of electrons within a coherence length, rather than a coherence volume. If one thinks of keeping the electron density constant and increasing the electron beam cross-section, P_{1D} would remain constant, because $P_e = S_e A$ and N_c are both proportional to the beam cross-section. This puzzling result is removed by the 3-dimensional treatment leading to Eq. (48), since N_c is replaced by $n_o V_c$, the number of electrons in a coherence volume. Hence the transverse coherence length $\sigma_\perp$ is important in the start-up process.

ACKNOWLEDGEMENTS

We have benefited from discussions with A. Gover, K.J. Kim and J.M. Wang.

REFERENCES

1. H.A. Haus, IEEE J. Quant. Electron. QE-17, 1427 (1981).

2. K.J. Kim, in Proceedings of Seventh International Free Electron Laser Confrence, Granlibakken, CA (1985).

3. J.M. Wang and L.H. Yu, ibid.

4. G.T. Moore, in Proceedings of the International Workshop on Coherent and Collective Properties of Electrons and Electron Radiation, Como, Italy, 1984, Nucl. Instrum. Methods, Sec. A.

5. E.T. Scharlemann, A.M. Sessler and J.S. Wurtele, Phys. Rev. Lett. 54, 1925 (1985).

6. R. Bonifacio, C. Pellegrini and L.M. Narducci, Opt. Comm. 50, 373 (1984).

7. The paraxial approximation has resulted in a plane wavefront $\tau - \zeta = 0$. An improved paraxial approximation (IPA) leads to $D_{IPA}(\Omega, q_\parallel, \vec{q}_\perp) = (1+q_\parallel) [\Omega^3 - (q_\parallel + q_\perp^2/(1 + q_\parallel)) \Omega^2 + \alpha\Omega - \alpha(1 + q_\parallel)]$, and a curved wavefront $\tau - \zeta - x^2/4\zeta = 0$.

8. For the one-dimensional Green's function this has been noted in the Appendix of ref. 4.

9. The approximate equality in Eq. (36) holds over a larger angular range in the improved paraxial approximation mentioned in ref. 7. For $\gamma >> 1$, $\theta << 1$, one sees that $q_\parallel + q_\perp^2/(1 + q_\parallel) = (\omega - \omega_1(\theta))/\omega_1(\theta)$.

LASER-INDUCED FLUORESCENCE SPECTRA OF Ar_2^*

P. R. Herman, P. E. LaRocque, and B. P. Stoicheff
Department of Physics, University of Toronto
Toronto, Ontario, M5S 1A7, Canada

ABSTRACT

A VUV-XUV source based on four-wave mixing in Hg vapor and tunable from ~120 to 86 nm has been used to investigate rovibronic spectra of Ar_2 in the region of 106 nm at a resolving power >3 x 10^5. Spectroscopic constants and potential energy curves have been obtained for the ground and three lowest excited states. The observation and analysis of three rotational branches in the A ↔ X band system has established that the first excited state has case (c) coupling and therefore the designation $A1_u$.

INTRODUCTION

Earlier work in this laboratory has shown that tunable, coherent, VUV radiation generated by four-wave frequency mixing in metal vapors is an ideal source for high-resolution spectroscopy and for radiative lifetime measurements. In previous conferences in this series, the sources have been described[1] along with their application to the study of fluorescence excitation spectra of Xe_2 and Kr_2[2]. These investigations were motivated by the interest of rare gas dimers as laser media, and by the realization that the spectroscopic constants and potential energy curves of the relevant upper states had not been determined experimentally.

The present work gives a preliminary account of recent results obtained from rovibronic spectra of $^{40}Ar_2$ and $^{36,40}Ar_2$ at ~106 nm. Analyses have yielded the vibrational numbering of several band systems for the first time, and thus enabled us to determine vibrational and rotational constants for the lowest excited states of Ar_2. Tanaka and Yoshino[3] have carried out extensive studies of the electronic spectra of Ar_2, and were successful in evaluating the vibrational constants for the ground (X) state and excited (C) state. The first rotational analysis was carried out by Colbourn and Douglas[4] for the B ↔ X band system. This was followed by a rotational analysis of the A ↔ X band system by Freeman, Yoshino, and Tanaka[5], who interpreted their observations in terms of Hund's case (b) coupling. The present spectra obtained at considerably higher resolution establish that case (c) coupling is dominant in the lowest excited state.

*Research supported by the Natural Sciences and Engineering Research Council of Canada, and the University of Toronto.

EXPERIMENTAL ARRANGEMENT AND OBSERVED SPECTRA

The experimental arrangement, shown in Fig. 1 is essentially the same as used in the earlier studies[1] of Xe_2 and Kr_2. A XeCl excimer laser (Lumonics TE-861M-2) was used to pump two dye-laser oscillators, each followed by two amplifiers. Linewidths as narrow as 0.03 cm^{-1} were achieved by incorporating two gratings and a four-prism beam expander in each dye laser resonator, and the output power was amplified up to 500 kW. The emission from one oscillator-amplifier system was doubled in KDP (the ν_1 beam) and tuned to be in two-photon resonance with the state $8s^1S_0$ of Hg. Tunable radiation from the second oscillator-amplifier system (beam ν_2) was then combined with the ν_1-beam and focused in the Hg vapor[6] to generate tunable coherent radiation at $2\nu_1 + \nu_2$, in the region of 106 nm. Argon dimers were formed at low temperatures (<10 K) by expanding either pure Ar or an Ar/He mixture (1/3) at a stagnation pressure of 5 atm, through a pulsed supersonic jet of 1 mm diameter.

Absorption spectra of three band systems A – X (108.2-107.4 nm), B – X (107.8-106.6 nm), and C – X (105.3-105.0 nm) were obtained by tuning the incident VUV radiation and detecting fluorescence emission with a photomultiplier (EMR-510G-08-013). Identical spectra of the weak A – X band system, but with improved signal-to-noise ratio were obtained by detecting Ar_2-ions formed by stepwise ionization (from $X \rightarrow A$ with $2\nu_1 + \nu_2$ radiation and $A \rightarrow$ ionization continuum with ν_1 or ν_2 radiation).

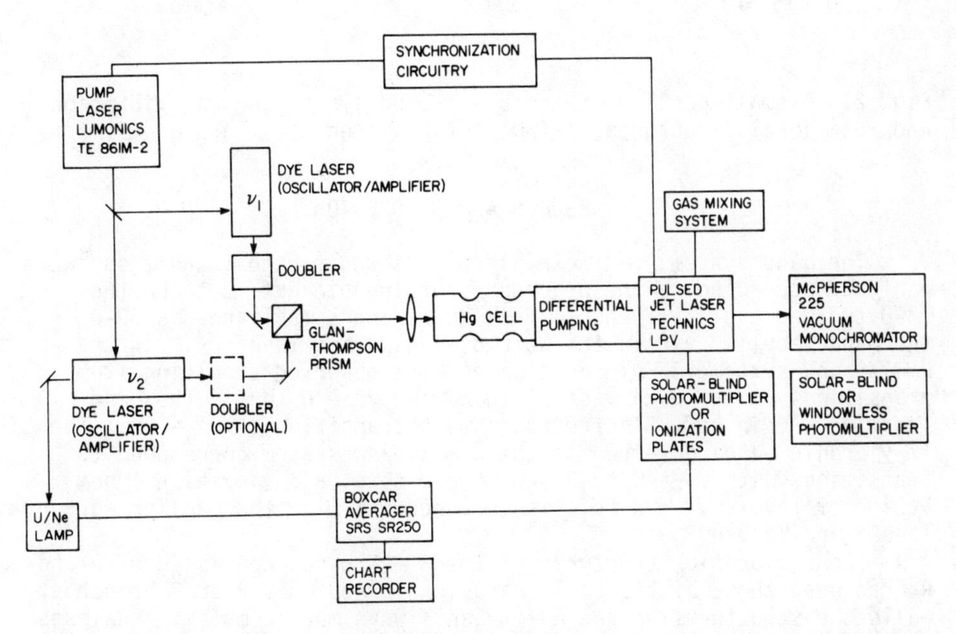

Fig.1. Experimental arrangement for generating tunable VUV radiation and for investigating spectra of Ar_2 formed by supersonic jet expansion.

As expected the observed spectra are simple in appearance because of the vibrationally and rotationally cold dimers produced in the jet. Nevertheless, by exciting the warmer dimers nearer the jet orifice, hot-bands originating from v" up to 5 were also recorded. In order to establish the vibrational numbering of the observed vibronic bands, spectra of the isotopes $^{40}Ar_2$ and $^{36,40}Ar_2$ were obtained for all three band systems. For this purpose, isotopic argon, ^{36}Ar, was mixed with ^{40}Ar and He in the ratio 1:9:30. A typical spectrum of the isotopic molecules is shown in Fig. 2.

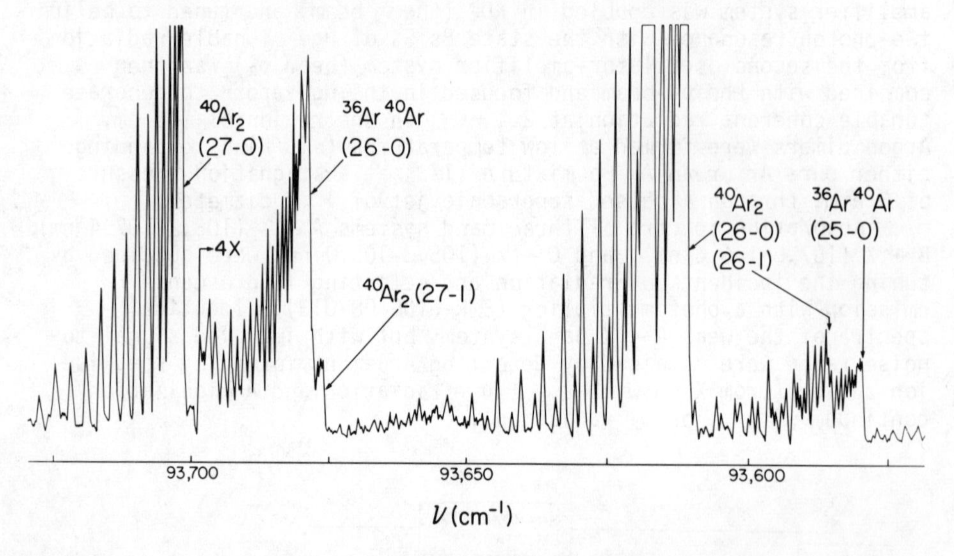

Fig. 2. A small portion of the B – X band system showing vibrational and rotational structures of two isotopic species, $^{40}Ar_2$ and $^{36,40}Ar_2$.

RESULTS AND DISCUSSION

The band system A – X exhibited 9 vibronic bands, and these were analyzed according to the procedure outlined in Ref. 2 to be the 23-0 to 30-0 and 31-1 bands. This numbering establishes as 30-0, the band labelled v-0 in the work of Tanaka and Yoshino[3]. In the B – X band system, 33 vibronic bands were observed resulting from transitions v' ← v" with v' = 20 to 28 and v" = 0 to 5. The band labelled v-0 in Ref. 3 corresponds to a transition to v' = 26. The 17 vibronic bands observed in the C – X band system correspond to transitions with v' = 0 to 9 and v" = 0 to 4, and vibrational numbering of levels v' = 0 to 4 is in agreement with the earlier work by Tanaka and Yoshino[3].

The rovibronic structures of the B – X bands consisted of P and R branches; those of the C – X bands exhibited only P or R branches, while the structures of the A – X bands were more complex. Analyses of these rotational structures for most of the observed bands were carried out using standard techniques. Here, only a summary of the more important findings is presented. For the ground state and the B state, the accuracy of known spectroscopic constants was improved. Rotational constants for the C state were derived for the first time. These were found to be similar to the rotational constants of the ground state, confirming that the potential energy curve of the C state is vertically above that of the ground state.

One of the most important results of the present work is the observation of three rotational branches P, Q, R in the A — X bands. This was made possible by careful adjustment of the oscillator cavities in order to achieve a resolving power of $\sim 5 \times 10^5$ and a linewidth of <0.2 cm^{-1} FWHM in the VUV. The resulting spectra revealed a clear splitting of the two rotational branches observed earlier by Freeman et al[5], into three branches. An example of one of the A — X bands is given in Fig. 3. In the subsequent analysis it was shown that the Q-branch transitions terminate on different rotational sublevels than the P- and R-branch transitions. This sublevel spacing varied as $J(J+1)$, as expected for Ω-type doubling. All of these results are consistent with Hund's case (c) coupling for the A state of Ar_2. With such strong evidence for case (c) coupling, we designate the electronic states as $X0_g^+$, $A1_u$, $B0_u^+$ and $C0_u^+$.

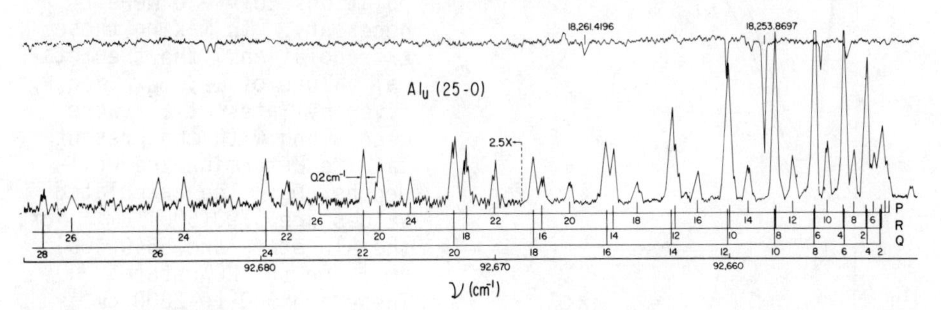

Fig. 3. Spectrum of the $A1_u - X0_g^+$ 25-0 band clearly showing 3 rotational bands, P, Q, and R. Upper trace is the opto-galvanic calibration spectrum of uranium.

Along with rotational constants, frequencies of band origins were obtained from the observed rovibronic bands. These values were then used to determine electronic energies T_e, vibrational frequencies ω_e, and anharmonic constants $\omega_e x_e$, $\omega_e y_e$... . For the ground state $X0_g^+$, values of $\omega_e = 30.676$, $\omega_e x_e = 2.420$, $\omega_e y_e = -0.0616$ and $\omega_e z_e = 0.00987$ cm^{-1} were derived from the data on levels v = 0 to 5. For the $C0_u^+$ state, values of $\omega_e = 66.83$ cm^{-1} and $\omega_e x_e = 3.997$ cm^{-1} were obtained using the energies for the levels v = 0 to 4. However, perturbations of the higher levels of the C state indicate departures from these values, and this will necessitate further study of the bands 6-0 to 9-0. While the rotational and vibrational constants for high v' bands of the A — X and B — X systems could be evaluated from the observed spectra, the long extrapolations to v' = 0 from v' = 23-31 for the A state and from v' = 20-28 for the B state, precluded the evaluation of reliable equilibrium values of R_e, ω_e, and $\omega_e x_e$.

Potential energy curves for all four states were calculated from the spectroscopic constants using an RKR technique devised by Vanderslice et al[7]. The results are displayed in Fig. 4. Here, the classical turning points for v" levels of the XO_g^+ state are indicated by horizontal lines ending in dots. These turning points (and therefore the shape of the potential energy curve) are in agreement with the potential curve calculated by Aziz and Slaman[8] (given by the solid line of Fig. 4) which is the most comprehensive calculation to date. Our value for the internuclear separation is $R_e = 3.761(3)$ Å.

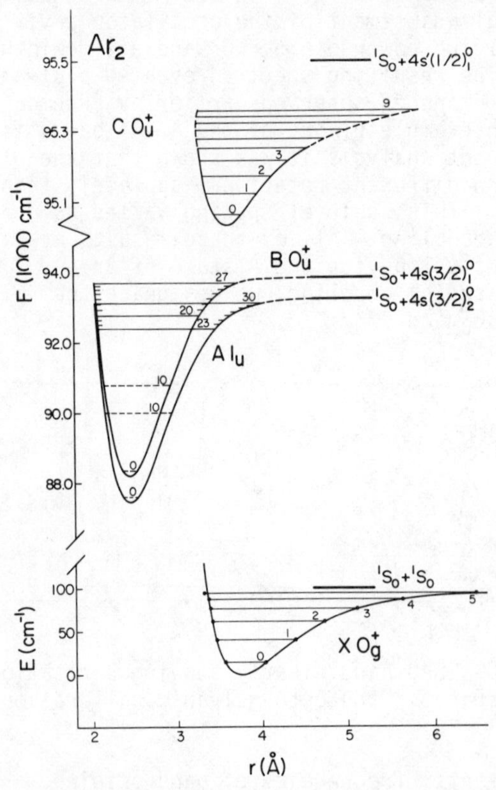

Fig. 4. Calculated potential energy curves showing observed vibronic levels and separated-atom states at dissociation limits.

The potential energy curves for the lowest stable states A and B are much less reliable since only high vibrational levels (>20) were accessible in the present experiments, and long extrapolations to v = 0 were necessary. In making these extrapolations, the theoretical values of ω_e, $\omega_e x_e$ and R_e given by Yates et al[9] were used along with the present data to determine the well-depths, D_e. The calculated values are 5790(500) cm^{-1} for the $A1_u$ state and 5640(400) cm^{-1} for the BO_u^+ state. These are 500 to 2000 cm^{-1} deeper than theoretical values calculated by Yates et al[9] and 500 to 1000 cm^{-1} shallower than values calculated by Michaelson and Smith[10]. The disagreement is not too serious considering the uncertainty in the present values and the complexity of theoretical calculations.

In Fig. 4, the potential curve for the CO_u^+ state is directly above that of the ground state. The low intensity of vibronic bands 5-0 to 9-4 and the lack of rotational features has limited the evaluation of spectroscopic constants to levels v = 0 to 4 for the C state. This added uncertainty is indicated by the dashed lines in the potential curve above v = 4. The internuclear separation was found to be R_e = 3.594(3)Å and well-depth D_e = 465.8(2) cm^{-1}. These values are in good agreement with the theoretical values of Castex et al[10]. Further details of the Ar_2 spectra and discussion are left to a future publication.

We are grateful to A. A. Madej for assistance with the Ar_2^+ detection.

REFERENCES

1. B. P. Stoicheff, J. R. Banic, P. Herman, W. Jamroz, P. E. LaRocque, and R. H. Lipson, in Laser Techniques for Extreme Ultraviolet Spectroscopy (T. J. McIlrath and R. R. Freeman, eds.) Amer. Inst. Phys., New York (1982), pp. 19-31; Can. J. Phys. 63, 1581 (1985).
2. R. H. Lipson, P. E. LaRocque, and B. P. Stoicheff, in Laser Techniques in the Extreme Ultraviolet (S. E. Harris and T. B. Lucatorto, eds.) Amer. Inst. Phys., New York (1984), pp. 253-258; J. Chem. Phys. 82, 4470 (1985).
3. Y. Tanaka and K. Yoshino, J. Chem. Phys. 53, 2012 (1970).
4. E. A. Colbourn and A. E. Douglas, J. Chem. Phys. 65, 1741 (1976).
5. D. E. Freeman, K. Yoshino, and Y. Tanaka, J. Chem. Phys. 71, 1780 (1979).
6. P. R. Herman and B. P. Stoicheff, Opt. Lett. 10, 502 (1985).
7. J. T. Vanderslice, E. A. Manson, W. G. Maisch, and E. R. Lippincott, J. Mol. Spectros. 3, 17 (1959), and 5, 83 (1960).
8. R. A. Aziz and M. J. Slaman, private communication (1986).
9. J. H. Yates, W. C. Ermler, N. W. Winter, P. A. Christiansen, Y. S. Lee, and K. S. Pitzer, J. Chem. Phys. 79, 6145 (1983).
10. R. C. Michaelson and A. L. Smith, J. Chem. Phys. 61, 2566 (1974).

Vacuum Ultraviolet Photolysis of Acetylene in the 110-135-nm Region

J J. Tiee, R. K. Sander, and C. R. Quick

Los Alamos National Laboratory, Los Alamos, NM 87545

R. Estler
Fort Lewis College, Durango, Colorado

Abstract

State-specific photofragmentation of acetylene in the 110- to 135-nm region has been studied using vuv laser and synchrotron sources. Investigations have been focused on learning the spectroscopic identity of the excited photoproducts by examining their time-resolved fluorescence. Results of the quenching of the excited photofragment emission and the emission polarization measurements are presented. An interpretation of these results in relating the observed photoproducts to the vuv photodissociation process is discussed.

Introduction

Vacuum ultraviolet photolysis of acetylene has been studied first by Stief et al.[1] at 1236 nm. An visible emission was reported by this work to be due to C_2 Swan band. Becker et al.[2] later attributed the emission to be from an excited state of C_2H_2 or that of the C_2H (designed here as C_2H^*) because the emission appeared to be broad covered the 400- to 580-nm region, unlike the diatomic spectrum of the C_2 radical. Similar works[3,4] on the vuv photolysis of C_2H_2 and C_2HBr have provided strong evidences that excited ethynyl radicals, C_2H^* are indeed the photoproducts. Very recently, additional experimental evidences were provided by other workers using a fast discharge vuv light source[5] and synchrotron radiation.[6] There seems to be little doubt that the broad emission is due to C_2H^*. However, a controversy has arisen over the mechanism of the observed long decay time: does it represent the radiative lifetime of the excited radical or a dark intermediate state that rapidly produces the emitting photoproducts. A proposal[7,8] for the dark intermediate state is that it is the lowest triplet state of vinylidene, CCH_2,[9-11] since it has been observed in flash photoexcitation of acetylene. The triplet state of the acetylene isomer,[9] CCH_2 has been calculated to be at 14,700 cm^{-1} above the ground state.[9]

With the availability of the laser-based high-resolution vuv light sources, a state-specific photofragmentation study of C_2H_2 can be implemented to address these issues. The intention is to examine the collision-free fluorescence lifetime and quenching efficiency of the observed emission in a state selective manner. In addition, using the fluorescence polarization measurements to learn about the excited state symmetry.

Experimental

Both a synchrotron light source and a laser-based coherent vuv source are used for these experiments. The synchrontron light source is used for providing a low-resolution survey of excitation spectra over a broad wavelength range (110 to 140 nm). With the large time-window and the high-spectral resolution of the laser source. The coherent light is used for obtaining time-resolved measurements of fluorescence decay and vibronically state-specific excitation spectra.

The synchrotron light source is a windowed (LiF) beam port (U9A) at the Brookhaven National Laboratory, National Synchrotron Light Source (NSLS). A 0.5-M monochromator is used to disperse the vuv light. Excited species emission is detected at right angles with a 0.2-M monochromator and a cooled photomultipler tube (PMT). The broadband synchrotron radiation is used to calibrate the spectral efficiency of the detection system. NIM photon counting electronics in conjunction with a LeCroy 3500 multichannel analyzer and a mini-computer (LS-11) are used for data acquisition and reduction.

The coherent vuv light is generated by frequency tripling in inert gas mixtures (Xe, Kr, and Ar). The apparatus is shown in Fig. 1. A Nd:YAG laser-pumped dye laser system is employed. The dye laser output of 40 to 70 mJ/pulse at the 550- to 600-nm region is mixed with residual 1.06-μm light from the YAG laser to produce 5 to 10 mJ of the 365- to 385-nm light. A 10-cm f.l. lens is used to focus the near uv light into the third harmonic generation (THG) section. A bandwidth of 1.5 cm^{-1} for the vuv light is approximated for the vuv light. A LiF lens (10 cm f.l. at 400 nm) is used to collimate or focus the vuv light to the sample cell section. A nitric oxide ionization cell with two plates biased at 200 V is utilized a detector for the vuv light.

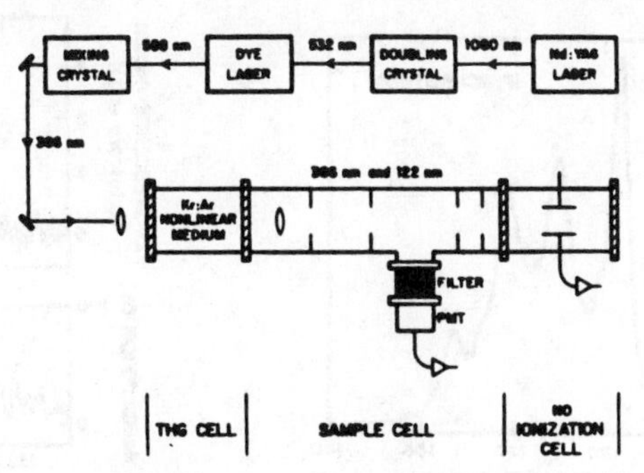

Figure 1. A schematic diagram of the experimental apparatus.

Since the THG process is only typically generated at a 10^{-6} efficiency, near uv scattered light is large compared to the desired photoproduct emission resulting from vuv photolysis. A number of light baffles are inserted along the beam path. A liquid filter (5-g/l hydrated $Fe_2(SO_4)_3$ and other interference filters are used to block the scattered fundamental laser light. The filtering system transmits emission from 400 to 940 nm with a transmission dip from 550 to 620 nm. In some experiments, a pinhole (~0.3 mm in diameter) is inserted into the midsection of the sample chamber. Because of the very large difference in the refractive index of the LiF lens, the vuv and near uv laser beams are focused at very different focal lengths. This allows the pinhole to discriminate much of the near uv light. A better than two orders of magnitude of reduction in the scattered light is attained. A gated PMT, which can be turn on/off for preset time delays and windows, is used for observing the emission. The output of the PMT is sent to either a boxcar averager or a transient digitizer/minicomputer system for signal averaging and data reduction.

Results and Discussion

The photoproduct emission spectra are obtained at different excitation wavelengths. The spectra are observed to be broad with no detectable structure, covering the entire visible region, similar to previous observations.[2,3,5] However, the observed spectra differ from the previous work in that the emission extending farther to the red, essentially to the cut-off wavelength of our PMT's, around 900 nm. This is most likely due to the broader spectral response of the PMT's used in the present work. The synchrotron excitation spectrum, and the laser absorption and excitation spectra are shown in Fig. 2. The synchrotron excitation spectrum shows pesks due to a combination of vibrational progression in Rydberg states and valence states. The spectrum agrees with the previous work,[6] which indicates a cut-off wavelength of emission at 136.5 nm, rather than 130 nm.[3] The laser spectra are scans of one vibronic band (Rydberg series, n = 4, v_2 = 1)

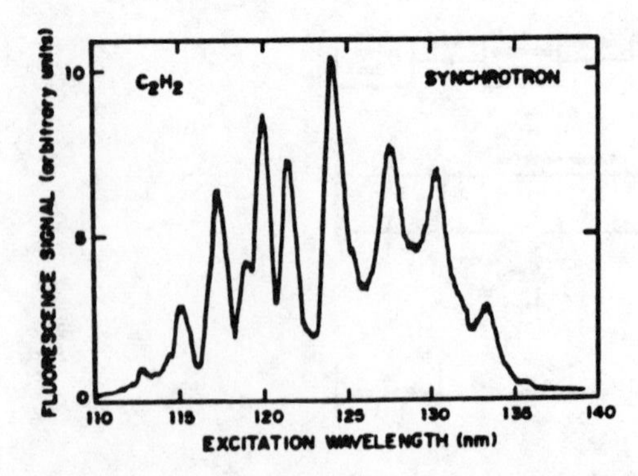

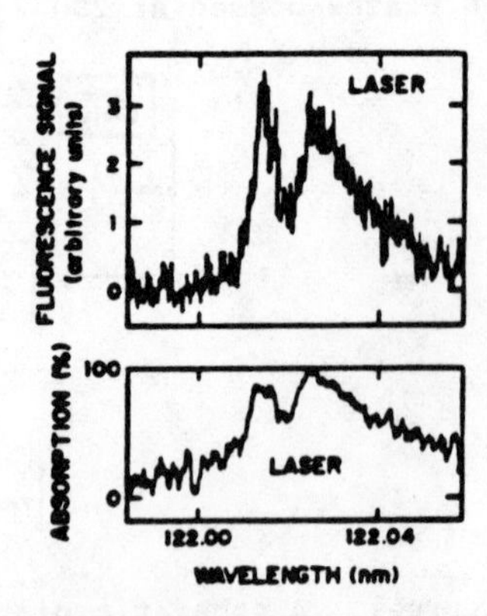

Figure 2. A synchrotron fluorescence excitation spectrum, a vuv laser fluorescence excitation spectrum, and a vuv laser absorption spectrum of C_2H_2.

which demonstrate the relatively high resolution of the laser source. However, since the laser linewidth is comparable to the rotational spacing of C_2H_2, it is not determined whether the actual bandwidth is smeared out due to the predissociation process.

The excitation spectra are also taken at different emission wavelength regions. Two excitation spectra obtained with bandpass filters (with a half-width of 40 nm) centered at 440 and 850 nm are shown in Fig. 3. There is a distinct shift in the amplitudes of vibronic peaks in the two spectra. This seems to indicate that higher excitation photon energy tends to produce more excited photoproducts. The temporal characteristics of the emission are also investigated in detail. It is found that single exponential decays give good fits to emission curves. It should be noted, however, since the emission is broad, the observing decay is not a state-selected decay. Collision-free fluorescence lifetimes of the emission are determined to be varied between 6 to 10 μs and are wavelength dependent. It appears that the shorter the emission wavelength the shorter the lifetime. The implication is then the more excited the photoproducts are, the shorter their fluorescence lifetimes are. This is consistent with the observation that the broadband emission lifetime is shortened at shorter excitation wavelengths.

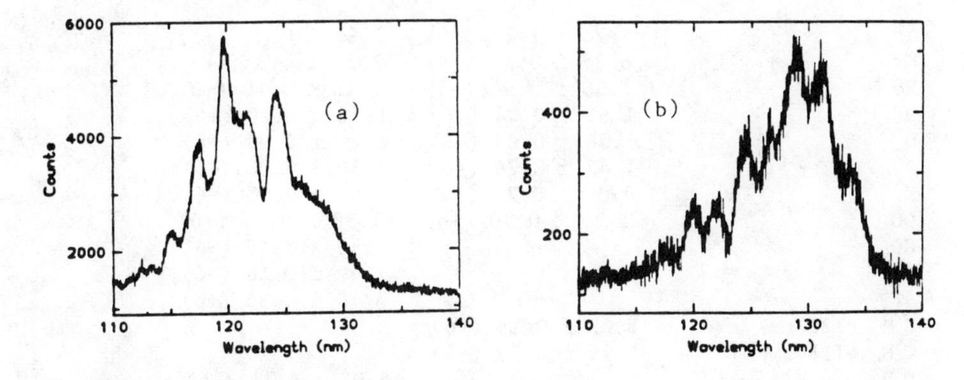

Figure 3. A fluorescence excitation spectrum of C_2H_2 obtained with (a) 440-nm and (b) 850-nm bandpass filters.

In the emission quenching measurements, a number of gases are used as quenchers. Several quenching rate constants are determined and listed in Table 1 for the emission resulted from the photolysis of either C_2H_2 or C_2D_2. Overall, no isotope effect is observed between these two sets of quenching rate constants. This is not surprising since the observed emission contains presumably large number of excited vibronic states, no state specificity is preserved within the photoproducts. The quenching data are fit to a simple quenching model, namely the harpoon model,[12] which assumes a crossing of a neutral repulsive potential energy curve with an attractive ionic potential energy curve to cause the quenching process. This model works well for molecules which have low ionization potentials or large electron affinities. In this case, C_2H does have a fairly large electron affinity (~3 ev). It is found that most of the calculated quenching cross-sections using this model to be in reasonable agreement to the measured cross-sections. The exceptions are the quenching cross-sections for C_2H_2, CF_4, and SF_6, which are measured to be much fast than the model predicts. In the C_2H_2 case, this can be explained to be the result of a reactive channel. As for CF_4 and SF_6, we believe that fast vibrational relaxation is probably the pathway, due to the high internal degree of freedom of the quenchers. One very important result is that no noticeably faster quenching rate constants are determined for heavy atoms or for fast triplet quenchers such as O_2. The implication is that the emitter can not be a triplet intermediate.

Table 1. Quenching Rate constants of C_2H and C_2D emission

	Rate Constants (sec^{-1} $torr^{-1}$)	
	C_2H	C_2D
He	4.71 ± 0.08 (+5)	4.9 ± 0.1 (+5)
Ar	7.7 ± 1.1 (+5)	7.9 ± 0.6 (+5)
Kr	6.25 ± 0.35 (+5)	7.3 ± 0.3 (+5)
Xe	1.30 ± 0.04 (+6)	1.13 ± 0.10 (+6)
H_2	1.65 ± 0.04 (+6)	1.36 ± 0.10 (+6)
D_2	1.36 ± 0.08 (+6)	1.26 ± 0.08 (+6)
O_2	1.43 ± 0.26 (+6)	1.30 ± 0.06 (+6)
N_2	9.8 ± 0.7 (+5)	1.01 ± 0.05 (+6)
CO	1.39 ± 0.90 (+6)	1.38 ± 0.20 (+6)
CO_2		2.58 ± 0.15 (+6)
CF_4		2.42 ± 0.18 (+6)
SF_6		3.70 ± 0.25 (+6)
C_2H_2 (122 nm)	5.03 ± 0.15 (+6)	
C_2H_2 (118.2 nm)	4.79 ± 0.15 (+6)	
C_2D_2		5.02 ± 0.14 (+6)

Since the excited C_2H_2 state symmetry is $^1\pi_u$, the exit channel of the predissociation must also be $^1\pi_u$; the H atom is ^2S, so the photoproduct must be produced in states correlating with π electronic symmetry. By examining the C_2H electronic state calculation of Shih et al.,[13] it is believed that $3^2A'$ is the most likely candidate emitting state of C_2H on the basis of energy spacing correlated with $^2\pi$ at its equilibrium C-C bond spacing. In addition, fluorescence polarization measurements are made with the vuv laser light polarized parallel with respect to the detector. A positive polarization ratio of ~6% is measured. This indicates a perpendicular transition to be the main contributor in the emission process and is in agreement with our tentative assignment of the $3^2A'$ --> $2^2A'$ transition as the one responsible for the observed emission.

Acknowledgment

Collaboration with J. Presses and R. E. Weston, Jr., in use of the Brookhaven National Laboratory Synchrotron beam line is grate-fully acknowledged, as well as technical support by R. Romero and M. Ferris. Financial support by DOE is also acknowledged.

References

1. L. J. Stief, V. J. Decarico, and R. J. Mataloni, J. Chem. Phys. 42, 3113(1965).
2. K. H. Becker, D. Haaks, and M. Schurgers, Z. Naturforsch. Teil A 26, 1770 (1971).
3. H. Okabe, J. Chem. Phys. 62, 2782 (1975).
4. H. Okabe, Photochemistry of Small Molecules, (Wiley, New York, 1978), p.262.
5. Y. Saito, T. Hikida, T. Ichimura, and Y. Mori, J. Chem. Phys. 80, 31 (1984).
6. M. Suto and L. C. Lee, J. Chem. Phys. 80, 4824 (1984).
7. A. H. Laufer, J. Chem. Phys. 73, 49 (1980).
8. A. H. Laufer and Y. L. Yung, J. Chem. Phys. 87, 181 (1983).
9. Y. Osamura and H. F. Schaefer III, Chem. Phys. Lett. 79, 412 (1981).
10. R. Krishnan, M. J. Frisch, J. A. Pople, and P. Von R. Schleyer, Chem. Phys. Lett. 79, 408 (1981).
11. K. A. White and G. C. Schatz, J. Phys. Chem. 88, 2049, (1984).
12. R. D. Levine and R. B. Bernstein, Molecular Reaction Dynamics (Oxford University Press, New York, 1974) pp.86-87.
13. S. K. Shih, S. D. Peyerimhoff, and R. J. Buenker, J. Mol. Spectrosc. 74, 124 (1979).

POSSIBILITY OF OBTAINING COHERENT SHORT WAVE RADIATION FROM A SOLID STATE FREE ELECTRON LASER

S.A. Bogacz and J.B. Ketterson
Department of Physics and Astronomy
Northwestern University, Evanston, IL 60201
and
Materials Science and Technology Division
Argonne National Laboratory
Argonne, IL 60439

ABSTRACT

The idea of using a crystal lattice or a superlattice as an undulator for a free electron laser is explored. A purely classical treatment of relativisitic positrons channeling through the proposed structure involving a self consistent solution of the wave equation for the radiating electromagnetic field and the kinetic equation for the positron distribution function leads to a positive gain coefficient for a forward radiating field. Matching the Kumakhov resonance to the undulator frequency further enhances the gain. This result, combined with a feedback mechanism arising from Bragg diffraction within the basic crystal lattice, leads to an instability of the radiation inside the crystal. Finally a numerical estimate of the Kumakhov-enhanced gain coefficient is made for the (110) planar channeling in a strain modulated Si superlattice.

INTRODUCTION

To scale the concept of the FEL to very short wavelengths e.g., X- rays, one must use either very high energy electron beams or short undulator periodicities. The first approach leads to a very small gain per unit length while the second one is limited by obvious geometrical restrictions in locating the opposing undulator magnets in close proximity.

In this paper we suggest using a solid state superlattice as an undulator in conjunction with Bragg diffraction (from the basic lattice periodicity) as a distributed feedback mechanism. The main idea can be sketched as follows.

---

*This work supported by the U.S. Department of Energy, under Contract W-31-109-ENG-38, and the Northwestern Materials Research Center through NSF grant DMR 82-16972.

0094-243X/86/1470322-7$3.00

Fig. 1 shows, schematically, a solid state superlattice. Such structures occur naturally in several alloy systems or may be prepared artificially with vapor deposition techniques. We consider a superlattice with an accompaning strain modulation, which is a natural consequence of the two constituents having different lattice spacings. A beam of relativistic particles while channeling through the crystal follows a well defined trajectory. Our treatment will be limited to positrons but could be modified for electrons. In Fig. 1, we show channeling paths (which for positrons would lie in low density regions of the crystal and conversely for electrons) parallel to the superlattice growth direction and also at a 45° angle to this direction. The latter are what interest us here because, as we see from the figure, the center of the channeling axis is modulated by the superlattice periodicity. This is the essence of the solid state undulator: i.e., the particles are continuously accelerated perpendicular to their flight path as they traverse the channel. The transverse motion is additionally enhanced by the Kumakhov resonance[2] - a transient oscillation of particles in the harmonic potential of the crystal field. Furthermore, the electric fields available for electrons involve the line averaged nuclear field and can be two or more orders of magnitude larger than the equivalent fields of macroscopic magnetic undulators.

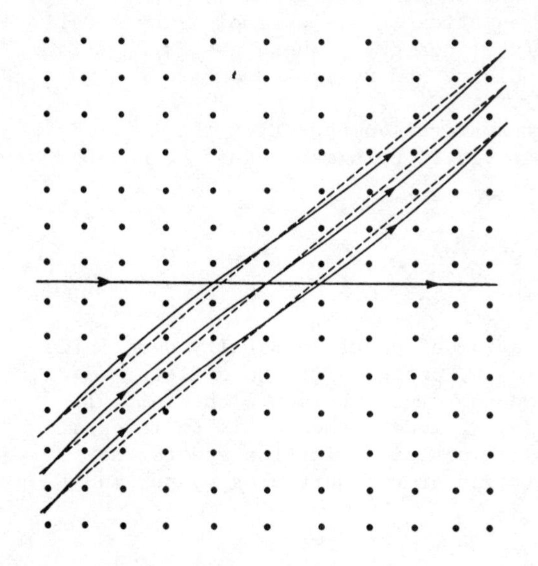

Fig. 1: Centers of the channeling trajectories for a (110) direction in a strain modulated superlattice. The (100) channeling direction yields no undulator effect.

Both of these factors hold the promise of greatly enhanced short wavelength undulator emission.

The feedback mechanism, necessary for the laser action is achieved through Bragg diffraction involving not the superlattice repeat distance, ℓ, but a basic lattice spacing, d. If the laboratory frame wavelength of the undulator emission satisfies $n\lambda = 2d_B$, where d_B is the spacing between neighboring sets of Bragg planes (perpendicular to the 45° channeling direction shown in Fig. 1), then counter propagating plane waves are set up; i.e., we have a standing wave just as with the mirrors of a conventional free electron laser. The Bragg condition is satisfied in practice by choosing the incident electron beam energy so as to satisfy the relation,

$$2d_B = n\lambda = n(\ell/2\gamma^2) ,$$

where γ is the Lorentz contraction factor.

THEORETICAL APPROACH

As was shown by F.A. Hopf et.al.[3] the free electron laser can be described in a purely classical way and the resulting gain is produced by a bunching of the particle density in the presence of a field.

A beam of relativistic positrons moving along the z-axis accelerated by the harmonic crystal field potential, ϕ, can be described in terms of a classical distribution function obeying the relativistic Boltzmann equation. Here we employ the model potential $\phi = \phi_0 + 1/2\ \phi_1\ (x-x_1 \cos gz)^2$, where $g = 2\pi/\ell$, ℓ is the strain modulation periodicity and x_1, ϕ_1 and ϕ_0 are parameters of the potential.

Assuming that only a transverse component of the A-field is present the linearized relativistic Boltzmann equation has the following form:

$$\frac{\partial}{\partial t} h + V_{\parallel} \frac{\partial}{\partial z} h - \frac{e}{c} V_{\perp} \frac{\partial A}{\partial z} \cdot \frac{\partial \Lambda}{\partial p_z} = - \frac{h}{\tau} \qquad (1)$$

where $h(t,z,p_z)$ is a fluctuation of particle density describing bunching of positrons due to the presence of the A-field. Here $\Lambda(p_z)$ is a longitudinal momentum distribution in the initial beam and the relaxation time, τ, models the collision integral accounting for small angle positron- lattice scattering.

The transverse acceleration of the particles by our model crystal field is given by

$$V_{\perp} = V_{\parallel} \frac{gx_1}{1-U^2} \sin gz \quad \text{where } V_{\parallel} = \frac{p_z}{m\gamma} \quad \text{and } U^2 = (gp_z)^2/e\phi_1 m\gamma. \qquad (2)$$

The bunching phenomenon leads to the generation of a transverse current which couples to the wave equation in the form

$$\left(\frac{\partial^2}{\partial z^2} - \frac{1}{c^2} \frac{\partial^2}{\partial t^2}\right) A = 4\pi n \int_{-\infty}^{\infty} dp_z \frac{e}{c} V_{\perp} \cdot h , \qquad (3)$$

where n is the concentration of particles in the beam.

Therefore our problem reduces to a self-consistent solution of Eqs. (1) and (3).

Using a procedure analogous to the Born approximation in scattering theory, we solve the above system of equations iteratively.[4] The resulting analytic solution for the A-field allows us to identity the amplitude gain coefficient for forward propagating radiation as follows:

$$\alpha = \frac{\pi}{4} n \int_{-\infty}^{\infty} dp_z \; Q^2 \; V_\| \; L \, \Gamma(\mu) \frac{\partial \Lambda}{\partial p_z} \quad ; \tag{4}$$

$$Q \equiv \frac{e}{c} V_\perp / V_\| \; , \; L\Gamma(\mu) = \mathrm{Im}\left[L \frac{e^{i\mu \, L/2} \sin \mu \, L/2}{(\mu \, L/2)^2} \right] \; ,$$

$\mu = \omega/V_\| - k - g$, $k = 2\pi/\lambda$, λ is the wavelength of the emitted radiation, L is the length of the undulator and now $\omega \to \omega + i/\tau$.

The integration in Eq.(4) can be carried out explicity in a limit describing position channeling through a superlattice, namely

$$L_C \ll L \ll L_p \quad \text{where } L_p \equiv \ell \left(\frac{\Delta p}{p}\right)^{-1} \text{ and } L_c \equiv \tau V_\| \quad . \tag{5}$$

Here L_c plays a role of the coherence length (is typically about 1 μm)[5] and the width of momentum distribution Λ is introduced in a standard way:

$$\frac{\Delta p}{p} \equiv \left[\left(p_z^2 \frac{\partial \Lambda}{\partial p_z} \right)_{max} \right]^{-1/2} \quad . \tag{6}$$

Our final result for the gain coefficient (for μ=0) is given by

$$\alpha = 2 \, n \frac{Q^2}{m} \frac{}{\gamma^2 (\gamma^2 - 1)^{1/2}} L_c^2 \quad , \tag{7}$$

and will be studied numerically in the last section.

DISCUSSION

According to the calculation presented, spontaneous emission of electromagnetic radiation results from a particular kind of particle density fluctuation, h, which has the form of a positron bunch propagating with the same frequency, ω, as the emitted electromagnetic wave and the phase velocity of the beam, $V_\|$.

Keeping in mind that the periodicity of the undulator represents a static driving force with a wavevector g, and k is the wave vector of the electromagnetic wave, we can analyze our results in the language of three wave mixing.

From this point of view, the solid state free electron laser reduces to a simple Umklapp process involving: 1) the propagating bunch in the positron density, 2) the emitted (or applied) electromagnetic wave and 3) the static periodic field of the undulator. Matching of the frequencies for the two "dynamic modes" assures "energy" conservation. Furthermore momentum conservation is given by the resonance condition (μ=0). The physical interpretation of the Umklapp process is illustrated in Fig. 2.

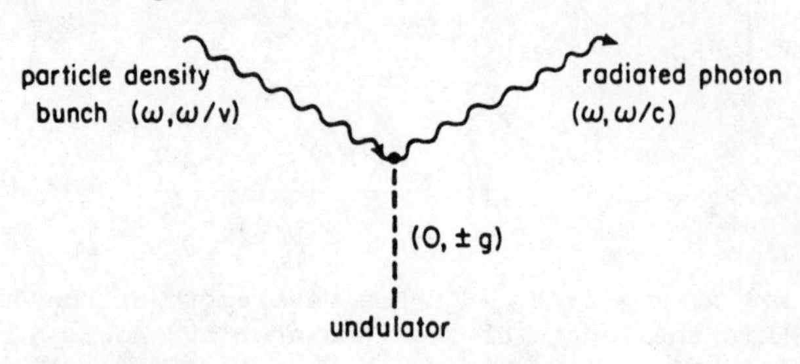

Fig. 2: Umklapp process - positron density bunch boosted by the momentum transfer from the undulator results in a spontaneous emission.

NUMERICAL ANALYSIS

We will discuss the feasibility of the proposed scheme by considering (110) planar channeling in a strain modulated Si crystal. We write the undulator period as ℓ = Nd, where d = 1.92 Å is the spacing between successive lattice planes. The strain modulation, of course, requires a second component, such as Ge; however, we will use the parameters of Si for convenience ($e\phi_1$ = 33 eV·Å^{-2})[6].

The relativistic particles, while channeling along the path, undergo transverse harmonic oscillations from the

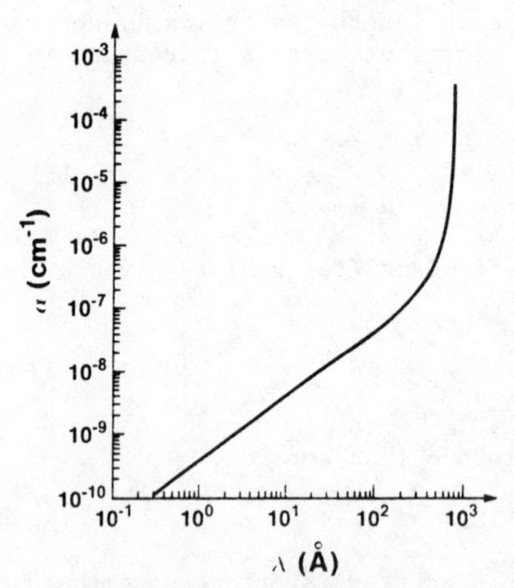

Fig. 3: Maximum value of the linear gain coefficient enhanced by the Kumakhov resonance as a function of wavelength of emitted radiation (λ). This calculation performed for Si, assumes that the particle oscillates with the largest allowed amplitude (half the channel diameter) and that the coherence length is energy independent; the latter assumption would not be valid as $\gamma \to 1$, where the graph shows a divergent gain.

crystal field - Kumakhov oscillations - with the characteristic frequency $\omega_k = (\phi_1/m)^{1/2}$. One can see from Eq. (2) that the pump parameter, Q, has a resonance ($U^2 \to 1$) if

$$gV_\parallel = \gamma^{-1/2}\,\omega_k \quad . \tag{8}$$

This enhances the gain coefficient, however, the excessive growth of Q would soon result in a rapid dechanneling of the particles. Indeed, dechanneling will occur if the transverse kinetic energy of the particle exceeds the binding energy of the harmonic potential. This condition fixes the maximum allowed value of the pump paramer, Q, as:

$$Q^{max} = \frac{1}{2}\, a\, \frac{e}{c} \left(\frac{e\phi_1}{mc^2}\right)^{1/2} \quad , \tag{9}$$

where a is a distance between adjacent channels. Finally the Kumakhov-enhanced gain coefficient given by Eqs. (7) and (9) is evaluated numerically in Fig. 3. As a source of high density particle current, we consider the state of the art 10,000 A Livermore pulsed electron-position facility, where the radius of the beam is about 1mm.[7] We have to keep in mind that two additional conditions have to be satisfied namely (μ=0) and Eq. (8). For a given wavelength λ they uniquely fix values of γ and superlattice modulation periodicity ℓ which is illustrated in Fig. 4 a) and b).

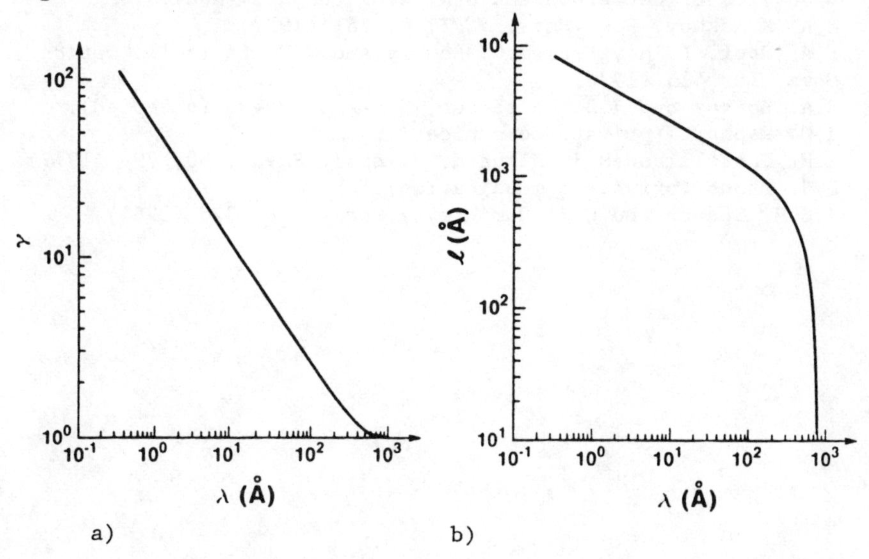

Fig. 4a,b: Positron beam energy (γ) and superlattice modulation length (ℓ) required for maximum of the gain coefficient at the vicinity of the Kumakhov resonance for a given wavelength.

CONCLUSIONS

In the X-ray region, our model calculation leads to a gain of the order 10^{-8} cm^{-1}. Possibly, by cooling the sample and using "stiffer" crystals (eg., diamond) one can bring the value of the coherence length to about 30 μm,[5] which would increase the gain by the factor of 10^3. To bring the gain to about 1 [cm^{-1}], to overcome X-ray attenuation, one has to hope that future particle accelerators will provide extremaly high particle densities (10^5 increase). In this case, our model FEL would make use of continuously distributed Bragg diffraction feedback mechanism.[4] The feasibility of the suggested source of coherent radiation rests on the availability of high current density sources (necessary to sustain spontaneous radiation) and the associated radiation damage and thermal heating of the sample.

Finally the quantum mechanical extension of this problem to the case of both electron and positron channeling should be attempted; this is especially relevant at low energies (i.e., when using the basic crystal itself as the undulator) where treating the particles as Bloch waves (the so called dynamical effect) is essential.[8]

REFERENCES

1. H. Ikezi, Y.R. Lin-Liu and T. Ohkawa, *Phys. Rev.* B30, 1567 (1984); J.B. Ketterson and G.K. Wong (unpublished).
2. M.A. Kumakhov, *Sov. Phys. JEPT,* 4, 781 (1977).
3. F.A. Hopf, P. Meystre, M.O. Scully and W.H. Louisell, *Optics Comm.* 18, 413 (1976).
4. S.A. Bogacz and J.B. Ketterson, *J. Appl. Phys.* (accepted).
5. J.O. Kephart (private communication).
6. R.H. Pantell and M.J. Alguard, *J. Appl. Phys.*, 50, 798 (1979).
7. D.S. Prono (private communication).
8. J.C.H. Spence and C.J. Humphreys. *Optik*, 66, 225 (1984).

LAMELLAR TRANSMISSION GRATINGS IN THE SOFT X-RAY REGION

R. C. McPhedran
A. Roberts
Department of Theoretical Physics,
University of Sydney,
Australia, 2006.

ABSTRACT

We calculate the behaviour of lamellar transmission gratings rigorously and with two approximate models, and show that all three theoretical models accord well with experiment.

INTRODUCTION

The development of transmission gratings for use in the x-ray and vacuum ultraviolet has led to applications in cosmic x-ray spectroscopy[1,2] and plasma diagnostics, as well as to their use as monochromator elements in soft x-ray synchrotron beams[3].

Here we study theoretically the diffraction by lamellar transmission gratings, using both a rigorous model and an approximate method combining plane-wave electromagnetism with Fourier optics. Another approximation, based on the solution for a grating composed of a perfectly conducting material, and appropriate in the case of radiation normally incident on the grating, is also developed. We investigate the validity of the approximations and compare our calculations with measurements.

RIGOROUS THEORY

The rigorous theory employed is based on the use of modal expansions for fields within the lamellar grating, and plane wave expansions in free space[4]. The two types of expansion are matched at the upper and lower surfaces of the grating giving a set of linear equations for unknown field quantities. Tests, such as conservation of energy and reciprocity, enable us to assess our calculated results to be accurate to better than ±1%. The two distinct cases of P and S polarization need to be treated separately and, in general, give different results. In the region of interest here, however, calculations have been made for both polarizations and the results found to be essentially the same.

0094-243X/86/1470329-7$3.00

GEOMETRIC OPTICS MODEL

One method which can be used to approximate the diffraction properties of a grating involves tracing the path of each ray of an incident plane wave through the grating. This technique is similar to that used by Tatchyn and Lindau[3] in modelling gratings compatible with synchrotron radiation. Refraction effects are ignored as are any rays which have been reflected twice. Representative ray paths and the parameters used in this theory are shown in Fig. 1.

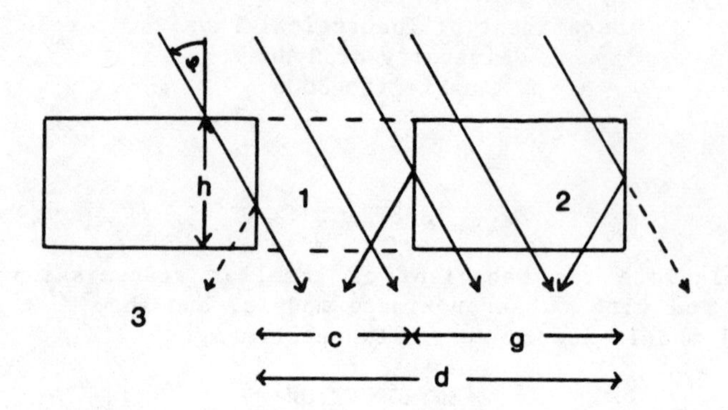

Fig. 1. Representative ray paths used in the development of equation (4). The refractive indices of regions 1, 2 and 3 are respectively r_1, r_2 and r_3, and the incident radiation has a vacuum wavelength, $\lambda = 2\pi/k_0$.

We find the field along the lower interface by multiplying the incident plane-wave field by phase and attenuation factors depending on the path taken by the ray, as well as by the appropriate transmission factors. Additional terms corresponding to rays reflected by the interfaces between regions 1 and 2 are also included. Fourier transforming, we find that the efficiency of the mth transmitted order is

$$\rho_m^T = \mathrm{Real}(\eta_m)\,|T_m|^2/\chi_0 \tag{1}$$

where

$$\chi_0 = k_0 \cos\varphi, \tag{2}$$

$$\eta_m = \sqrt{k_0^2 r_3^2 - (k_0 \sin\varphi + 2\pi m/d)^2} \tag{3}$$

and

$$T_m = \frac{t_{02}t_{21}t_{13}e^{ik_0 r_2 h/\cos\varphi}\left(e^{i(k_0(r_1 - r_2)/\sin\varphi - 2\pi m/d)a} - 1\right)}{(k_0 d(r_2 - r_1)/\sin\varphi - 2\pi m)} - \frac{t_{01}t_{13}e^{ik_0 r_1 h/\cos\varphi}\left(e^{-i2\pi mc/d} - e^{-i2\pi ma/d}\right)}{2\pi m}$$

$$+ \frac{t_{01}t_{12}t_{23}e^{i(k_0 r_1 h/\cos\varphi + (k_0(r_2 - r_1)\sin\varphi - 2\pi m/d)c)}}{(k_0 d(r_2 - r_1)/\sin\varphi - 2\pi m)} \times (e^{i(k_0(r_2 - r_1)/\sin\varphi - 2\pi m/d)a} - 1)$$

$$- \frac{t_{02}t_{23}e^{ik_0 r_2 h/\cos\varphi}(1 - e^{-i2\pi m(a + c)/d})}{2\pi m}$$

$$- \frac{t_{01}r_{12}t_{13}e^{i(k_0 r_1 h/\cos\varphi - 2\pi mc/d)}(1 - e^{i(2\pi m/d + 2k_0\sin\varphi)a})}{(2k_0 d\sin\varphi + 2\pi m)}$$

$$- \frac{t_{02}r_{21}t_{23}e^{ik_0 r_2 h/\cos\varphi}(1 - e^{i(2\pi m/d + 2k_0\sin\varphi)a})}{(2k_0 d\sin\varphi + 2\pi m)} \, . \qquad (4)$$

(Note that if $\varphi = 0$, then only the second and fourth terms contribute.) In (4) $a = h\tan\varphi$ and t_{ab} and r_{ab} are, respectively, the complex transmission and reflection amplitude coefficients for a ray going from medium a into medium b. If necessary, polarization effects can be included in these last two quantities.

This expression is valid for angles of incidence, φ, such that $\tan\varphi < \min\{c,g\}/h$. If φ is increased beyond this, more terms in the expression for T_m need to be considered.

Results obtained from the exact theory and this approximate model when φ equals 0° and 30° are compared in Table I, the limit of validity of equation (4) for this example being 60°. We used optical constants of gold from Ref. 5. As the scale size of the grating is reduced to the same order as the wavelength, the approximation remains valid into the region where the structure is effectively transparent.

φ (degrees)	Order	Transmitted Efficiencies (%)	
		Exact	Approximate
0	+1	13.442	13.101
	+2	1.641	1.310
	+3	0.455	0.583
30	+1	13.594	13.588
	+2	1.902	1.911
	+3	0.389	0.383

Table I Exact and approximate values for transmission efficiencies. The grating and incidence parameters are: d=1.0 μm, c=0.4 μm, h=0.25 μm, $=9.9\times10^{-4}$ μm, $r_1=1.0$, $r_2=0.99885+i4.42\times10^{-4}$, $r_3=1.0$.

INFINITE CONDUCTIVITY MODEL

Another approximation, of theoretical interest, although restricted to normal incidence, is based on the solution for a perfectly conducting grating. One parameter involved in the development of the exact theory is η, where

$$\eta = \min\{c,g\} \times k_0\sqrt{r_2^2 - r_1^2}. \tag{5}$$

We have shown that as $|\eta|$ and, in particular, the imaginary part of η become large, the problem approaches that of the corresponding perfectly conducting grating. The case where $r_2 \to \infty$ has been investigated in detail. A similar situation arises for a lossy structure if $dk_0 \to \infty$, i.e. the scale size of the grating is much larger than the wavelength. However, since r_1 and r_2 are both close to unity in the soft x-ray region, both the P and S polarization boundary conditions approach those for a perfectly conducting grating with P polarized radiation. This can be seen in Fig. 2, where the plot of the electric field predicted by the exact electromagnetic theory shows that the magnitude of the field approaches zero at the interface between regions 1 and 2.

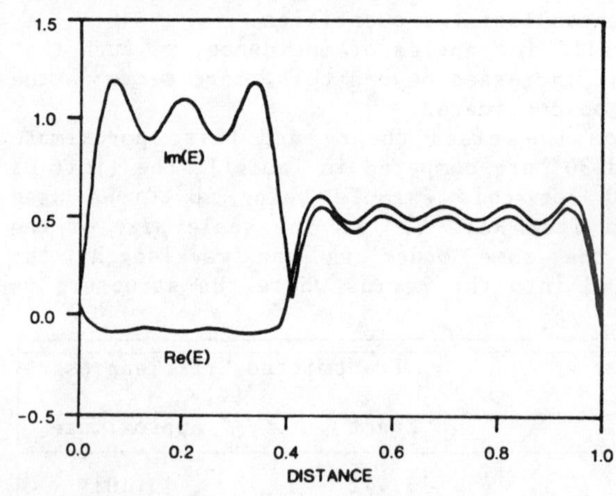

Fig. 2. Electric field at the mid-depth of a gold transmission grating, plotted over one period as a function of the distance along the grating. The grating parameters are the same as those given with Table I.

At normal incidence the incident and diffracted rays strike interfaces within the grating at grazing angles, since the angular dispersion of the grating is low. Since rays approaching the interface from either medium are strongly reflected at such angles of incidence the fields in medium 1 "see" medium 2 as being highly conducting and vice-versa. Under these circumstances, the field produced by a finitely conducting grating can thus be regarded as being the superposition of the solutions for two perfectly conducting gratings. One grating is composed of material of refractive index r_1 in region 1 and a perfectly conducting material in region 2, while the complementary grating has this material in region 1 and material of refractive index r_2 in region 2. This approximate method remains valid as long as $|\eta|$ remains large. (We have found that estimates of the first order transmission efficiencies to within a relative error of 5% can be obtained with $|\eta|$ greater than 50.)

A comparison of exact theory, (c), and the two approximate models, (d) and (e), appears in Table II. It can be seen that the estimates for the transmission efficiencies are in good agreement with the rigorous model. The values obtained using the second technique are, however, in significantly better agreement with the rig-

λ(Å)	EFFICIENCIES (%) COMBINED ORDERS ±n		
n =	1	2	3
5.4			
(a)	9.9	0.6	0.6
(b)	13.1	0.8	0.8
(c)	11.4	1.8	0.3
(d)	11.8	1.1	0.5
(e)	11.2	1.9	0.3
7.0			
(a)	19.4	1.5	1.1
(b)	25.8	2.0	1.5
(c)	21.2	2.7	0.8
(d)	21.3	2.0	0.9
(e)	20.7	2.7	0.7
8.3			
(a)	22.7	1.2	1.7
(b)	30.3	1.6	2.3
(c)	26.6	3.2	0.9
(d)	27.1	2.6	1.1
(e)	26.3	3.3	0.8
9.9			
(a)	23.2	1.5	1.5
(b)	30.9	2.0	2.0
(c)	26.8	3.3	0.9
(d)	27.5	2.6	1.2
(e)	26.6	3.4	0.8
13.3			
(a)	20.9	1.7	1.2
(b)	27.9	2.3	1.6
(c)	25.6	3.2	0.8
(d)	26.3	2.5	1.1
(e)	25.3	3.4	0.6
14.4			
(a)	18.7	2.1	1.0
(b)	24.9	2.8	1.3
(c)	24.2	3.1	0.7
(d)	24.9	2.4	1.1
(e)	23.8	3.3	0.6

orous model than those obtained using the simpler method. This is most evident in the values obtained for the (much lower) second and third order efficiencies.

Table II Transmission efficiencies of a lamellar grating with the parameters given with Table I. The five values correspond to: (a) experimentally determined values from Ref. 1., (b) same as (a) adjusted to take into account the presence of a supporting grid, (c) rigorous electromagnetic theory, (d) approximate geometric optics theory and (e) approximate theory based on the solution for a perfectly conducting grating.

In Table II we also compare the theoretical calculations with the experimental values of Bräuninger et al. both uncorrected, (a), and corrected, (b), for the effects on transmission of a supporting grid. The three theories agree well with the measured values, although errors including uncertainties in the correction factor and deviations from a rectangular cross-section for the wires have been introduced.

OPTIMIZATION OF THE FIRST ORDER TRANSMISSION EFFICIENCY

At normal incidence equation (4) is considerably simplified and the transmission efficiencies become

$$\rho_0^T \simeq |t_1 c + t_2 g|^2/d^2 \qquad (6)$$

and

$$\rho_m^T \simeq |t_1 - t_2|^2 \sin^2(m\pi c/d)/m^2\pi^2 \quad , \; m \neq 0, \qquad (7)$$

where $t_1 = e^{ik_0 h}$ in vacuo and $t_2 = e^{ik_0 rh}$, for a material with refractive index r in region 2. By equating to zero the differential coefficient of (7) with respect to h, the first order transmitted efficiency can be optimized by solving for h. For h=0.36 μm, the combined first order efficiencies are 32%, up 5% on the value of Table II.

Also, as noted by Bräuninger et al., the first order efficiency can be improved by increasing the mark-space ratio to 0.5. This is reflected in the argument of the sin term of equation (7).

By summing (6) and (7) over all possible values of m, we find that the total transmitted energy is approximately

$$\text{E.T.} \simeq |ct_1 + gt_2|^2/d^2 + |t_1 - t_2|^2 cg/d^2. \qquad (8)$$

Since the energy reflected is negligible, the energy absorbed is given by

$$\text{E.A.} \simeq 1 - \text{E.T.} \,. \qquad (9)$$

Thus, as h becomes large, $t_2 \to 0$ and E.T. $\to$ c/d, while the energy absorbed approaches its complement.

CONCLUSION

In this paper we have outlined two approximate techniques for the calculation of diffraction properties of lamellar transmission gratings in the soft x-ray region. These approximations agree well with a rigorous modal theory although at normal incidence results obtained from a technique based on the solution for a perfectly conducting grating are in better agreement than those given by the simpler theory. The latter model, however, remains quite accurate in its estimates of first order efficiencies, is valid at off normal angles of incidence and is computationally less complex than the other technique. This method can also be used to optimize efficiencies by varying grating parameters.

REFERENCES

1. H. Bräuninger, H. Kraus, H. Dangschat, K. P. Beuermann, P. Predehl and J. Trümper, Appl. Opt., 18, 3502 (1979).
2. H. W. Schnopper, L. P. van Speybroeck, J. P. Delvaille, A. Epstein, E. Källne, R. Z. Bachrach, J. Dijkstra and L. Lantward, Appl. Opt., 16, 1088 (1977).
3. R. Tatchyn and I. Lindau, Nucl. Inst. and Meth., 172, 287 (1980).
4. L. C. Botten, R. C. McPhedran, J. L. Adams, J. R. Andrewartha and M. S. Craig, Optica Acta, 28, 413 (1981).
5. D. W. Lynch and W. R. Hunter, in Handbook of Optical Constants of Solids, ed. E. D. Palik (Academic Press, Orlando, 1985) p. 290.

SOFT X-RAY EMISSIONS FROM VARIOUS TARGETS IRRADIATED BY LONG WAVELENGTH LASERS

H. Daido, Y. Kitagawa, H. Matsunaga, H. Shiraga, F. Miki,
M. Fujita, T. Yabe, H. Fujita, Y. Kato, S. Nakai,
and C. Yamanaka

Institute of Laser Engineering, Osaka University
Suita, Osaka 565 JAPAN

ABSTRACT

We observed efficient emission of soft x-rays ($h\nu$<1.5 keV) from 10 μm laser wavelength laser irradiated targets in which the inhibition of thermal conduction and expansion loss was expected to occur.

INTRODUCTION

Many papers have been published on intense pulsed x-ray emission from laser produced plasmas.[1-10] At short laser wavelength (<0.5 μm), the physical process of the x-ray conversion from laser produced hot and dense plasma have been investigated experimentally[2-4] and theoretically.[5] The computed predictions based on the non LTE model agree with the experimental data.[4]

On the other hand, at long laser wavelengths such as CO_2 laser whoes efficiency is very high compared to the solid state lasers, a few papers have been published.[8-10] At 10.6 μm wavelength, Nishimura et al. have obtained the conversion efficiency of only 1 % from the gold plane target at laser intensity of 10^{14} W/cm^2, while Rockett et al. showed the conversion efficiency of more than 10 % from the gold coated spherical target at same laser intensity. Measurement and calculation methods based on the black body assumption are almost same in both works. The reason of this discrepancy is attributed the different target geometries such as plane geometry of the infinite boundary in Nishimura's experiment while the finite boundary spherical target in Rockett's experiment. At long wavelength (10μm) laser, the resonantly absorbed laser energy goes into hot electrons, which go towards the under dense corona region[11]. The strong self-generated magnetic field is generated in the corona region and the electrons spread over the target surface outside the focal region due to E x B drift[12-15]. Significant amount of the absorbed energy is deposited into large area of the target and is lost by fast ions[16]. The energy coupling between the laser and the dense plasma is thereby weak, leading to the small conversion efficiency compared to that at short wavelengths especially for plane target geometry. In this report, we show the improved radiation conversion at long wavelengths using novel target in which the inhibition of thermal conduction and expansion loss is expected to occur.

BASIC CONCEPT

Figure 1(a) shows the schematic diagram of the lateral spreading of hot electrons due to E x B drift and the fast ion expansion. These are the dominant loss mechanisms for x-ray conversion. Figure 1(b) shows the cross sectional view of the novel target in which the hot electrons are trapped inside the target. The ablated plasma converges to the center of the target and form the central hot-core with the strong self-generated magnetic field of more than several MGauss, which then inhibits strongly the thermal conduction from the hot plasma to cold target wall.[17] Another loss mechanism is the plasma expansion[7] from the laser focal region to vacuum, which is usually observed in the plane geometry. The expansion loss is inhibited by the novel geometry of the target as shown in Fig. 1(b). These two inhibition mechanisms against the plasma cooling are expected to give larger conversion efficiency than plane targets.

CO_2 LASER EXPERIMENT

One arms of the LEKKO VIII CO_2 laser system[18] delivers 100 J in 1 ns and the average intensity at the focal point is 1.3×10^{14} W/cm^2. The hot electron temperature is estimated to be around 15 keV[12] and the self-generated magnetic field[14] is expected to be more than 1 MGauss (100 Tesla) at this laser intensity. Two types of targets, shown schematically in Fig. 2(a) and (b) are irradiated to inverstigate the plasma life time for various target radii and irradiation geometries (one beam or two beam irradiation). The corresponding x-ray pinhole images and typical diode signals of the soft x-rays which escape from the hole of the target are also shown in Fig. 2. The tamper shell made of 15 μm thick parylene (C_8H_7Cl) is transparent enough to observe hot central core. The duration of the soft x-ray signal is much longer than that with a direct irradiation target, where the typical x-ray duration is comparable to that of the laser pulse.[19] Figure 3 shows the characteristic time of the x-ray emission as a function of the cavity diameter when the target is irradiated by a 100 J CO_2 laser. The different lines correspond to the entire width (solid circles), the width of the secondary peak (triangles) and the separation between the two peaks (squares) of the x-ray signals. The first peak is due to the direct impact of the laser on the inner surface, while the second peak is due to the plasma formed at the center. These data agree well with the simulation code HISHO in which the radial heat transport is inhibited by the assumed magnetic field of 10MG[17]. The laser absorption on these targets are 40-60%

measured by 23 ch. optical calorimeter array. In our novel target, the x-ray conversion efficiency is expected to be high with the good absorption and thermal insulation.

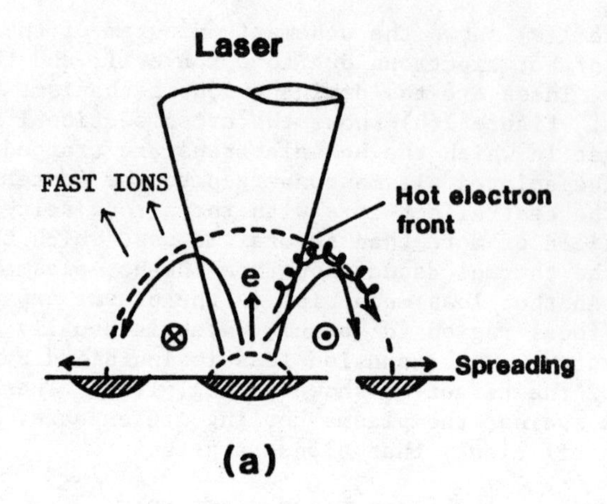

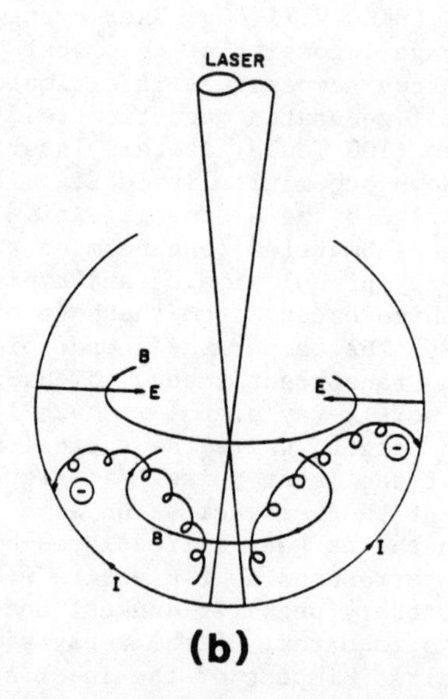

Fig.1 Schematic diagram of the magnetic field generation and related physical processes in the plane geometry (a), and the novel geometry (b).

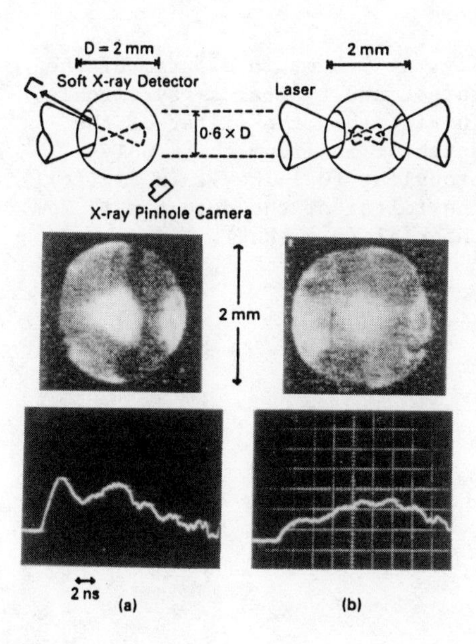

Fig. 2 Top figures show the schematic diagram of the target with one hole (a) and two holes (b). Middle pictures show typical x-ray pinhole images (white regions) of the targets whose spectrum is higher than 1keV. Bottom figures show typical soft x-ray temporal signals (100eV-1keV) detected by a soft x-ray detector and Tektronix 7104 oscilloscope. The sweep time is 2ns/div.

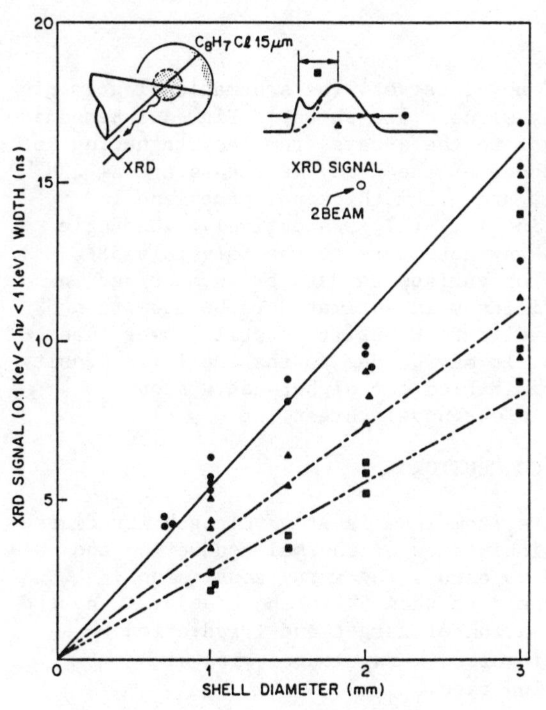

Fig. 3 Observed plasma lifetime as a function of the target diameter when the target is irradiated by a 100-J CO_2 laser.

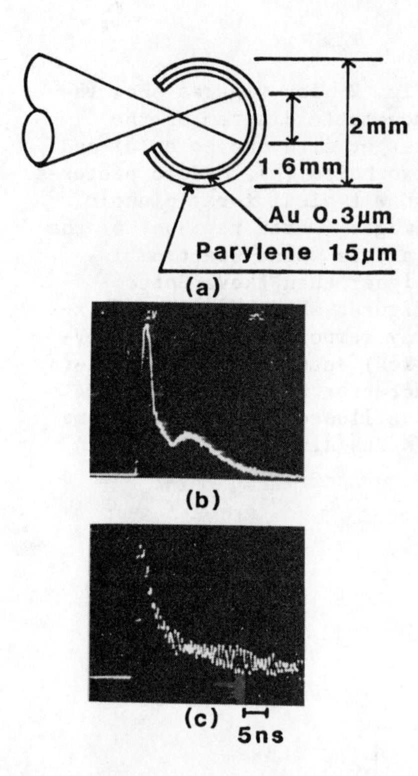

Fig. 4 Schematic diagram of the target and typical x-ray signals. In Fig. (b), the filter of the detector is 1.2μm thick polypropylene (0.1-1keV). In Fig.(c), the filter of the detector is 2μm thick Al (0.5-1keV).

We test the Au coated novel target. The schematic diagram of the target and typical x-ray signals are shown in Fig. 4. Secondary peak in Fig. 4(b) corresponds to the x-rays from the stagnating hot central core. The photo-cathode of the x-ray detectors are Al and the filters of the detectors are 1.2μm thick polypropylene in Fig. 4(b) and 2μm thick Al in Fig. 4(c), respectively. The angle between the laser and the x-ray detectors to the target is 38°. The x-ray radiation energy for various targets are summarised in Fig. 5. The conversion efficiency is estimated to be more than 5% in the case of Au coated novel target but it is still lower than that at shorter wavelengths. It may be due to the small ion density in the target[5]. Futher optimization for higher conversion efficiency is investigated in our novel target.

CONCLUSIONS

We show the soft x-rays from 10μm laser wavelength irradiated novel targets in which the inhibition of thermal conduction and expansion loss was expected to occur. The x-ray conversion efficiency is estimated to be more than 5% in the case of Au coated novel target. Futher optimization of target and irradiation geometries gives a practical pulsed x-ray source with high efficiency and high repetition rate.

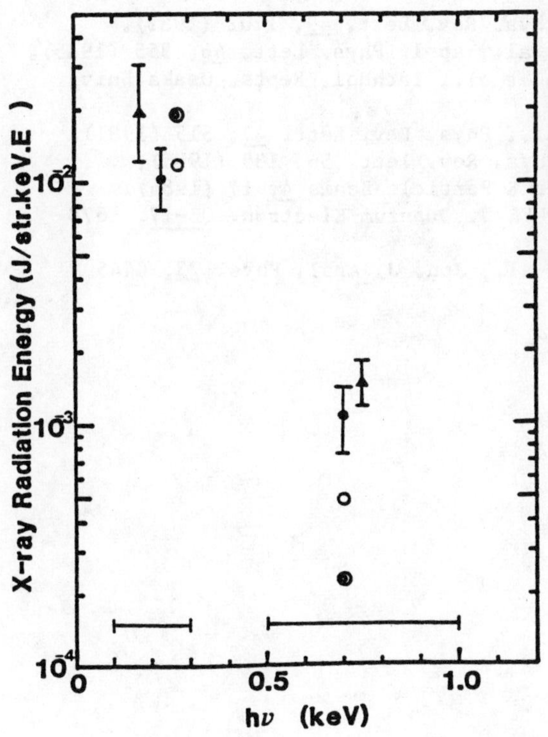

Fig. 5 X-ray radiation energy as a function of photon energy. ▲ denotes Au coated novel target with diameters of 1.5 and 2mm. •,○,◎ denote the parylene shell novel target with diameters of 1, 2 and 3mm, respectively.

ACKNOWLEDGEMENT

The authors wish to thank Professor A. Hasegawa, Professor K. Nishihara and Professor K. Mima for their fruitful discussions and also to thank Dr. T. Norimatsu and M. Takagi for their technical support of this experiment.

REFERENCES

1) W. C. Mead et al., Phys. Rev. Lett. 47, 1289 (1981).
2) W. C. Mead et al., Phys. Fluids 26, 2316 (1983).
3) K. Okada, T. Mochizuki et al., Jpn. J. Appl. Phys. 22, L671 (1983).
4) T. Mochizuki, T. Yabe et al., Phys. Rev. A33, 525 (1986).
5) S. Kiyokawa, T. Yabe et al., Jpn. J. Appl. Phys. 22,L772(1983).
6) T. Yamanaka et al., Laser & Particle Beams 4, 43 (1986).
7) P. H. Y. Lee et al., Laser & Particle Beams 2, 303 (1984).
8) H. Nishimura et al., Phys Fluids 26, 1688(1983).
9) P. D. Rockett etal., Phys. Fluids 25, 1286 (1982).
10) Y. Kitagawa et al., Jpn. J. Appl. Phys. 25, No.3 (1986).
11) T. H. Tan et al., Phys Fluids 27, 296 (1984).
12) P. A. Jaanimagi et al., Appl. Phys. Lett. 49, 1702 (1982).

13) M. A. Yates et al., Phys. Rev. Lett. 49, 1702 (1982).
14) K. Terai, H. Daido et al., Appl. Phys. Lett. 46, 355 (1985).
15) R. Tateyama, H. Daido et al., Technol. Repts. Osaka Univ. 32, 305 (1982).
16) D. M. Villeneuve et al., Phys. Rev. Lett. 47, 515 (1981).
17) A. Hasegawa et al., Phys. Rev. lett. 56, 139 (1986); H. Daido et al., Laser & Particle Beams 4, 17 (1986).
18) C. Yamanaka et al., IEEE J. Quantum Electron. QE-17, 1678 (1981).
19) K. Terai, H. Daido et al., Jpn. J. Appl. Phys. 23, L445 (1984).

PRODUCING A STEADY-STATE POPULATION INVERSION*

R. K. Richards
Physics Division, Oak Ridge National Laboratory
Oak Ridge, TN 37831

D. C. Griffin
Department of Physics, Rollins College
Winter Park, FL 32789

ABSTRACT

An observed steady-state transition at 17.5 nm is identified as the $2p^5 3s3p\ ^4S_{3/2} \rightarrow 2p^6 3p\ ^2P_{3/2}$ transition in Na-like aluminum. The upper level is populated by electron inner shell ionization of metastable Mg-like aluminum. From the emission intensity, the rate coefficient for populating the upper level is calculated to be approximately 5×10^{-10} cm^3/sec. Since the upper level is quasi-metastable with a lifetime 22 times longer than the lower level, it may be possible to produce a population inversion if a competing process to populate the lower level can be reduced.

INTRODUCTION

Quasi-metastable levels of Na-like ions and atoms have previously been proposed as appropriate upper levels in producing population inversions.[1] In this previous study, the inversion between the $2p^5 3s3p\ ^4S_{3/2}$ upper level and the $2p^6 3p^2P_{3/2}$ lower level in Na and Mg^+ was investigated. In this paper, an investigation of this same transition in Al^{2+} is presented. Also included is a technique for populating the upper level, an observation of the transition, and an experimental estimate of the population rate.

INNER SHELL IONIZATION

The inner shell (2p) ionization rate of Al^+ for energies well above threshold (100 eV) may exceed the outer shell ionization rate.[2] For Al^+ in the metastable states ($2p^6 3s3p\ ^3P_{0,1,2}$), the 2p ionization will populate the $2p^5 3s3p$ levels in Al^{2+}. Using a multiconfigurational code to calculate the autoionizing and radiative decay rates and assuming a 2J+1 weighting factor in populating

*This work supported by the Office of Fusion Energy of the U.S. Department of Energy under contract No. DE-AC05-84OR21400 with Martin Marietta Energy Systems, Inc.

the upper levels, a radiative emission spectrum can be predicted. Figure 1 shows such a spectrum which includes a 0.05 nm broadening and a 1% increase in the predicted energy levels to slightly decrease the wavelength of the lines, which improves the fit to the data and is well within the uncertainty of the calculation.

An experimental observation of these transitions is shown in Fig. 2. This was taken with a scanning 2.2 m monochromator observing radiation from the ELMO Bumpy Torus experiment.[3] This device produces a steady-state plasma with aluminum as the dominant impurity and has an electron temperature typically 300 eV. These same transitions have been observed in emissions from Penning discharges.[4,5]

Because the $^4S_{3/2}$ level has the lowest energy of the $2p^53s3p$ levels the $2p^53s3p\ ^4S_{3/2} \rightarrow 2p^63p\,^2P_{3/2}$ transition will occur at the longest wavelength, in Fig. 2 this is at 17.5 nm. The intensity of this transition may be used to estimate the population rate coefficient by comparison to another intensity with a known excitation rate. Since the 16 nm pair is produced by inner shell ionization of Al^{2+} in this plasma[3] with a rate coefficient approximately 5×10^{-9}

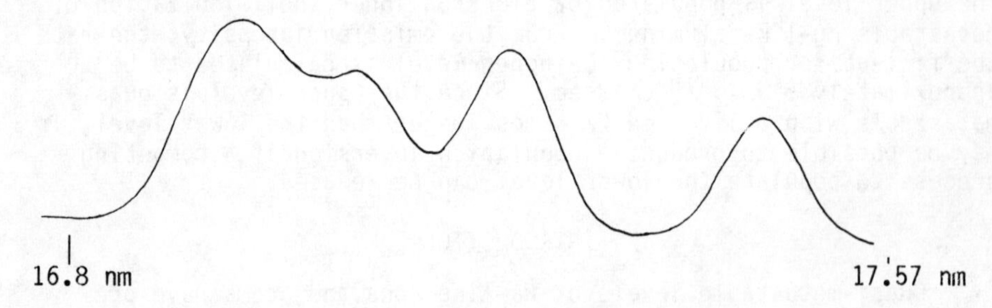

Fig. 1. Predicted spectrum of the $2p^53s3p$ - $2p^63p$ transitions.

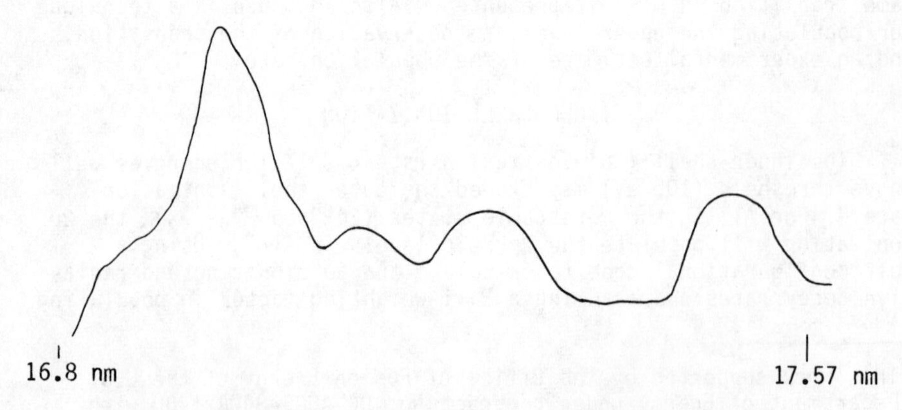

Fig. 2. Observed spectrum between 16.8 and 17.0 nm.

cm^3/sec[6], the density ratio of Al^{2+} to metastable Al^+ is typically 3, and the intensity ratio of the 17.5 nm line to the 16 nm pair is approximately 0.03; then the rate coefficient for populating the $2p^5 3s3p\ ^4S_{3/2}$ level will be $5 \times 10^{-10}\ cm^3/sec$.

THE POPULATION INVERSION

The lifetime of the $2p^5 3s3p\ ^4S_{3/2}$ state is 39 nsec which is 22 times longer than the $2p^6 3p\ ^2P_{3/2}$ state. This large difference will permit a steady-state population inversion to be maintained. The major problem in keeping the inversion comes from the ground state population in Al^{2+}. Because of the large electron collisional excitation rate from the ground state to the $2p^6 3p\ ^2P_{3/2}$ level, the ground state density must be kept well below that of the metastable Al^{1+} density.

REFERENCES

1. S. E. Harris, D. J. Walker, R. G. Caro, A. J. Mendelsohn, and R. D. Cowan, Opt. Lett. 9, 168 (1984).
2. W. Lotz, Z. Phys. 216, 241 (1968).
3. J. W. Swain, J. A. Cobble, D. L. Hillis, R. K. Richards, and T. Uckan, Phys. Fluids 28, 1922 (1985).
4. E. S. Warden and H. W. Moos, Appl. Opt. 16, 1902 (1977).
5. D. S. Finley, F. Bowger, F. Paresce, and R. F. Malina, Appl. Opt. 18, 649 (1979).
6. S. M. Younger, Phys. Rev. A 24, 1272 (1981).

PARTIALLY COHERENT RADIATION FROM LASERS, UNDULATORS, AND LASER PRODUCED PLASMAS

Nasif Iskander* and Nadine Wang**
Center for X-Ray Optics, Lawrence Berkeley Laboratory
University of California
Berkeley, CA 94720

Abstract

The coherence properties of several existing and proposed sources of soft x-rays are compared. Average and peak values of spectral brightness and coherent power are calculated and plotted.

Definitions

Spectral brightness is defined as the photon flux or power per unit area per unit solid angle in a specified bandwidth. Spectral brightness differs from brightness only in its consideration of bandwidth. We have chosen as our units for spectral brightness:

$$B = \frac{\text{photons/sec}}{\text{mm}^2\text{mrad}^2(0.1\%\ \text{BW})}$$

The spectral brightness of the LLNL Se laser and the laser produced plasma sources were approximated using simple geometrical models. For the synchrotron sources, we used approximations describing the flux distribution formulated by K-J. Kim.[1]

The portion of the flux that exhibits full transverse or spatial coherence is that part satisfying an area-angle product $(d\cdot\theta)^2 = (\lambda/2)^2$, where d is the diameter of the source, and θ is the half-angle of the radiation

---

* Also Department of Physics, University of California at Berkeley, Berkeley, CA 94720
** Also Department of Electrical Engineering and Computer Science, University of California at Berkeley, Berkeley, CA 94720

cone. Such radiation is said to be "diffraction limited".

The longitudinal or temporal coherence of the radiation is related to its spectral purity. We can define the coherence length, ℓ_c, as:

$$\ell_c = \lambda^2/\Delta\lambda$$

The coherent flux with a coherence length ℓ_c can be calculated as

$$F_c = B(\lambda/2)^2(\lambda/\ell_c)$$

Sources

In our calculations for the LLNL Se laser, we used a simple geometrical model of a cylindrical plasma source, whose diameter and length match the dimensions of the driving laser focus. We used the parameters of the published highest gain case, with 2.2 cm long line focus, 200 μm wide.[2] A peak radiated power of 360 W and a peak spectral brightness of 2 X 10^{11} were reported.[3,4] This value for the brightness agrees with our calculated value.

The laser produced plasma source calculations were also based on very simple geometrical models. We assumed that the plasma was a spot the same size as the driving laser focus, and that it radiated evenly into 4π steradians. We assumed that the lines in the emission were Doppler broadened, which results in a bandwidth of about 0.01%. both targets considered here were radiated with 5320 Å laser light. the chlorine target was irradiated for 124 ps at an incident intensity of 2 x 10^{15} W/cm^2, focussed to a 130 μm diameter spot. This resulted in a 126 ps emission, with a conversion efficiency to the He α line of ξ_x = 35 x 10^{11} photons/(J-sphere).[5] The titanium target was irradiated for 57 ps at an intensity of 1.1 x 10^{16} W/cm^2, in a 61 μm diameter spot. This resulted in a 48 ps emission with a conversion efficiency into the He α line of $\xi_x = 2.6 \times 10^{11}$ photons/(J-sphere).[5]

The VUV ring at NSLS (Brookhaven) is a 750 MeV ring, with a period of 170 ns. Our calculations are based on a single bunch operation with an average current of 500 mA. We used a σ value for the bunch duration of 600 ps.[6] The values for emittance that we used are: $\sigma_x = 0.3$ mm $\sigma_y = 0.1$ mm, $\sigma_{x'} = 0.2$ mrad, and $\sigma_{y'} = 0.01$ mrad.[7]

The proposed soft x-ray undulator for beam line XI of the x-ray ring at NSLS will have 38 periods, 8 cm long each, with a maximum field strength of 3 kGauss. This corresponds to a maximum K value of 2.3. The ring itself operates at 2.5 GeV with an average current of 200 mA. The circumference of the ring is 150 m, and the σ length per bunch is 5-10 cm. The horizontal emittance is about 10^{-9} meter-radians.[8] In our calculations we assumed one bunch per operation.

The PEP ring at SSRL is a 15 GeV machine, currently running a 4 mA beam. This will eventually be increased to 30 mA. Curves for both modes of operation are plotted in figures 1 through 3. We assumed a 3 bunch operation with $\sigma = 2$ cm per bunch. the circumference of the ring is 2.2 km.[9] The emittance values we used are: $\sigma_x = 0.3$ mm, $\sigma_y = 0.1$ mm, $\sigma_{x'} = 0.2$ mrad, and $\sigma_{y'} = 0.01$ mrad.[7] The PEP undulator has 30 periods of 7.3 cm, with a maximum field strength of 0.244 T. This corresponds to a maximum K value of 1.52.[9]

The 54-pole wiggler at SSRL has a period length of 7 cm with a maximum field strength of 1.35 T. This corresponds to a maximum K of 8.8. The ring runs at 3 GeV with an average current of 60 mA. It has a repetition rate of 1.38 MHz, and we considered its 16 bunch mode of operation with $\sigma = 2$ cm per bunch.[9] The emittances are given by: $\sigma_x = 3.2$ mm, $\sigma_y = 0.15$ mm, $\sigma_{x'} = 0.18$ mrad, $\sigma_{y'} = 0.03$ mrad.[7]

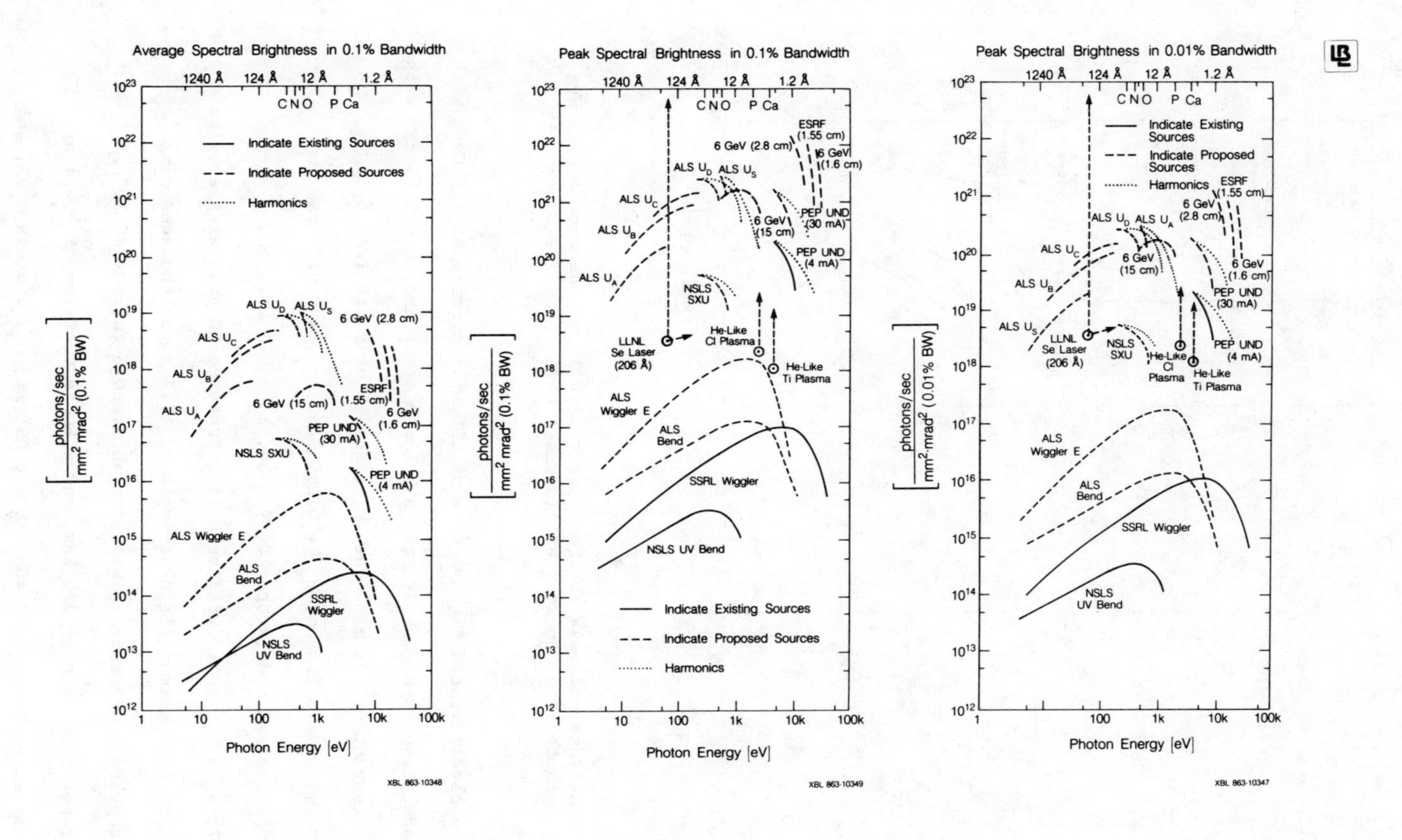

<u>Figure 1</u> Average and peak values of spectral brightness in 0.1% and 0.01% bandwidths are plotted for the sources discussed here.

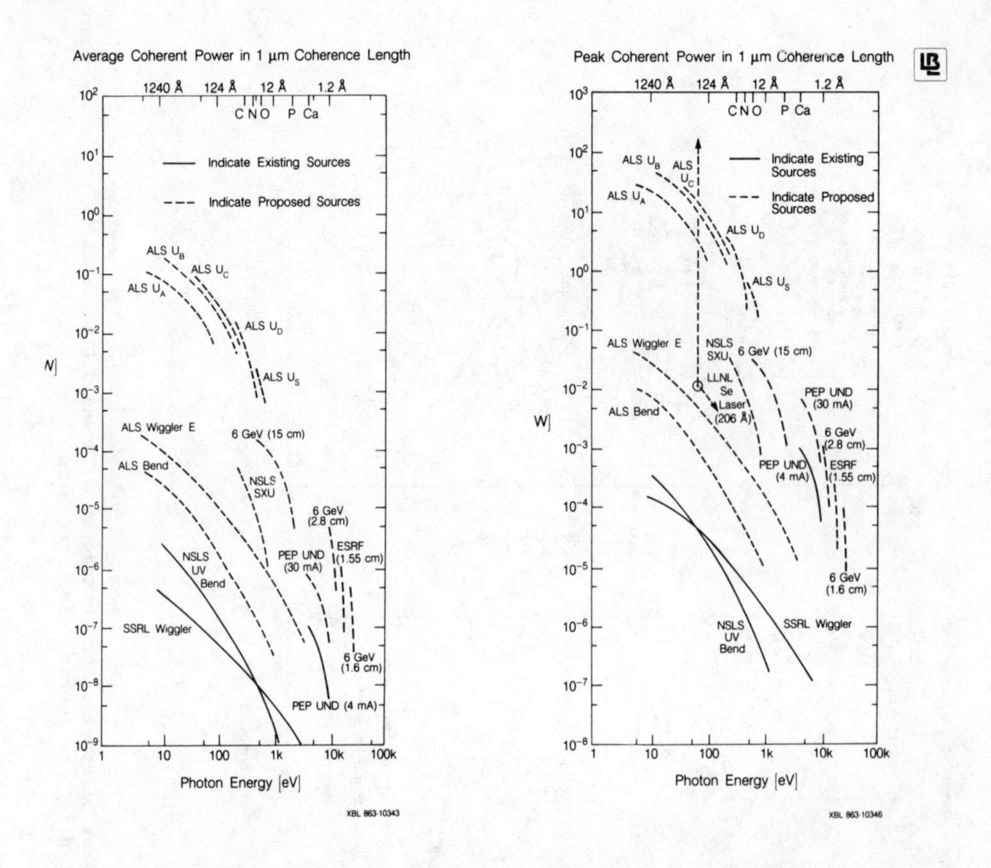

Figure 2 Average and peak coherent power in a 1 μm coherence length are plotted for the sources discussed here.

The proposed Advanced Light Source at LBL will operate at 1.5 GeV, with a 2.4 MHz repetition rate. We assumed a 1 bunch operation, σ = 0.55 cm, with an average current of 7.6 mA. The emittance is characterized by: σ_x = 0.173 mm, σ_y = 0.053 mm, $\sigma_{x'}$ = 0.055 mrad, $\sigma_{y'}$ = 0.017 mrad. There are 5 proposed undulators: U_A(N = 30, λ_u = 3.5 cm), and U_S(N = 167, λ_u = 3.0 cm). Wiggler E is a proposed 50 pole wiggler with 10 cm periods and maximum field strength of 1.77 T for the same ring.[10,11]

The proposed 6 GeV machine will run an average current of 100 mA. We assume a repetition rate of 375 KHz, and a bunch duration of σ = 1 ns. The emittance is given by: σ_x = 0.405, σ_y = 0.098 mm, $\sigma_{x'}$ = 0.018 mrad, and

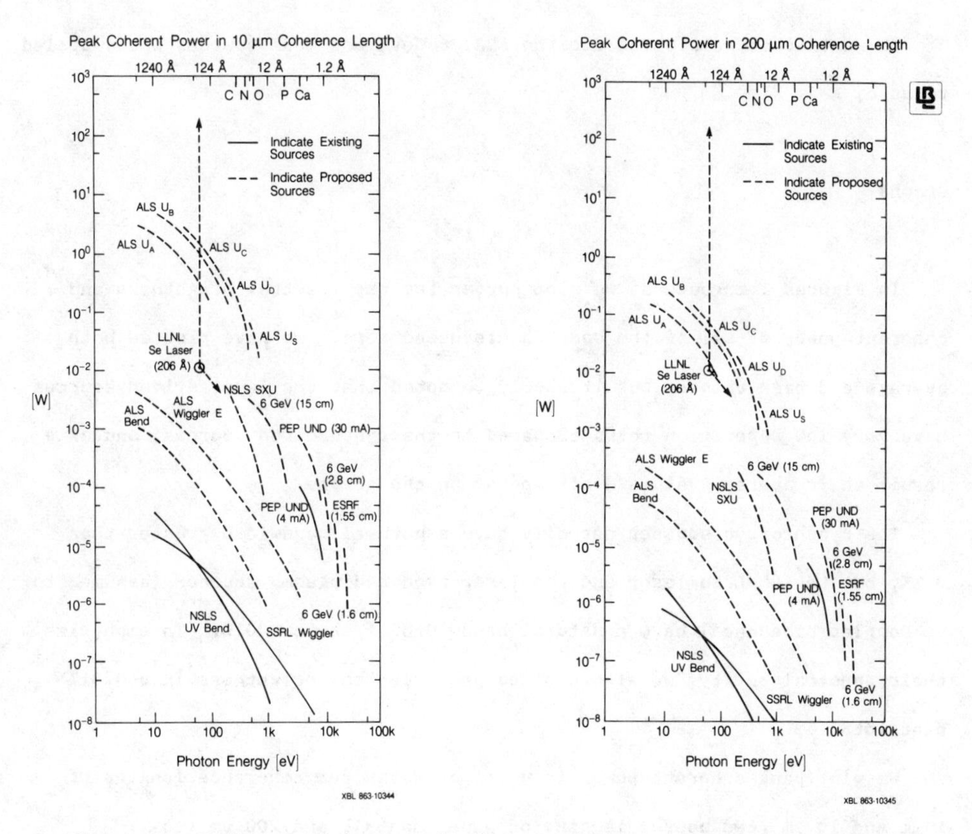

Figure 3 **Peak coherent power in 10 μm and 200 μm coherence lengths is plotted for the sources discussed here.**

$\sigma_{y'} = 0.017$ mrad. We show in the graphs curves for three of the proposed undulators. The 15 cm undulator has 33 periods with a maximum K of 2.67. The 2.8 cm period undulator has 178 periods, and a maximum K of 0.85. The 1.6 cm period undulator has 312 periods, with a maximum K of 0.35.[11]

The ESRF is a proposed European 5 GeV ring. We assume that it will run a 100 mA current, with a time structure similar to that of the 6 GeV machine discussed above. The emittance of the ring is characterized by: $\sigma_x = 0.410$ mm, $\sigma_y = 0.052$ mm, $\sigma_{x'} = 0.016$ mrad, $\sigma_{y'} = 0.013$ mrad. We plotted a curve for one undulator on this ring, with 322 periods, 1.55 cm long each, with a maximum K value of 0.35.[11]

The data that we plotted for the ALS, 6 GeV, and ESRF sources was supplied to us by K-J. Kim.

Graphs

In Figures 1 through 3, we plot curves for the spectral brightness and coherent power of all of the sources discussed here. We have plotted both average and peak values, but it should be noted that the laser driven sources have very low repetition rates compared to the synchrotron sources, and as a result their average values don't appear on the graphs.

The synchrotron sources normally have a natural bandwidth greater than 0.1%, but the LLNL Se laser and the laser produced plasma sources (assumed to be Doppler broadened) have a natural bandwidth of about 0.01%. To emphasize their spectral purity, we also plotted peak spectral brightness in a 0.01% bandwidth.

We plot peak coherent power in units of Watts for coherence lengths of 1 μm and 10 μm (two useful lengths of experiments), and 200 μm (to illustrate the naturally long coherence length of the LLNL Se laser).

REFERENCES

1. K-J. Kim, "Characteristics of Synchrotron Radiation", X-Ray Data Booklet, ed. D. Vaughan, Lawrence Berkeley Laboratory, Berkeley, California, PUB-490, 1985.
2. M.D. Rosen et al., Phys. Rev. Letters, 54, 106, 1985.
3. D.L. Matthews et al., Phys. Rev. Letters, 54, 110, 1985.
4. J.F. Holzrichter, E.M. Campbell, J.D. Lindle, and E. Storm, Science, 229, 1045, 1985.

5. D.W. Phillion, E.M. Campbell et al., to be published; D.T. Attwood, N.M. Ceglio, E.M. Campbell, J.T. Larson, D.M. Matthews, and S.L. Lane, "Compression Measurements in Laser-Driven Implosion Experiments", Laser Interaction and Related Plasma Phenomenon, (Plenum, New York, 1981), eds. H.J. Schwarz, H. Hora, M.J. Lubin, and B. Yaakobi; B. Yaakobi et al., Optics Communications, 38, 196, 1981.
6. Personal Communication between D.T. Attwood and G. Vignola, NSLS, Brookhaven National Laboratory.
7. D.T. Attwood, K-J. Kim, K. Halbach, and M.R. Howells, "Undulators as a Primary Source of Coherent X-rays", paper presented at the International Conference on Insertion Devices for Synchrotron Sources, October 1985; also published as Lawrence Berkeley Laboratory, Berkeley, California PUB-20569, 1985.
8. Personal Communication with S. Krinsky, NSLS, Brookhaven National Laboratory.
9. Personal Communications with A. Hofmann, SLAC Stanford.
10. K-J. Kim and N. King, "Photon Performance of a High Brightness, Soft X-Ray and VUV Synchrotron Radiation Facility", Lawrence Berkeley Laboratory, Berkeley, California, PBU-487, 1985.
11. Personal Communication with K-J. Kim, Center for X-Ray Optics, Lawrence Berkeley Laboratory.

LOW POWER PULSED LASERS AS PLASMA SOURCES OF SOFT X-RAYS

F.O'Neill, Y.Owadano and I.C.E.Turcu
Rutherford Appleton Laboratory, Chilton, Didcot, Oxon. OX11 0QX, U.K.
A.G.Michette, C.Hills and A.M.Rogoyski
Physics Dept., King's College, Strand, London WC2R 2LS, U.K.

ABSTRACT

The development of plasma sources of soft x-rays, using pulsed KrF (600mJ) and Nd:YAG (500mJ) lasers, is described. The intensities and spectral properties of the sources have been investigated for various target materials. The use of the sources in soft x-ray contact microscopy will be described.

INTRODUCTION

In recent publications,[1-4] the successful use of high power pulsed lasers for soft x-ray contact microscopy of both wet and dried biological material was reported. The lasers used in these experiments were the VULCAN and SPRITE systems of the Central Laser Facility at the Science and Engineering Research Council Rutherford Appleton Laboratory. VULCAN is a Nd:glass laser giving about 100J per beam at a wavelength of 1053nm, or about 30J of frequency doubled radiation, in a 1ns pulse every 20 minutes or so. SPRITE is a KrF excimer laser giving about 100J of radiation at a wavelength of 249nm in a 50ns pulse with a similar repetition rate. When the laser energy is focused to a small spot ($\leq 300\mu m$ diameter) on a target, a plasma is formed which emits soft x-rays in both continuum and line spectra. Typically the emitted x-ray intensity in the water window (2.3-4.4nm) over all angles is a few joules. The recording medium used in these experiments was a copolymer resist PMMA-MAA, consisting of poly(methylmethacrylate) and methacrylic acid. At the x-ray wavelengths predominant in the spectrum from a carbon target (around 2-4nm) this required an exposure of about $100J\ m^{-2}$, which could be achieved by positioning the specimen, in contact with the resist, about 50mm from the target at an angle of about 50^0 to the incoming laser beam. A difficulty in these experiments was controlling the exposure, and hence the development procedure, because of the variability of the incident laser energy.

Smaller pulsed lasers, with about 1J per pulse, can have much better energy stability from pulse to pulse, leading to more constant exposure. They are also much cheaper and easier to operate than the higher power lasers, and therefore more suitable for individual laboratories. By moving the specimen chamber closer to the target, and reducing the angle, it is still possible in principle to achieve the necessary level of exposure in a single pulse. This can, however, result in damage to the specimen chamber window (possibly breaking it) and to the resist, caused by debris and radiation from the target. Alternative ways of obtaining suitable exposures include the use of more sensitive resists and the inclusion of x-ray focusing optics.

In this paper, the use for contact x-ray microscopy of two low power lasers, one a KrF excimer laser with 600mJ per pulse and the other a Nd:YAG laser with 500mJ per pulse, is reported. Various target to specimen distances were used, with both PMMA-MAA and EBR-9, a very sensitive resist used in electron beam lithography.

EXPERIMENTAL ARRANGEMENTS

The experimental arrangement for the work with the excimer laser is shown in figure 1. A Lambda-Physik EMG150 KrF laser, with an injection locked unstable

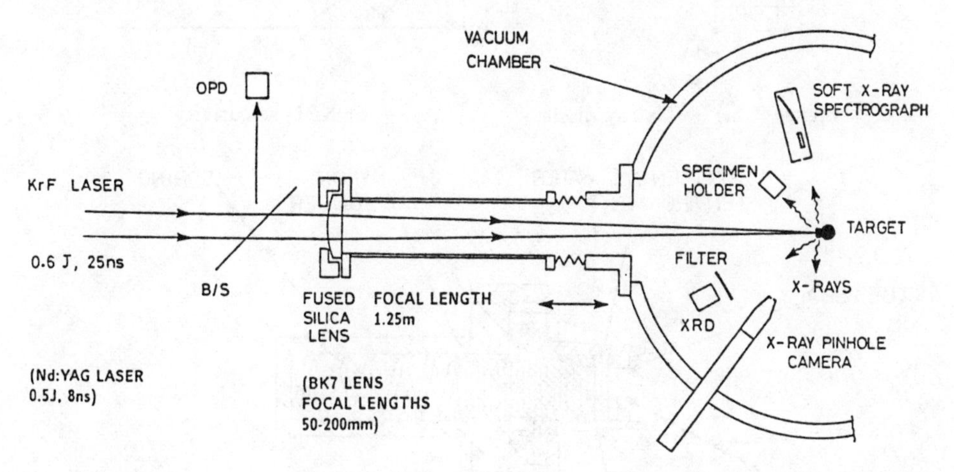

Figure 1. Experimental arrangement.

resonator cavity, was used. The diffraction-limited beam size was 10x25mm, and each 25ns long pulse gave 600mJ of radiation at a wavelength of 249nm. The beam was focused onto a pre-aligned target. A similar arrangement, using a glass lens, was used for the work with the Nd:YAG laser. This was a JK Lasers HY750 with 500mJ in each TEMoo pulse lasting 8ns, a beam diameter of 2.3mm and a wavelength of 1064nm. In this case optical alignment was carried out with a He-Ne laser on the same optical axis as the HY750, using specular reflection from the target to find the best focus. The offset due to the different focal lengths of He-Ne and Nd:YAG laser light is calculable, and the re-positioning was done using a stepper motor drive which moved the target to within 20μm of the correct focus. Each of the lasers can be pulsed several times each second, but in the experiments reported here single pulses were used for the exposures, except for the carbon target spectra for which several pulses were used. The x-ray output was monitored by a pinhole camera, a soft x-ray diode (XRD, figure 2a), and a spectrograph. By use of suitable filters the XRD output allowed estimates of the total intensity in particular energy ranges to be obtained, as indicated in figure 2b. In order to allow the specimen to be positioned closer to the target, a specimen chamber design different to that used in the previous experiments was used. This is shown in figure 3. For images of hydrated specimens, a small amount of quick drying insulating varnish was used to seal round the silicon nitride window, in order to maintain the specimen environment when the vacuum chamber was evacuated. Although it was not necessary to have a high vacuum in order to allow x-ray image formation, the path length of soft x-rays in air at atmospheric pressure being several millimetres, a vacuum af about 5×10^{-5} torr was needed to prevent electrical breakdown of the x-ray diode.

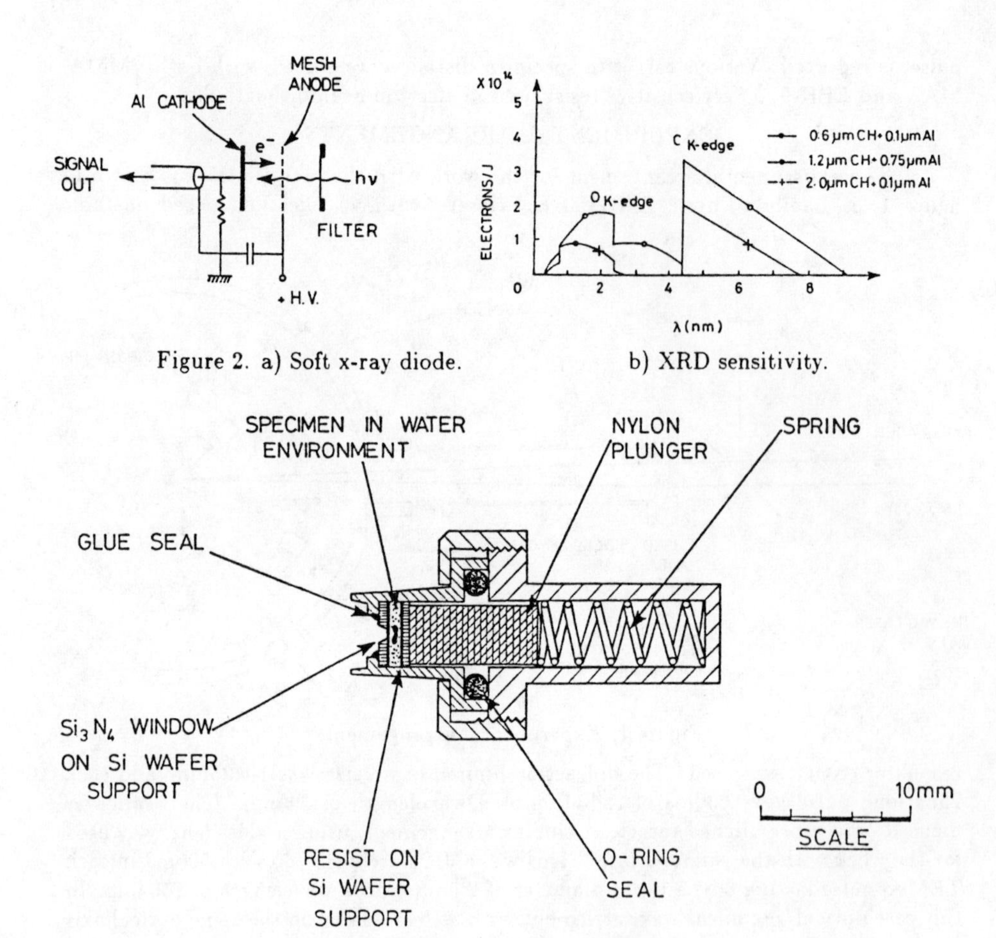

Figure 2. a) Soft x-ray diode. b) XRD sensitivity.

Figure 3. The specimen chamber.

RESULTS - DIAGNOSTICS

Targets of carbon and brass were used in both sets of experiments. In the KrF laser experiments, the x-ray diode measurements show that, at an angle of 45^0 to the laser beam, the soft x-ray intensity was about 1.4mJ per steradian for carbon targets and about 8mJ per steradian for brass targets for irradiances on target of about $3x10^{15} W\ m^{-2}$. The x-rays were concentrated in a pulse (figure 4b) approximately the same length as the laser pulse (figure 4a). For carbon targets the x-ray spectra (figure 5a), obtained using eight laser pulses, show lines at wavelengths of 4.027nm (strongest), 3.50nm and 3.37nm, due to, respectively, the $1s^2$-1s2p and $1s^2$-1s3p lines in C V (helium-like ions) and the Lyman-α line in C VI (hydrogen-like ions). The level of the trace between the lines is consistent with microdensitometer noise. From the relative intensities of the C V lines and C VI lines, and assuming coronal equilibrium,[5] the electron temperature of the emitting region is estimated to be 63eV, while the pinhole camera image shows the size of the region to be about 100μm (figure 4c). For brass targets, more continuum emission was obtained, most of this being in the useful, for x-ray microscopy, water-window wavelength range.

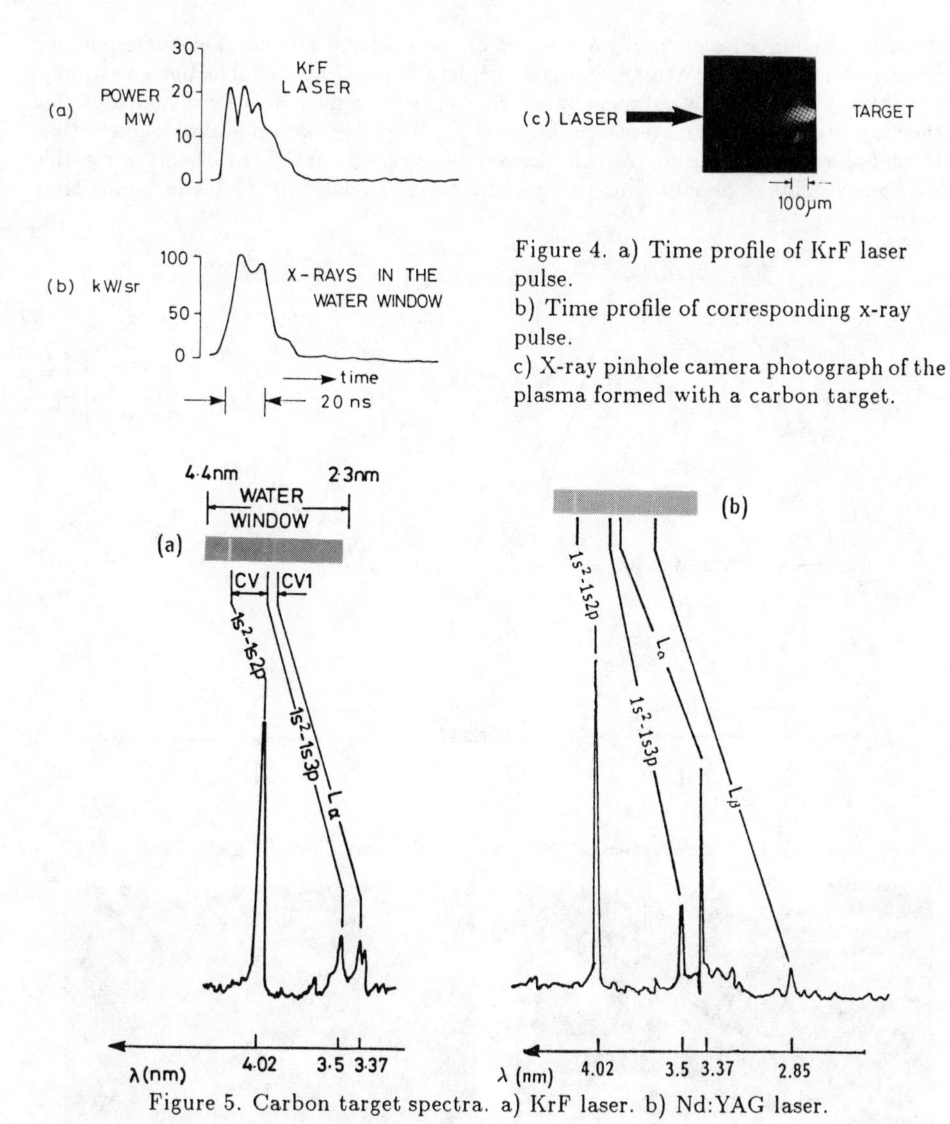

Figure 4. a) Time profile of KrF laser pulse.
b) Time profile of corresponding x-ray pulse.
c) X-ray pinhole camera photograph of the plasma formed with a carbon target.

Figure 5. Carbon target spectra. a) KrF laser. b) Nd:YAG laser.

A spectrum obtained from a carbon target in the Nd:YAG laser experiments is shown in figure 5b. Five pulses were used to obtain this spectrum, and the intensity ratio of the lines shows that the electron temperature of the emitting region is higher than for the KrF laser. A spectrum (single pulse) from a brass target is shown in figure 6a, demonstrating the greater amount of continuum emission. Other targets were also investigated with this laser, for example calcium chloride, a single pulse spectrum from which is shown in figure 6b. Both these spectra show emission outside the water window range, the long wavelength part of which would be absorbed by a silicon nitride window. Electron micrographs (figure 7) of the target after irradiation show craters about 100μm diameter, giving a lower limit on the size of the emitting region. The corresponding irradiance is about 10^{16}W m^{-2}. X-ray diode data are not yet available, but an estimate of the x-ray intensity was obtained by imaging an electron microscope grid and adjusting the target to resist distance until the exposed resist required a similar development time to that of a similar exposure made in the KrF laser experiment. For carbon targets, this gave an estimate of about 0.2mJ per steradian at an angle of 50^0 to the incoming laser beam.

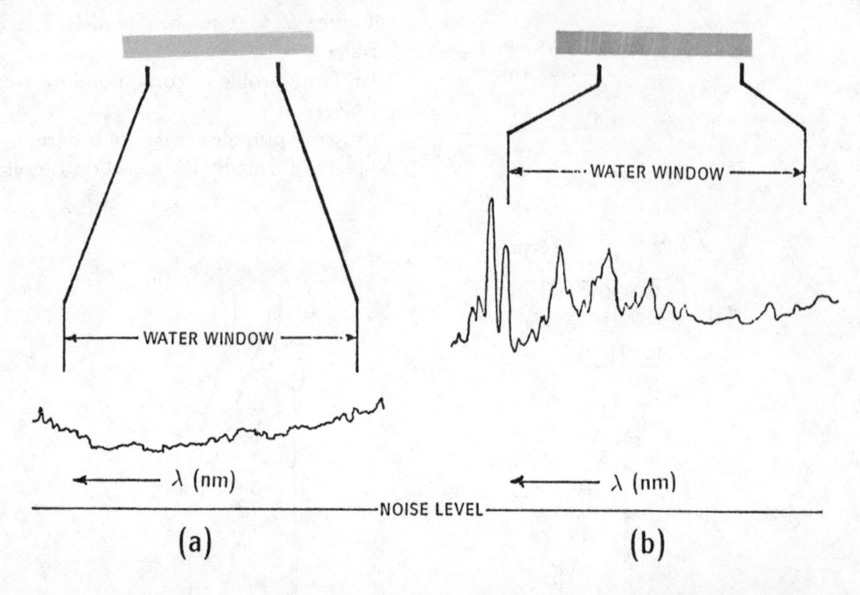

Figure 6. Soft x-ray spectra (Nd:YAG). a) Brass target. b) CaCl target.

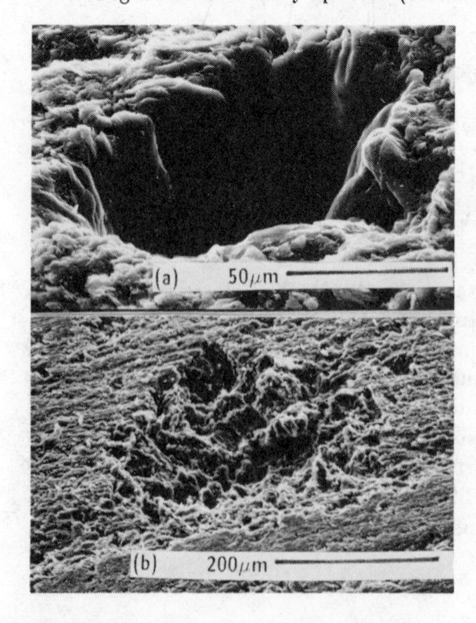

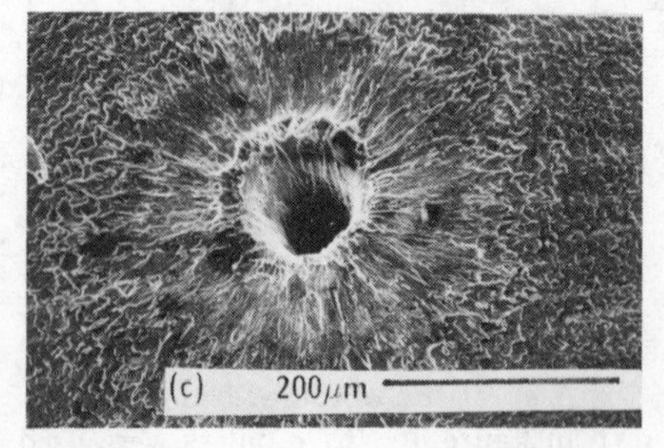

Figure 7. Craters formed on targets.
a) Carbon (laser correctly focused).
b) Carbon (defocused).
c) Brass (focused).

RESULTS - EXPOSURES

In both series of experiments exposures were made on PMMA-MAA and EBR-9 resists. In all cases, it proved simple to obtain exposures of electron microscope grids as test objects. Although no systematic biological study was carried out in these experiments, several types of material were imaged in order to try assess the potential of the low power lasers as sources for contact x-ray microscopy. The specimens included air-dried and wet blood, algae, foxglove epidermal hairs, fatty spreads, flagellae, and tissue cultures. In order to obtain good exposures of these specimens, it was necessary when using carbon targets and PMMA-MAA resist to move the specimen chamber to within 10mm of the target. Because of the specimen plus resist thickness, typically about 1.5μm, the resolution was still limited by diffraction (to about 80nm for a wavelength of 4nm) rather than penumbral blurring for these source distances and the exposures obtained in the experiments. The proximity of the specimen chamber to the target almost invariably led to the silicon nitride window being destroyed, which could have allowed the resist to be exposed by scattered ultraviolet radiation in the KrF laser experiments. That this was not so was checked by replacing the silicon nitride window by a 6mm thick calcium fluoride filter which was transparent to ultraviolet radiation but opaque to soft x-rays. The resist then remained unexposed. To prevent window destruction, the specimen chamber was moved further away and a ten times more sensitive resist, EBR-9, was used. Although it proved possible to obtain images in this way, they showed very little range of contrast. Clearly some compromise is needed, i.e. a resist mid-way in sensitivity between PMMA-MAA and EBR-9 which still allows imaging without window breakage.

Figure 8. Cyanidium algae.

Optical micrographs of one of the developed EBR-9 resists are shown in figure 8. The specimens were cyanidium algae, which are typically 5μm in diameter, and the image was obtained using the KrF laser. The lack of detail in the image is not due to inherent poor resolution but to the poor range of contrast given by the resist.

CONCLUSIONS AND FURTHER WORK

It has been demonstrated that low power pulsed lasers can generate plasmas that emit soft x-rays with sufficient intensity in single pulses to expose x-ray resists. Some further developments, for example in the choice of more suitable resists, in the investigation of a wider range of target materials, and in the use of x-ray focusing optics are needed before the evaluation of these sources for soft x-ray contact microscopy can be completed.

ACKNOWLEDGEMENTS

The financial support of the Science and Engineering Research Council and of the Royal Society Paul Instrument Fund is gratefully acknowledged. The experiments would not have been possible without the support of technical staff at both King's College and the Rutherford Appleton Laboratory, and we should also like to thank Professor P.J.Duke (King's College and the Daresbury Laboratory) for supplying the vacuum chamber for the Nd:YAG laser experiments. Dr. R.Feder (IBM Yorktown Heights) for supplying resists and windows, and Dr A.D.Stead (Botany Department, Royal Holloway and Bedford New College) for assisitance and advice with biological specimen preparation. This manuscript was typeset with TEX.[6]

REFERENCES

1. A.G.Michette, P.C.Cheng, R.W.Eason, R.Feder, F.O'Neill, Y.Owadano, R.J.Rosser, P.Rumsby & M.Shaw, To be published in J. Phys. D, (1986).
2. R.W.Eason, P.C.Cheng, R.Feder, A.G.Michette, R.J.Rosser, F.O'Neill, Y.Owadano, P.Rumsby & M.Shaw, To be published in Optica Acta, (1986)
3. P.C.Cheng, R.Feder, D.Shinozaki, K.H.Tan, R.W.Eason, A.G.Michette & R.J.Rosser, To be published in Nucl. Instrum. Methods in Physics Research, (1986)
4. R.W.Eason, P.C.Cheng, R.Feder, A.G.Michette, F.O'Neill, Y.Owadano, R.Rosser, M.Shaw, A.D.Stead & E.Turcu, submitted to Science, (1986).
5. T.F.Stratton, in Plasma Diagnostic Techniques (R.Huddlestone & S.L.Leonard, eds.), Academic Press, New York, (1965).
6. D.E.Knuth, The TEXbook, Addison-Wesley, Reading, Mass.,(1984).

X-RAY AND ELECTRON HOLOGRAPHY USING A LOCAL REFERENCE BEAM

Abraham Szöke[a]
High Energy Physics Laboratory
Stanford University
Stanford, California 94305

ABSTRACT

Three dimensional image of the close vicinity of an atom can be obtained using characteristic x-rays, photoelectrons or Auger electrons emitted by it. These methods (some new, some old) are shown to be a natural extenstion of holographic microscopy. An algorithm is proposed for the reconstruction of the image by computer.

INTRODUCTION

It was the search for ways to produce three dimensional images of atomic resolution that led Gabor to the discovery of holography.[1] He realized that if an unknown wave is made to interfere with a known wave and the intensity of the resulting interference pattern is recorded, this "hologram" contains most of the information needed to reproduce the unknown wave. The second part of his discovery was that if the known wave, called the "reference" wave is simple, its modulation by the recorded interference pattern reproduces the unknown wave. If the hologram is recorded on film, this second step can be achieved by shining an appropriate reference wave on the film. Holography has since grown into a large field of science with many important applications but Gabor's original goal still remains to be achieved. The aim of this article is to discuss x-ray and electron holography of microobjects that can be well oriented in space and are clustered in large numbers. Some of the techniques proposed are new but most of them are already practiced; the emphasis of this article is on a unified treatment of these methods.

As an introduction, consider a metallo-organic molecule with a single heavy metal atom in a well defined orientation. Irradiate the molecule with X-rays slightly above the K-edge of the heavy metal and observe the angular distribution of the characteristic (K) X-rays emitted by it. In the absence of surrounding atoms, this distribution is isotropic: the electric field of the emitted radiation at the point of observation, at a distance R, is given by

$$E_o = A \frac{e^{i(kR-\omega t)}}{R} \quad (1)$$

where $k=2\pi/\lambda \approx \omega/c$ is the wave vector of the emitted radiation, assumed to be monochromatic. The electric field is treated as a classical and scalar quantity in order to simplify notation.

In the presence of the surrounding atoms some of the emitted photons get scattered. An integral expression for the scattered field is given by[2]

$$E_1 = \frac{ik}{c} \int J(\bar{r}') \frac{e^{ik|\bar{R}-\bar{r}'|}}{|\bar{R}-\bar{r}'|} d^3r' \quad (2)$$

where $J(\bar{r}')$ is the current density induced by the radiated electromagnetic field at $\bar{r}'$. Assuming the presence of discrete atoms at positions $\bar{r}_i$ in the "radiation zone" and weak scattering (see Fig. 1), the total field can be written as

$$E = F \quad 1 + \sum_i \frac{\ell_i f_i(\theta_i)}{r_i} e^{ikr_i(1-\cos\theta_i)} \quad (3)$$

where ℓ_i is the scattering length (possibly complex), $f_i(\theta_i)$ is an angular distribution, and $|F|^2 = |A|^2/R^2$.

This is the same as the fundamental equation of holography[1]. In words, the angular distribution of the characteristic radiation contains a fringe pattern, which is a hologram of the surroundings of the emitting molecule. The fringe visibility is approximately

$$V = \frac{\ell_i}{r_i} f_i(\theta_i) \quad . \quad (4)$$

It decreases rapidly with the distance of the scatterer from the emitter. The fringe pattern has an angular width of $\approx(\lambda/r_i)^{1/2}$ in the direction of $\bar{r}_i$, and $\approx \lambda/r_i$ in the perpendicular direction.

The experiment can be done on a large number of similarly oriented molecules; each metal atom that emits a photon produces a hologram of its environment. The waves emitted by different atoms are incoherent and therefore do not interfere. Thus, if the object is a crystal of the metallo-organic molecule and the

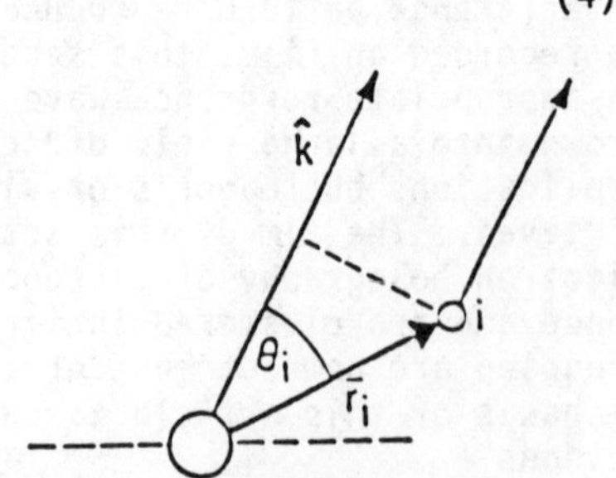

Figure 1: Scattering geometry; emitter is at the origin, $\bar{r}_i$ points to the scatterer i, $\bar{k}$ is the direction to the observer, and θ_i is the angle between $\bar{k}$ and $\bar{r}_i$.

molecules (crystallites) are oriented within an angle δ (typically ~1°) all the individual holograms are in register within δ; therefore their incoherent superposition produces a single hologram that resolves the environment within $r_i \ll \lambda/\delta$. Also, the crystal size, L, has to be smaller than $R\lambda/r_i$. Neither of these conditions imposes any practical limitations on the method. Note that the periodicity of the crystal is immaterial, so the experiment could be done equally well on a liquid crystal, a stretched fiber, or a well oriented membrane.

Quite general considerations show that approximately 200 photons have to be scattered and observed per resolution element in order to detect it with reasonable confidence.[3,4] Therefore the time of exposure and the consequent damage are inversely proportional to the volume of the crystal. In this method, not unlike in X-ray crystallography, Debye-Waller factors, absorption corrections, etc. have to be applied. Rough estimates show that the sensitivity of this method is comparable to that of EXAFS.[5] Note that, compared to "ordinary" holography, the coherence requirement of the source is easy to satisfy. Also this holographic method has intrinsically high magnification: the fringes are of macroscopic dimensions being always in the far field. (The experts call them Fourier-Frauenhoffer holograms.)

Photoelectron diffraction[6] is a similar experiment using photoelectrons. Suppose the same molecule is irradiated by monochromatic X-rays, with energy $\hbar\omega = I_p + E_1$, where I_p is the metal atom's ionization potential. This results in the ejection of photoelectrons of energy E_1 and energy width $\Delta E_1 = \hbar\Delta\omega + \hbar\Gamma$, where $\hbar\Delta\omega$ is the energy width of the incident X-rays and Γ is the decay rate of the inner shell vacancy. The angular pattern of the ejected photoelectrons, usually a dipole pattern, is well described in the literature[5]. Some of the ejected electrons get scattered by the atom's environment; therefore the angular distribution of these electrons is described by an equation similar to Eqs. 2 and 3 where the wave function of the electron, ψ, is written instead of E. Therefore, if the angular distribution of the photoelectrons is recorded, an electron hologram of the emitting atom's close environment is obtained.

When discussing electron holograms, or photoelectron diffraction notice that Eqs. 1 and 2 are similar to the fundamental EXAFS equations of Lee et. al[7]. As the scattering is strong, the integral equation cannot be simplified as in Eq. 3; it can be expanded into multiple scattering diagrams, and corrected for extinction of the coherent electron wave (mostly by inelastic scattering) and atomic disorder (similar to a Debye-Waller factor). As a consequence, similarly to LEEDS,[8] only very thin layers permit a simple interpretation of the electron image. For the same reason, the use of 1 -2 keV photoelectrons is attractive. Their deBroglie wavelength is 0.4 - 0.3A, their range is ~150A. The angular distribution of their coherent scattering is strongly forward peaked; in our arrangement this actually simplifies the hologram. Similar experiments can be done using an appropriately chosen Auger-electron instead of photoelectron. An interesting variation on this theme is to emit a high energy photoelectron from a light atom (in the vicinity of the heavy atom), at an energy where the heavy atom has a bound resonance. The resulting hologram(s) locate the heavy atom with reference to the emitting light atom, providing chemical selectivity.

Two more "conventional" methods for holographic microscopy will be described for completeness. Consider a thin, but regular array of molecules, for example a lipid layer on some thin substrate. "Decorate" this layer with heavy atoms by deposition of a sub-monolayer quantity of it. We assume that the heavy atom is always deposited in the same position relative to the lipid molecules (X-ray crystallographers sometimes do this successfully in protein crystals.) Irradiate the layer by monochromatic X-rays, or monoenergetic electrons. There are three kinds of interference patterns produced: ordinary Bragg scattering (LEEDS) of the crystal, the same for the "decorating" overlayer, and the interference between the radiation scattered from the heavy atom and its surroundings. The last mentioned produces a hologram, as discussed above.

The intensity of Bragg reflections can be minimized by minimizing the coherence area of the incident radiation, while the intensity of the hologram remains largely unaffected. This is surprising, as conventional holography always fights for larger coherence length, but the fact is well established[8]. In conventional X-ray tubes, each atom emits incoherently; therefore the minimum transverse coherence length of $\approx f\lambda$ and longitudinal coherence length of $\approx f^2\lambda$ are achieved if a shallow ($<f^2\lambda$) X-ray emitting region of an anticathode is imaged onto the sample with optics of effective angular aperture of 1/f (aperture/focal length). Electrons can be emitted by very small cathodes, the minimum coherence area is limited by the aberrations of electron optics. These last methods are similar to those proposed by Pendry[9] and carried out by Heinz et. al.[10,11]

Let us discuss now the holographic reconstruction problem, i.e. how to obtain the likeness (image) of the "object," given the hologram. Consider first a very simple case: a (scalar) spherical wave emanates from the reference atom at the origin and is scattered isotropically from a single scatterer located at r_1. The electric field of the wave, on a sphere of radius R, in the far field, is given by Eq. 3',

$$E = A\,\frac{e^{ikR}}{R} + N(\bar{r}_1)\,\frac{e^{ik|\bar{R}-\bar{r}_1|}}{|\bar{R}-\bar{r}_1|} \qquad (3')$$

where $N(\bar{r}_1)$ is a complex quantity (pure number). This can be written, using outgoing spherical Hankel functions, $h_\ell^{(1)}$, spherical Bessel functions j_ℓ, and spherical harmonics $Y_{\ell m}(\theta,\varphi)$ as[2,13]

$$E = A\,[ik\, h_o^{(1)}\,(kR)\,Y_{oo}(\theta,\varphi) \tag{5}$$

$$+ N(\bar{r}_1)\,4\pi ik \sum_{\ell=0}^{\infty} j_\ell(kr_1) h_\ell^{(1)}(kR) \sum_{m=-\ell}^{\ell} Y^*_{\ell m}(\theta_1,\varphi_1)\,Y_{\ell m}(\theta,\varphi)]$$

Assuming a strong reference wave (or a weak scatterer), we get for the intensity on the sphere

$$I(\theta,\varphi) = \frac{c}{8\pi}\,|E|^2 \approx I_o[1 + 2\quad(\sum_{\ell,m} O_{\ell m} Y_{\ell m}(\theta,\varphi))] \quad , \tag{6}$$

where

$$O_{\ell m} = 4\pi N(\bar{r}_1)(-i)^{\ell+2} j_\ell(kr_1) Y^*_{\ell m}(\theta_1,\varphi_1) \tag{7}$$

If a photographic material is exposed to this intensity, we assume it has transmission $T = T_o + aI(\theta,\varphi)$. Let us irradiate the sphere with a spherical incoming wave $E_{in}(R^+) = E_o h_o^{(2)}$ $(kR)\,Y_{oo}(\theta,\varphi)$, and calculate the wave amplitude on the inside surface of the sphere $E_{in}(R^-) = E_{in}(R^+)\,T$. The solution of Maxwell's equations inside the now empty sphere, satisfying these boundary conditions, can be found after some algebra

$$E(\bar{r}) = \frac{E_o T_o}{4\pi}\quad 2j_o(kr)\; -i\; N(\bar{r}_1)a\;\frac{\sin k|\bar{r}-\bar{r}_1|}{|\bar{r}-\bar{r}_1|} \tag{8}$$

This is the mathematical expression of Gabor's holographic reconstruction: the light intensity inside the sphere reproduces the reference beam and the scatterer. If the first term is filtered out, the intensity of the second term reproduces the "object", within the limitations of the uncertainty principle. Note that if the hologram is recorded on the full sphere, there is no "doubling" of the image and the resolution is $\approx\lambda$ in all three dimensions.

A more general holographic reconstruction problem is the following. Assume that a complicated, but <u>known</u> wavefront interferes with an unknown wave; the intensity of the interference pattern is recorded on a sphere. Find an algorithm to reconstruct the source distribution of the unknown wave. In parallel to Eq. 6 the intensity distribution on the sphere is

$$\begin{aligned} I(\theta,\varphi) &= \frac{c}{8\pi}\,|E_R + E_O|^2 \\ &= \frac{c}{8\pi}\,|\sum_{\ell,m} (R_{\ell m} + O_{\ell m}) h_\ell^{(1)}(kR) Y_{\ell m}(\theta,\varphi)|^2 \end{aligned} \tag{9}$$

where the components $R_{\ell m}$ are assumed to be known, and $O_{\ell m}$ unknown. Define the spherical harmonic components of $I(\theta,\varphi)$ as

$$I_{\ell m} = \int I(\Omega) Y^*_{\ell m}(\Omega) d\Omega \tag{10}$$

Using the formula from Messiah[12] (p. 1057, Eq. C-16), we get

$$I_{\ell_3 m_3} = \frac{c}{8\pi R^2} \sum_{\ell_1 m_1} \sum_{\ell_1 m_2} (R_{\ell_1 m_1} + O_{\ell_1 m_1})(R^*_{\ell_2 m_2} + O^*_{\ell_2 m_2})(-i)^{\ell_1 - \ell_2} \tag{11}$$

$$(-1)^{m_2+m_3} \left[\frac{(2\ell_1+1)(2\ell_2+1)(2\ell_3+1)}{4\pi}\right]^{1/2} \begin{pmatrix} \ell_1 & \ell_2 & \ell_3 \\ 0 & 0 & 0 \end{pmatrix} \begin{pmatrix} \ell_1 & \ell_2 & \ell \\ m_1 & -m_2 & -m_3 \end{pmatrix}$$

Where the ( ) are Clebsch-Gordan coefficients, and the summation is restricted to $m_1-m_2-m_3 = 0$; ℓ_1,ℓ_2,ℓ_3 are all positive and also satisfy the triangular inequalities.

This is an (infinite) set of quadratic matrix equations for the unknowns, $O_{\ell m}$. It can be written symbolically as

$$I = C\,(R + O)\,(R^* + O^*) \tag{12}$$

The three principal terms, RR*; RO* + R*O; OO* can be recognized as the reference, the hologram, and the oject's self interference terms respectively. It is expected that in the presence of noisy experimental data, (imperfect knowledge of $I_{\ell m}$, $R_{\ell m}$) these equations constitute an ill posed problem; therefore they can only be solved by least squares, maximum likelihood or maximum entropy algorithms. If Eq. 12 can be solved, the three dimensional brightness distribution of the "object" can be determined by a linear algorithm which follows from Eq. 7. Note the essential difference between the appearance of the singular function exp(ix)/x in the source (Eq. 3) and of the regular function sin(x)/x in the reconstruction (Eq. 9). This corresponds to the following statement: If it is known _a priori_ that the object is a single scatterer, it can be located to arbitrary accuracy even when it is smaller than the wavelength of the imaging radiation; but if this knowledge is lacking, the wavelength of the radiation limits the resolution obtainable.

Finally, there is a very close parallel between the holographic reconstruction problem and the phase problem of X-ray crystallography. An X-ray diffraction pattern is a "sampled" $I(\theta,\varphi)$: it is different from zero only when the Bragg conditions are satisfied. The original, "unphased" pattern is analogous to OO* in Eq. 13; as the structure becomes better known, more and more of the intensity is "shifted" to R. Thus our procedure is completely analogous to the refinement procedure for electron density maps[13]. This observation points to an interesting possibility. If the three dimensional image of the close neighborhood of a metal atom in a biological macromolecule can be obtained, the Fourier transform of its electron density can be calculated with the proper phases. It can then be used as a starting point for the phasing of an x-ray diffraction pattern. This method may supplement heavy atom substitution, isomorphons replacement and anomalous dispersion techniques for the solution of crystal structures. Alternately it can be viewed as a starting point for direct methods.

ACKNOWLEDGEMENTS

This research was supported by a grant from the Air Force Office of Scientific Research. It was carried out while the author was on personal research leave from Lawrence Livermore National Laboratory.

REFERENCES

(a) Permanent address: Lawrence Livermore National Laboratory
Livermore, CA 94550

1. D. Gabor, Nature, 161, 777 (1948); Proc. Roy. Soc. (London) A197, 454 (1949); Proc. Phys. Soc. (London) B64, 449 (1951).
2. J. D. Jackson, Classical Electrodynamics, Wiley 1975. pp 392.
3. J. W. Goodman, Statistical Optics, Wiley 1985, Ch. 9.
4. Malcolm R. Howells, private communication.
5. K. O. Hodgson, ed. EXAFS and Near Edge Structure III, Springer, 1984.
6. Angle-resolved X-ray Photoelectron Spectroscopy, C. S. Fadley, Prog. Surf. Science 16, 275 (1984); J. J. Barton and D. A. Shirley Phys. Rev. B32, 1892, 1906 (1985); M. Sagurton et al. Phys. Rev. B33, 2207 (1986).
7. P. A. Lee et al., Revs. Mod. Phys. 53, 769 (1981).
8. J. B. Pendry, Low Energy Electron Diffraction, Academic, London, 1974.
9. E. Leith in Handbook of Optical Holography, H. J. Caulfield ed. Academic, 1979.
10. J. B. Pendry, D. K. Saldin, Surface Sci. 145, 33 (1984).
11. K. Heinz, D. K. Saldin, J. B. Pendry, Phys. Rev. Letters 55, 2312 (1985).
12. The author with collaboration of A. Hawryluk did some model experiments on a similar scheme in 1975 but never published them.
13. A. Messiah, Quantum Mechanics, Wiley, 1958.
14. A. Bricogne, Acta Cryst, A40, 410 (1984).

SOFT X-RAY NANO LITHOGRAPHY OF SEMITRANSPARENT MASKS FOR THE GENERATION OF HIGH-RESOLUTION HIGH CONTRAST ZONE PLATES

C.J.Buckley (KC), R.Feder (IBM), M.T.Browne (KC)

KC: Department of Physics, Kings College, Strand, London WC2R 2LS, UK.
IBM: IBM T.J.Watson research Centre,PO Box 218, Yorktown Heights, NY10598 , USA.

ABSTRACT

Contamination E-Beam lithography has been used to fabricate accurate carbon zone plates to an outer zone width of 40 nm. X-ray lithography using synchrotron radiation has been employed to generate gold zone plate replicas from semi-transparent Carbon ZP masks which have an outer zone width of 75 nm.

INTRODUCTION

During the past few years increasing effort has been applied to making Fresnel Zone plates (FZP's) suitable as focusing elements for soft X-ray based microscopes. 1,2,3,4,5. The factor which ultimately limits the spatial resolution of images formed using a circular FZP is the outer zone spacing.6 Zone plates with the finest outer zone spacing made to date have been fabricated by E-beam lithography (EBL) and it is likely that EBL will continue to produce the finest zone plates for the foreseeable future.

This paper discusses the conditions necessary to fabricate ZP's which are capable of working to the diffraction limit, outlines the fabrication of ZPs by Contamination EBL and describes how X-ray lithography has been used to replicate Carbon ZP's in a higher contrast medium.

USEFUL ZONE PLATES

For a ZP to be useful for soft x-ray microscopy it must be capable of focusing incident x-rays to a small well defined circularly symmetric spot.7 (see fig 1). For effective microscopy the spot should be smaller than that which can be produced optically and a sub hundred nano meter spot size is considered desirable. The zone plate must also have a suitably long (>0.4 mm) focal length. This places a constraint on the minimum diameter of the ZP as the focal length (and consequently the working distance) of the ZP is proportional to its radius.8

As well as having fine outer zone spacings, the ZP pattern must be highly accurate if the theoretical resolution is to be obtained. Simpson 7 calculates that the zone pattern must be accurate to about one third of the finest zone period over the whole of the zone plate diameter if the ZP focus is to remain within the Strehl limit. For example, a ZP with a finest zone width of 50 nM (100 nM period) and a radius of 35 uM, (giving a focal length of 1 mm at 3.5 nM wavelength) the positioning of the zones must be accurate to about 30 nM over a 70 uM field.

Focusing efficiency is an important consideration for realistic imaging times and photon statistics. The major factor effecting the focusing efficiency of accurate FZP's is the transmission contrast of the odd and even zones.

The following section briefly reviews the employment of electron beam lithography for the production of useful ZP's and outlines the production of ZPs by Contamination EBL.

ZONE PLATE FABRICATION BY E-BEAM LITHOGRAPHY

The main factors which efect the quality of ZP's fabricated by conventional static-substrate E-beam lithography are:

1) Forward scattering of the electron beam in the resist
2) Distortion of the E-beam scan
3) Drift of the substrate during writing

Forward scattering of the electron beam by the resist substrate has the effect of limiting the the fineness of lines exposed in conventional resists. This makes it difficult to obtain practical lines of 100nM and smaller with a 1:1 mark space ratio 9. Zone plates made with unequal odd/even zone areas have reduced first order efficiency, and have an increased background flux due to the presence of second order radiation 8.

Scan distortions are present in all E-beam machines and can have a serious effect on the quality of the final ZP. As mentioned in the previous section, the distortion of the ZP pattern should not be more than one third of the outermost zone period across the diameter of the ZP. Attempts to correct for beam distortion in conventional EBL have been reported by Shaver et al 10, though the distortion correction reported was only down to 250nM across the drawing field. Drift of substrate during pattern writing will also result in pattern distortion, and steps must be taken to either compensate for it, or reduce its affect by minimizing the writing time.

The above realities make the fabrication of useful ZP's capable of sub 100 nM resolution by conventional EBL very difficult. The following section outlines how Contamination writing has been employed to overcome much of the above problems to create useful ZP's for X-ray microscopy.

ZP FABRICATION BY CONTAMINATION EBL

Contamination writing 11 has been employed to fabricate high resolution ZP's 4. The Contamination EBL is aptly suited to the this task, as the problem of forward scattering is greatly reduced - enabling lines of width 20 nM to be drawn with an aspect ratio of 5:1 and a mark space ratio,of 1:1 (fig 2).

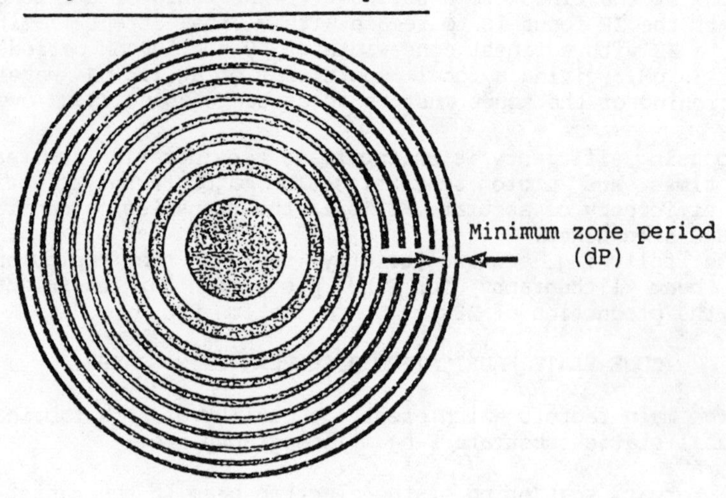

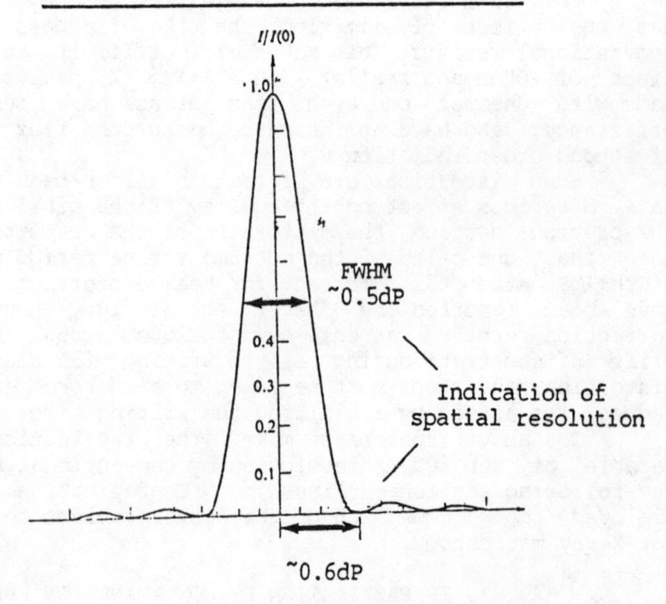

Fig 1. The spot produced by an accurate Fresnel zone plate consisting of one hundred zones or more is a close approximation to an Airy pattern. The radius of the first minimum of the pattern is approximately 0.6 times the minimum zone period (or 1.2 times the outer zone width) and is a good indication of the spatial resolution obtainable in zone plate microscopy.

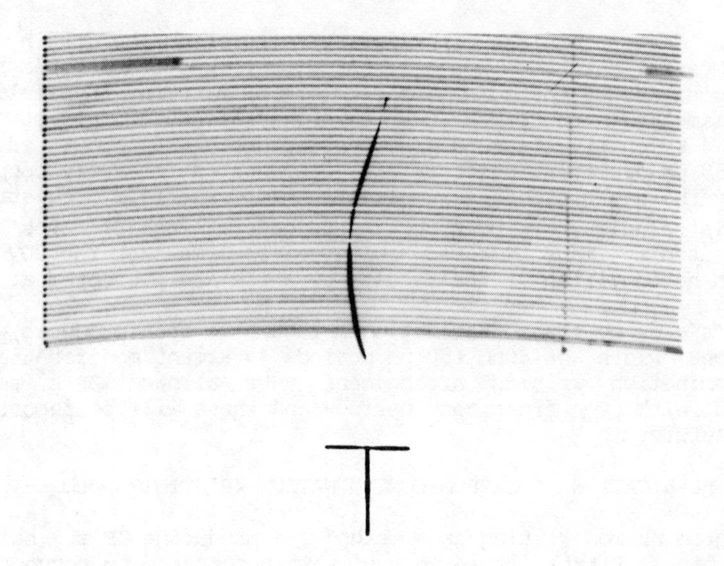

Fig 2. TEM of 20 nm wide zone arcs written by STEM contamination EBL.

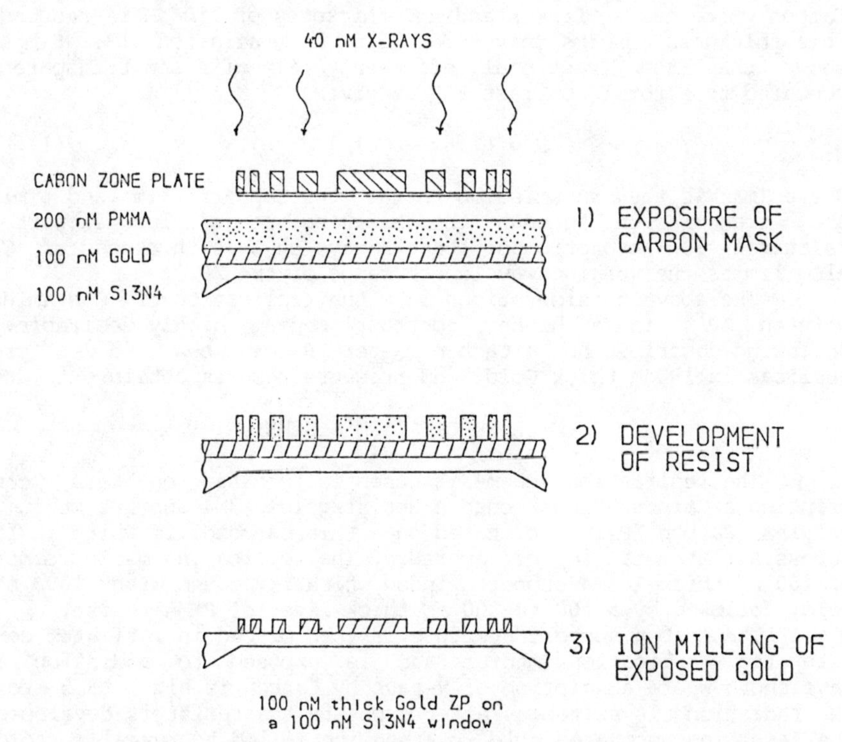

Fig 3. Steps for the replication of Carbon ZP's by x-ray lithography.

Scan distortions are minimized by restricting the drawing field to a 5 uM axial patch - where relevant distortions reduce to a simple non orthogonality of the scan. The complete ZP pattern (typically 50 um in diameter) is formed by a highly accurate field stitching technique 4 which gives a zone placement accuracy of about 12.5 nM over a 50 uM diameter. Drift is also compensated for every couple of seconds by the repetitious scan location of in-field registration marks. This technique has been applied to generate useful ZP's which have been used to image hydrated biological specimens at the VUV ring/ Brookhaven national lab - the results of which are presented at this meeting 12.

ZP's made by the contamination method have been produced with a finest zone width of 40nM (80 nM period) 4. Recent modifications to the Contamination writing arrangement have allowed ZP's to be fabricated with even finer zone periods and these will be reported in the near future.

REPLICATION OF CARBON CONTAMINATION ZP'S INTO GOLD

Contamination writing as a method for producing ZP's has two main draw backs. Firstly it is very slow when compared to conventional resist EBL. The production of a Contamination written ZP takes one to two days - during which time problems may arise 4 causing degradation of the ZP. Secondly, the efficiency of a ZP which is comprised of Carbon which has a "free standing" thickness of 110 nM is reduced - as this thickness absorbs only 56% of 3.5 nM radiation 13. Simpson 7 shows that the fractional efficiency (F) of a semitransparent FZP compared to a total contrast FZP is given by

$$F = [T_{max} - T_{min}]^2 \qquad ..(1)$$

Where Tmax is the transmission through the support film and Tmin is the transmission through the blocking zones plus support. The relatively low absorption of these X-rays by this thickness of Carbon also limits the working wavelength range of the ZP.

The above considerations make the replication of Contamination written ZP's in a higher contrast medium highly desirable. The following describes how a Carbon master ZP can be used to produce replicas in 100nM thick Gold, and presents results obtained to date.

ZP REPLICATION SCHEME

The replication scheme is essentially based on X-ray contact printing a Carbon ZP mask onto a Resist/gold/Si3N4 support medium. The original Carbon ZP is fabricated on a thin Carbon film which is formed across a flat metal support aperture. The replicating medium comprises of 100 nM thick Si3N4 support window which is coated with 100 nM of gold, followed by a 100 to 200 nM thick layer of PMMA resist.

The Carbon support aperture is then placed in intimate contact with the replication medium and is exposed to radiation at a wavelength where absorption of x-rays by Carbon is high. (3.5 to 4.4 nM radiation is suitable) After exposure the resist is developed and the resulting uncovered gold is Argon ion milled to reveal a gold ZP pattern on a Si3N4 support membrane. (see fig 3).

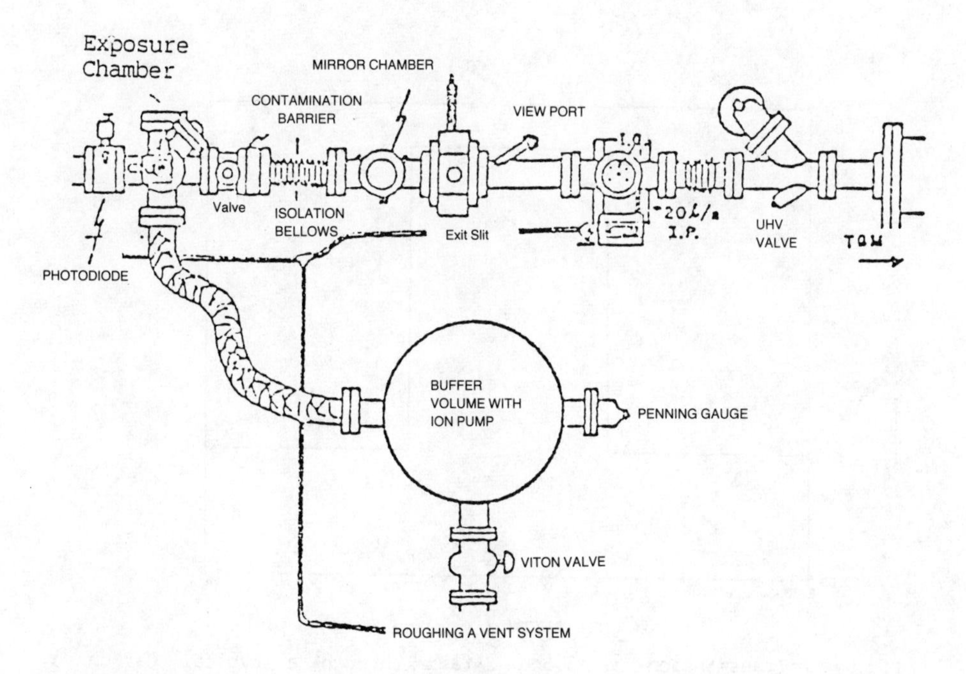

Fig 4. Diagram of part of U15 beam line from UHV valve to exposure chamber.

EXPOSURE

The U15 beam line of the NSLS storage ring at BNL was used for exposure. Synchrotron radiation is ideal for this type of lithography as it combines high X-ray intensity with the good collimation necessary to avoid resolution degradation by penumbral blurring. The arrangement of the beam line is shown (fig 4) with the exit slit of the monochromator 40 cm from the IBM exposure mount. The maximum divergence of the beam at U15 was 2 milli rads (two sigma value). If the ZP/resist contact is better than 3um, then penumbral blurring is less than 10 nM at the resist.

DOSE MANIPULATION

The X-ray dose used for this replication was relatively high (about 1 -> 5 J/cm2). A high dose is used to provide a large differential in resist development dissolution rate, with a relatively low incident contrast differential. This is employed to counter the effects of two limiting problems. These are :

1) Poor mask contrast

&

2) Diffraction effects

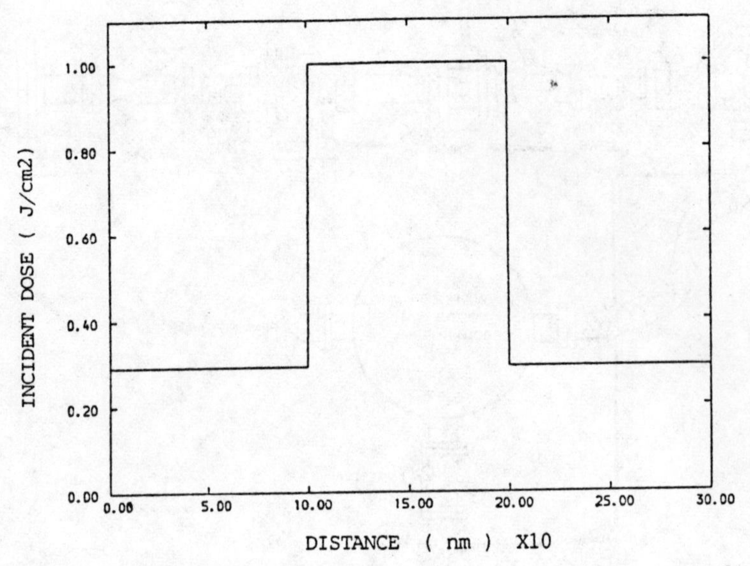

Fig 5a Transmission of 3.5nm X-rays through a typical Carbon contamination written zone pair.

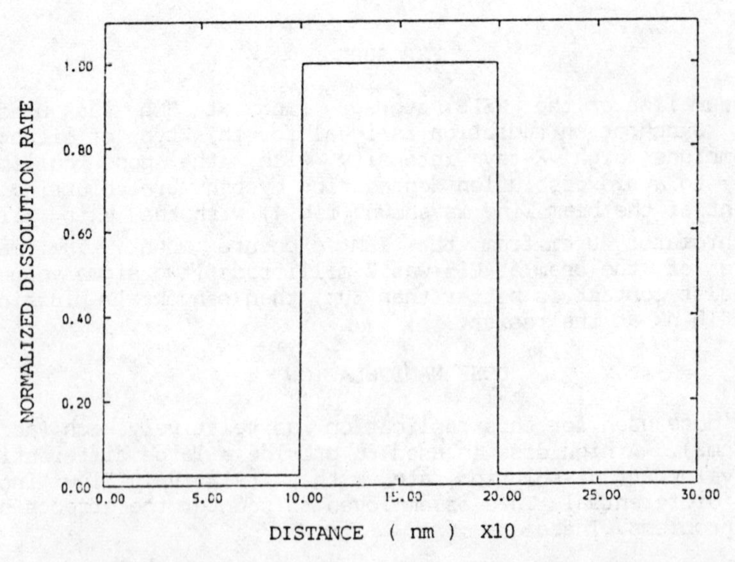

Fig 5b Normalized PMMA dissolution rate for incident dose profile in 5a (max dose 1 J/cm2).

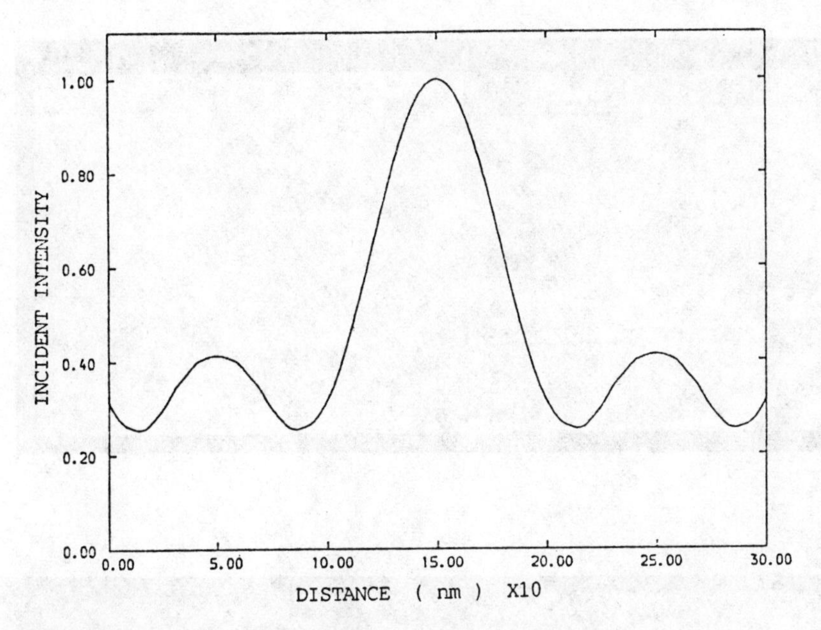

Fig 6a Intensity profile of 3.5 nm flux transmitted through carbon zone pair at surface of resist (2um mask/resist separation).

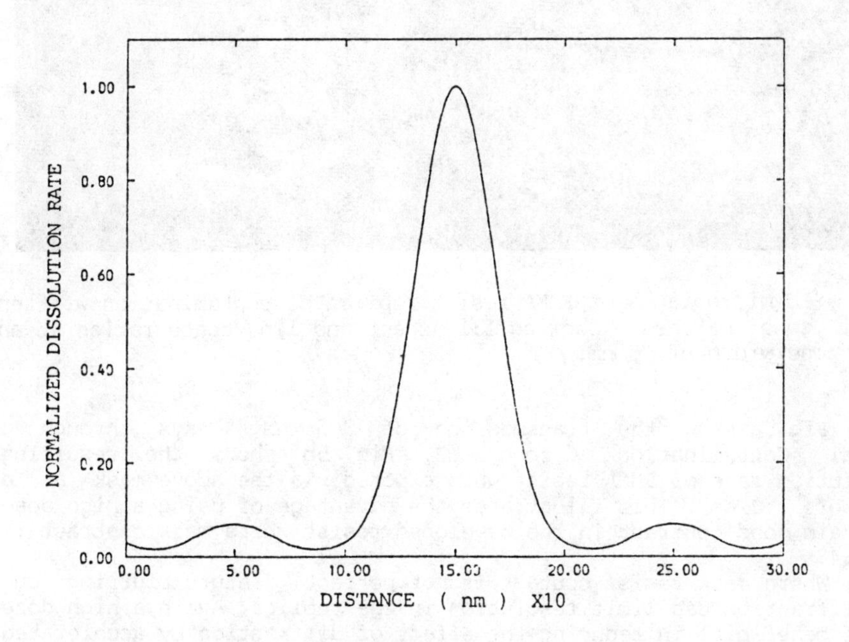

Fig 6b Normalized dissolution rate for incident dose profile in 6a.

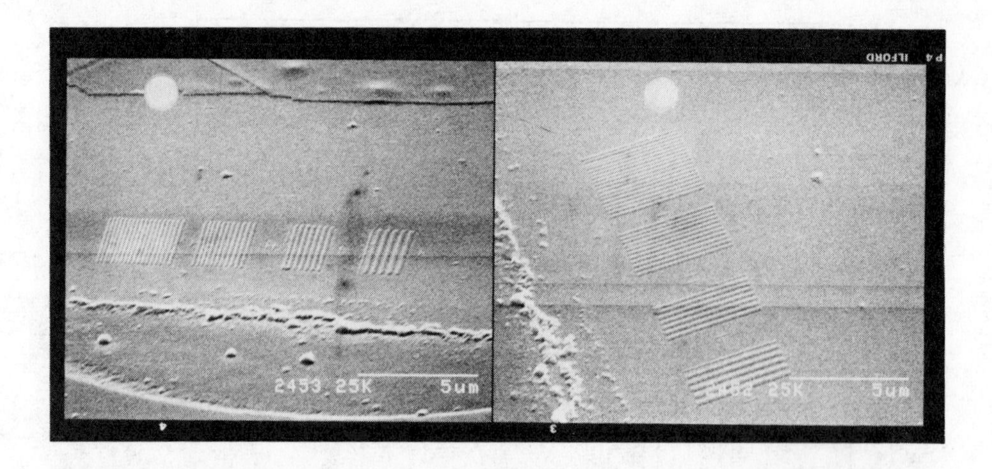

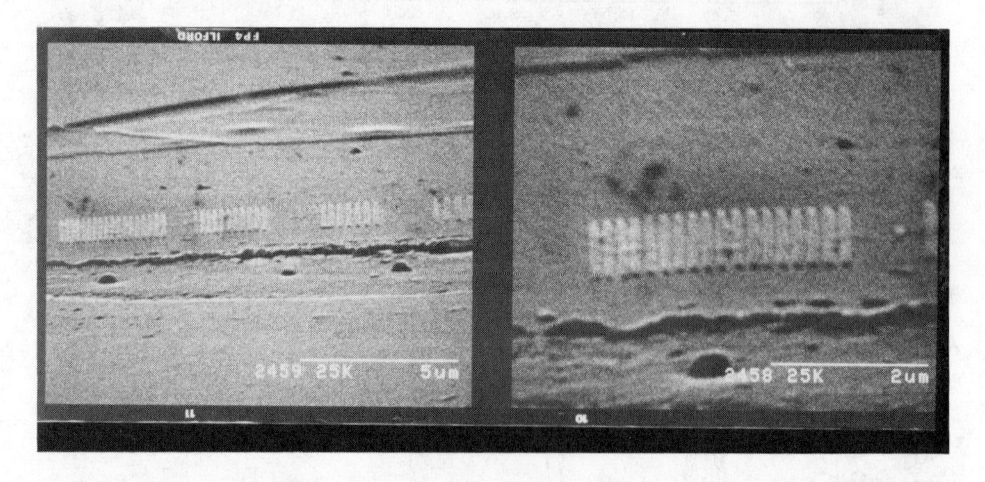

Fig 7 SEM micrographs of PMMA resist replica of contamination written linear zone patches - showing 1:1 aspect and line/space ratios at an outer zone width of 75 nm.

Fig 5a shows the transmission of 3.5 nM X-rays through a typical Contamination Carbon ZP. Fig 5b shows the resulting dissolution rate of PMMA resist when exposed via the above mask at a dose of 1 J/cm2. This illustrates the advantage of using a high dose to obtain good contrast in the developed resist where mask contrast is poor 14.

Where mask/resist contact is not perfect, feature blurring due to diffraction can limit resolution in the replica. Again a high dose can be beneficial in reducing the effect of diffraction by accelerated resist development where the incident intensity is highest. This

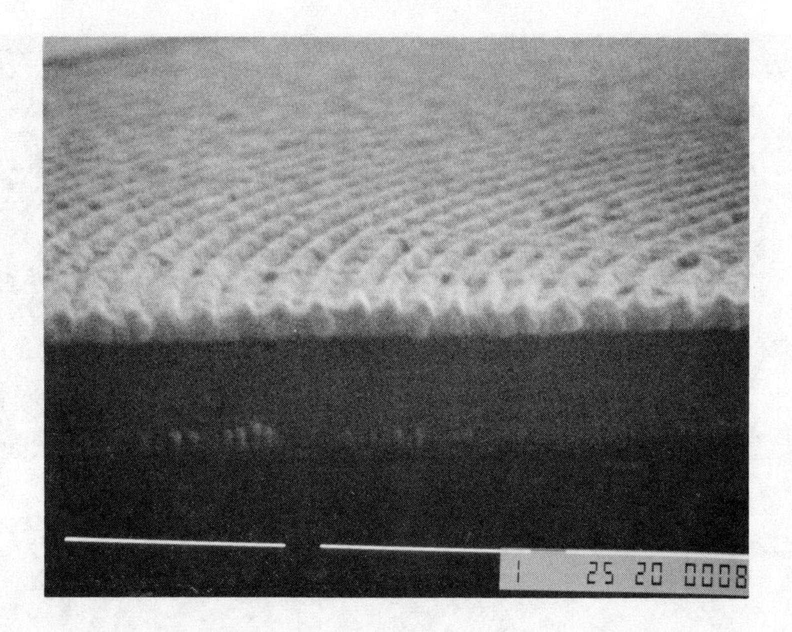

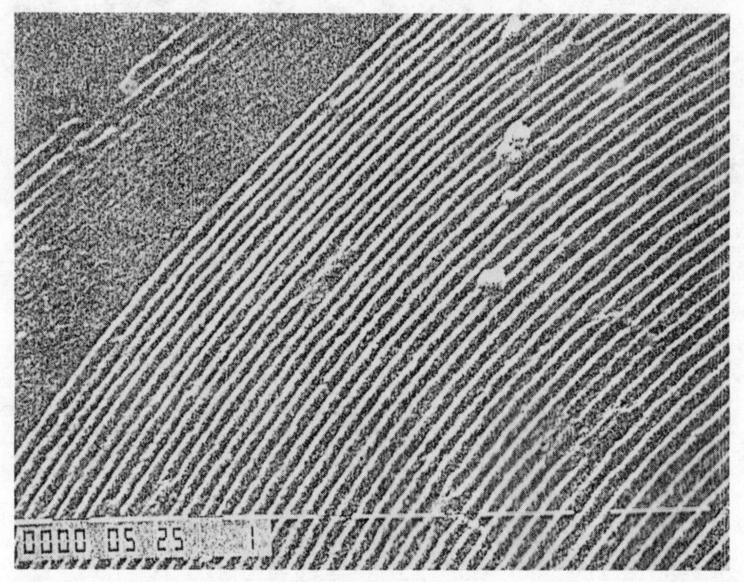

Fig 8. SEM micrographs of relief profile of a gold replica (75 nm outer zone width) on a fractured piece of silicon after ion milling.

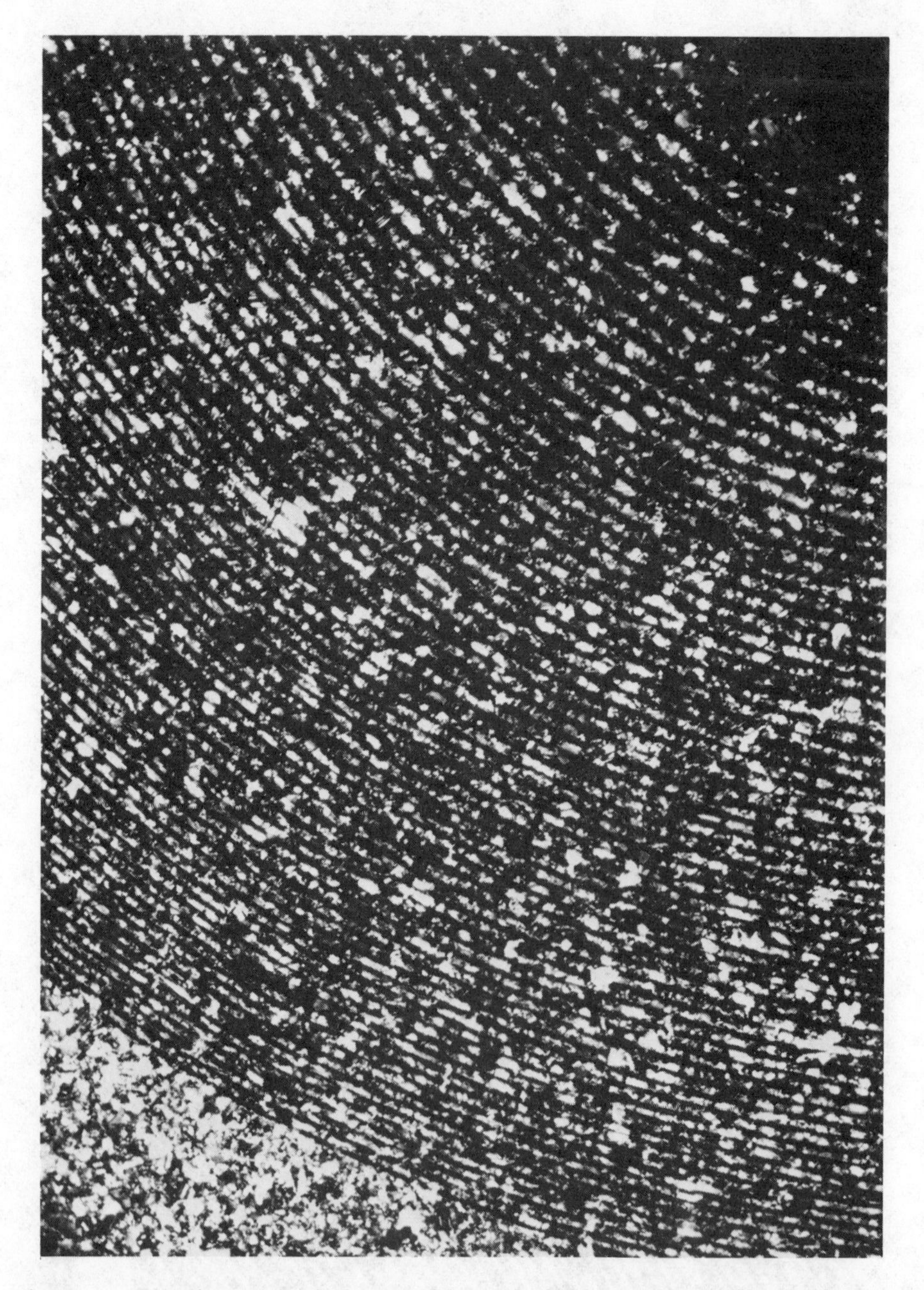

Fig 9 TEM micrograph of a replica of the same ZP as fig 8, but on a silicon nitride X-ray transmission window. The gold shows a high degree of granularity and the mark-space ratio is not 1:1, note however, that the finest zone spacings have been reproduced (75 outer zone width in this case).

reduces the feature spread caused by diffraction (see figs 6a & 6b).

Where high differential resist dissolution rates are employed for contrast enhancement, self absorption of X-rays by the resist must be considered 15. For example; If 3.5 nM radiation is used to expose 1 uM thick PMMA resist, then the dose received at the base of the resist would be less than one tenth of that received by the surface. This would result in very low dissolution rate at the base of the resist compared to that at the top. This would lead to overcutting (side wall attack) and loss of resolution. The effect of resist self absorption can be minimized by using a resist thickness of 100 -> 200 nM where the surface/base dose difference is only about 30%.

Resist of this thickness was used in ZP replication experiments, the details and results of which are presented below.

REPLICATION RESULTS

Initial experiments involved the replication of linear zone patterns to asses the viability of replication of ZP's with similar characteristics. The conditions under which replication was performed are as follows:

Mask substrate thickness	35 nm Carbon film
Zone material + substrate thickness ...	135 nm Contamination Carbon
Minimum mask/resist separation	2 um
Unexposed PMMA resist thickness	200 nm
Resist pre-bake temperature	160 C
Exposure	1 J/cm2 at 3.5 nm wavelength
Resist development	15 secs in IPA at 20 C

Fig 7 shows SEM micrographs of linear zone patches replicated in PMMA. The resist profile obtained is a faithful replica of the Carbon original and shows an aspect and mark space ratios of about 1:1 at a line width of 75nM.

Complete zone plate patterns were exposed in a similar manner, and following development were given a brief exposure to oxygen plasma followed by normal incidence argon ion milling. Milling was performed at 500 volts 20 ma for 10 mins. Fig 8 shows the relief profile of a gold replica ZP (75nm outer zone width) on a fractured piece of Silicon after ion milling. Fig 9 is a TEM micrograph of a replica of the same ZP in fig 8 but on a Silicon nitride x-ray transmission window. There are obviously problems with the granularity of the gold, and the mark space ratio of the zones is not 1:1 The replication technique has reproduced the finest zone spacings of the ZP (75nm outer zone width in this case) however, and one of the replicas of the above ZP was used to form images on the Stony brook/BNL scanning x-ray microscope.

CONCLUSIONS

ZP's suitable for x-ray microscopy have been made by Contamination writing to an outer zone width of 40 nm. X-ray lithography has been employed to create improved contrast replicas of Carbon contamination ZP's. The results presented in this publication represent initial experiments performed in October 1985. In the time available the technique was tested to an outer zone width resolution of 75 nm. Zone plates of this outer zone width have been replicated and used to form images in a scanning x-ray microscope.

Future experiments will concentrate on improving the granularity of the gold, the line mark space ratio and will attempt the replication of higher resolution zone plates.

ACKNOWLEDGEMENTS

The authors would like to acknowledge the following persons for direct or indirect advice and support.

IBM T.J.Watson Research centre Yorktown heights:
P.C.Cheng, D.Shinozaki, M.Calderoldo, R.E.Acosta, U. Parazchec, and U.Vladamerski.

Brookhaven National Lab/Stony Brook University:
C.Jacobsen, J.Kirz, I.MacNulty, R.Rosser, B.X.Yang, H. Rarback and the staff at the NSLS, BNL, and Stony brook University Physics department.

Kings College London:
R.Burge, P.Charalambous, J.Kenney, K.Ogawa, W. Luckhurst and P.Duke (SRS Daresbury Laboratorty).

The research at Kings College is funded in part by the SERC and the research at Stony Brook and Brookhaven is supported by the NSF and the DOE.

REFERENCES

1. G.Schmahl et al, "Zone plates for x-ray microscopy" Proc. 1983 Symp. X-ray microscopy. Springer 1984.

2. D.Kern et al, "Electron Beam Fabrication and characterization studies of Fresnel zone plates...", Proc. Advances in Soft X-ray Science and Technology. SPIE 1984.

3. H.Aritome, K.Nagata and S.Namba, "Fabrication of ZP's with a minimum zone width smaller than 100 nm by EBL", Microelectronic Engineering 3, Elsevier Science 1985.

4 C.J.Buckley, M.T.Browne & P.Charalambous, "Contamination lithography for the fabrication of Zone Plate X-ray lenses", E-Beam,X-Ray &Ion Beam Techniques 5. SPIE 1985.

5. E.Kratschmer, D. Stephani, H.Beneking, "High resolution 100 keV E-Beam lithography", Microcircuit Engineering 83. Academic Press 1983.

6. M.Born and E.Wolf, "Principles of Optics" 5th Edn. Pergamon Press 1980.

7. M.Simpson, Doctor of philosophy Thesis (unpublished), Queen Elizabeth College, London University, P37,111, 1984.

8. C.J Buckley, "Scanning Soft X-ray microscopy with Kings College STEM ZP's on the Brookhaven/Stony Brook microscope" (internal report), KC/BNL P20, 1985.

9. C.Wilkinson, "High resolution EBL", Proc. New Ways of Examining the Sub Micron world, 1984, to be published.

10. D. Shaver and D.Flanders, N.Ceglio and H.Smith, "X-ray zone plates fabricated using E-Beam and X-ray lithography", J.Vac.Sci.Tec. Nov 1979.

11. M.Browne, P.Charalambous and R.Burge, "Uses of contamination writing in STEM: Projection electron lithography", Inst Phys. Conf. Ser. No 61, Ch 2. 1981.

12. C. Jacobsen et al, "Scanning Soft X-Ray Microscopy at the National Synchrotron Light Source", Proc. Short wavelength coherent radiation Conf. Opt.Soc.America. 1986.

13. B. Henke (soft x-ray data), Dept of Physics & Astronomy, University of Hawaii, Honolulu, Hawaii 96822 (Last revision: January 1982).

14. R.Feder et al, Science 197 (1977) 259.

15. Halebich, J.Silverman and J.Warlaumont, "Synchrotron Radiation X-ray Lithography", Nuc. Instr. Methods. 222, 1984, Elsevier Press.

TUNABLE XUV RADIATION GENERATED BY RESONANT THIRD- AND FIFTH-ORDER FREQUENCY CONVERSION

R. Hilbig, G. Hilber, A. Lago, B. Wolff
and R. Wallenstein
Fakultät für Physik, Universität Bielefeld,
4800 Bielefeld, FRG

A B S T R A C T

Third- and fifth-order sum- and difference frequency conversion of pulsed dye laser radiation generates coherent tunable radiation in the vacuum ultraviolet at wavelengths λ_{vuv} = 60 - 200 nm. The generated VUV light is of narrow spectral width (ΔE = 0.01 - 1 cm^{-1}) and high spectral intensity (0.03 - 10^4 W; 10^8 - $3\cdot10^{13}$ photons/pulse). Because of their spectral brightness these VUV light sources are a powerful tool for VUV spectroscopy of atoms, molecules and solid states.

INTRODUCTION

Nonlinear frequency mixing in gases is a well established method for the generation of optical radiation in the spectral region of the vacuum ultraviolet (VUV) at wavelengths λ_{vuv} = 100 - 200 nm and in the extreme ultraviolet (XUV) at λ_{xuv} < 100 nm. The pioneering work in this field has been published more than ten years ago by J.F.Ward and G.H.C. New[1] and by S.E. Harris, J.F. Young and coworkers[2]. During the past decade the results of a large number of theoretical and experimental investigations demonstrated that frequency mixing of powerful laser light generates intense VUV and XUV radiation with tunable frequency and high spectral brightness.

The principles of frequency mixing and the experimental progress towards the generation of coherent VUV radiation have been reviewed by W. Jamroz and B.P. Stoicheff[3] and by C.R. Vidal.[4]

In this contribution we present a few examples of more recent investigations of third- and fifth harmonic generation ($\omega_{vuv} = 3\ \omega_1$ or $5\ \omega_1$) and third- and fifth order sum frequency mixing ($\omega_{vuv} = 2\ \omega_1 + \omega_2$ or $4\ \omega_1 + \omega_2$) of dye laser radiation (with the frequencies ω_1 and ω_2) in rare gases. These frequency conversions are simple, reliable methods for the production of tunable XUV radiation of narrow spectral width ($\Delta E = 0.01 - 1\ cm^{-1}$) and high spectral intensity ($0.03 - 10^2$ W; $10^8 - 2 \cdot 10^{11}$ photons/pulse). The results of current investigations are very promising for a further, considerable increase of the XUV pulse power and for an extension of the tuning range to even shorter wavelengths.

THIRD-ORDER FREQUENCY CONVERSION

Nonresonant third-order sum- and difference frequency mixing ($\omega_{vuv} = 2\ \omega_1 \pm \omega_2$) in the rare gases Xe and Kr and in Hg vapor generated broadly tunable VUV radiation in the wavelength range λ_{vuv} = 110 - 200 nm.[5-12] Frequency tripling in Ar and Ne produced XUV light in spectral regions between 72 and 105 nm.[13,14]

In these experiments laser pulse powers of 1 to 5 MW provided conversion efficiencies of typically 10^{-5} to 10^{-6}. The power of the generated VUV light pulses was in the range of 1 to 20 W ($0.3 - 6 \cdot 10^{10}$ photons/pulse). Main limitations on the efficiency are caused by dielectric gas-breakdown in the focus of the laser light and by nonlinear intensity dependent changes of the refractive index[4,15] which destroy the phase-matching.

The VUV intensity generated by nonresonant conversion methods is sufficient for most investigations in linear (absorption or fluorescence) spectroscopy. Other applications (like multiphoton excitation and ionization or photodissociation) require more powerful VUV light pulses.

The pulse power can be increased by several orders of magnitude using resonant conversion methods.[16]

Tuning the laser frequency, for example, to a two-photon resonance the induced polarization is resonantly enhanced. This enhancement provides conversion efficiencies of $\eta > 10^{-4}$ even at input powers of only a few kilowatts.

The two-photon resonant conversion is usually of the type $\omega_{vuv} = 2\,\omega_1 \pm \omega_2$ where ω_1 is tuned to a two-photon transition, and ω_2 is a variable frequency.

In the past this type of frequency conversion has been investigated in detail in metal vapors (like Sr[17,18], Mg[19-21], Cs[22], Ba[23], Hg[24-26] and Zn[27]) and in the rare gases Xe and Kr.[11,28-31] Experimentally rare gases are easy to handle and are thus a very appropriate medium for the construction of a simple and reliable VUV source. In Xe, for example, the resonant frequency mixing has been investigated using the 5p-6p and 5p-6p' resonances.[28,31] Among the metal vapors Hg requires the lowest temperatures. In this metal vapor the two-photon resonant conversion is very efficient.[24-26]

The measurements performed in Xe[28,31] and Hg[24-26] provided broadly tunable VUV of KW peak power and detailed information on different saturation phenomena.

A very efficient resonant frequency conversion which generates broadly tunable VUV is the two-photon resonant mixing in Kr.[11]

In Kr the lowest two-photon resonance is the 4p-5p [5/2,2] transition. This transition requires UV radiation of $\lambda_R = 216.6$ nm. Radiation at this wavelength can be generated by doubling the output of a Fl.27 dye laser ($\lambda_L = 543.9$ nm) and mixing the UV with the infrared (ω_{IR}) of the Nd-YAG laser. In this way the maximum available pulse power is close to 1 MW.

Tuning the dye laser in the range $\lambda_2 = 216 - 900$ nm the sum frequency $\omega_{vuv} = 2\,\omega_R + \omega_2$ is in the XUV ($\lambda_{xuv} = 96.6 - 72.1$ nm); the difference frequency should be continuously tunable in the range of $\lambda_{vuv} = 123 - 217$ nm.

Tuning, for example, the dye laser in the range $\lambda_L = 540 - 730$ nm the conversions $2\omega_R - \omega_L$, $2\omega_R - (\omega_L + \omega_{IR})$ and $2\omega_R - (2\omega_L)$ generated VUV at λ_{vuv} = 127.5 - 135.5 nm, 145.5 - 155 nm and 155 - 181 nm, respectively. Radiation at λ_{vuv} = 135 - 145 nm was produced by $2\omega_R - \omega_L$ with λ_L = 428 - 548 nm, the tuning range of Coumarin dye lasers. At optimum conditions an input of P_R = 200 KW and P_L = 1 MW generated VUV pulses close to 0.5 KW.

While the difference frequency generates continuously tunable VUV in the range of 120 to 200 nm the corresponding sum frequency produces light at wavelengths λ_{xuv} < 100 nm. First experimental results are shown in Figure 1.

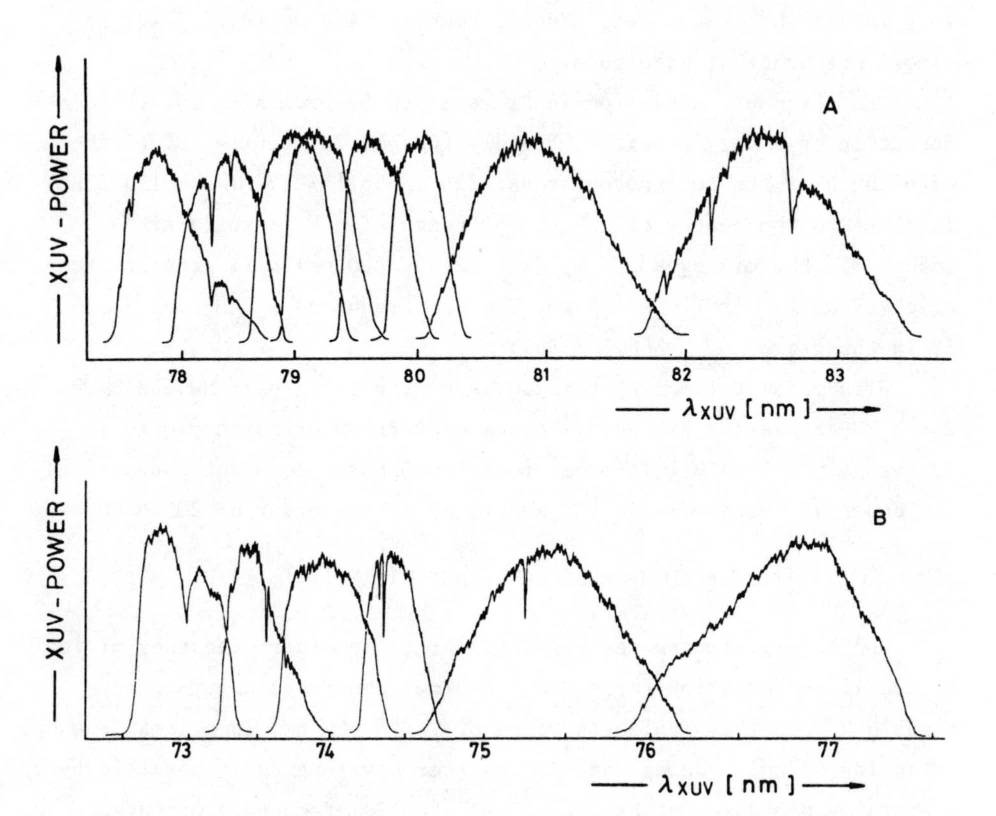

Figure 1: Sum frequency mixing $\omega_{xuv} = 2\,\omega_R + \omega_{uv}$ in Kr. A: radiation with $\omega_{uv} = 2\omega_L$ (λ_L = 540-730 nm) is generated with the dyes Fl 27, different Rhodamine dyes, DCM and Pyridine I; B: $\omega_{uv} = 2\omega_L + \omega_{IR}$ with λ_L = 550-730 nm.

With λ_2 = 218.5 - 365 nm the XUV light is continuously tunable in the region of λ_{xuv} = 72.5 - 83 nm. The attenuations observed at certain wavelengths are caused by resonance absorption of Kr ions created in the focus of the UV laser radiation.

In these experiments the laser light is focused into a pulsed free Kr jet;[32,33] the VUV is detected by a differentially pumped spectrometer-detector arrangement. The pulse power of the generated XUV light has been measured by a windowless vacuum photo diode (with gold cathode). According to these measurements the pulse power P_{xuv} generated in the focus of the laser light should exceed 100 W. Absorption in the Kr gas jet reduces, however, the detected power by almost one order of magnitude.

The resonant conversion in Kr seems to be promising for the construction of a very powerful, broadly tunable VUV source. Since the wavelength of the two-photon transition 4p-6p [3/2,2] (λ_R = 193.5 nm) is close to the center of the tuning range of a narrowband ArF*-laser[34,35] the mixing $2\omega_R - \omega_L$ (λ_L = 216 - 800 nm) will generate tunable VUV at λ_{vuv} = 110 - 175 nm. The sum frequency $\omega_{xuv} = 2\omega_R + \omega_L$ is in the range λ_{xuv} = 66.8 - 86.3 nm.

With present laser systems pulse powers of P_R = 15 MW and P_L = 2 - 5 MW are easily achieved. If the conversion efficiency η is in the range $\eta = 1\text{-}3 \cdot 10^{-3}$ (typical for a two-photon resonant conversion) the power of the generated VUV should be on the order of 20 to 60 KW.

FIFTH-ORDER FREQUENCY MIXING

In the experiments described so far third order frequency mixing of dye laser radiation (λ_L = 216 - 800 nm) generated continuously tunable VUV in the wavelength range of 72 to 200 nm. In principle an extension of this tuning range to shorter wavelengths is possible by conversion processes of higher order.[29,36] Sum-frequency mixing of fifth order, for example, should produce coherent VUV at wavelengths λ_{xuv} = 42 - 72 nm.

In the past fifth harmonic generation has been investigated with powerful fixed frequency solid state (Nd-YAG or Nd-Glass) and gas

(XeCl, KrF) lasers.[29,36-38] In one of these experiments[38] input powers of more than 300 MW (mode-locked Nd-YAG fourth harmonic, $\lambda = 266.1$ nm) provide, for example, conversion efficiencies of 10^{-5} to 10^{-6}. Since the pulse power of most dye laser systems is lower by one or two orders of magnitude nonresonant fifth-order frequency mixing of this radiation would produce intensities below a useful level.

A considerable increase of the generated VUV power is expected, however, from resonant six-wave mixing. Recently resonant fifth-order conversion has been demonstrated in Ar. In these investigations the UV radiation ($\lambda_{uv} = 318.9$ nm) of a frequency doubled dye laser($\lambda_L = 637.8$ nm) is resonant with the four-photon Ar-transition 3p-9p [1/2, 0]. Simultaneously $3\,\omega_{uv}$ is close to the transition frequency ω_{res} of the first Ar resonance ($3p^1S_o$ - 4s[3/2,1]). The energy difference $\Delta E = 3\omega_1 - \omega_{res}$, of 91 cm^{-1} is sufficiently small so that the fifth-order conversions $\omega_{xuv} = 5\omega_{uv}$ and $\omega_{xuv} = 4\omega_{uv} + \omega_L$ are not only four-photon but also (almost) three-photon resonant. The conversion $\omega_{xuv} = 3\omega_{uv} + 2\omega_L$ is five- and nearly three-photon resonant.

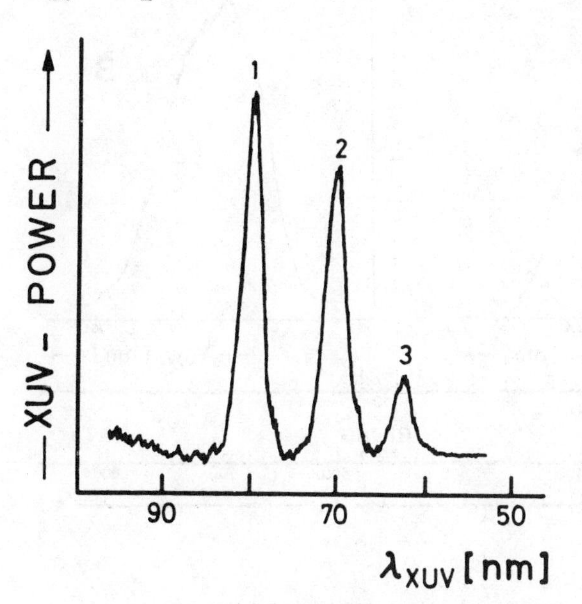

Figure 2: XUV radiation generated by resonant fifth order sum frequency mixing in Ar;
1: $\omega_{xuv}=3\omega_{uv}+2\omega_L$;
2: $\omega_{xuv}=4\omega_{uv}+\omega_L$;
3: $\omega_{xuv}=5\omega_{uv}$ with $\omega_{uv} = 2\omega_L$ and $\lambda_L=637.8$ nm.

Figure 2 shows the relative intensities of the VUV light observed at these fixed frequencies. (The width of the recorded lines is determined by the low resolution of the used 0.2 m VUV spectrometer.)

Tunable radiation is generated by $\omega_{xuv} = 4\omega_{uv} + \omega_v$. With the radiation of a second dye laser (λ_v = 216 - 800 nm) the six-wave mixing $\omega_{xuv} = 4\omega_{uv} + \omega_v$ should cover the whole range of 58 to 72 nm (Figure 3).

In the two examples displayed in this figure λ_v is tuned in the range of 550 - 580 nm (dye: Rhodamine 6G) and of 275 - 290 nm (the frequency doubled dye laser radiation). The generated XUV is tunable in the wavelength regions λ_{xuv} = 69.85 - 70.1 nm (Fig. 3 B) and λ_{xuv} = 62.0 - 62.6 nm (Fig. 3 A).

Of special interest is of course the XUV pulse power obtainable by this method. Measurements of the dependence of the power P_F of the fifth harmonic on the laser power P_L (P_L = 0.2 - 1.5 MW) confirmed that P_F is proportional to P_L^5. The value of P_F could be estimated by comparison with the known pulse power of the VUV generated by four-wave mixing processes. The measured ratio of P_F and of the tripled

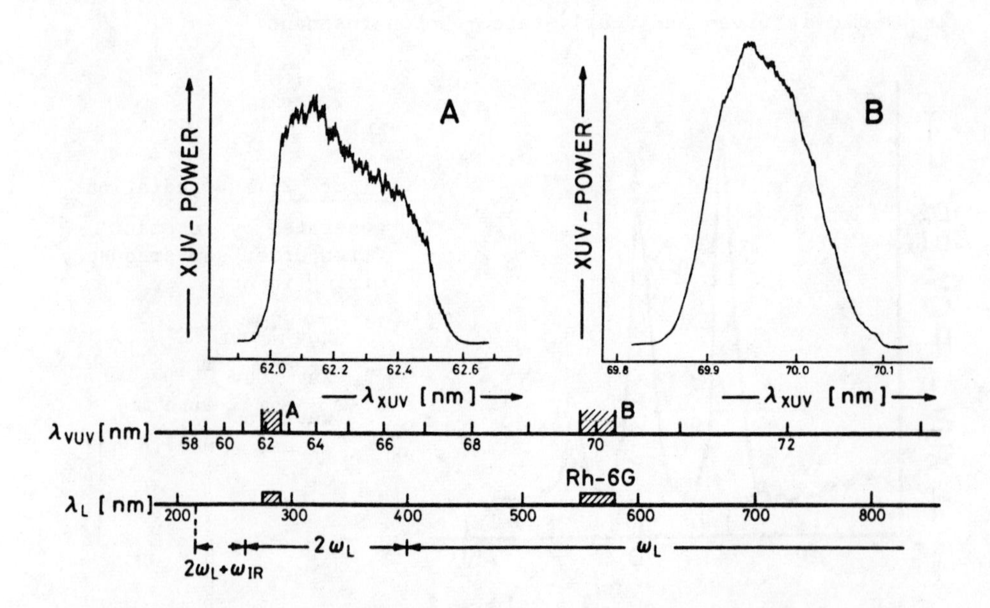

Figure 3: Tunable radiation generated by resonant six-wave mixing $\omega_{xuv} = 4\omega_{uv} + \omega_v$ in Ar; λ_{uv} = 318.9 nm, λ_v = 550 - 580 nm (B) and λ_v = 275 - 290 nm (A).

radiation P_T is about one hundred. Since for the laser pulse power P_{uv} = 1.5 MW (used in the present investigation) P_T is on the order of 10 W the power P_F should be about 0.1 W or $3\cdot10^8$ photons/pulse.

Analogous to the results in Ar six-wave mixing in Ne should generate VUV at even shorter wavelengths. In the case of Ne the UV radiation of a frequency doubled dye laser (λ_{uv} = 282.1 nm) and of the sum frequency $\omega'_{uv} = \omega_{uv} + \omega_{IR}$ (λ'_{uv} = 223.0 nm) is four-photon resonant with the 2p-6p[1/2,0] transition ($\omega(2p-6p) = 3\,\omega'_{uv} + \omega_{uv}$). The frequency $3\omega'_{uv}$ is close to ω_{res} of the resonance transition $2p\,^1S_o \rightarrow$ 3s[3/2,1]. The difference $\Delta E = 3\omega_1 - \omega_{res}$ is only 69.8 cm^{-1}. The (almost) three- and four-photon resonant conversion $\omega_{xuv} = 3\omega'_{uv} + \omega_{uv} + \omega_v$ (with λ_v = 216 - 800 nm) should thus be well suited for the production of tunable XUV in the range λ_{xuv} = 46.2 - 54.8 nm.

At present the resonant six wave mixing in Ar and Ne is subject of detailed investigations. The results obtained so far indicate that the mixing schemes described above are appropriate for the desired extension of the tuning range of the VUV light generated by nonlinear frequency up conversion.

REFERENCES

1 G.H.C. New and J.F. Ward, Phys. Rev. Lett. 19, 556 (1967)
J.F. Ward and G.H.C. New, Phys. Rev. 185, 57 (1969).
2 S.E. Harris and R.B. Miles, Appl. Phys. Lett. 19, 385 (1971)
R.B. Miles and S.E. Harris, IEEE J. Quantum Electron. QE-9, 470 (1973)
A.H. Kung, J.F. Young, G.C. Bjorklund and S.E. Harris, Phys. Rev. Lett. 29, 985 (1972).
3 W. Jamroz and B.P. Stoicheff, in Progress in Optics, E. Wold, ed., North Holland, Amsterdam 1983, vol. 20, pp. 326-380.
4 C.R. Vidal, in Tunable Lasers (I.F. Mollenauer and J.C. White, eds., Springer Verlag, Heidelberg 1984).
5 G.C. Bjorklund, IEEE J. Quantum Electron. QE-11, 287 (1975).
6 R. Mahon, T.J. McIlrath and D.W. Koopman; Appl. Phys. Lett. 33, 305 (1978).
7 D. Cotter, Optics Commun. 31, 397 (1979).
8 R. Wallenstein, Optics Commun. 33, 119 (1980).
9 R. Hilbig and R. Wallenstein, IEEE J. Quantum Electron. QE-17, 1566 (1981).
10 R. Hilbig, PhD thesis, 1984 (to be published in Appl. Phys.).
11 R. Hilbig, G. Hilber, A. Timmermann, and R. Wallenstein; "Laser Techniques in the Extreme Ultraviolet", in Proc AIP Conf. vol. 119, p. 1 (1984).

12 R. Hilbig and R. Wallenstein; Appl. Optics 21, 913 (1982).
13 R. Hilbig and R. Wallenstein; Optics Commun. 44, 283 (1983).
14 R. Hilbig, A. Lago and R. Wallenstein, Optics Commun. 49, 297 (1984).
15 H. Langer, H. Puell, and H. Röhr; Opt. Commun. 34, 137 (1980).
16 D.C. Hanna, M.A. Yuratich, and D. Cotter, Nonlinear Optics of Free Atoms and Molecules; Berlin: Springer Verlag, 1979.
17 R.T. Hodgson, P.P. Sorokin, and J.J. Wynne; Phys. Rev. Lett. 32, 343 (1974).
18 H. Scheingraber, H. Puell, and C.R. Vidal; Phys. Rev. A 18, 2585 (1978).
19 D.M. Bloom, J.T. Yardley, J.F. Young, and S.E. Harris; Appl. Phys. Lett. 427 (1974).
20 S.C. Wallace and G. Zdaziuk; Appl. Phys. Lett. 28, 449 (1976).
21 H. Junginger, H.B. Puell, H. Scheingraber, and J.R. Vidal, IEEE J. Quantum Electron. QE-16, 1132 (1980).
22 K.M. Lang, J.F. Ward, and B.J. Orr, Phys. Rev. A 9, 2440 (1974).
23 J. Heinrich and W. Behmenburg; Appl. Phys. 23, 333 (1980).
24 F.S. Tomkins and R. Mahon; Opt. Lett. 6, 179 (1981), IEEE J. Quantum Electron. QE-18, 913 (1982).
25 J. Bokor, R.R. Freeman, R.L. Panock, and J.C. White, Opt. Lett. 6, 182 (1981);
M. Jopson, R.R. Freeman, and J. Bokor; Proc. XIIth IQEC, App. Phys.. B 28, 203 (1982); see also "Laser Techniques in the Extreme Ultraviolet" in Proc. AIP Conf., vol. 90.
26 R. Hilbig and R. Wallenstein, IEEE J. Quantum Electron. QE-19, 1759 (1983).
27 W. Jamroz, P.E. La Rocque, and B.P. Stoicheff; Opt. Lett. 7, 617 (1982).
28 R. Hilbig and R. Wallenstein, IEEE J. Quantum Electron, QE-19, 194 (1983).
29 J. Bokor, P.H. Bucksbaum and R.R. Freeman; Opt. Lett. 8, 217 (1983).
30 K.D. Bonin and T.J. McIlrath, JOSA B2, 527 (1984).
31 A. Lago, PhD thesis (to be published).
32 A.H. Kung, N.A. Gershenfeld, C.T. Rettner, D.S. Bethune, E.E. Marinero, and R.N. Zare, in "Laser Techniques in the Extreme Ultraviolet", Proc. AIP Conf. vol. 119, p. 10 (1984), and references therein.
33 A. Lago, G. Hilber, R. Hilbig and R. Wallenstein, Laser u. Optoelektronik, vol. 17-4, p. 357 (1985).
34 H. Schomburg, H.F. Döbele, and B. Rückle; Appl. Phys. B20, 201 (1982).
35 J. Bokor, J. Zavelovich and C.K. Rhodes; Phys. Rev. A 21, 1453 (1980).
36 J. Reintjes, Appl. Opt. 19, 3889 (1980) and references therein.
37 J. Reintjes, L.L. Tankersley and R. Christensen; Opt. Commun. 39, 355 (1981).
38 J. Reintjes, R.C. Eckhardt, C.Y. Shen, N.E. Karangelen, R.C. Elton, and R.A. Andrews; Phys. Rev. Lett. 37, 1540 (1976).

GENERATION OF TUNABLE INTENSE COHERENT RADIATION AROUND 126NM WITH AN ARGON EXCIMER LASER

Wataru Sasaki, Yoichi Uehara and Kou Kurosawa
Department of Electronics, University of Osaka Prefecture,
Mozu-Umemachi, Sakai, Osaka 591, JAPAN

Etsuo Fujiwara, Yoshiaki Kato and Chiyoe Yamanaka
Institute of Laser Engineering, Osaka University,
Yamada-oka, Suita, Osaka 565, JAPAN

Masanobu Yamanaka
Course of Electromagnetic Energy Engineering, Osaka University,
Yamada-oka, Suita, Osaka 565, JAPAN

Junji Fujita
Institute of Plasma Physics, Nagoya University,
Chikusaku, Nagoya 464, JAPAN

ABSTRACT

Generation of intense coherent radiation with an electron beam pumped argon excimer laser is presented. A tunable oscillation was established in the wavelength range from 124nm to 127.5nm, and the maximum output power of 2.2MW was obtained at the line center. The spectral region is extended using stimulated Raman scattering in hydrogen molecules. The conversion efficiency is as high as about 50% in 8atm hydrogen gas.

1. INTRODUCTION

Recently a great deal of attention has been paid to a tunable intense coherent radiation in the VUV region of the spectrum with respect to applications to a diagnostics for fusion plasmas and a semiconductor processing as well as a linear and a nonlinear laser spectroscopy. To obtaine the tunable coherent radiation in the VUV region there are two ways, direct laser oscillations and indirect generation. Rare gas excimer lasers are one of the attractive candidates of the intense VUV source partly because of potentiality to the high power operation and partly because of considerable wide range of tunable oscillation. Among rare gsa excimer lasers an argon excimer laser has oscillated the shortest wavelength at 126nm.[1] The high power operation in argon excimer was not established when we started to develop them. The most serious problem faced to obtaining a high power oscillation of argon excimer laser was an optical damage of the cavity mirror, aluminum thin film mirror with MgF_2 coating.[2] To overcome this problem we have used an uncoated quartz plate or an uncoated silicon single crystal as one of cavity mirrors and an uncoated MgF_2 plate as an output mirror.[3] These cavities have fairy high optical loss so that we have to use intense electron beam pumping to overcome the loss.[4]

In this paper, we have reported that a tunable high power operation was obtained around 126nm based on an electron beam pumped argon excimer laser. When an intense electron beam pumping with long pulse duration was employed, high power tunable oscillation was established over the range from 124nm to 127.5nm with a spectral width of 0.3nm.[5] The output power was 2.2MW at the line center. A stimulated Raman scattering(SRS) in molecular hydrogen gas was successfully used to expand the spectral region. A very high conversion efficiency up to 50% and a low threshold power for the SRS were established for the first Stokes line in hydrogen gas at a pressure of 8atm. In addition to the argon excimer laser a potentiality of generation of intense VUV radiation with other rare gas excimer lasers is also discussed and the predicted spectral region are described.

2. EXPERIMENTAL APPARATUS

An experimental setup is illustrated in Fig.1. An electron beam device used in this experiment was composed of a Marx generator(700kV, 5kJ), a coaxial Blumlein-type pulse forming line and a coaxial diode(Fig.1(a)). In Fig.1(b) is shown the laser part. The laser diode consisted of a 40-cm-long cathode and a concentric anode pipe. The stainless steel anode pipe was 5mm in diameter and 70μm thick. The voltage applied to the diode was 700kV with 90nsec duration(FWHM). The electron dose distribution along the anode pipe measured with Radcolor film(Nitto Electric Industrial Company No.381) indicated that the energy deposition was very homogeneous over 38cm along the anode pipe. The ends of the anode pipe opened into high-pressure chambers, where the optical elements of the laser cavity were placed. The cavity length was approximately 80cm. An uncoated MgF_2 plate of 6mm thickness as an output mirror with 60% transmittance and 5% reflectivity at 126nm and a MgF_2 prism backed by a quartz plate as a dispersive element were used to compose the laser cavity. The MgF_2 prism was found to be highly resistive to an optical damage with intense VUV radiation. Since the dispersion of refractive index of MgF_2 became very large in this spectral region, the apex angle of the prism of 16° was enough to tune the oscillation wavelength of the argon excimer laser with narrow spectral width. It should be noted that the MgF_2 prism having such advantages for the VUV laser was used in this experiment instead of a diffraction grating.

For the stimulated Raman scattering experiments the output laser beam was focused into a Raman cell of 1m length with a 50cm

MgF_2 lens. Before hydrogen gas was filled in the Raman cell, it was evacuated with a vacuum pump and then purged several times by high purity helium gas. The system for measuring the laser characteristics was similar to that reported previously. The spectra were observed using a Seya-Namioka-type 0.5m sptectrometer with a 1200-line/mm grating, which has a spectral resolution of 0.15nm in 100μm slit width. For wavelength calibration, a capillary discharge Kr tube, emitting atomic lines at 116nm and 123nm, was used. The output beam from the Raman cell was divided with a MgF_2 beam splitter. The transmitted beam was introduced to a calibrated phototube(HTV R1328U-04) to measure the intensity and the waveform, and the reflected beam was detected with a photomultiplier tube after passing through the spectrometer. The transmittance of the spectrometer was measured to be ~0.1% at 126nm. There might be a small difference between the transmittance at 126nm and that at the first Stokes line, but we neglected it. When the spectrometer was used as

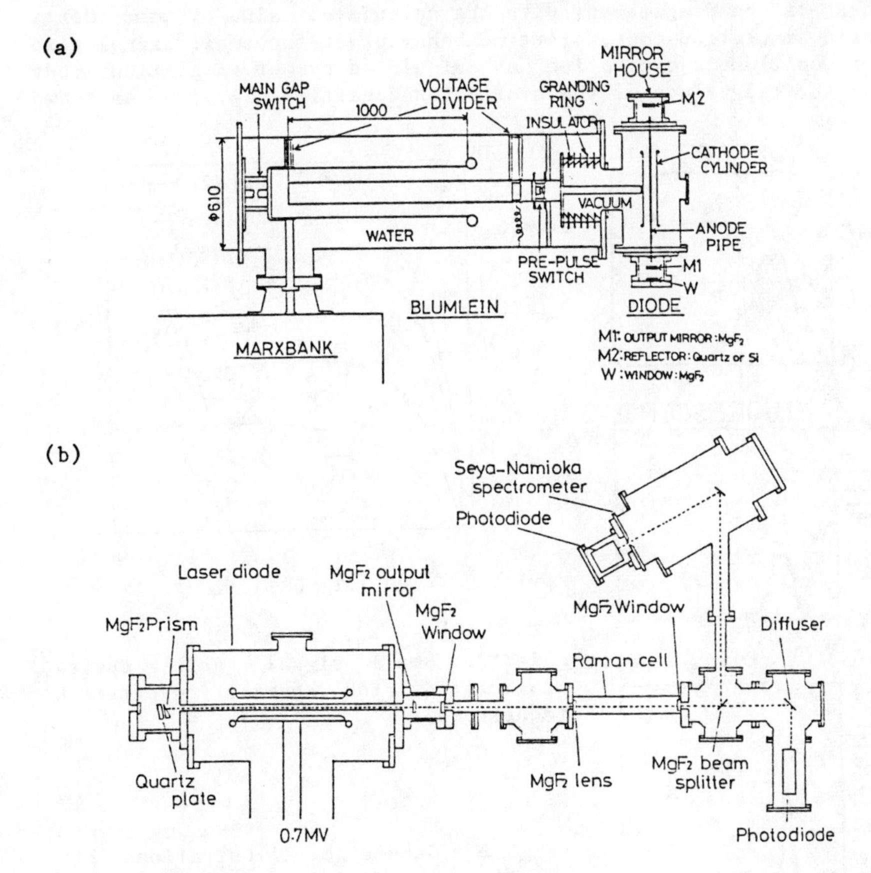

Fig. 1 Schematic drawing of experimental setup: electron beam device (a) and laser part (b).

a spectrograph, the photomultiplier tube was replaced by a Kodak SWR film. All the optical paths in the detection system were in vacuo to avoid the absorption of the laser radiation by air.

3. RESULTS AND DISCUSSION

Figure 2 shows measured waveforms of the diode voltage, the fluorescence at 126nm and the laser pulse. According to our Monte Carlo code for simulating behavior of electrons through the anode pipe, the argon gas is excited by electrons having a kinetic energy greater than 400keV. As shown in the figure it takes about 20nsec to build up the population inversion for starting laser oscillation. It is found that the duration of effective excitation is 60nsec in which the diode voltage exceeds 400kV. Such a long pulse excitation, therefore, has advantage to overcome the high cavity loss owing to the low reflectivity mirrors. Total energy deposited to the argon gas was measured to be 30J, which corresponded to excitation density of 62.5MW/cm^3. This is good agreement with the calculated value by the Monte Carlo simulation code. A time behavior of the small signal gain for continuous excitation was calculated by our simulation code for several values of the excitation densities(Fig.3). As shown

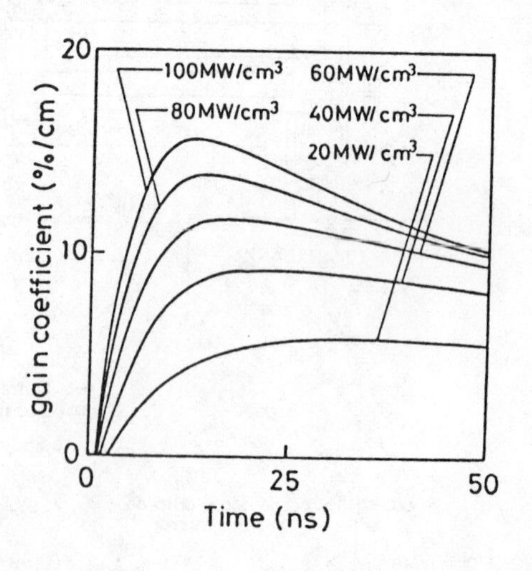

Fig. 3 Small signal gain curves calculated for several excitation densities.

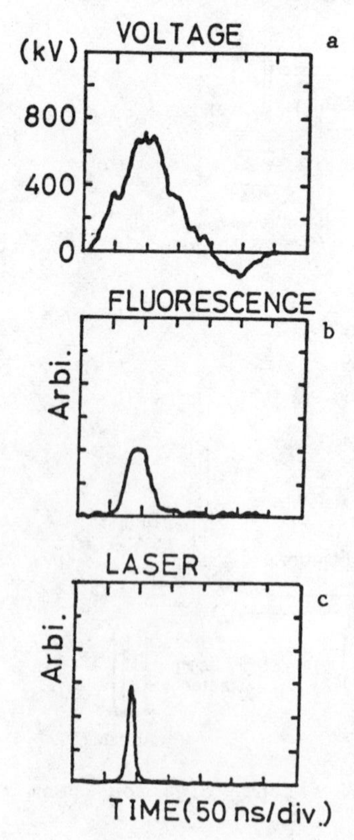

Fig. 2 Schematic illustrations of waveforms of diode voltage (a), fluorescence (b), and laser pulse (c).

Table I. Output power of argon excimer lasers with various cavities.

Reflector		Output Mirror			Output Power without Damage(MW)
Material	Reflectivity at 126nm (%)	Material	Reflectivity (%)	Transmittance (%)	
Al/MgF_2 Thin film	80	MgF_2	5	60	<1.5
Synthesized Quartz	14	MgF_2	5	60	>2.4
Si single Crystal	20	MgF_2	5	60	>3.4
MgF_2 Prism +Syn.Quartz	8.5*	MgF_2	5	60	>2.1

* See the text.

in Fig.3 the rise time of the gain coefficient about 15nsec for the excitation density of $60MW/cm^3$ is comparable to the measured value.

Table I shows the comparision of output power for various kinds of cavity compositions. As shown in the Table I, Al/MgF_2 mirror was damaged at the output power near 1.5MW. The estimated damage threshold was $100mJ/cm^3$, which was considerablly smaller than that in the visible spectral region. On the contrary an uncoated quartz mirror and a Si mirror worked very well with a stable high power operation of 2.2MW and 3.4MW, respectively, without any damage. According to the theoretical calculation from the refractive index n and the absorption coefficient k, the reflectivity of quartz and Si are 20% and 52%, respectively. Whereas the measured values of quartz and Si were 14% and 20%, respectively. It seems that the differences between theoretical values and the experimental ones are due to the scattering loss from the surfaces of the mirrors. These facts show that the development of high resistive mirror to the optical damage with high reflectivity is strongly required for the further high power operation of the argon excimer laser. One of the candidates will be a multilayer dielectric mirror with CaF_2 as a material for high reflective index and MgF_2 as a low one, though it has not been developed yet.

Tuned oscillation was obtained from 124nm to 127.5nm with the spectral width of 0.3nm, as shown in Fig.4. It should be noted that the output power was close to that of the untuned oscillation near the line center, in spite of the high loss cavity. We made the laser beam take the short optical path in the prism, but the rather high absorption (about 20%) of the prism was not avoidable in this spectral region (about 50% of transmittance). It seemed that the intense excitation compensated the additional optical loss.

A wavelength dependence of the optical gain is shown in Fig.5. The maximum gain exceeded 10%/cm at the line center of 126nm. In general an oscillation condition is described by

$$g > \frac{1}{L_{exc}} \ln\left(1/(R_1 \cdot R_2)\right) \qquad (1)$$

where g is the gain coefficient, L_{exc} is the excitation length, and R_1 and R_2 are the reflectivity of the cavity mirrors. In our experimental setup, $R_1=0.05$(for the output mirror) and the effective reflectivity (R_2) of dispersive elements consited of the MgF_2 prism and quartz mirror is 0.035, in which the absorption of MgF_2 prism is taken into account. The right hand side of Eq.(1) becomes $7.9\times 10^{-2}\ cm^{-1}$. The range in which the gain exceeds this value is indicated by the dashed line in Fig.5. It is nearly equal to the tuning range in Fig. 4. The gain exceeds 5% at 122nm so that if we could use a high reflectivity mirror, the direct laser oscillation at the Lyman α line would be posible.

Taking advantage of the high power output in this spectral region, we can expand the wavelength region of VUV spectra by a stimulated Raman scattering. H_2 gas and D_2 gas are good Raman media in this spectral region. A densitometer trace of Raman spectra recorded by a SWR film shown in Fig.6 at the 8atm of hydrogen gas in 1m long Raman cell. The output beam of the argon excimer laser with 1mrad beam divergence was focused into the Raman cell with the MgF_2 lens of 50cm focal length. The laser power was 450kW in the Raman cell and the power density at the focal point was estimated to be $225MW/cm^3$. Intence first Stokes(133.6nm) and second Stokes lines(141.4nm) are observed for the fundamental pump light at 126.5nm. The lines at 116.5nm and 123.6nm are the spectra from krypton discharge lamp for calibration. The frequency shift of the first Stokes line agrees with the reported frequency, $4160cm^{-1}$, of the first vibrational state in the ground state of hydrogen molecules. Peak powers of Raman line and fundamental line were calibrated by the photomultiplier and the phototube as mentioned preceeding section. In Fig.6 no anti-Stokes line was observed because the dispersion of refractive index of hydrogen molecules gave rise to the mismatch of the phase velocity of Stokes wave and anti-Stokes wave in the Raman medium. A detailed discussion about the anti-Stokes scattering will appear in this proceeding. The peak power of the first Stokes line is shown in Fig.7(a) in terms of

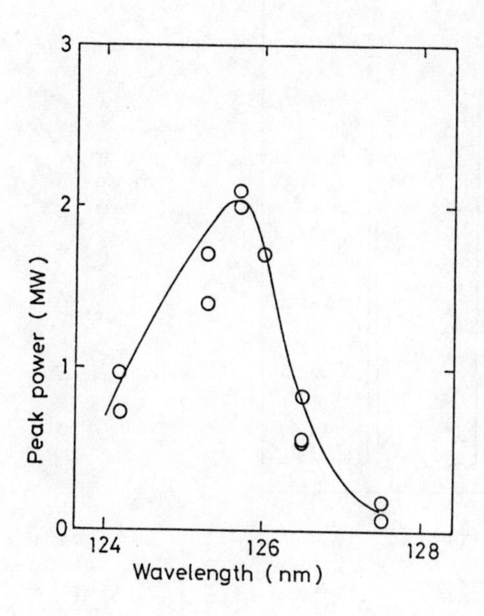

Fig. 4 Peak power of laser output. The argon pressure is 40atm.

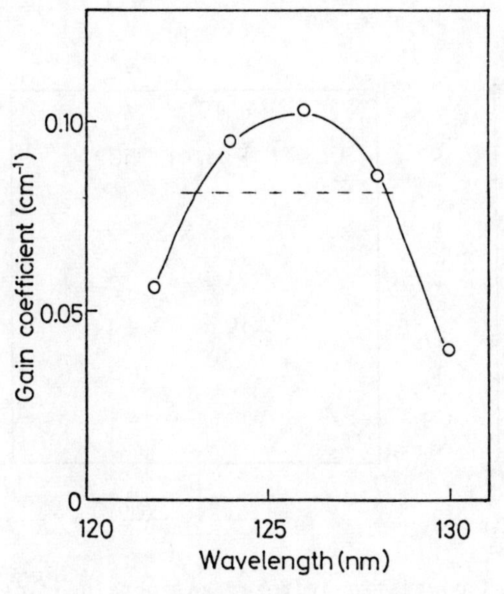

Fig. 5 Laser gain coefficient. The argon pressure is 40atm.

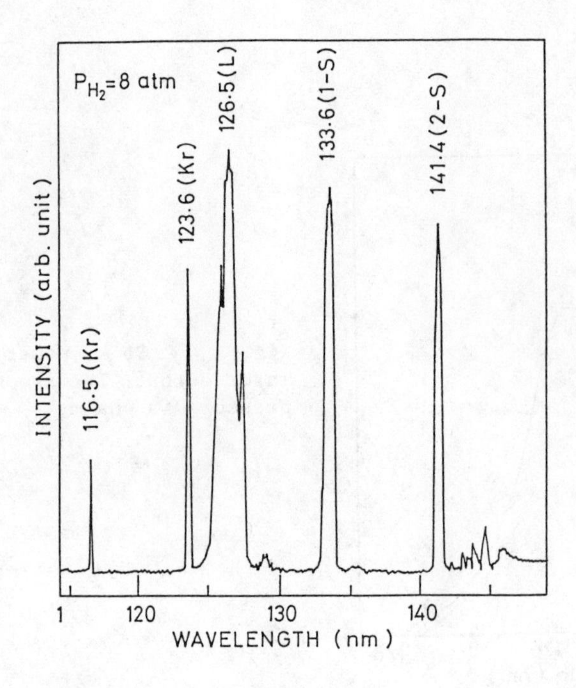

Fig. 6 Typical spectrum of stimulated Raman scattering in hydrogen gas at 8atm.

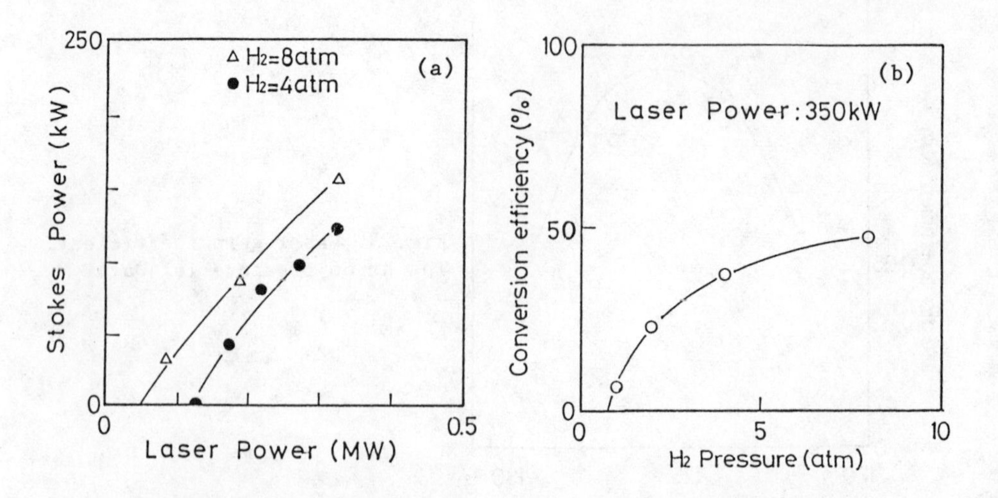

Fig. 7 Stimulated Stokes peak powers in hydrogen gas obtained with various pumping powers (a) and conversion efficiencies of the first Stokes lines for various hydrogen gas pressures (b).

the input peak power for 4atm and 8atm of hydrogen pressure. It is increased linearly with the input peak power. The threshold power of stimulated Raman scattering for the first Stokes was found to be 100kW at 8atm. This value corresponds to the power density of $50MW/cm^3$ at the focal point.

In Fig.7(b) measured conversion efficiency to the first Stokes line was illustrated as a function of hydrogen pressure. The conversion efficiency tends to saturate with increasing the gas pressure.

The high conversion efficiency and the low threshold power makes the SRS to be a great advantage for the generation of coherent VUV radiation. As mentioned above the threshold power of SRS is about 100kW so that frequency conversion might be possible over the entire wavelength region of the tuned oscillation of the argon excimer laser, since the output power of the tuned oscillation is exceeds 100kW over almost whole wavelength region obtained as shown in Fig.4. Although we tried to the stimulated Raman scattering in deuterium gas, no SRS was observed with input power of few handred kilowatts. Since the cross section of Raman scattering of deuterium molecule is a half of hydrogen molecule, more input power seems to be required.

There are three excimer lasers, Xe excimer laser with 172nm, Kr excimer laser with 145nm and Ar excimer laser with 126nm. We can obtain the oscillation at 145nm and 172nm with changing argon gas to krypton or xenon gases. Using these excimer lasers combined with Raman shifter we might obtain a considerablly wide range of the VUV spectra as shown in Fig.8. The stopping power of the krypton is in between the values of argon and xenon. The oscillation wavelength of the krypton excimer laser is longer than that of argon excimer laser, so that the situation of optical elements is better than that for the argon excimer laser. Although the krypton excimer laser has not been studied well, it is prusible to estimate the output power level between that of the argon excimer laser and the xenon excimer laser.

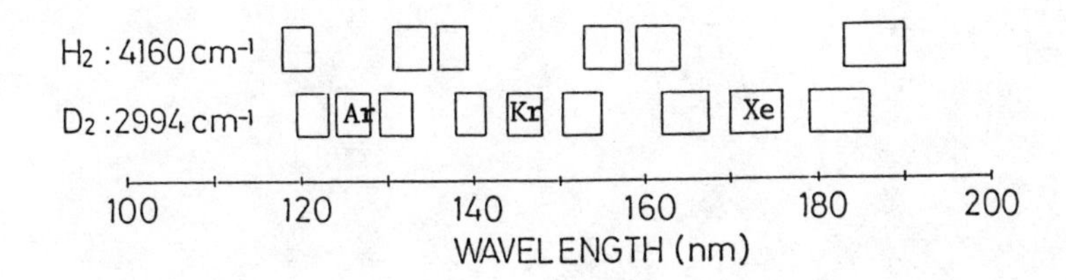

Fig. 8 Pictorial illustration of wavelength extension obtained by stimulated Raman scattering in H_2 and D_2 gases pumped with three different kinds of excimer lasers.

CONCLUSIONS

High power tunable oscillation of the electron beam pumped argon excimer laser was obtained with the spectral range between 124nm and 127.5nm. Use of MgF_2 prism and quartz mirror made it stable operation of the order of several MW without any mirror damage.

The frquency conversion by stimulated Raman scattering was obtained for the first time in this spectral region. It has been found that the stimulated Raman scattering is a very useful technique for converting the wavelength in the VUV region of spectrum because of high conversion efficiency.

This research was partly supported by a grant-in-aid for a special research project of nuclear fusion from the Ministry of Education, Science and Culture of Japan.

REFERENCES

1.W.H.Hughes, J.Shannon and R.Hunter, Appl. Phys. Lett. 24, 488(1974).
2.W.-G.Wrobel, Rep.IPP 1/186, Max Planck Institut fur Plasmaphysik, Euratom Association, D-8046 Garching, Federal Republic of Germany(1981, unpublished).
3.Y.Uehara, W.Sasaki, S.Saito, E.Fujiwara, Y.Kato, M.Yamanaka, K.Tsuchida and J.Fujita, Opt.Lett. 9, 539(1984).
4.Y.Uehara, et al., submitted to IEEE.
5.Y.Uehara, W.Sasaki, S.Kasai, S.Saito, E.Fujiwara, Y.Kato, C.Yamanaka, M.Yamanaka, K.Tsuchida and J.Fujita, Opt. Lett. 10, 487(1985).

Time-resolved vuv photoemission spectroscopy

P. H. Bucksbaum, J. Bokor, R. Haight[a], R. R. Freeman
AT&T Bell Laboratories
Murray Hill, NJ 07974 and Holmdel, NJ 07733

ABSTRACT

Time- and angle-resolved vuv photoemission spectroscopy is an important application of laser-produced vuv radiation. This paper reviews the techniques used to produced time-resolved photoemission spectra, and discusses the physical problems which can be explored by this new technique. Two experiments are reviewed in detail: spectroscopy of the photoexcited InP (110) surface; and electron dynamics on the cleaved Si (111) 2x1 reconstruction.

INTRODUCTION

Angle-resolved vuv photoemission spectroscopy (ARUPS) is a well established technique for studying the electronic structure of condensed matter systems and their surfaces[1]. The conventional sources of vuv photons in ARUPS studies are resonance lamps or synchrotrons. Picosecond lasers and nonlinear optics technology can provide laser-based ultraviolet radiation sources capable of producing ultrashort pulses up to 35 eV.[2] These sources provide time-averaged total photon flux comparable to synchrotron bending magnet beam lines, with substantially higher peak spectral brightness per pulse. Also, because the vuv pulses can be precisely synchronized to powerful short pulsed lasers operating in the near visible wavelength region, laser-based sources for ARUPS offer the opportunity to study transient excitations of solids and surfaces. This paper reviews the new field of time-resolved ARUPS. The requirements of the vuv source are discussed, with particular attention to systems which have already been used to produce time-resolved spectra. New physical problems which may be addressed by this technique are reviewed, and two examples are explored in detail: spectroscopy of the photoexcited InP (110) surface; and electron dynamics on the cleaved Si (111) 2x1 reconstruction.

Conventional ARUPS can be used to determine the occupied bands in a solid or solid-vacuum interface. An incident photon with frequency ν promotes an electron in the solid to an excited state. If the state lies above the vacuum level, the electron may travel out of the surface into the vacuum, where it appears with kinetic energy equal to $h\nu$ minus the binding energy. A measurement of the kinetic energy and vector momenta of electrons photoemitted from a crystal surface by monochromatic ultraviolet radiation can be used to determine the binding energy and crystal momenta of those electrons before the photon absorption event. Energy versus momentum band dispersions can be obtained in this way for normally occupied bands, i.e. those lying below the Fermi level. Still more information comes from examining the electron intensity, normalized to the incident photon flux, as a function of photon energy and polarization. The intensity is affected by the density and location of the final states, as well as relative symmetries of the initial and final states.

0. To be published in the Proceedings of the Third Topical Meeting on Short Wavelength Coherent Radiation: Generation and Applications; Monterey, CA, March 24-27, 1986.

In time resolved ARUPS, the surface is prepared prior to photoemission by a pulse of visible or infrared "pump" radiation. The pump promotes some of the electrons into excited states which are not normally occupied. Photoemission from these states then may be observed with a pulse of vuv "probe" radiation, as shown in figure 1. Synchronization of the pump and probe pulses is achieved by means of a variable path length optical delay. Time resolution is comparable to the pulsewidth of the laser, and transient effects may be followed over many nanoseconds, limited only by available table space for the delay line optics.

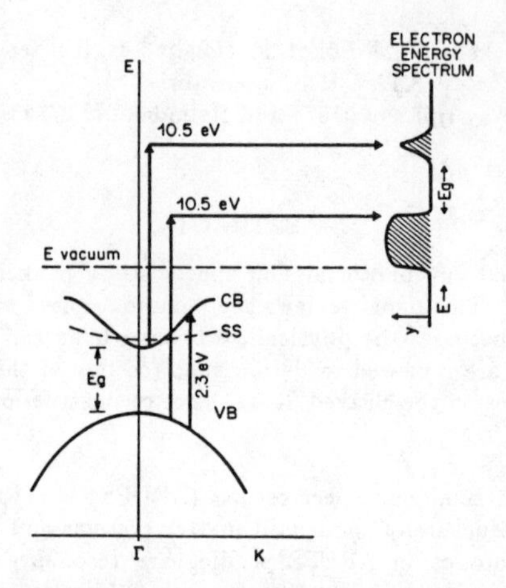

Figure 1: Schematic of a typical time-resolved ARUPS experiment, showing the excitation (2.3 eV pump in this example) and subsequent photoemission (10.5 eV probe) from a model semiconductor. A normally unoccupied surface state is shown as dashed curve.

The spectroscopy of the normally unoccupied states is one of the most direct applications of this new technique. In addition, the high density of excited states in the solid or at the interface may actually alter the electronic band structure during excitation. A simple example is the potential energy shift of the entire spectrum due to the photovoltage effect, which occurs when a dipole depletion or accumulation layer at the surface is removed by the diffusion of photoexcited free carriers away from the interface. The lattice may also be altered during photoexcitation, by heating, for example, thereby shifting the positions of the electron energy bands. If enough energy is transferred to the lattice, ultimately phase transitions will occur, and these may also be studied using photoemission.

Perhaps the most important application of time resolved ARUPS is the study of carrier dynamics. Through choice of pump laser frequency and polarization, it is possible to selectively populate a particular region of an excited band. The system then returns to the ground state through inter- and intraband relaxation processes, often involving transitions between surface and bulk energy bands or diffusion into the bulk crystal. Every step of the relaxation process may be followed using a short photoemission probe to take a "snapshot" of the electronic structure before, during, and after excitation.

APPARATUS

Time-resolved ARUPS experiments require a short pulsed laser; the means of converting laser light to the pump and probe wavelengths; a variable optical delay between the pump and probe; and an ultrahigh vacuum chamber, containing a port for the vuv radiation to enter, another window for the pump radiation, an electron spectrometer, and the sample under investigation. In addition the chamber must have the normal surface preparation and characterization instruments, such as LEED, Auger analysis, and surface cleaning equipment.

The laser used in the experiments described below is an actively mode-locked and Q-switched Nd:YAG operating at 1064 nm. This produces a train of about 20 pulses of 80 picosecond duration, separated by 12 nanoseconds. The laser operates at a repitition rate of 10 Hz. A single pulse is switched out of the train with a pockels cell and amplified to 50-100 mJ in external Nd:YAG amplifiers. It can then be converted easily into a number of wavelengths for the vuv probe pulse, and the visible, infrared or ultraviolet pump pulse, through nonlinear harmonic generation in crystals and gases, or stimulated Raman scattering. The use of a single laser for both the pump and the probe is crucial: this is the single best way to insure that the two light pulses incident on the sample are precisely timed with respect to each other.

Ideally, one would like to have as short a pulse as possible, since some electronic relaxation phenomena occur on femtosecond timescales. Rhodes at this meeting has reported producing harmonics of a sub-picosecond excimer laser in the vacuum ultraviolet, and this is a possible direction in future source development[3]. It should be pointed out, however, that the short pulse excimer laser technology is really designed for high power applications, and as such, is not well matched to the requirements of time-resolved photoemission. This is because all ARUPS experiments are limited in the amount of peak light intensity they can use by space charge. In a time of flight spectrometer which utilizes a field free drift region, the maximum practical photon fluence which may be used without observing space charge broadening of the spectrum is approximately 10^6 vuv photons per laser shot, focused onto a 0.1 cm diameter spot on the sample surface. A more promising direction may be high repitition rate amplification of dye lasers which operate in the 30-200 femtosecond range.

In the work performed to date, the vuv wavelength is 118.2 nm, the ninth harmonic of the Nd:YAG frequency. Vuv conversion is accomplished in two steps. First, the 1064 nm radiation is tripled in KD*P, producing typically 3 mJ of 355 nm light. This is then focused to a waist at the object slit of a 0.3 m vacuum monochrometer (MacPherson model 218) which has been back-filled with 5 - 7 Torr of xenon. This is well below the optimum density for phase matched third harmonic conversion; the xenon density is adjusted to limit the vuv flux on the sample to $\sim 10^6$ photons per pulse to avoid distortions in the energy and angular distributions of the photoemitted electrons arising from space charge effects. The monochromator filters out the unwanted 355 nm fundamental, and the 118.2 nm probe beam is refocused through the monochromator exit slit (see Fig. 2). The waist which appears at the image plane is then refocused again onto the sample, with an ellipsoidal mirror used at 70° incidence angle. The sample chamber is isolated from the Xe filled beam line with a LiF window. For shorter wavelengths, windows can no longer be used, and linear absorption in the nonlinear medium becomes a serious problem. In this case, a pulsed gas jet is used to produce a "windowless cell" of nonlinear medium. The light is refocused into the chamber with a dichroic mirror, which transmits most of the fundamental wavelength, and reflects much of the vuv[4]. A thin foil of indium or aluminum serves as a final light block for scattered fundamental, as well as a contamination shield between the source region and the ultra-high vacuum chamber.

The ultrahigh vacuum chamber maintains a base pressure of 5×10^{-11} torr and includes low energy electron diffraction (LEED) and a retarding field Auger electron spectroscopy system. Sample surfaces are prepared by cleaving oriented single crystals *in situ*. Photoelectron spectra are measured by a time-of-flight spectrometer, consisting of a mu-metal and copper magnetically and electrically shielded 24 cm drift tube. Electrons are detected by a pair of microchannel plates with an angular acceptance of $\pm 2.5°$. The signals are digitized using a 500 MHz Tektronix 7912AD transient digitizer interfaced to a laboratory computer. Single photoelectrons produce nearly symmetric pulses of 1.5 nanoseconds full width at half maximum. On every laser shot, the computer system determines the flight time of each of the 5 to 10 arriving photoelectrons on each laser pulse. The energy resolution of the system is 105 meV at an electron energy of 6.5 eV.

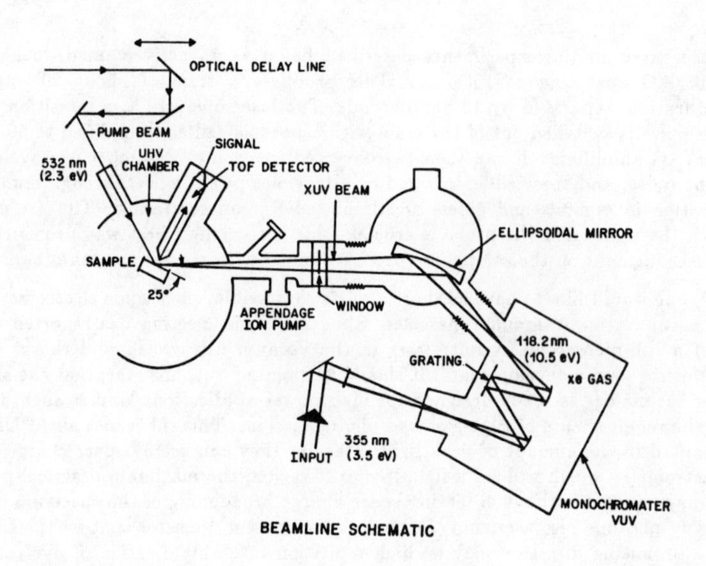

Figure 2: Schematic of the beamline, UHV photoemission chamber and apparatus used to generate the vuv probe light. The pump and probe beams shown were used in the InP experiment described below.

SURFACE SPECTROSCOPY OF INDIUM PHOSPHIDE (110)

The first demonstration of time-resolved ARUPS was a study of electron dynamics in the picosecond time domain on photoexcited InP(110) surfaces[5]. Transient population in a normally unoccupied surface state was observed in the photoemission spectra following excitation of a bulk carrier plasma by 532 nm 50 picosecond pump laser pulses.

On the cleaved (110) surface of all the III-V semiconductors (except GaP), the bandgap is free of intrinsic surface states.[6] A normally unoccupied surface resonance lies at an energy slightly above the conduction band minimum (CBM). If carriers are excited into the bulk conduction band, this surface resonance becomes populated. Using time-resolved ARUPS, we can directly observe the transient population of electrons in this surface resonance, determine the position of its energy minimum in momentum space, and measure the surface band dispersion.

The pump laser employed to excite the InP was the second harmonic of Nd:YAG at 532 nm. The fluence used was 0.5 mJ/cm^2, well below the melting threshold for InP which is of the order[6] of 100 mJ/cm^2. Figure 3 shows a representative set of photoemission spectra taken with p-polarized probe radiation and the detector aligned along the <110> direction, i.e., normal to the sample surface. Figure 3(a) displays a spectrum for the unexcited surface in which the probe pulses actually arrived at the sample 133 psec before the pump pulses (t = -133 psec). The energy zero is chosen as our best determination (±50meV) of the valence band maximum (VBM). The spectrum shown in Fig. 3(b) was taken with the pump and probe pulses temporally overlapped (t = 0) and clearly shows a new feature centered at +1.47 eV. When the probe was delayed by 266 psec the spectrum in Fig. 3(c) was obtained. Here the 1.47 eV peak is significantly reduced in intensity as well as width. This can be more clearly seen in the insets of Fig. 3(b) and (c) which show a magnified view of this feature.

Figure 3(d) displays a spectrum obtained at t = 0 after the surface has been exposed to atomic hydrogen. The disappearance of the new feature upon hydrogen exposure was observed for both n- and p- type samples, and is evidence for its assignment as a surface state. The peak is believed to be the C_3 unoccupied surface state, which is predicted but has never been directly

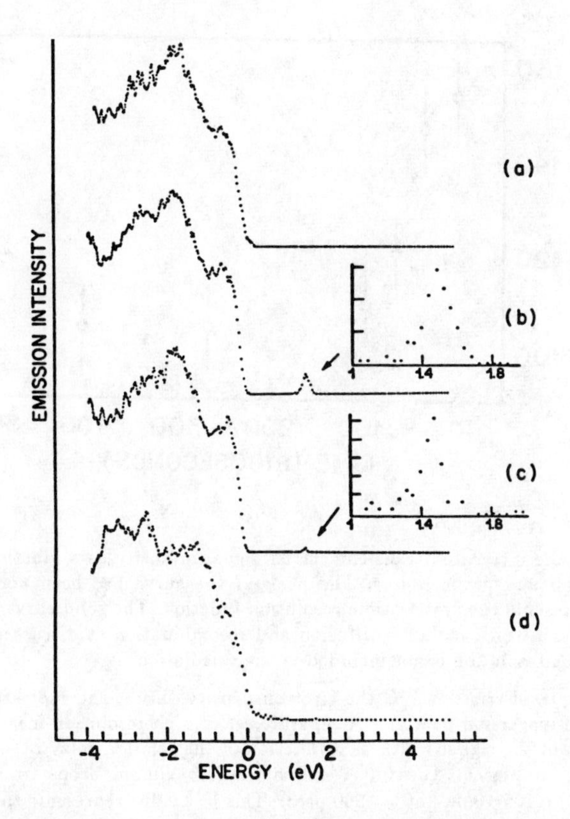

Figure 3: Pump-probe ARUPS spectra from InP (110) for time delays of: a) t=-133 psec, b) t=0, c) t=266 psec and d) t=0 where the surface has been exposed to hydrogen. The insets in b) and c) are magnified views of the 1.5 eV region where the vertical scale has been expanded by a factor of 4.7 and 28 respectively.

observed previously. We were able to determine the energy position of the bottom of the C_3 band to be 120±50 meV above the CBM. Our direct measurement of this value is in good agreement with that inferred by van Laar, *et al.*[7] based on measurements of the "surface exciton" state associated with the indium 4d core level.

Since the C_3 state lies within the bulk conduction band strong coupling between the surface and bulk bands may be expected. In particular, the surface band becomes occupied in the presence of a degenerate bulk electron population. Evidence for this should be seen in the dependence of the integrated C_3 peak intensity on the relative time delay, t, between the pump and probe pulses as shown in Fig. 4. This should reflect the time dependence of the bulk carrier density near the surface. A numerical calculation of the spatial and temporal evolution of the bulk plasma density supports this hypothesis. The dominant decay mechanism at the surface for times up to approximately 1 nsec is diffusion of electrons into the bulk due to the large density gradient associated with the short absorption depth of the 532 nm pump radiation (100nm). The fact that the decay curve of the C_3 surface state peak follows what we expect for the bulk carrier density at the surface leads to the conclusion that the C_3 surface state population is equilibrated with the bulk.

The insets of Fig. 3 show that the energy width of the C_3 peak varies with time. The peak is observed to narrow from the high energy side. This behavior is interpreted as variation of the

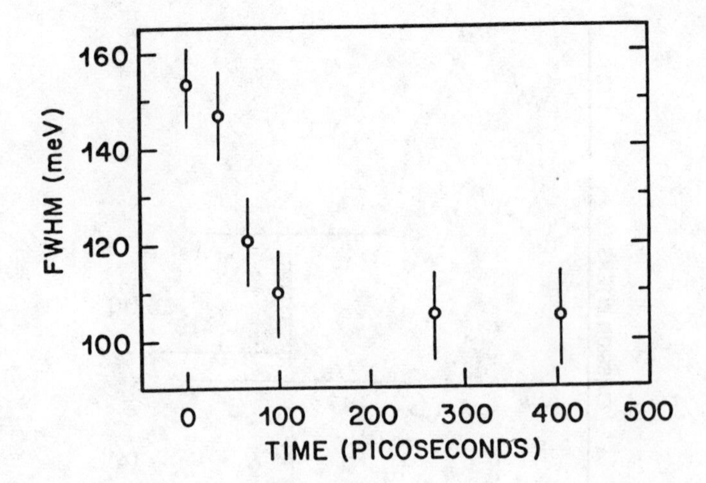

Figure 4: Plot of the integrated transient carrier signal intensity as a function of time delay, t, between the pump and probe pulses. The peak of the curve has been normalized to 1. The dashed curve represents the system time resolution function. The solid curve is the result of the 1-dimensional model which includes diffusion and recombination as discussed in the text. The effect of finite time resolution is not included in the calculation.

surface electron quasi-Fermi level as the electron density varies, and indicates that the surface band has positive dispersion near $\bar{\Gamma}$. A separate set of high resolution measurements designed to follow changes of the signal width as a function of the relative delay between the pump and probe are shown in Fig. 5. The full width at half maximum drops from its maximum of 154±7meV at t=0 to 105±10meV at t= 260 psec. This last value represents the energy resolution of the electron spectrometer.

The dependence of the C_3 peak as a function of electron emission angle was studied to obtain information on dispersion and the momentum dynamics of the excited electrons. Dispersion can be observed directly in Fig. 6, where the energy distribution taken at $\theta = 0°$ is compared with that taken at $\theta = 6°$ along the $\overline{\Gamma X}$ direction. The centroid of the peak shifts to higher energies with increasing angle, consistent with a positively dispersing band. These energy and angular distribution measurements can be used to determine the energy versus $k_{\|}$ dispersion relation for the C_3 surface state. We obtain $0.20 \leq m^*/m_e \leq 0.24$ along $\overline{\Gamma X}$, and $0.10 \leq m^*/m_e \leq 0.15$ along $\overline{\Gamma X'}$, where m_e is the free electron mass. The anisotropic effective mass values differ significantly from the nearly isotropic bulk conduction band effective mass of 0.07 m_e, providing additional confirmation of our identification of this peak as a surface state.

SILICON (111) 2X1 SURFACE ELECTRON DYNAMICS

Electron diffraction from a (111) silicon surface cleaved in ultrahigh vacuum exhibits a 2x1 pattern: that is, the surface appears to reconstruct to form a unit cell with unequal sides in the ratio 2 to 1. This surface reconstruction is unusual in that it is metastable, returning to the thermodynamically favored 7x7 reconstruction after heating. A strong infrared absorption band localized on the surface of the silicon is known to accompany the reconstruction[8]. This absorption is polarization dependent, occuring only for light which is polarized in the surface direction orthogonal to the longer periodicity. In recently completed time-resolved ARUPS experiments, this absorption has been shown to be due to excitation between occupied and empty surface bands which have been seen directly in the transient electron spectra[9]. The decay history of the

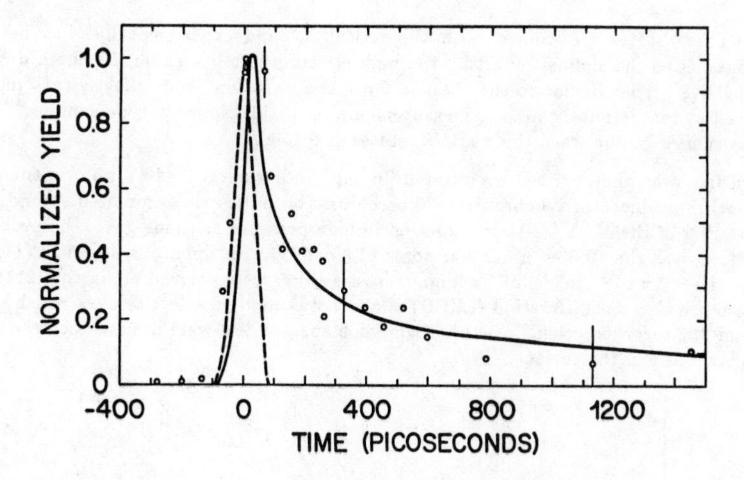

Figure 5: Transient carrier signal full width at half maximum (FWHM) as a function of the delay between the pump and probe pulses. The dashed curve represents the system time resolution function.

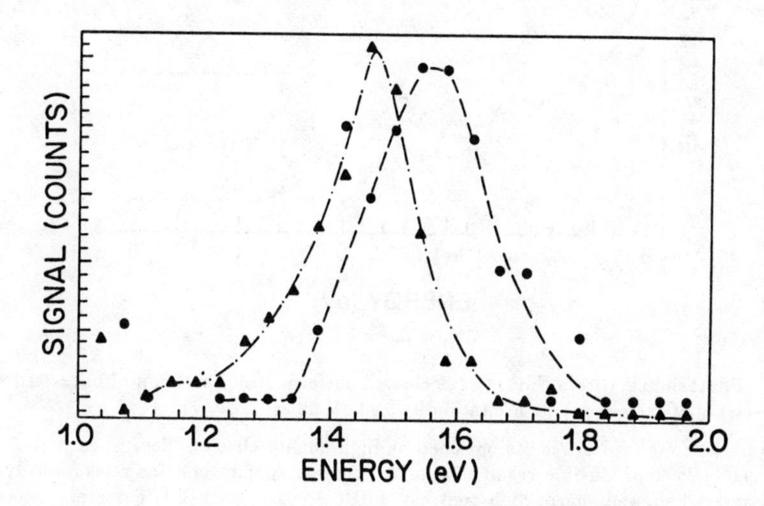

Figure 6: Comparison of the transient carrier signal at t=0 for photoelectrons measured at an exit angle of 0° (filled triangles) and 6° (filled circles) along the $\overline{\Gamma X}$ direction respectively. The dashed curves are to guide the eye.

excited surface state shows the effects of recombination and interdiffusion between surface and bulk states. The decay rates were found to be cleavage dependent, strongly implicating the role of steps and/or defects in the relaxation dynamics.

A wide variety of experiments aimed at elucidating the surface crystallography and electronic structure of the 2x1 surface support the π-bonded chain model originally proposed by Pandey.[10] In this model, the surface reconstructs by forming long chains along the 110 direction. Electrons move along the chains, forming an occupied band and an empty band with a gap inbetween. These states are known both theoretically[11] and experimentally[12] to be highly dispersive parallel to the chains, reflecting the high electron mobility along the chain direction, and nondispersive perpendicular to the chains. Optical absorption[8] and reflectivity[13] measurements, as well as recent photoemission measurements on highly doped n-type samples[14] give a value for the energy gap between the π and π^* states as 0.45 eV.

The surface was photoexcited by infrared "pump" radiation at 0.44 eV photon energy (2.8 μm wavelength), produced in synchronized 80 psec duration pulses via stimulated second Stokes Raman scattering of the Nd:YAG laser radiation in high pressure methane gas. The probe radiation, as before, was the 10.5 ev ninth harmonic of Nd:YAG. Phosphorous doped Si $(\underline{111})$ bars ($\rho \sim 8\ \Omega-$cm, $N_D \sim 3\times10^{14}$ cm^{-3}), of 3×3 mm^2 cross section were cleaved along the $[\overline{2}11]$ direction. Low-energy electron diffraction (LEED) analysis was used to select cleaves which showed single domain 2x1 reconstruction. The photoemission spectra also were used as a cross-check of the domain structure of the surface.

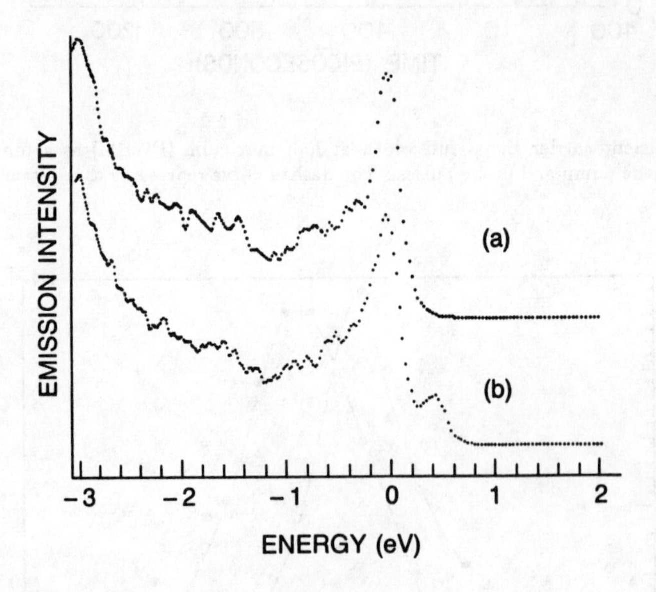

Figure 7: Photoemission from Si (111) 2x1 cleaved surface. 7(a): Spectrum of unexcited surface. 7(b): Spectrum of surface during excitation by 0.44 eV pulse.

Figure 7 shows typical results obtained on high quality cleaves. Several such cleaves were studied that gave reproducible results for photoemission spectra and for relaxation dynamics. These cleaves all showed sharp, high contrast, single domain 2x1 LEED patterns. The spectra shown in Fig. 7 were taken at an emission angle of 45° corresponding to the $\overline{J}$ point in the surface Brillouin zone (SBZ). In Fig. 7(a), a photoemission spectrum for the unexcited surface is shown. This spectrum agrees well with that obtained by Uhrberg, *et al.*[12] In Fig. 7(b), we show the spectrum obtained from the photoexcited surface at exact time coincidence of the pump and probe pulses. It is remarkably similar to the photoemission spectrum of highly doped n-type silicon recently published by Martensson, *et al.*[14] The additional peak at 0.5 eV arises from transient population in the π^* surface state. Its angular distribution is also the same as that shown in Ref. 16. Since the π^* population in the experiment of Martensson, *et al.*[14] is produced by doping, they were able to estimate the absolute population as 0.01 electrons per surface atom. By comparing the intensity of the π^* peak in our spectra to that shown in Ref. 16, we can estimate to population which we produce by photoexcitation to be also of the order of 0.01 electrons per surface atom.

The relaxation dynamics for the excited surface was studied by measuring the signal intensity in the 0.5 eV π^* peak as a function of the relative time delay between the infrared excitation pulse and the ultraviolet probe pulse. In Fig. 8, the time decay curve obtained from the same cleave used to obtain Fig. 7 is shown. Initially, the π^* signal rises rapidly with the excitation laser pulse, and begins to decay rapidly, essentially following the fall of the laser pulse. However, the rapid decay appears to cease, and the signal then decays much more slowly.

Figure 9 shows the results of performing the same experiment on a cleave of "lesser" quality. The LEED pattern for this cleave was single-domain, but the sharpness of the spots was slightly degraded compared to the LEED pattern taken from the cleave used to obtain Fig. 7.

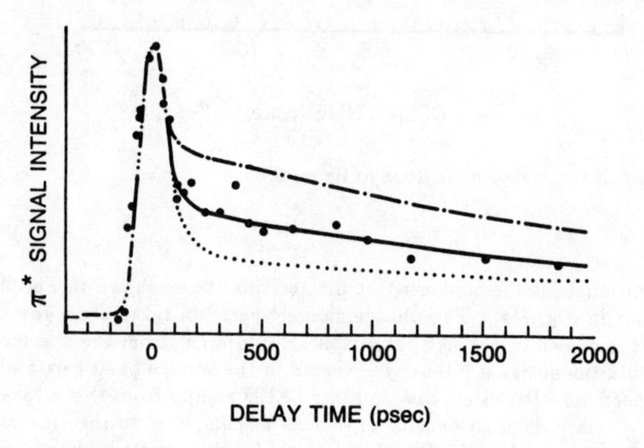

Figure 8: Time decay of the transient feature shown in figure 7.

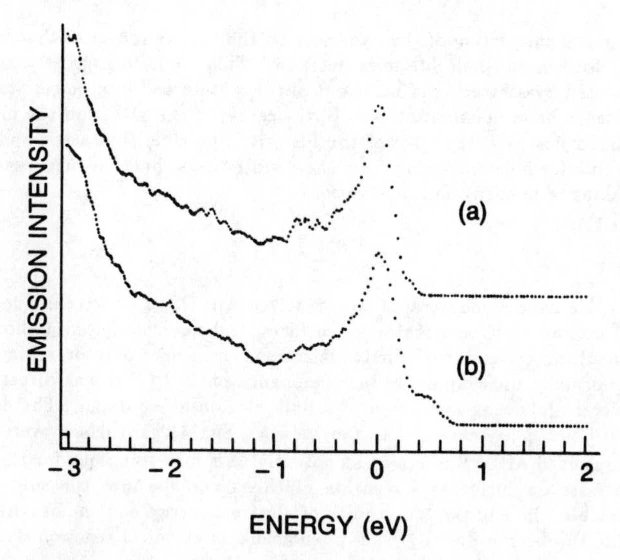

Figure 9: Photoemission from a poor cleave, Si (111) 2x1. 9(a): Unexcited spectrum. 9(b): Photoexcited spectrum.

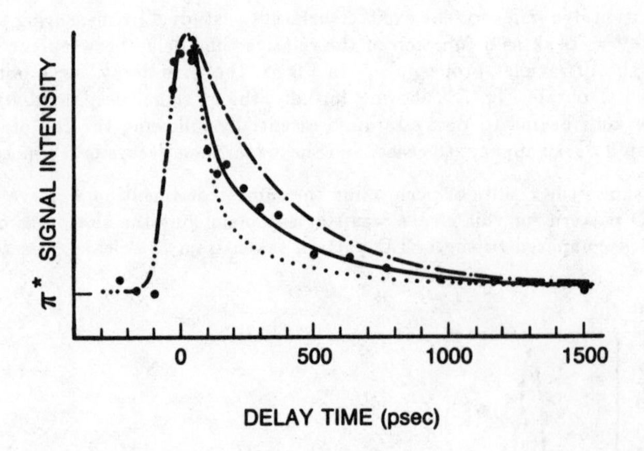

Figure 10: Time decay of the transient feature in figure 9.

Note that the conventional photoemission spectrum for this cleave shown in Fig. 9(a) is essentially identical to that in Fig. 7(a). The photoemission spectrum taken from the photoexcited surface of this cleave is shown in Fig. 9(b) and is clearly different from the spectrum shown in Fig. 7(b). Note the additional signal intensity observed in the surface band gap, and the general broadening of the spectrum. Based on this and our LEED results from this cleave, we believe that this cleave has a relatively high density of defects as compared to the cleaves typified by photoemission spectra shown in Fig. 7. The decay curve for this cleave is shown in Fig. 10 and is clearly different from the decay curve shown in Fig. 8. These data strongly suggest that cleavage defects play a major role in the relaxation dynamics of the electronic excitations of the π-bonded chains.

A detailed model calculation of the dynamics on the surface indicates that on the defected cleave, both the electron and hole lifetimes decrease. This tends to suggest that the hole lifetime, τ_h, is dominated by scattering of holes out of the π state and into defect states. However, since the bulk valence band maximum lies at $\overline{\Gamma}$, the center of the SBZ, and the maximum of the π surface state band lies at $\overline{J}$, the edge of the SBZ, it is possible that steps and defects could also increase the rate for hole scattering from the π state to the bulk valence band by providing the required breaking of momentum conservation.

SUMMARY

In summary, we have demonstrated time-resolved ARUPS, a new technique for measuring the properties of excited electronic states on surfaces, and have used the method to study the energy and momentum dynamics of excited electrons on semiconductor surfaces. Transient population in a normally unoccupied surface resonance on InP (110) was directly observed in photoemission spectra following excitation of a bulk electon-hole plasma. The decay dynamics of electrons and holes photoexcited on the cleaved Si(111)2x1 surface were also studied. Picosecond time-resolved ARUPS offers great potential as a new technique for the general study of transient processes on surfaces. Extension of this technique into the subpicosecond time regime seems possible. In addition to studies of electron energy and momentum dynamics, it should be possible to observe directly such phenomena as chemical reaction dynamics on surfaces and ultrafast phase transitions.

REFERENCES

a. Present address: IBM Watson Research Center, Yorktown Heights, NY

1. Reviews on this subject: R. H. Williams, G. P. Srivastava, and I. T. McGovern, Rep. Peog Phys. *43*, 1357-1413 (1980); E. W. Plummer and W. Eberhardt, Adv. Chem. Phys. **49**, I. Prigogine and S. A. Rice, eds. (1982).
2. J. Bokor, P. H. Bucksbaum, and R. R. Freeman, Opt. Lett. **28**, 21 (1983); see also *Laser Techniques in the Extreme Ultraviolet,* S. E. Harris and T. B. Lucatorto, eds. (Optical Society of America, Boulder, CO, 1984).
3. C. K. Rhodes, et. al. in this Proceedings; also see recently published accounts of subpicosecond excimer lasers by J. H. Glownia, P. P. Sorokin, Opt. Lett. **11**, 21 (1986).
4. R. W. Falcone and J. Bokor, Opt. Lett. **8**, 21 (1983).
5. R. Haight, J. Bokor, J. Stark, R. H. Storz, R. R. Freeman, and P. H. Bucksbaum, Phys. Rev. Lett. **54**, 1302 (1985).
6. J. Bokor, R. Haight, J. Stark, R. H. Storz, R. R. Freeman, and P. H. Bucksbaum, Phys. Rev. **B32**, 3669 (1985).
7. W. Gudat and D. E. Eastman, J. Vac. Sci. Technol. **13**, 831 (1976); J. R. Chelikowsky and M. L. Cohen, Solid State Commun. **29**, 267 (1979).
8. J. van Laar, A. Huijser, and T. L. van Rooy, J. Vac. Sci. Technol. **14**, 894 (1977).
9. M. A. Olmstead and N. M. Amer, Phys. Rev. Lett. **52**, 1148 (1984).
10. J. Bokor, R. H. Storz, R. R. Freeman, and P. H. Bucksbaum, submitted to Phys. Rev. Letters
11. K. C. Pandey, Phys. Rev. Lett. **47**, 1913 (1981).
12. J. E. Northrup and M. L. Cohen, Phys. Rev. Lett. **49**, 1349 (1982).
13. R. I. G. Uhrberg, G. V. Hansson, J. M. Nocholls, and S. A. Flodström, Phys. Rev. Lett. **48**, 1032 (1982).
14. P. Chiaradia, A. Cricenti, S. Selci, and G. Chiarotti, Phys. Rev. Lett. **52**, 1145 (1984).
15. P. Martensson, A. Cricenti, and G. V. Hansson, Phys. Rev. **B32**, 6959 (1985).

LASER-XUV EXCITED STATE SPECTROSCOPY

B. F. Sonntag
Universität Hamburg, II
Institute of Experimental Physik
Luruper Chausse 149
2000 Hamburg 52
Federal Republic of Germany

C. L. Cromer, J. M. Bridges, T. J. McIlrath, and
T. B. Lucatorto
National Bureau of Standards
Gaithersburg, MD 20899

ABSTRACT

For Ca, the localization of the 3d-orbital critically depends on the effective potential. In order to probe this in detail we have studied the 3p-absorption spectra of Ca, Ca*, and Ca^+ in the wavelength range 25 nm to 45 nm. The Ca was contained in a heat pipe operated at temperatures >700°C. The VUV radiation of a (ruby) laser-produced plasma transmitted through the heat pipe was dispersed by a 1.5 m grazing incidence spectrometer. A multichannel photoelectric detector covering a spectral range of approximately 4 nm detected the dispersed VUV. Excitation and ionization was achieved by pumping the Ca $4s^2\ ^1S_0 \rightarrow$ Ca $4s4p^3P$ transition with a dye laser. Collisions convert the initial excitation into ionization. Various population densities have been probed by delaying the VUV continuum pulse relative to the initiation of the dye laser pulse. All spectra are dominated by a strong $3p \rightarrow 3d$ resonance. Towards higher photon energies, and the Ca and Ca^+ spectra display sharp Rydberg lines whereas there are only broad structures in the Ca* spectrum.

INTRODUCTION

Calcium (Z=20) is at a position near the beginning of the transition metals Sc(Z=21) to Zn(Z=30), which are characterized by the progressive filling of the 3d shell. For the elements proceeding Ca, the 3d orbital is generally concentrated near its

0094-243X/86/1470412-11$3.00

hydrogenic position and is associated with excited levels. In contrast, for the transition elements the 3d-orbital is collapsed near the inner core region of the atom, becomes significantly more tightly bound, and is thus associated with the ground state of the atom.

The occurrence of orbital collapse is associated with striking changes in atomic structure. The marked and usually sudden decrease in orbit size results in the aforementioned higher binding energy and a large increase in overlap with the inner shell electrons. This latter change is manifested in the redistribution of oscillator strength. For example an uncollapsed 3d orbital has little overlap with a collapsed np orbital ($n<3$) so that photoabsorption from the np shell is expected to be strongest in the $np \rightarrow \epsilon d$ continuum transition and not the $np \rightarrow 3d$ transition. When 3d collapse occurs, a dramatic increase of oscillator strength in the $np \rightarrow 3d$ is observed.

Our study of Ca was motivated by this desire to better understand the characteristics of 3d-orbital collapse. With the technique of exciting or removing a 4s valence electron, we can produce slight modifications in the effective potential and study the effect on the 3p photoabsorption spectra. This is exactly what we have done with the observations of photoabsorption in the 27-50 eV range of Ca, Ca* ($4s4p\,^3P$), and $Ca^+(4s)$ which will be explained below.

EXPERIMENTAL

The experimental apparatus shown in Fig. 1 was designed to optimize the time-resolved measurements necessary to observe the absorption spectra of the laser-excited or laser-ionized sample. The major components of the experimental apparatus have been described in detail elsewhere, and consist of:

- a laser-plasma VUV continuum light source
- heat-pipe oven
- a 1.5 meter grazing incidence spectrometer
- a VUV optical multichannel analyzer consisting of a channel electron multiplier array (CEMA) image intensifier coupled to a linear self-scanned diode array
- flashlamp-pumped dye laser

The laser plasma continuum source is well suited for time-resolved measurements because it provides an intense pulse of quasi-continuum radiation with a pulse duration similar to that of the focused laser. For our source we used a ruby laser with a 3-5 joule output in 25 ns, which gives us a 25 ns temporal resolution in our VUV absorption measurements. The emission spectra for a variety of targets has been published elsewhere;[2] samarium was chosen as the optimum target material for the wavelength range of interest for Ca. The samarium continuum spectrum is provided in Fig. 2.

The VUV radiation, after passing through the sample cell and a capillary window, was focused onto the spectrometer entrance slit with a grazing incidence toroidal mirror. The dispersed VUV was detected with the CEMA positioned tangent to the Rowland circle of the spectrometer. Photoelectrons produced at the front surface of the CEMA are multiplied in the microchannel plate and proximity focused onto a phosphor-coated fiber optic bundle. The visible photons are then detected with the linear diode array.

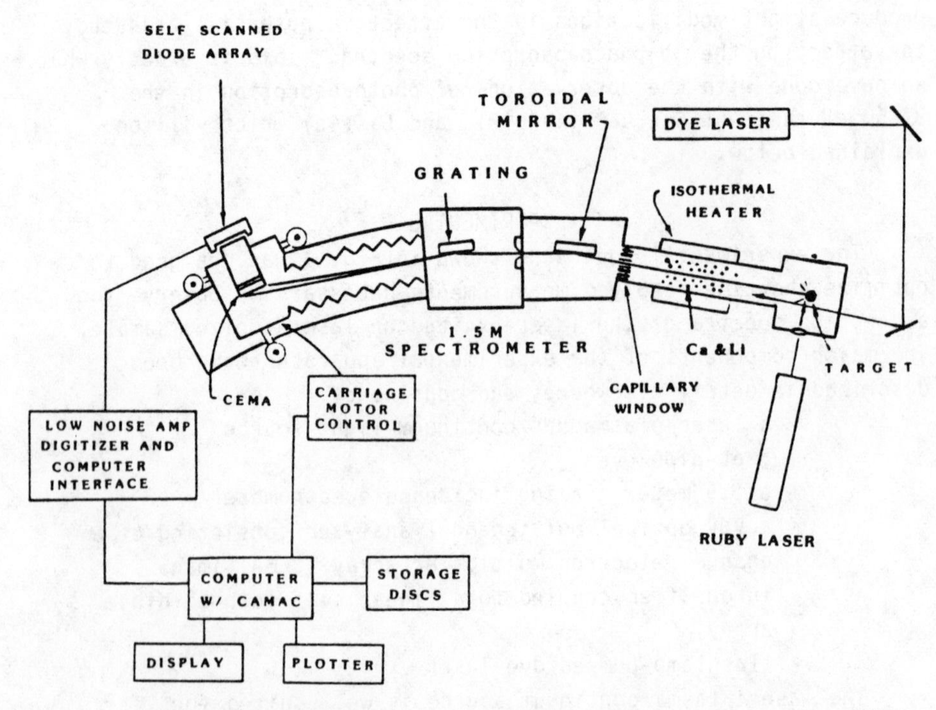

Fig 1. Layout of spectrometer for time-resolved absorption studies of dye laser excited or ionized atoms.

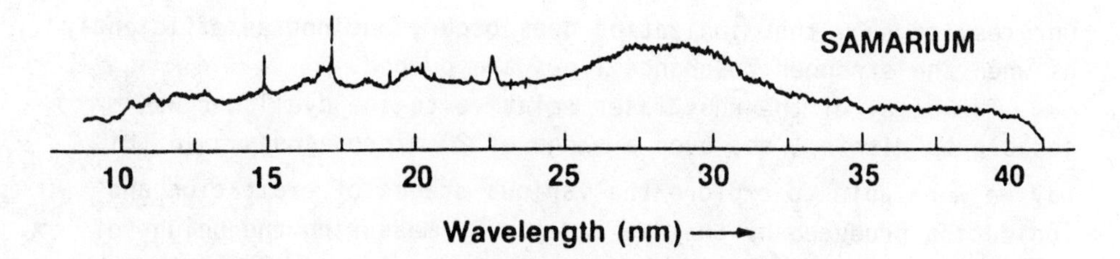

Fig. 2 Spectrum of the laser-induced plasma emission of a samarium target.

The CEMA detector was specially designed for sensitivity and high resolution. Our measurements show our resolution to be better than 0.015 nm, with the CEMA's spatial resolution better than 30 microns. The large dynamic range and linearity of the detection make it easy to determine the quantities of real physical interest, such as $\log(I/I_0)$.

A coaxial flashlamp-pumped dye laser was used to excite and ionize the Ca vapor in the heat pipe. As shown in Fig. 1 the laser was focused past the target and into the heat pipe, filling the field of view of the spectrometer. The heat pipe used was specifically designed to minimize the solidification of the Ca at the cold ends of the oven. A concentric, welded sodium jacket was used on the outside of the cell to isothermaly heat the central region of the oven. A mixture of Ca and Li (10:1) was used to lower the overall melting point, and to improve the flow of the sample in the heat pipe. With these modifications the Ca/Li heat pipe could be operated up to 950 C for indefinite periods, without any sign of solid deposits at the ends of the oven. Li was chosen as the companion metal as it has a very small cross-section (< 1 Mb) in the region of interest and forms on alloy with Ca with a much reduced melting point.

Calcium excitation and ionization

Our previous experiments on the absorption spectra of laser-excited and laser-ionized atomic vapors utilized a pulsed dye laser tuned to the resonance line of the atom.[3,4,5] The Ca resonance line is in the UV and thus is not easily accessible to high energy, long pulse dye lasers. The intercombination line at 657.2 nm is readily accessible, but has an oscillator strength approximately 10^{-5} times smaller than that of the resonance line.

Our results show that ionization does occur, but not as efficiently as when the stronger resonance lines are pumped.

The delay of the ruby laser relative to the dye laser was setable to within 1 ns, over a range of 20 microseconds. In this way we were able to explore the various stages of excitation and ionization produced by the dye laser. By measuring the height of the strongest absorption feature in each species, we can infer the population densities of the neutral, excited, and ionized species as a function of delay time; Fig. 3 shows the relative population densities of each at 700 C and at 730 C.

We observe that the peak excited state density exceeds the ground state density during the dye laser pulse, a phenomenon which was not observed under similar conditions in previous experiments on Li, Na, and Ba. This is probably due to the much larger

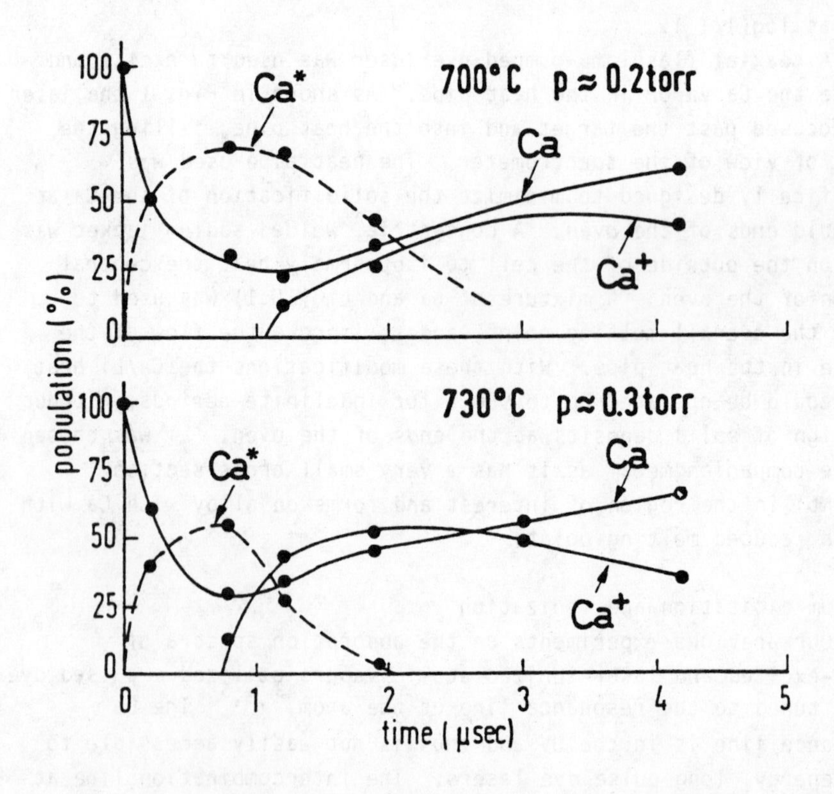

Fig. 3 Evolution of the population of Ca, Ca* and Ca^+ as a function of decay time after initiation of the dye laser.

statistical weight of the excited ^3P_J levels combined with the metastability of the intercombination line used here. Another interesting point is that, unlike previous observations in which complete ionization was obtained with a laser fluence of 5 MW/cm^2, the Ca vapor reaches a maximum degree of ionization of only 50% long after the dye laser has turned off. Again this is probably associated with the small transition probability of the intercombination line.

Comparing the behavior at 700 C with that at 730 C, one can see that the higher Ca density at 730 C results in a faster deexcitation of the Ca metastable state, and a more efficient conversion of the excitation to ionization. This vapor density dependence is clear evidence for the collisional nature of the ionization mechanism behind the resonantly driven process.

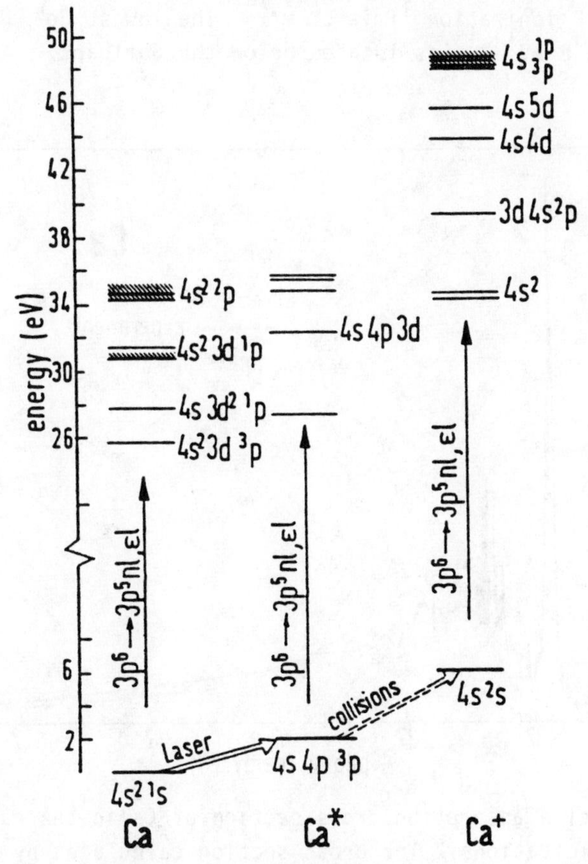

Fig. 4 Schematic energy diagram of Ca, Ca* and Ca$^+$.

RESULTS AND DISCUSSION

A schematic energy level diagram for Ca, Ca* and Ca^+ is presented in Fig. 4. The laser excitation and the collisions populating the Ca* 4s4p ^3P and Ca^+ 4s ^2S states are indicated at the bottom. The VUV radiation excites 3p electrons into bound $n\ell$ or continuum states $\varepsilon\ell$. Some assignments are made for the autoionizing core-excited states and some 3p ionization limits are given.

3p-photoabsorption of Ca

The photoabsorption and photoionization of Ca in the region of the 3p-excitation has been studied in great detail both in experiment[6-9] and theory.[10-12] The two-step autoionization mechanism resulting in Ca^{2+} has attracted much interest.[7-9] The Ca^{2+}/Ca^+ ratio increases steplike from zero to unity above the $3p^5 3d4s^2 {}^2P$, ${}^4P$ ionization limits.[7-9,13] The lowest $3p^5 3d4s$ ${}^4P$ limit in Ca^+ at 30.8 eV[13,14] is located below the dominant

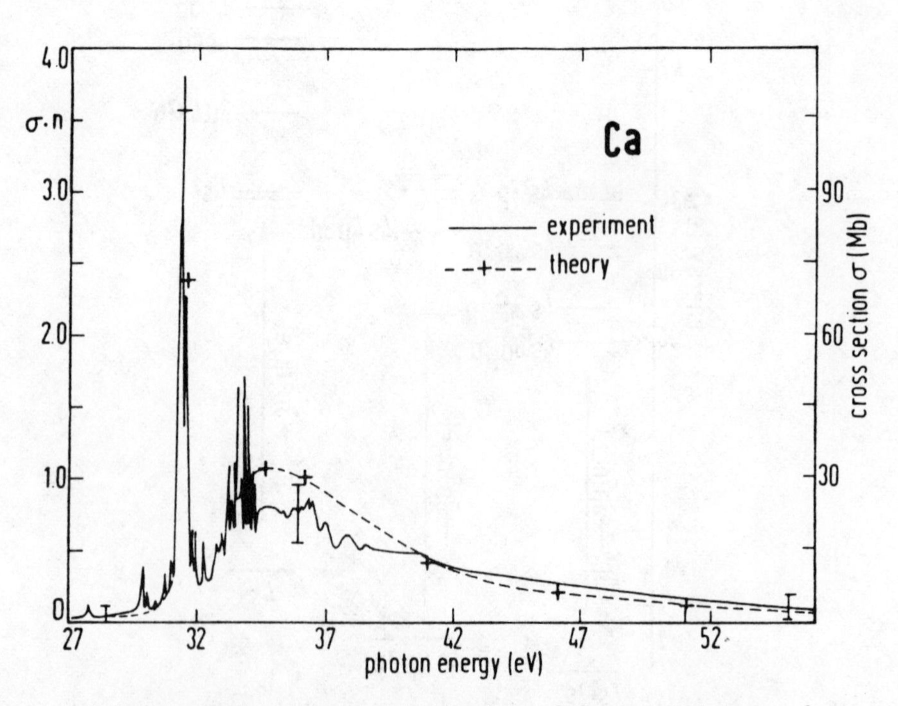

Fig. 5 Relative absorption cross section of Ca in the range of the 3p excitations. The cross section calculated by Altun et al. is given by the dashed line.

$3p^6 4s^2\ ^1S \rightarrow 3p^5 4s^2 3d\ ^1P$ transition at 31.40 eV which can thus decay to a $3p^5\ 3d4s^4P\ \varepsilon\ell$ state followed by a second autoionization of the $3p^5\ 3d4s^4P$ ion. Our Ca 3p absorption spectrum is shown in Fig. 5.

The energy positions of the numerous absorption lines are in good agreement with those reported by Mansfield and Newsom.[6] The strong $3p^6 4s^2\ ^1S \rightarrow 3p^5 4s3d\ ^1P$ line is clearly to be seen. Toward higher energies there are well developed Rydberg series converging towards the $3p^5 4s^2\ ^2P_{3/2,}$ (34.31 eV; 34.66 eV) ionization limits. The broad structures are caused by two-and three-electron excitations.[6,14] In contrast to the earlier photographic measurements[6] we were able to determine the relative absorption cross section with reliability. There is good agreement between our relative cross sections and the total photoionization cross section recently obtained by Schmidt.[9]

For comparison the cross section calculated by Altun et al.[11] is included. The experimental curve has been normalized to the

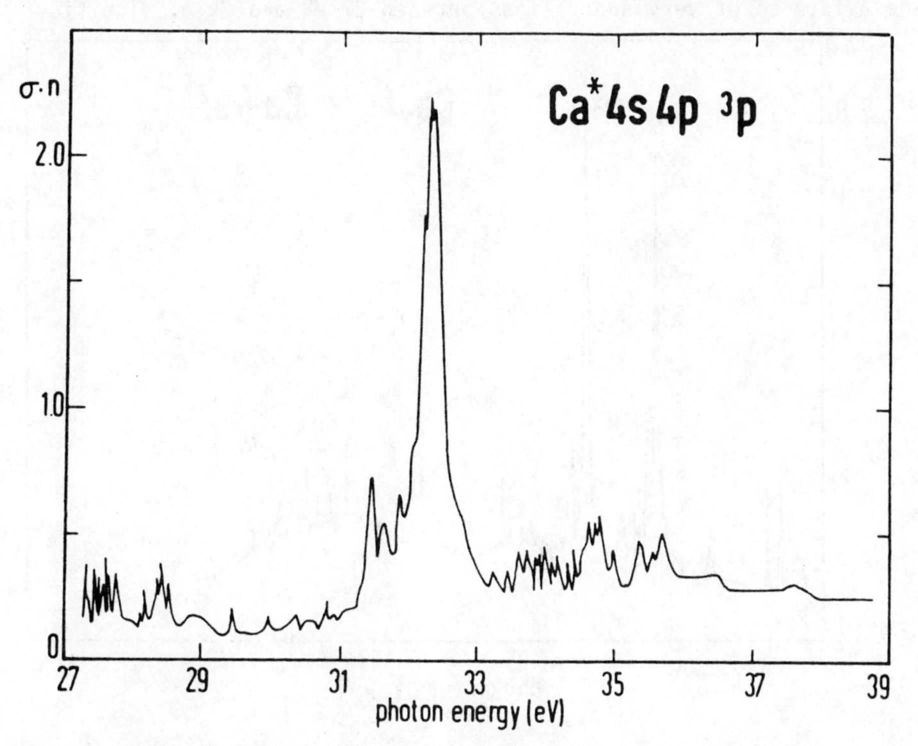

Fig. 6 Relative absorption cross section of Ca* in the range of the 3p excitations.

theoretical cross section at the high energy end. The general behavior is well described by the calculations.

3p-photoabsorption of Ca*

The $3p^6 4s4p \rightarrow 3p^5 3d4s4p$ resonance at 32.3 eV dominates the spectrum of Ca* presented in Fig. 6. The vapor even for optimal timing contains a fraction of Ca $4s^2$ ^1S atoms (see Fig. 2). Therefore the spectrum represents a superposition of the Ca and Ca* absorption. The lines in the ranges 31 - 31.7 eV and 33 - 34.5 eV are due to Ca. The most puzzling finding is the complete absence of Rydberg series. There are two mechanism which one could invoke for an explanation:

i) the oscillator strength is distributed over many lines thus diminishing the strength of each;

ii) the rapid Auger decay of the $3p^5 4s4p$ core broadens the lines thus smearing the lines.

The existence of very sharp lines between 27 eV and 29 eV signifies

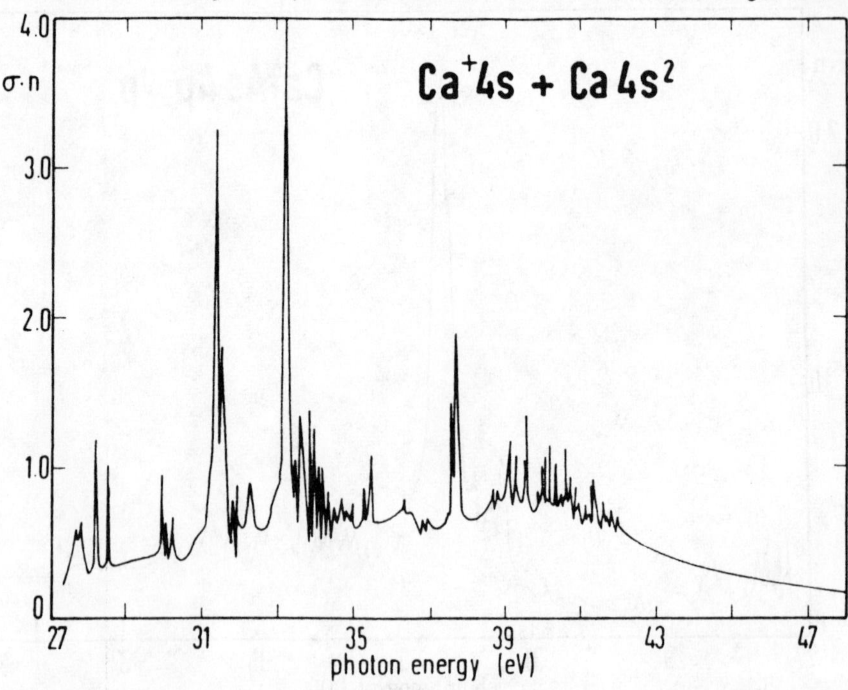

Fig. 7 Absorption of a mixture of Ca^+ 4s and Ca $4s^2$ in the region of the 3p excitations.

a limitation of the latter argument.

3p-absorption of Ca^+

Fig. 7 shows a superposition of the Ca and Ca^+ absorption spectra in the range of the 3p excitations. By comparison with the Ca spectrum, the Ca^+ features can be easily identified. The $3p^64s$ $^2S_{1/2} \rightarrow 3p^54s^2\ ^2P_{3/2,1/2}$ transitions give rise to the two lines at 28.21 eV and 28.55 eV. These energies are in good agreement with tabulated values.[15] The prominent $3p^64s\ ^2S_{1/2} \rightarrow 3p^53d(^1P)4s\ ^2P$ line is located at 33.21 eV which agrees well with the value of 33.20 eV calculated by Hansen.[17] The energy of 33.38 eV reported by Mansfield and Olney[13] is slightly higher. The spectrum of Ca^+ shows Rydberg series converging towards the $3p^54s\ ^3P_{2,1,0}\ ^1P_1$ (41.944 eV, 42.116 eV, 42.324 eV, 42.582 eV) ionization limits, but the analysis is severly hampered by the great number of overlapping series. From published energy values for Ca I and Ca III[15] a crude estimate of 0.6 has been extracted for the quantum defect of nd states. Based on this the lines around 37.6 eV are assigned to transitions to $3p^54s$ 4d and $3p^54s5d$ final states. For the $3p^54s4d$ states this assignment is corraborated by Hartree-Fock calculations.[16] According to these calculations the line at 35.42 eV can also be attributed to $3p^64s \rightarrow 3p^54s4d$ transitions. The lowest $3p^53d\ ^3P_0$ (37.082 eV) ionization limit[15] lies well above the strong $3p^64s^2S \rightarrow 3p^53d(^1P)4s^2\ ^2P$ line. This precludes the $3p^53d4s \rightarrow 3p^53d\ \varepsilon\ell$ autoionization. This is different from the situation encountered for Ca.

Work supported in part by Air Force Office of Scientific Research (Contract No. ISSA 85-0033) and a grant from the NATO Scientific Programme.

REFERENCES

1. C. L. Cromer, J. M. Bridges, J. R.Roberts and T. B. Lucatorto, Appl. Opt. 24, 2996 (1985)
2. J. M. Bridges, C. L. Cromer and T. J. McIlrath, Appl. Opt. (in press).
3. T. J. McIlrath and T. B. Lucatorto, Phys. Rev. Lett. 38, 1390 (1977).
4. T. B. Lucatorto, T. J. McIlrath, J. Sugar and S. M. Younger, Phys. Rev. Lett. 47, 1124 (1981).

5. T. B. Lucatorto, T. J. McIlrath and G. Mchlman, Appl. Opt. 18, 2916 (1979).
6. M. W. D. Mansfield and G. H. Newsom, Proc. R. Soc. Lond. A357, 77 (1977).
7. D. M. P. Holland and K. Codling, J. Phys. B14, 2345 (1981).
8. Y. Sato, T. Hayaishi, Y. Itikawa, Y. Itoh, J. Murakami, T. Nagata, T. Sasaki, B. Sonntag, A. Yagishita and M. Yoshino, J. Phys. B18, 225 (1985).
9. M. Schmidt (private communication).
10. P. C. Deshmukh and W. R. Johnson, Phys. Rev. A27, 326 (1983).
11. Z. Altun, S. L. Carter and H. P. Kelly, Phys. Rev. A27, 1943 (1983).
12. Z. Altun and H. P. Kelly, Phys. Rev. A31, 3711 (1985).
13. M. W. D. Mansfield and T. W. Ottley, PRoc. Roy. Soc. Lond. A365, 413 (1979).
14. J. M. Bizau, P. Gérard, F. J. Wuilleumier and G. Wendin, Phys. Rev. Lett. 53, 2083 (1984).
15. J. Sugar and C. Corliss, J. Phys. Chem. Ref. Data 8, 865 (1979).
16. A. W. Weiss (private communications).
17. J. E. Hansen, J. Phys. B8, 2759 (1975).

LIFETIME MEASUREMENTS OF PERTURBED TRIPLET STATES OF THE CO MOLECULE

K.Strobl and C.R.Vidal
Max Planck Institut für Extraterrestrische Physik
D-8046 Garching, F. R. G.

ABSTRACT

Lifetimes of individual rotational vibrational levels of the perturbed $a'\,^3\Sigma^+$ (v=14) and the $e\,^3\Sigma^-$(v=5) states of the CO molecule have been measured using state selective excitation and the results are compared with theory.

INTRODUCTION

The CO molecule has already been the subject of several radiative lifetime measurements. Most recently, lifetime measurements have been performed on the $A\,^1\Pi$ state of the CO molecule[1] using state selective excitation. The measurements clearly showed increased lifetimes due to the interaction with neighbouring triplet states. The unperturbed lifetimes of the $A\,^1\Pi$ state have by now been pretty well settled[2]. However, with regard to the neighbouring singlet and triplet states which have a very small transition moment to the ground state and whose lifetimes may vary by several orders of magnitude, rather large uncertainties still exist in the literature. The large variations between the lifetimes of neighbouring rotational vibrational levels give rise to a very intricate pressure dependence which defies a simple Stern Vollmer analysis. Therefore accurate lifetimes can only be obtained under truely collisionfree conditions and by pumping individual levels as done in this contribution. The most extensive analysis of perturbations in the CO molecule which is a key information for the lifetimes of perturbed singlet or triplet states, has so far been reported by *Field* and coworkers[3,4].

THEORY

In the limit of a pure two-level system the wave function of a triplet level interacting with a singlet level can be given by a linear combination of the unperturbed states

$$|\psi_{3,1}\rangle = \alpha|\psi_3^o\rangle + \beta|\psi_1^o\rangle \tag{1}$$

where the normalization of the wave function requires $|\alpha|^2 + |\beta|^2 = 1$. With this relation the transition moments from the perturbed triplet or singlet levels to a pure triplet level $|\overline{\psi_3^o}\rangle$ or a pure singlet level $|\overline{\psi_1^o}\rangle$ are given by

$$\langle\overline{\psi_3^o}|\mu|\psi_{3,1}\rangle = \alpha\langle\overline{\psi_3^o}|\mu|\psi_3^o\rangle \qquad \text{and} \qquad \langle\overline{\psi_1^o}|\mu|\psi_{3,1}\rangle = \beta\langle\overline{\psi_1^o}|\mu|\psi_1^o\rangle\,. \tag{2}$$

The radiative lifetime due to spontaneous emission from some level $|p\rangle$ is defined to be

$$\tau_p = \left(\sum_{q(<p)} A_{pq}\right)^{-1} . \tag{3}$$

For a two-level system the lifetime of a perturbed triplet (T) or singlet (S) state can therefore be given by

$$\frac{1}{\tau_T} = \frac{1-|\beta|^2}{\tau_T^o} + \frac{|\beta|^2}{\tau_S^o} \qquad \text{and} \qquad \frac{1}{\tau_S} = \frac{1-|\beta|^2}{\tau_S^o} + \frac{|\beta|^2}{\tau_T^o}, \tag{4}$$

respectively where τ_T^o and τ_S^o are the lifetimes of the unperturbed triplet and singlet state. In our case of interest we typically have $\tau_S^o \approx 10\,nsec$ and $\tau_T^o \approx 5\,\mu\text{sec}$. Since the mixing coefficient $|\beta|^2$ defining the interaction between the singlet and triplet levels has to be smaller than 0.5, we immediately see that τ_S can increase at the most by a factor of 2, whereas τ_T can be shortened by as much as a factor of 250. It is for this reason that the radiative lifetimes of the triplet states are much more sensitive to the mixing coefficients $|\beta|^2$ than the lifetimes of the interacting levels of the $A^1\Pi$ state which change at the most by a factor of two as observed by *Maeda* et al.[1]. A more general discussion for the multi-level system is in preparation[5].

EXPERIMENTAL

A pulsed, tunable coherent vacuum uv source which uses the method of two-photon resonant sum frequency mixing[6] in a Mg Kr mixture, was emloyed for pumping through one of the weak intercombination lines individual rotational vibrational levels of different perturbed triplet states in the CO molecule[7]. The details of the coherent VUV source have already been described[7]. The source has a pulse duration of about 2 *nsec*, an energy of typically 10^{11} to 10^{12} photons per shot over the desired spectral region from 141 to 149 *nm* and a repetition rate of up to 20 *Hz*.

In order to obtain genuine radiative lifetimes all measurements had to be done under collision free conditions using either a pulsed nozzle beam or a low pressure gas cell operated at about 5 *mTorr* and lower. The fluorescence cell looked very similar to a system recently used by *Klopotek* et al.[8] where the gas cell could be replaced by a pulsed nozzle beam. Due to the adiabatic expansion giving rise to CO molecules with a rotational temperature of about $20^o\,K$, the pulsed nozzle beam was used for the rotational levels with small J, whereas the gas cell was employed for the radiative lifetime measurements of levels with J typically larger than 10.

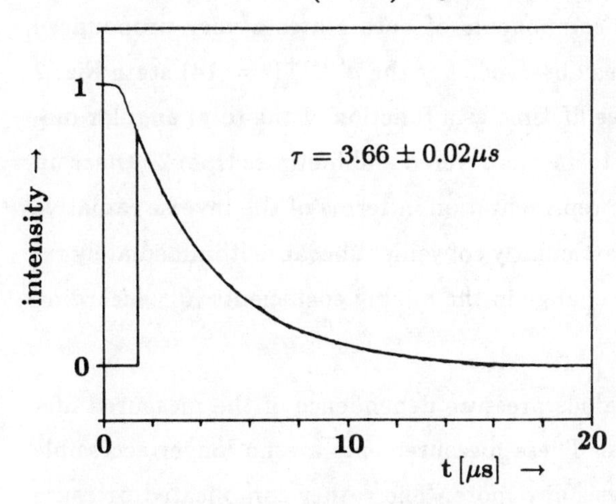

Fig. 1. Typical fluorescence signal averaged over 500 pulses. The smooth curve shows the corresponding least squares fit. The time constant τ and its one σ standard error have been obtained from an online fit of the measurement.

Suppressing the scattered light of the incident VUV source by using an appropriate filter the subsequent visible fluorescence[7] was measured with a fast photomultiplier which is followed by a fast charge sensitive integrator and a transient digitizer. Averaging over typically 500 pulses a very good signal to noise ratio was obtained for individual levels within typically 3 *min* of measuring time. The effective time constant τ was obtained from an online least squares fit fitting a function of the form $y = a \cdot e^{-t/\tau} + b$ to the data as shown in Fig. 1. For this purpose the time dependent detection probability of the photomultiplier due to the geometry of the fluorescence cell had to be taken into account. In the range of overlap there was excellent agreement between measurements taken with the pulsed nozzle beam and those taken with the low pressure gas cell.

RESULTS

Lifetimes have been measured on many individual rotational vibrational levels of the perturbed $a'\,^3\Sigma^+$(v=14) and the $e\,^3\Sigma^-$(v=5) states of the CO molecule. The first state had been selected because it allows a detailed comparison with the mixing coefficients $|\beta|^2$ of *Field* and coworkers[3,4]. The second state was chosen to demonstrate the superior sensitivity of the lifetime measurements. For the latter state the mixing coefficients $|\beta|^2$ were so small that they could not be determined from the energy level perturbations.

For all three fine structure components of both states a very pronounced J-dependence of the lifetimes was observed. For the $a'\,^3\Sigma^+(v = 14)$ state Fig. 2 shows the inverse of the radiative lifetime as a function of the total angular momentum J. A range in J from 1 to 14 was covered and lifetimes from 279 *nsec* up to 3.20 μsec were obtained. The representation in terms of the inverse radiative lifetimes as chosen in Fig. 2, is particularly convenient because it immediately reveals the corresponding relative change in the mixing coefficients $|\beta|^2$ according to Eq. (4).

Fig. 2 also shows the anomalous pressure dependence of the measured lifetimes for a selected set of levels. These measurements are no longer accessible to a simple Stern Vollmer analysis any more. The rather complicated pressure dependence originates from the collisional coupling between levels of widely different lifetimes. Levels with longer (shorter) lifetimes than the lifetime of the

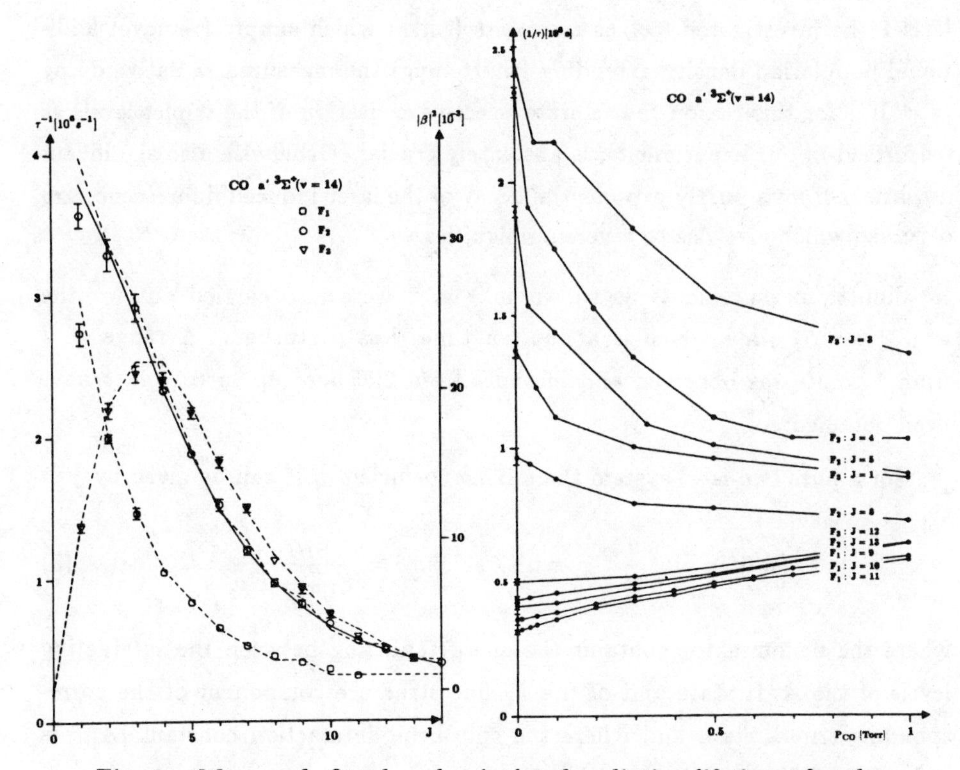

Fig. 2. Measured, fitted and calculated radiative lifetimes for the three different fine structure components of the $a'\,^3\Sigma^+$(v=14) state. The rotational levels are perturbed by the $A\,^1\Pi$(v=4) state. The standard errors of the measurements are indicated by the vertical bars and amount to typically 2.7% . The solid line is from a least squares fit of the measured radiative lifetimes of the F_2 component. The dashed lines are obtained using the mixing coefficients $|\beta|^2$ of *Field* et al.[3,4] and the unperturbed radiative lifetime τ_T^o from the least squares fit. The absolute scale of the mixing coefficients $|\beta|^2$ is given for the unperturbed radiative lifetimes $\tau_S^o = 9.4\,nsec$ and $\tau_T^o = 4.59\,\mu sec$. The figure on the right shows the inverse of the effective lifetime for several levels shown in the left figure as a function of the CO gas pressure.

level to be investigated, act as reservoirs (sinks) which supply (remove) additional population density extending (shortening) the measured radiative decay rate. It is for this reason that a state selective excitation of the triplet levels as performed in our experiments, is absolutely crucial. Otherwise also significant deviations from a purely exponential decay of the laser induced fluorescence are observed which give rise to severe ambiguities.

Similar measurements as shown in Fig. 2 were also carried out for the $e\,^3\Sigma^-(v = 5)$ state which is about ten times less perturbed. A range of J from 1 to 19 was observed and lifetimes from 2.55 *μsec* up to 6.14 *μsec* have been obtained[5].

For a pure two-level system the mixing coefficient $|\beta|^2$ can be given by

$$|\beta|^2 = \sin^2\frac{\theta}{2} \qquad \text{with} \qquad \sin\theta = \frac{2|A_{10}^{\pm}|}{E_{\Pi^{\pm}} - E_{F_2}} \tag{5}$$

where the denominator contains the energy splitting between the interacting levels of the $A\,^1\Pi$ state and of the F_2 fine structure component of the corresponding triplet state and where the spin orbit interaction constant $|A_{10}^{\pm}|$ is defined according to *Wicke* et al.[9]

$$A_{10}^{\pm} = \langle ^1\Pi|H_{so}|^3\Sigma_1^{\pm}\rangle \tag{6}$$

with H_{so} being the operator for the spin orbit interaction. A least squares fit of the measured lifetimes for the F_2 component of the two different $^3\Sigma$ states was performed using the unperturbed lifetimes of the $A^1\Pi$ state, $\tau_S^o = 9.4\,nsec$ for v=4 and $\tau_S^o = 9.6\,nsec$ for v=3 [2] and using the energy level splitting $E_{\Pi^{\pm}} - E_{F_2}$. The latter quantity has been given by several authors in the literature. The values differ typically by 1% which according to Eq. (5) amounts to less than 2% in the mixing coefficients. In this manner the unperturbed lifetime of the triplet states τ_T^o and the spin orbit interaction constants $|A_{10}^{\pm}|$ were obtained. The numbers in brackets are the standard errors in units of the last digit. The spin orbit interaction constant for the $a'\,^3\Sigma^+$ state is in good agreement with calculations of *Field* et al.[3]. For the $e\,^3\Sigma^-$ state the spin orbit interaction constant could not be given before. The overall accuracy of the measurements was typically 3 to 5%. The results of the final least squares fit were:

System	$\tau_T^o[\mu sec]$	$\|A_{10}^{\pm}\|[cm^{-1}]$
$A^1\Pi(v=4) - a'\,^3\Sigma^+(v=14)$	4.59(6)	4.381(17)
$A^1\Pi(v=3) - e\,^3\Sigma^-(v=5)$	7.53(9)	5.071(31)

Tab. 1. Results from a least squares fit of the F_2 component of the $^3\Sigma$ states.

SUMMARY

Lifetimes of individual rotational vibrational levels of the perturbed $a'\,^3\Sigma^+$ (v=14) and the $e\,^3\Sigma^-$(v=5) state of the CO molecule have been measured under collision free conditions. Pumping individual levels of the triplet states with different total angular momenta J, widely different lifetimes have been measured with an acuracy of 3 to 5%. It is also shown that lifetime measurements provide a more sensitive measure of the mixing coefficients $|\beta|^2$ than energy level perturbations.

REFERENCES

1. M. Maeda and B. P. Stoicheff, AIP Conf. Proc. **119**, 162 (1984).
2. R.W. Field, O. Benoist d'Azy, M. Lavollée, R. Lopez-Delgado and A. Tramer, J. Chem. Phys. **78**, 2838 (1983).
3. R.W.Field, R.G.Wicke, J.D.Simmons and S.G.Tilford, J. Mol. Spectr. **44**, 383 (1972) and private communication.
4. A. Le Floch, F. Launay, J. Rostas, R. W. Field and K. Yoshino, to be submitted.
5. K. H. Strobl and C. R. Vidal, J. Chem. Phys. to be submitted.
6. C. R. Vidal, *Four wave frequency mixing in gases* in *Tunable Lasers*, edited by L. Mollenauer and J. C. White, Topics in Applied Physics (Springer, Heidelberg, to be published).
7. P.Klopotek and C.R.Vidal, Can. J. Phys. **62**, 1426 (1984).
8. P. Klopotek and C. R. Vidal, J. Opt. Soc. Am. B **2**, 869 (1985).
9. B.G. Wicke, R.W. Field and W. Klemperer, J. Chem. Phys. **56**, 5758 (1972).

EXPERIMENTS IN PHOTOFRAGMENTATION DYNAMICS USING COHERENT VACUUM ULTRAVIOLET FOR PRODUCT DETECTION

I.M. Waller, H.F. Davis and J.W. Hepburn,
Center for Molecular Beams and Laser Chemistry,
University of Waterloo, Waterloo, Ontario N2L 3G1 Canada.

ABSTRACT

The photofragmentation dynamics of CS_2, $Fe(CO)_5$ and $Co(CO)_3NO$ at 193nm have been studied under molecular beam conditions. The S and CO products of these photofragmentations have been probed by laser-induced fluorescence spectroscopy, using tunable VUV generated by four wave sum mixing in Mg vapour. As well as determining product quantum state distributions, doppler spectroscopy has been used to probe product translational energy and the anisotropy of the product recoil.

INTRODUCTION

The quest for a precise, microscopic understanding of chemical reactions has given rise to an area of research known as "state-to-state" chemistry. The goal of experiments in state-to-state chemistry is to initiate a chemical reaction in a system where the initial state is as thoroughly characterized as possible, and to analyze the products of such a reaction, determining such properties as product internal energies, orientations of products, and correlations between properties of different products. The complexity of doing these experiments on bimolecular reactions means that currently, the most detailed experiments are being carried out on unimolecular photochemical reactions[1]. By studying photofragmentation reactions, the problem of defining the initial state for the reaction is greatly simplified. The reaction is initiated by exciting molecules, which can be internally cooled by supersonic expansion, to a precisely defined dissociating excited state. By polarizing the dissociation laser, an initial direction can be defined for the subsequent dissociation. Once the dissociation occurs, the resulting fragments are detected spectroscopically, in order to determine the product internal energy distribution in detail. The simultaneous use of doppler spectroscopy on the photofragments can provide a very detailed picture of the products, giving information on translational energy for a given product quantum state, dissociation direction with respect to photolysis laser polarization, and orientation of products. In order to ensure collision-free conditions, these experiments are best performed on

0094-243X/86/1470430-12$3.00 Copyright 1986 American Institute of Physics

supersonic molecular beams, where the molecule of interest is diluted in a large excess of a carrier gas, typically He. To avoid contamination of the results due to photochemistry of van der Waal's clusters[2] dilute expansions (on the order of 1% of the target gas in He) under mild expansion conditions should be used. In some cases, multiple photon absorption (sequential or concerted) can complicate the observed photochemistry so it is best to operate at the lowest possible photolysis laser power. All of these considerations mean that the method used to probe the products must be able to detect very low densities of products in a small volume: typical total product densities are less than $10^{10}cm^{-3}$,and the dimensions of the photolysis zone less than 1 cm.

The two most commonly employed methods for detecting products under these conditions are laser-induced fluorescence spectroscopy (LIF) and resonant two-photon ionization spectrosopy. Both methods rely on resonant excitation of a well characterized electronically excited state of the product being detected. The tuning range of commercial pulsed dye laser systems restricts state-to-state experiments to molecules where the products absorb in the visible or near-ultraviolet. However, we have demonstrated in previous work[2,5-7] that by taking advantage of four-wave mixing to generate tunable coherent vacuum ultraviolet, it has been possible to extend these studies to systems where the products absorb only in the vacuum ultraviolet region of the spectrum. These include chemically important species such as: H_2, CO, N_2, hydrogen halides and a wide range of light atom products (H, C, O, etc.)[3]. One technique which has proven very useful is four-wave mixing in Mg vapour[4], which generates radiation tunable from 1400-1700Å, and has been used to detect Br[5], S[2] and CO[2,6-7] products from a variety of processes.

In this paper, the usefulness of coherent VUV for state-sensitive product detection is illustrated by some recent results on the following photofragmentations:

$$CS_2 + 193\text{ nm} \rightarrow CS + S \qquad (1)$$

$$Fe(CO)_5 + 193\text{ nm} \rightarrow Fe(CO)_x + yCO \qquad (2)$$

$$Co(CO)_3NO + 193\text{nm} \rightarrow yCO + Co(CO)_xNO \qquad (3)$$

For reaction (1), there are two possible product channels:

$$CS_2 + 193\text{ nm} \rightarrow CS(^1\Sigma^+) + S(^3P) + 45\text{ kcal/mole} \qquad (1a)$$

$$CS_2 + 193\text{ nm} \rightarrow CS(^1\Sigma^+) + S(^1D) + 19\text{ kcal/mole} \qquad (1b)$$

For CS_2 in the 2000Å region, there is a very strong structured absorption system due to transitions from the linear $^1\Sigma^+_g$ ground state to a bent $^1\Sigma^+_u(^1B_2)$ excited state. The photochemistry of this excited state has been the subject of a great deal of study[8], which has resulted in a controversy over the relative yields of the singlet and triplet dissociation channels. In three recent studies, the relative yields of $S(^3P):S(^1D)$ were determined to be 0.25:1[9], 6:1[10], and 0.66:1[8], a range of more than an order of magnitude. As well, there is some uncertainty about the energy release for each channel in the fragmentation. By probing the products of the dissociation with tunable coherent VUV, it is possible to directly address these problems and to provide a measurement of the spatial anisotropy of the dissociation caused by photolysis using polarized light.

The study of large molecule photochemistry is also an active area in chemical physics, as it addresses the issue of energy flow and statistical behaviour in isolated molecules. Organometallic compounds provide attractive model systems for such studies for several reasons. These compounds are often quite volatile and they absorb throughout the ultraviolet and into the visible. A wide variety of ligands can be attached to the central metal, and compounds with clusters of metal atoms at their core can be made. Studying such compounds should provide insight into catalysis by metal clusters, a very important chemical process.

In order to probe the photofragmentation of these larger molecules, coherent VUV is again used for product detection, this time to detect the CO fragment, a common product in organometallic photochemistry[11].

In the 193nm photolysis of $Co(CO)_3NO$ and $Fe(CO)_5$, the energy of the photon is enough to remove all of the ligands from the central metal atom. Thus the excitation is well above the lowest dissociation thresholds, and one is studying the photofragmentation of highly excited molecules. In order to probe the dynamics of such processes, the energy distribution of the fragments and the dissociation anisotropy (if it exists) are measured.

The case of $Co(CO)_3NO$ is particularly interesting, as there are two different leaving groups that can be probed by LIF (NO can be detected by excitation of the $A^2\Sigma$-$X^2\Pi$ transition around 220nm). Although the thermochemistry of even these simple organometallics is not well known[11], it is believed that the Co-NO bond is stronger than the Co-CO bond, thus measurements of the relative yields and energy distributions for the CO and NO products are important.

The photochemistry of $Fe(CO)_5$ has received a lot of attention recently[12], but little is known about the detailed dynamics of this photodissociation. The results we report here are the beginning of a detailed study on the photofragmentation dynamics of $Fe(CO)_5$ at several wavelengths in the ultraviolet and visible.

EXPERIMENTAL

A schematic diagram of the experimental apparatus is shown in Figure 1. The photolysis laser used was an ArF excimer laser (Questek 2000) which provided 15nsec wide pulses of 193nm radiation at a 10hz repetition rate. The excimer laser was run in the "powerlok" mode, providing for a constant pulse energy of about 25mJ/pulse throughout an experiment. The laser output was attenuated to 1mJ/pulse by layers of stainless steel mesh, collimated and reduced by a suprasil lens telescope and spatially filtered by an iris diaphragm to a 3mm spot. The unpolarized 193nm light was then passed through a stacked plate polarizer consisting of 12 suprasil windows at brewster's angle, arranged so

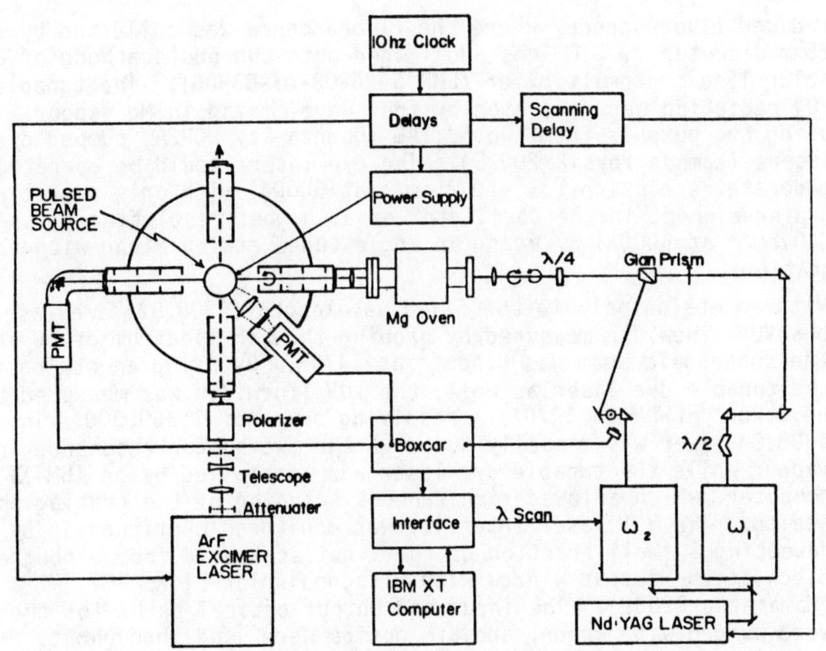

Figure 1: Schematic of experimental apparatus.

the 193nm beam was not deflected by insertion of the polarizer. The photolysis laser pulse energy was measured by a Gentek ED200 energy meter which could be inserted before the input window. The 193nm beam was sent through a series of optical baffles and intersected the pulsed molecular beam at a distance of 3-4cm from the nozzle.

The fixed frequency dye laser is labelled ω_1, the tunable dye laser ω_2. Generated VUV has frequency $\omega = 2\omega_1 + \omega_2$. The axis of the supersonic beam is perpendicular to the page.

The molecule under study was introduced to the photolysis region by a pulsed supersonic nozzle of the design described by Adams et al.[19] The nozzle orifice was 0.16mm and the gas pulse was 200μsec wide. The CS_2, $Fe(CO)_5$ and $Co(CO)_3NO$ samples were commerically obtained, purified by vacuum distillation and degassed thoroughly before use. The resulting reagents were held at constant temperature and then their vapour pressure diluted in a flow of Helium to about 1% of the total 900 torr that was expanded from the source. By measuring the fluorescence excitation spectrum of the CO impurity present in a 1% $Fe(CO)_5$ in He beam, the rotational temperature of the CO under these molecular beam conditions was determined to be 10K.

At a variable time delay after the photolysis laser pulse (typically 0.5-1.0μsec) the VUV probe laser was sent through the photolysis volume and the S or CO products were detected by laser-induced fluorescence, where the fluorescence was collected by a 25mm diameter f1 LiF lens and imaged onto the photocathode of a solarblind photomultiplier (EMR 542G-08-17-03900). The tunable VUV radiation was generated by four-wave mixing in Mg vapour, using the outputs from two Nd:YAG (Quanta-Ray DCR2A) pumped dye lasers (Lambda Physik 2002E). The dye lasers could be operated at moderate resolution ($\Delta\nu \approx 0.25cm^{-1}$ at 5000Å) with only a grating tuning element in the oscillator or at higher resolution ($\Delta\nu \approx 0.07cm^{-1}$ at 5000Å) by using an angle tuned etalon along with the grating.

With an etalon only in the fixed wavelength (4308.8Å) dye laser, the VUV linewidth measured by probing the LIF spectrum of CO in the supersonic beam was $0.4cm^{-1}$ at 1470Å. By using an etalon in the tunable dye laser as well, the VUV linewidth was measured to be $0.2cm^{-1}$ FWHM at 1470Å, a resolving power of 1:350,000. The 4308.8Å laser was manually tuned to the two-photon resonance in Mg vapour while the tunable dye laser was controlled by an IBM-XT computer, which allowed simultaneous scanning of the grating and etalon. The VUV laser intensity was monitored continuously by detecting a small fraction of the light scattered from a roughened elbow at the output window with a second solar blind PMT (Hamamatsu R2032). The input and output optical paths for the VUV were purged with argon, and LiF optics were used throughout.

The signals from both photomultipliers were averaged by boxcar averagers (SRS SR250 or PAR 162) and the boxcar outputs passed to the controlling computer through a multiplexed 12 bit A/D converted. The observed LIF signal was divided by the VUV laser power, and standard reference signals (either a single S atom line or a series of CO lines) were recorded frequently during the experiments to correct for any long-term drift in sensitivity. Signals due to interference from CO impurities in the beam or photolysis due to the probe laser were measured by recording LIF spectrum with the excimer laser blocked. There was no background atomic S signal observed, and interference from the background CO was restricted to v=0, $J<10$ because of the internal cooling due to the supersonic expansion. Above J=10, no significant background CO was observed. Below J=10, the background signal was subtracted from the observed total signal. It should be mentioned that the observation of background CO was due to the high sensitivity of the VUV LIF method. No impurity CO could be detected in the metal carbonyl samples used when the infrared absorption spectrum was recorded by FTIR, which means that the CO impurity was present in less than 1% concentration.

RESULTS

The S atom products from CS_2 photolysis were detected using the $^1P_1 \leftarrow {}^1D_2$ transition at 1448.2Å and the $^3D_{J'} \leftarrow {}^3P_{J''}$ transitions between 1474Å and 1487Å. From the relative areas of the spectral peaks due to $^1D_2$ and ^3P_J S atoms, and using relative line strength measured independently[13], we determined that 69 ± 5% of the atomic S product is formed in the ^3P state, which means the ^3P:^1D branching ratio is 2.2. The relative

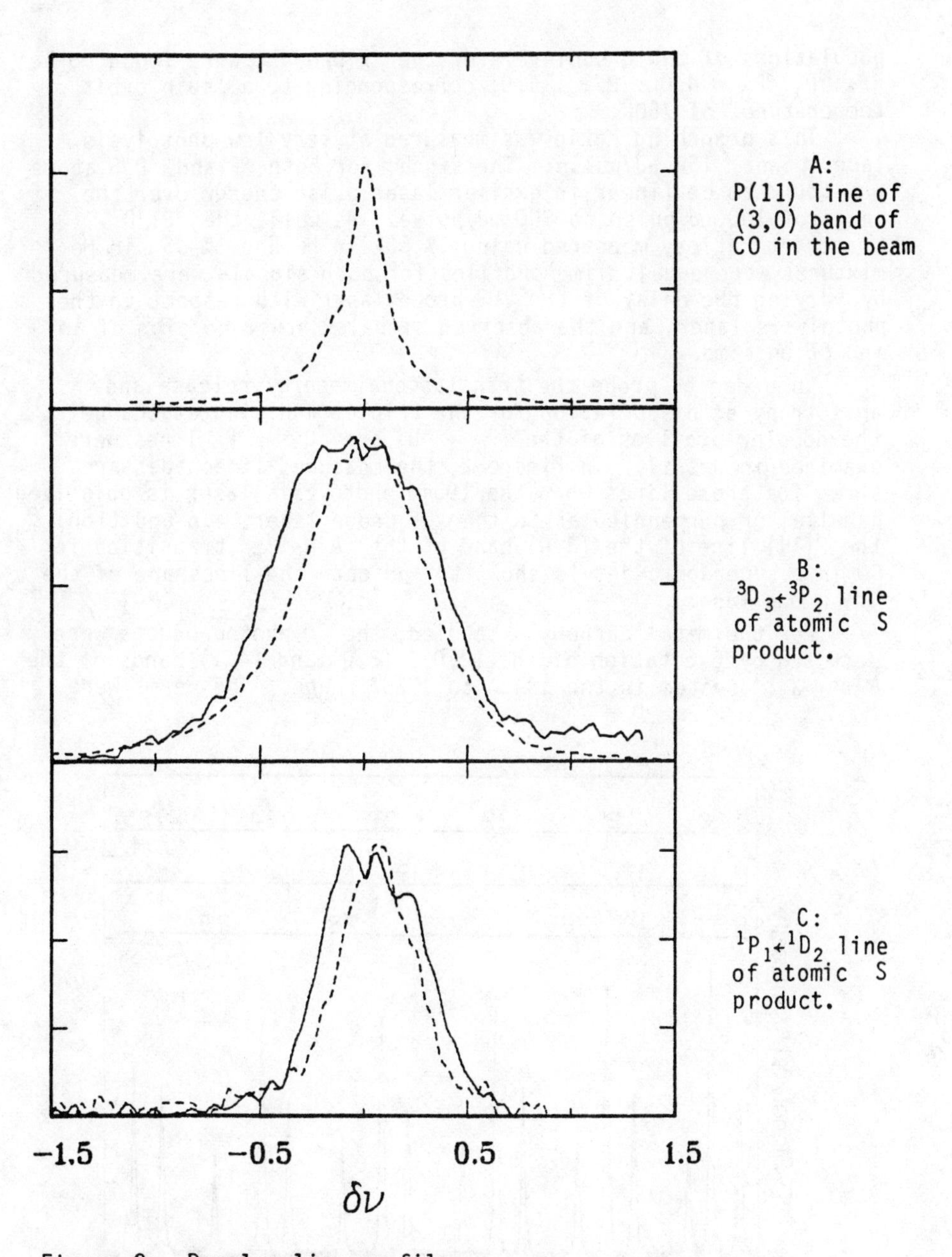

Figure 2: Doppler line profiles.

The lineshapes for jet cooled CO, and the $^3D_3 \leftarrow ^3P_2$ and $^1P_1 \leftarrow ^1D_2$ lines of the S product from CS_2 + 193nm → CS + S. For the atomic S lines, the solid curves are the Doppler profile measured with the photolysis laser polarized parallel to the VUV probe and the dashed curves are the line and profiles for the photolysis laser polarization perpendicular to the probe laser.

populations of the J sublevels of the ^3P product were found to be $^3P_2 : {}^3P_1 : {}^3P_0 = 4.0 : 2.1 : 1.0$, corresponding to a "spin orbit temperature" of 750K.

This branching ratio was measured at very low photolysis laser power, 150 μJ/pulse. The signal for both ^3P and ^1D S atoms was found to be linear in excimer laser pulse energy over the range of 100 μJ/pulse to 800 μJ/pulse. As well, the $^3P : {}^1D$ branching ratios, measured using 1% CS_2 in He and 5% CS_2 in He mixtures were equal. Time profiles for both signals were measured by varying the delay of the VUV probe laser with respect to the photolysis laser, and the observed signals showed no sign of an induction time.

In order to probe the translational energy release and anisotropy of dissociation for the triplet and singlet channels, the doppler profiles of the $^1P_1 \leftarrow {}^1D_2$ and $^3D_3 - {}^3P_2$ lines were examined in detail. In figure 2, the measured lineshapes are shown for these lines when the 193nm photolysis laser is polarized parallel or perpendicular to the VUV probe laser. In addition, the P(11) line of the (3,0) band of the $A^1\Pi - X^1\Sigma^+$ transition for CO in a supersonic jet is shown to indicate the lineshape of the VUV probe laser.

For the metal carbonyls studied, the CO photoproducts were detected by excitation of the (2,0), (3,0) and (4,1) bands of the $A^1\Pi \leftarrow X^1\Sigma^+$ system in the 1447Å to 1500Å range. The beams were

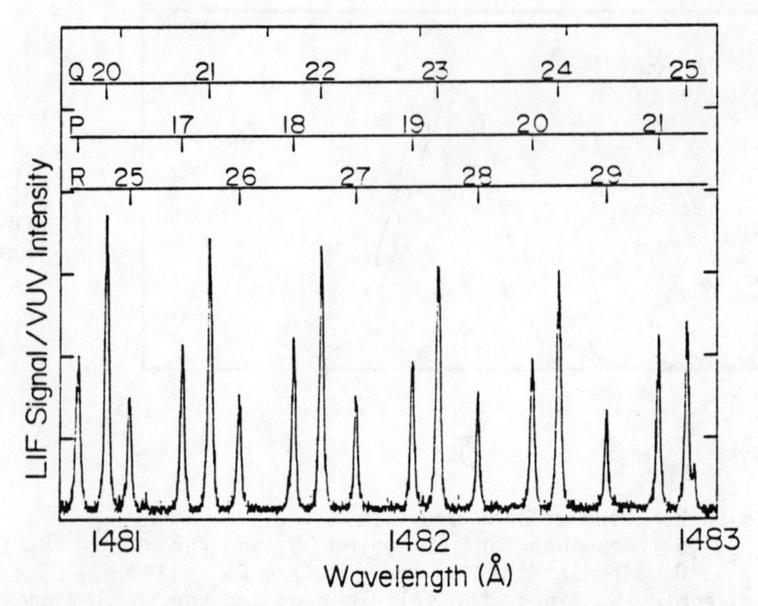

Figure 3. LIF spectrum of CO product from $Fe(CO)_5$. A portion of the (2,0) band of the CO spectrum is shown, with the rotational assignments indicated.

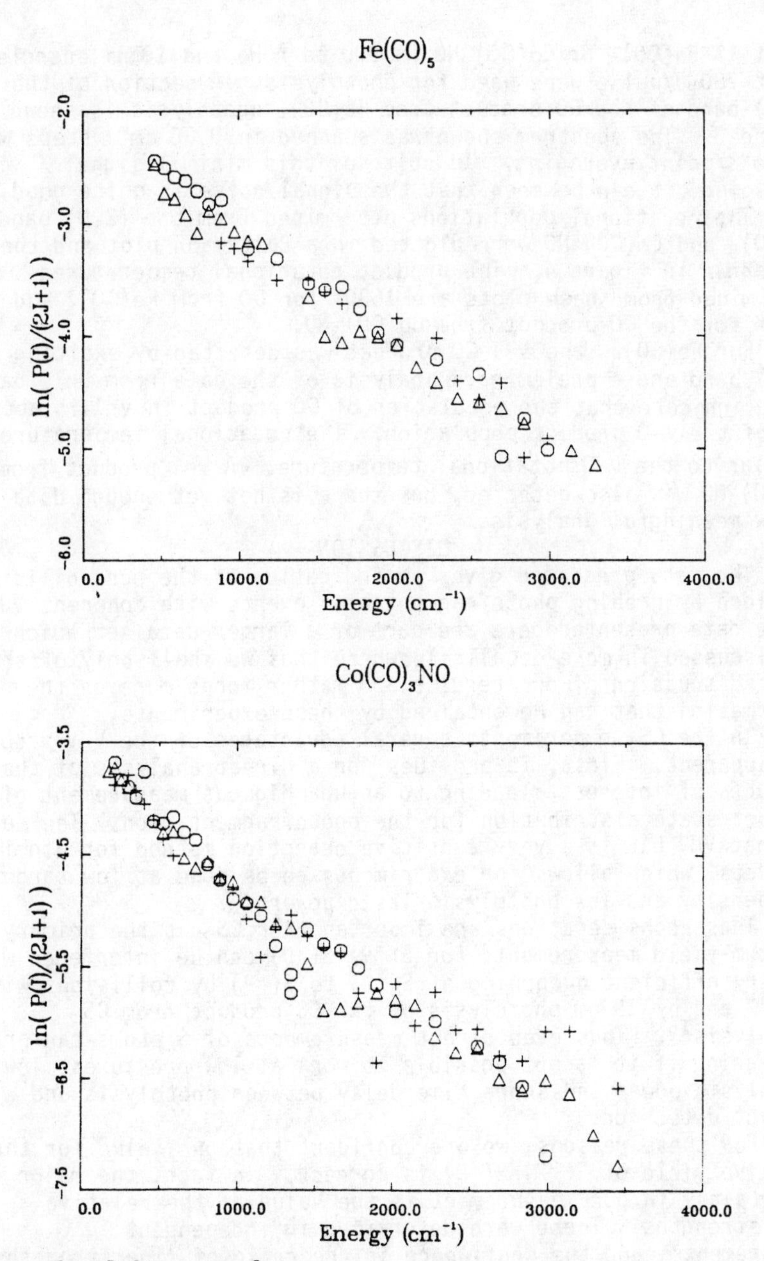

Figure 4. Boltzmann plots.

The rotational populations of CO were measured from the rotational line intensities of the (2,0) band of the CO products from $Fe(CO)_5$ and $Co(CO)_3NO$. All three branches in the band were analyzed and the symbols refer to rotational state populations measured from different branches: crosses - R branch lines, triangles - Q branch lines, circles - P branch lines.

about 1% $Fe(CO)_5$ or $Co(CO)_3NO$ in 900 torr He and 193nm energies of about 700μJ/pulse were used for photolysis. A section of the (2,0) band of the CO product from $Fe(CO)_5$ photolysis is shown in figure 3. The spectrum shown was scanned in 0.08 cm^{-1} steps with 3 shots/point averaging. In spite of this minimal signal averaging, it can be seen that the signal/noise is quite good.

The rotational populations determined from the (2,0) band for $Fe(CO)_5$ and $Co(CO)_3NO$ were plotted in a Boltzmann plot and these are shown in figure 4. The product rotational temperatures determined from these plots are 1530K for CO from $Fe(CO)_5$ and 1400K for the CO product from $Co(CO)_3NO$.

For $Fe(CO)_5$, the v=1 CO product was detected by exciting the (4,1) band and a preliminary analysis of the data from this band would indicate that the population of CO product in v=1 is about 12% of the v=0 product population. The rotational temperature is similar to the v=0 rotational temperature. A v=1 product from $Co(CO)_3NO$ was also detected, but there is not yet enough data to allow meaningful analysis.

DISCUSSION

The data presented gives an indication of the possibilities provided by probing photofragmentation events with coherent VUV. These data presented here are part of a larger data set which will be discussed in more detail elsewhere thus we shall only offer a brief discussion of our results and rather focus more on the information that can be obtained by these experiments.

In the CS_2 experiments several advantages of the VUV probe are apparent. First, it provides for a direct analysis of the products of interest, leading to an unambiguous measurement of the product state distribution for the photofragmentation. The second is that VUV LIF is a very sensitive detection method for atomic products, which allows for experiments to be done at low target gas density and low photolysis laser power.

These considerations are important for CS_2 as the primary quantum yield measurements for $S(^3P)$:$S(^1D)$ can be interfered with by very efficient quenching of $S(^1D)$ to $S(^3P)$ by collisions with CS_2 [14] and by 193nm photolysis of the CS product from CS_2 photolysis[15]. Thus even direct measurements of S atoms can prove misleading if it is not possible to work at low pressures, low photolysis power and short time delay between photolysis and product detection.

For these reasons, we are confident that our value for the relative yield of $S(^3P)/S(^1D)$ is correct. In fact, the major uncertainty in our measurement is the value of the relative line strengths. These were determined in independent measurements, and the confidence in the ratio of line strengths is about ±10%, compared with the ±5% reproducibility in $S(^3P)/S(^1D)$ signal ratios under a wide range of conditions.

Along with the high sensitivity of the VUV LIF detection method, the high resolution of the probe laser makes it possible to use doppler spectroscopy, even when probing a relatively heavy product such as atomic S, as our results demonstrate. Although the velocity resolving power of this method when applied to S atoms is not as good as photofragment translational spectroscopy[16], doppler spectroscopy has the advantage that it can separate different outgoing channels (i.e. singlet vs. triplet) unambiguously, while time of flight spectroscopy with a mass spectrometer detector must necessarily detect all possible pathways simultaneously. This is also important when the dissociation anisotropy is taken into account, as again the effect of the photolysis laser polarization on the product doppler lineshapes provides a direct measure of the anisotropy of each channel independently of the other possible channels.

As an illustration of this, a preliminary analysis of the data presented in figure 2 would indicate an average product translational energy of 10 kcal/mole for the $S(^1D)$ channel, and 23 kcal/mole for the $S(^3P)$ channel. These results are in good agreement with the published values of McCrary et al.[8], who determined the translational energy distribution for the total products using time of flight spectroscopy. By combining these results with those from LIF of the CS product, McCrary et al.[8] postulated average translational energies of 9 kcal/mole for CS + $S(^1D)$ and 23 kcal/mole for CS + $S(^3P)$. The data in figure 2 indicates an anisotropy in the dissociation resulting from the alignment of the transition dipole along the dissociation axis, the S-C-S bond. Such an anisotropy was not observed in previous experiments[9], due to the fact that in the current study, the CS_2 molecules are rotationally cooled by supersonic expansion to ≈10K, while in the work of Yang et al.[9] a thermal, effusive beam of CS_2 was used. This rotational cooling accentuates any anisotropy, and for both $S(^1D)$ and $S(^3P)$ the observed anisotropy is consistent with the expected excited state lifetime of about 1 psec, estimated from the fluorescence quantum yield for excitation at 201nm[17].

A recent photofragment translational spectroscopy experiment carried out by Y.T. Lee and coworkers[18], which measured the time of flight spectrum of the CS product from 193nm photolysis of CS_2 is consistent with our results. Although the time of flight experiments do not in themselves provide a clear separation between the singlet and triplet channels, it is possible to use our lower resolution doppler profiles, and our measured $S(^3P)/S(^1D)$ branching ratio to separate the two contributions to the observed time of flight spectrum. The recent work of Lee also shows an anisotropy in the dissociation, again enhanced by the supersonic cooling of the CS_2 in the beam.

In the metal carbonyl experiments, the high sensitivity for detection of CO is demonstrated. For the beam conditions used in these experiments, we estimate the $Fe(CO)_5$ density in the photolysis region was on the order of 10^{-5} torr. With the low photolysis laser powers used here only on the order of 1% of those molecules were dissociated. This means that about 10^{-7} torr total pressure of hot CO product was produced by the dissociation. Under the conditions used to record the spectrum shown in figure 3, a density of 10^7 cm^{-3} of CO in a given quantum state can be detected with a signal/noise of 1. Relatively simple improvements in the light collection optics used and more signal averaging would improve this detection limit by about a factor of 10.

For $Fe(CO)_5$, in spite of the fact that the photon energy of 6.41eV is far above the first dissociation threshold ($Fe(CO)_5 \rightarrow Fe(CO)_4 + 5CO$) of about 2eV, and in fact is above the energy required for complete dissociation of the molecule ($Fe(CO)_5 \rightarrow Fe + CO$) the energy release appears statistical. The rotational energy distribution, although quite excited, can still be described by a single rotational temperature. Our preliminary result on the CO v=1 population corresponds to a vibrational temperature of 1400K, in close agreement with the 1530K rotational temperature measured. These results would indicate a rapid equilibrium of the initial excitation in the molecule, followed by a statistical dissociation.

Similarly, for CO from $Co(CO)_3NO$ we can see Boltzmann behaviour for the rotational populations, with a temperature very close to the $Fe(CO)_5$ result. In figure 4 it can be seen that there may be a non-linearity in the Boltzmann plot in this case, but more data must be obtained before any conclusions of this sort are possible.

FUTURE WORK

The results presented in this paper provide only a glimpse of the possibilities provided by using coherent VUV to probe photofragmentation. This method can be applied to a wide range of reactions, those generating the CO and atomic product described here and, with extension of the VUV probe to shorter wavelengths, reactions producing H_2 and H.

For CS_2, we are in the process of doing a detailed analysis of the doppler profiles. Further work will involve using different photolysis wavelengths within the same absorption system to investigate how the dynamics are affected by exciting different vibrational states in the excited electronic state.

The work on the photochemistry of organometallics will be pursued in several ways. To complete the study on 193nm photolysis of $Fe(CO)_5$ and $Co(CO)_3NO$, more data needs to be collected on the vibrationally excited CO products. The product translational energy and dissociation anisotropy will be probed by measuring the doppler lineshapes for CO products using a polarized photolysis laser. Finally, for the $Co(CO)_3NO$ product, the NO product can be probed by LIF using a tunable ultraviolet laser in the 2200Å region.

Additional experiments can be carried out at different photolysis wavelengths, starting with the 249nm, 308nm and 351nm outputs of the excimer laser. Further study of different organometallics, especially those based on metal atom clusters is planned, to investigate the effect of cluster size on the dissociation dynamics.

ACKNOWLEDGEMENTS

This research has been supported by NSERC (Canada) with partial support provided by the Research Corporation (#10127) and the Petroleum Research Fund (#17811-AC6). JWH wishes to acknowledge support of NSERC in the form of a University Research Fellowship, IMW acknowledges the support of NSERC in the form of a graduate scholarship and HFD thanks Bruker Spectrospin (Canada) for an undergraduate research scholarship.

The authors wish to thank Dr. J. Frey for communicating results prior to publication, and Professor R. Bersohn for helpful discussions. D. Hart of our research group is thanked for help in setting up the VUV laser system.

REFERENCES

1. M. Shapiro and R. Bersohn, Ann. Rev. Phys. Chem., 33, 409 (1982).
   J.P. Simons, J. Phys. Chem., 88, 1287 (1984).
   R. Bersohn, J. Phys. Chem., 88, 5145 (1984).
2. N. Sivakumar, I. Burak, W.Y. Cheung, P.L. Houston and J.W. Hepburn, J. Phys. Chem., 89, 3609 (1985).
3. J.W. Hepburn, Israel J. Chem., 24, 273 (1984).
4. S.C. Wallace and G. Zdasiuk, Appl. Phys. Lett., 28, 449 (1976).
5. a) J.W. Hepburn, D. Klimek, K. Liu, R.G. Macdonald, F.J. Northrup and J.C. Polanyi, J. Chem. Phys., 74, 6226 (1981).
   b) J.W. Hepburn, K. Liu, R.G. Macdonald, F.J. Northrup and J.C. Polanyi, J. Chem. Phys., 75, 3353 (1981).
6. J.W. Hepburn, F.J. Northrup, G.L. Ogram, J.M. Williamson and J.C. Polanyi, Chem. Phys. Lett., 85, 127 (1982).
7. D.J. Bamford, S.V. Filseth, M.F. Foltz, J.W. Hepburn and C.B. Moore, J. Chem. Phys., 82, 3032 (1985).
8. V.R. McCrary, R. Lu, D. Zakheim, J.A. Russell, J.B. Halpern and W.M. Jackson, J. Chem. Phys., 83, 3481 (1985) and references quoted in the introduction.
9. S.C. Yang, A. Freedman, M. Kawasaki and R. Bersohn, J. Chem. Phys., 72, 4058 (1980).
10. M.C. Addison, R.J. Donovan and C. Fotakis, Chem. Phys. Lett., 74, 58 (1980).
11. G.L. Geoffroy and M.S. Wrighton, Organometallic Photochemistry, Academic Press, Inc., N.Y. (1979).
    V. Balzani and V. Carassiti, Photochemistry of Coordination Compounds, Academic Press, Inc., N.Y. (1970).
12. Y. Nagano, Y. Achiba and K. Kimura, J. Chem. Phys., 84, 1063 (1986).
    R.L. Whetten, K.-J. Fu and E.R. Grant, J. Chem. Phys., 83, 4899 (1983).
    D.V. Horak and J.S. Winn, J. Phys. Chem., 87, 265 (1983).
13. N. Sivakumar, I. Burak, G. Hall, P.L. Houston and J.W. Hepburn, paper in preparation.
14. D.J. Little, A. Dalgleish and R.J. Donovan, Faraday Disc. Chem. Soc., 53, 211 (1972).
15. G. Dorhofer, W. Hack and W. Langel, J. Phys. Chem., 88, 3060 (1984).
16. J.W. Hepburn, R.J. Buss, L.J. Butler and Y.T. Lee, J. Phys. Chem., 87, 3638 (1983).
17. K. Hara and D. Phillips, Trans. Faraday Soc., 74, 1441 (1978).
18. J. Frey, A. Wodtke and Y.T. Lee, unpublished data.
19. T.E. Adams, B.H. Rockney, R.J.S. Morrison and E.R. Grant, Rev. Sci. Instrum., 52, 1469 (1981).

RADIATIVE LIFETIME MEASUREMENTS OF THE $A1_u$ STATES OF Ar_2 AND Kr_2*

A. A. Madej, P. R. Herman, and B. P. Stoicheff
Department of Physics, University of Toronto
Toronto, Ontario, M5S 1A7, Canada

ABSTRACT

Time-resolved fluorescence studies of Ar_2 and Kr_2 formed by supersonic jet expansion and excited by monochromatic coherent VUV radiation have been used to measure radiative lifetimes of high vibronic levels of the $A1_u$ states. Values of 160 and 56 ns, were found for Ar_2 and Kr_2, respectively. These differ significantly from lifetimes of 3 μs and 0.26 μs obtained for $v' \sim 0$ levels using high gas pressures and excitation by charged particles or synchrotron radiation. Such large variations with vibrational level probably arise from changes of the electronic transition moment with changes in internuclear separation.

INTRODUCTION

The combined use of tunable coherent VUV laser sources and a pulsed supersonic jet has permitted the investigation of fluorescence excitation spectra of Xe_2, Kr_2, and Ar_2 at high resolution. The resonant four-wave mixing techniques using metal vapors (Mg, Zn, Hg)[1] and the spectra of Xe_2 and their analysis[2] were reviewed in the first two conferences in this series. Here, we present preliminary results of radiative decay measurements for specific rovibronic levels of the $A1_u$ states of Ar_2 and Kr_2. These and nearby 0_u^+ and 0_u^- states participate in the operation of electron-beam pumped excimer lasers[3], and knowledge of their radiative lifetimes is of considerable interest. Several earlier experiments of lifetime measurements have been reported. These investigations used either charged particles[4-6] or synchrotron radiation[7,8] for fluorescence excitation of Ar_2 and Kr_2 excimers formed at high pressures ($P > 100$ torr). At these pressures, the collisional frequency is sufficiently high that rapid vibrational relaxation occurs, resulting in observed fluorescence from low vibrational levels ($v' \sim 0$) of the $A1_u$ states. In the present experiments, radiative lifetimes were measured for high vibronic levels of the $A1_u$ states of Ar_2 and Kr_2, in a pressure regime ($<10^{-2}$ torr) where lifetimes are shorter than mean collision times. These results differ substantially from the values obtained for the levels $v' \sim 0$, and are considered to be due to large changes in electronic transition moment with internuclear distance for the transitions $A1_u \leftrightarrow X0_g^+$, in the rare gas dimers.

*Research supported by the Natural Sciences and Engineering Research Council of Canada, and the University of Toronto.

EXPERIMENT AND RESULTS

The experimental arrangement was essentially the same as that employed in the spectroscopic studies of Xe_2, Kr_2, and Ar_2. Tunable and monochromatic ($\Delta\nu \sim 0.3\,cm^{-1}$) VUV radiation was generated by the process of resonantly enhanced four-wave sum mixing in Hg vapor[9]. This pulsed radiation was incident on a beam of rare gas dimers formed in their ground states (at < 2 K) by expansion through a pulsed supersonic nozzle, with the gas pressure in the interaction region being typically 1 - 10 millitorr. Pure argon at a back pressure of ∿5 atm was used in the jet expansion to form the Ar_2 molecules and a Kr/He mixture of 1/6 at ∿5 atm for the Kr_2 molecules. Fluorescence emission was detected by a solar-blind photomultiplier tube (EMR-510G-08-013), with signal averaging and storage provided by a programmable transient digitizer (Tektronix 7912 AD). The time response of the detection system was ∿2.5 ns, with the incident VUV excitation pulse being ∿3 ns FWHM.

The $A1_u - X0_g^+$ band systems of Ar_2 and Kr_2 occur at 107.5 and 125.7 nm, respectively. Only high vibrational levels of the $A1_u$ states are accessible from the ground states because of the relative positions of the corresponding potential energy curves shown in Fig. 1. Thus the spectroscopic and present lifetime studies were limited to v' = 24 to 30 for Ar_2, and to v' = 32 to 38 for Kr_2. For Ar_2 with its single isotope and small moment of inertia, the rovibronic spectra were well-resolved, and it was possible to select and to excite specific rovibronic levels. For Kr_2 with its several isotopes and relatively large moment of inertia the vibronic bands appeared as shaded band heads, so that excitation for the present lifetime measurements was limited to the band heads.

The VUV wavelength was tuned to each selected rovibronic line or band head, and the fluorescence signal was detected and accumulated. For each pulse of incident radiation, ∿10 to 100 photons were observed, and measurements of 1400 pulses were averaged for Ar_2, and 800 for Kr_2. Typical fluorescence decay curves are shown in Fig. 2. These

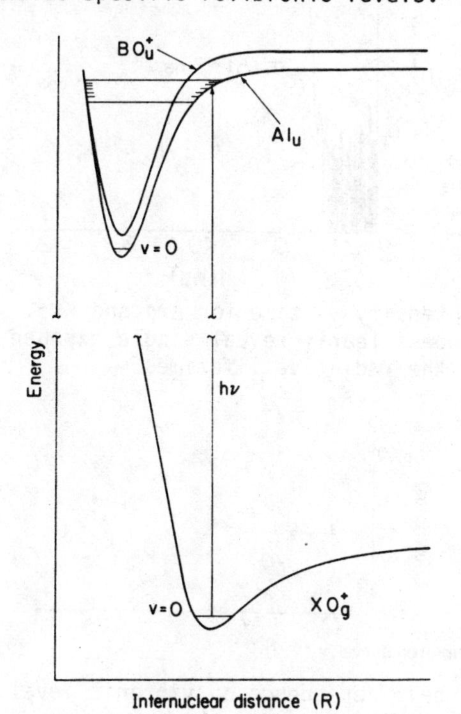

Fig. 1. Schematic potential energy curves for rare gas dimers showing transitions from v = 0 of the ground state to high vibrational levels of the $A1_u$ state.

clearly revealed single exponential decays with time, yielding radiative lifetimes for each level investigated. Measured lifetimes were independent of pressure over the small pressure range of these studies (1 to 10 millitorr). For Ar_2, the lifetimes were also found to be independent of rotational quantum number within a given vibronic level. Graphs of measured lifetimes showing their dependence on vibronic level are presented in Fig. 3. For Ar_2, the radiative life-

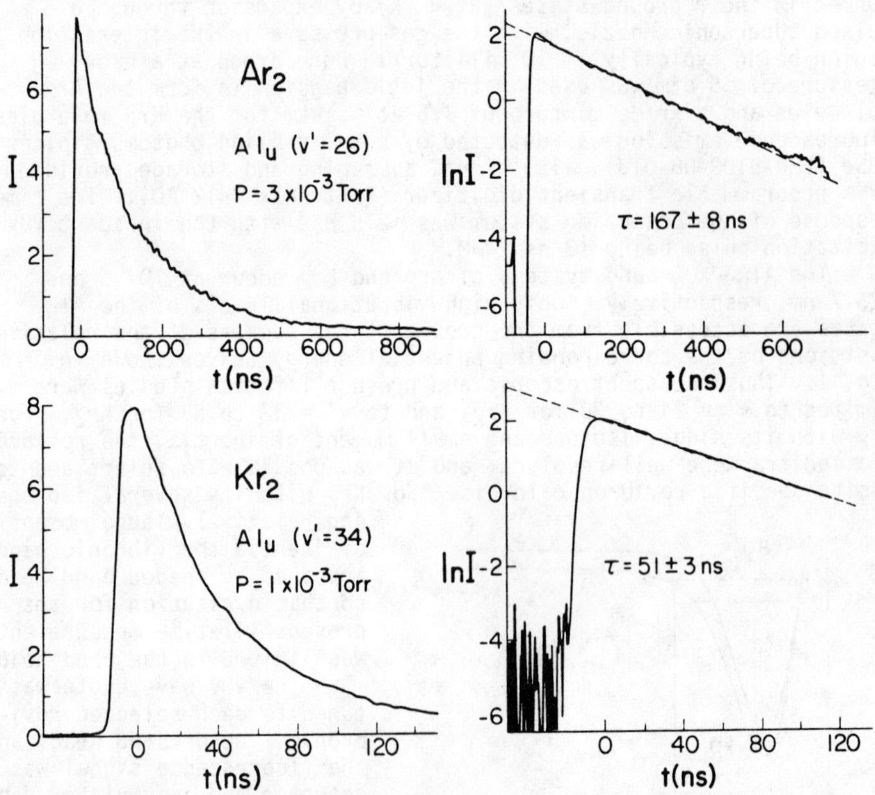

Fig. 2. Graphs of fluorescence intensity vs time for Ar_2 and Kr_2, including semi-log plots whose slopes clearly reveal single exponential decays and yielded values of the radiative lifetimes.

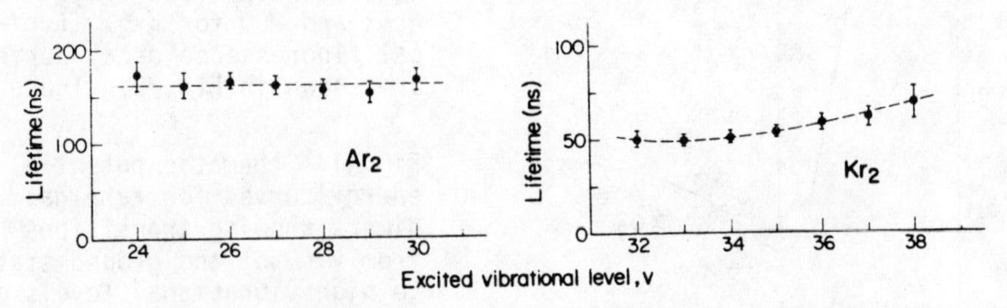

Fig. 3. Radiative lifetimes and their dependence on vibronic level.

times were observed to be constant in the range v' = 24 to 30, giving an average value of $\tau = 160 \pm 10$ ns. For Kr_2, however, a noticeable variation with v' was found: a minimum lifetime of $\tau = 50 \pm 3$ ns was obtained at v' = 33, and a maximum value of $\tau = 69 \pm 8$ ns at v' = 38, the highest vibronic level for which measurements could be made in the present experiment.

DISCUSSION

The significance of the present results is the large difference in radiative lifetimes found here for high vibronic levels in comparison with earlier values of 3 μs[5] and 0.26 μs[7] obtained for v' ∿0 levels of Ar_2 and Kr_2, respectively. These results imply a reduction in lifetime by factors of ∿20 and ∿5 for Ar_2 and Kr_2 in going from v' ∿0 to high vibronic levels. Such a decrease in radiative lifetime with increasing vibronic level has been predicted for Ne_2 by Schneider and Cohen[10], but to date, there has been no experimental confirmation. Large variations in radiative lifetimes may arise from changes in the electronic transition moment $\mu(R)$ with changes in internuclear distance R. For the 1_u states of the rare gas dimers, an electric dipole transition to the ground state 0_g^+ is not allowed in the separated-atom or united-atom limits. However, at intermediate distances, spin-orbit coupling enables the transition to be weakly allowed. Theoretical results[10,11] show that this causes a strong R dependence in $\mu(R)$, and this is indicated qualitatively in Fig. 4. The transition moment is essentially zero at very small and

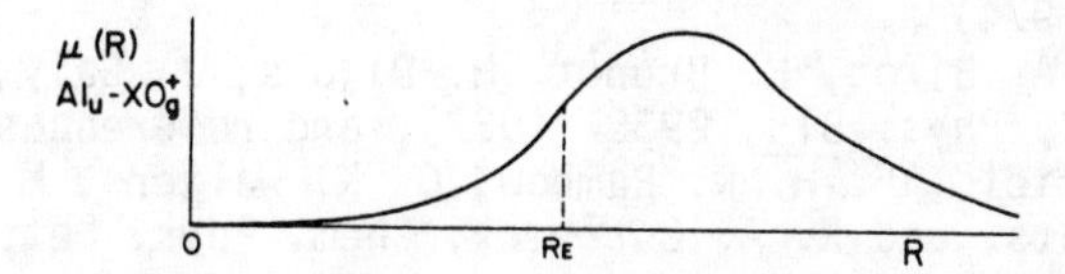

Fig. 4. Schematic diagram indicating the dependence of electronic transition moment $\mu(R)$ on internuclear distance, R.

very large distances, and smoothly approaches a maximum at R > Re (the equilibrium internuclear separation of the 1_u state). If this relation for $\mu(R)$ is known, the value of the radiative lifetime (τ) and its variation with R can be estimated from the usual relation for the spontaneous transition probability $A = \tau^{-1} = \text{constant} \times |\Delta E|^3 |\mu(R)|^2$. Here ΔE is the vertical energy between two states. In this way, Schneider and Cohen[10] have calculated a change in τ for the Ne_2 1_u state from 11.9 μs at v' = 0 to 1.0 μs at v' = 10. For Xe_2, we find a reduction by a factor of ∿3 in τ over the range v' = 0 (R ∿3.1Å) to v' = 40 (R ∿4.3Å) using the data on $\mu(R)$ given by Ermler et al[11]. These calculations give some validity to the above explanation for the large variation in radiative lifetimes obtained for Ar_2 and Kr_2.

Recently, we have received calculated values of $\mu(R)$ for Ar_2 from Professor Ermler. It now remains to carry out the corresponding calculations for the spontaneous transition probabilities for

Ar_2, and to compare the derived lifetimes with the experimental values. Also, we plan to continue the present set of lifetime measurements of the rare gas dimers with experiments on Xe_2 for comparison with the above calculations.

We thank Professor H. M. van Driel for the use of the transient digitizer and Professor G. K. Walters of Rice University for his helpful remarks and suggestions. This research was supported by the Natural Sciences and Engineering Research Council of Canada (NSERC) and the University of Toronto (U. of T.). A. A. Madej is grateful for a NSERC Postgraduate Scholarship and P. R. Herman for a U. of T. Open Scholarship.

REFERENCES

1. B. P. Stoicheff, J. R. Banic, P. Herman, W. Jamroz, P. E. LaRocque, and R. H. Lipson, in Laser Techniques for Extreme Ultraviolet Spectroscopy (T. J. McIlrath and R. R. Freeman, eds.) Amer. Inst. Phys., New York (1982), pp. 19-31; Can. J. Phys. 63, 1581 (1985).
2. R. H. Lipson, P. E. LaRocque, and B. P. Stoicheff, in Laser Techniques in the Extreme Ultraviolet (S. E. Harris and T. B. Lucatorto, eds.) Amer. Inst. Phys., New York (1984), pp. 253-258; J. Chem. Phys. 82, 4470 (1985).
3. M. H. R. Hutchinson, Appl. Phys. 21, 95 (1980).
4. L. Colli, Phys. Rev. 95, 892 (1954).
5. J. W. Keto, R. E. Gleason, Jr., and G. K. Walters, Phys. Rev. 33, 1365 (1974).
6. P. Millet, A. Birot, H. Brunet, H. Dijols, J. Galy, and Y. Salamero, J. Phys. B15, 2935 (1982), and references therein.
7. T. D. Bonifield, F. H. K. Rambow, G. K. Walters, M. V. McCusker, D. C. Lorents, and R. A. Gutcheck, Chem. Phys. Lett. 69, 290 (1980).
8. T. D. Bonifield, F. H. K. Rambow, G. K. Walters, M. V. McCusker, D. C. Lorents, and R. A. Gutcheck, J. Chem. Phys. 72, 2914 (1980).
9. P. R. Herman and B. P. Stoicheff, Opt. Lett. 10, 502 (1985).
10. B. Schneider and J. S. Cohen, J. Chem. Phys. 61, 3240 (1974).
11. W. C. Ermler, Y. S. Lee, K. S. Pitzer, and N. W. Winter, J. Chem. Phys. 69, 976 (1978).

COMPLEMENTARY BRANCHING RATIOS BY SATELLITE EXCITATION

L.D. Van Woerkom and W.E. Cooke
Physics Department
University of Southern California
Los Angeles, Ca 90089-0484

ABSTRACT

It is shown that if a state autoionizes to produce a fraction, B, of products in an excited ionic state, then one can obtain the complementary fraction, 1-B, by exciting the character of a different autoionizing state, but at the location of the first state's resonance. This structure will appear as a satellite to the second state with a branching ratio that depends on the first state only. This behavior is illustrated using the 6pns and 6pnd autoionizing states of barium.

INTRODUCTION

In the continuing attempt to develop short wavelength lasers, many schemes have centered on using inversions in atomic ions as the lasing medium. Atomic ions have the natural advantages of high energy, line emissions, so that the difficulty becomes only to create a sufficiently dense inversion. Shake-up photoionization has been used to create visible lasers[1], and has been proposed for some vuv lasers[2]. In this process, an atom is photoionized with a sufficiently high energy photon, that the remaining ion is "shaken" into an excited state, creating an inversion. Such a process can be dramatically enhanced if an autoionizing state lies in that energy region, and if that state naturally decays into excited ionic states, primarily. This process has already been used to demonstrate an autoionization pumped Ba^+ ion laser[3] in the visible spectrum. More recently, amplified spontaneous emission at 93 nm from Kr^+ ions has been produced, presumably by a similar process.[4] However, the autoionizing resonance itself can inhibit the production of an inversion if the state naturally produces primarily ground state ions, and consequently, the general usefulness of this process has been restricted. Here, we demonstrate that one usually has the ability to tailor dramatically the final product composition, by a judicious choice of the excitation route by which the autoionizing states are produced. The process which we describe, satellite excitation, is not limited only to autoionizing states, but can be used to modify the branching ratio of any short-lived resonance.

THEORY

We have studied the case of two different autoionizing states coupled to two distinct continua. It is assumed that the two bound states do not interact directly and that the two continua do not interact directly. For a two continuum problem there will always exist two valid wavefunctions with orthogonal continua for each energy.[5] If the two independent solutions are labelled by $|+\rangle$ and $|-\rangle$, the total wavefunctions may be written as

$$|+\rangle = A_1^+|1\rangle + A_2^+|2\rangle + \cos\theta|a\rangle + \sin\theta|b\rangle$$

$$|-\rangle = A_1^-|1\rangle + A_2^-|2\rangle + \cos(\theta+\pi/2)|a\rangle + \sin(\theta+\pi/2)|b\rangle \quad (1)$$

where the subscripts 1,2 refer to the two bound states and the angle, θ, determines the particular linear combination of the two continua for a given problem. In general, θ and the A coefficients all depend on energy. The wavefunctions are such that only one bound state ($|1\rangle$ or $|2\rangle$) couples strongly to one solution ($|+\rangle, |-\rangle$ respectively). Furthermore, near a bound resonance, the structure of the wavefunctions is determined by the resonant state's coupling to the continua and thus fixes the value of θ. As the energy is tuned from one resonance to the next, θ will change in general since the dominant resonant state is different. If ϕ_i defines the value of θ near the ith resonance, the following properties of the coefficients hold:

$$\text{on resonance 1:} \quad \theta=\phi_1 \; ; \; A_1^+ \gg 1 \; ; \; A_2^+ \sim 0$$

$$\text{on resonance 2:} \quad \theta=\phi_2 \; ; \; A_2^- \gg 1 \; ; \; A_1^- \sim 0 \quad (2)$$

In general, the first wavefunction will have very little admixture of any bound state <u>not</u> located at resonance $|1\rangle$; the second wavefunction <u>will</u> have very little of the bound character of the state which <u>is</u> located at resonance $|1\rangle$. This analysis will be valid provided that the bound states are separated by more than their widths so that overlapping resonance effects are not important. In this case, excitation of a character off resonance occurs through the wings of the lineshape.

However, for short-lived states, such as autoionizing states, this process can nevertheless be dominant in many energy regions.

We will consider branching ratios in the case that only one bound state character is excited and no continuum character (direct photoionization) is excited at all. The general energy dependent branching ratio for exciting only character |1> and observing only continuum |a> is

$$B(E) = \sigma_{1a}(E)/\sigma_{1t}(E) \tag{3}$$

where the numerator is the cross section for exciting character |1> and observing continuum |a>, and the denominator is the cross section for observing all decay products when exciting only character |1>. Since two wavefunctions exist for each energy, a coherent superposition is always excited; thus, the cross sections must be obtained by summing the appropriate transition moments before squaring. For our case of only exciting character |1>[6,7]

$$\sigma_{1a} \propto \mu_1^2[(A_1^+\cos\theta)^2+(A_1^-\cos(\theta+\pi/2))^2 \\ +2A_1^+A_1^-\cos\theta\cos(\theta+\pi/2)\cos\pi(\nu^+-\nu^-)]$$

$$\sigma_{1t} \propto \mu_1^2[(A_1^+)^2 + (A_1^-)^2] \tag{4}$$

where μ_1 is the transition moment from some initial state to the |1> character. The extra cosine factor in the cross term arises because the continuum functions have different phase shifts (ν^+,ν^-) in the two solutions, and this affects their overlap. The first two terms in the partial cross section are just the product of the excitation moment (A_1) into a given solution (|+> or |->) times the probability of that solution producing continuum |a>. Thus, this picture is appealing in that the probabilities are found from the admixture coefficients.

As evident in equation (4), the general branching ratio can be a complicated function of energy. If, however, one only considers energies near a resonance, expression (3) simplifies. By using equations 2 and exciting only the |1> character at the |1> resonance, ($E=E_1$), the |+> solution is primarily populated. This yields a branching ratio of $B_1(E_1) \approx \cos^2\phi_1$. (This is not exact since a small amount of the other solution is still excited.) Similarly, if the |2> character is excited at the |2> resonance, ($E=E_2$), one finds $B_2(E_2) \approx \cos^2(\phi_2+\pi/2) = \sin^2\phi_2$. These are the normal branching ratios for resonances |1> and |2> and reflect the fact that each bound state is coupled predominantly to one wavefunction ($|+\rangle$ or $|-\rangle$ respectively). It is possible, however, to excite a different character at a given resonance by making use of the finite admixture in the wings as mentioned earlier. For example, excitation of bound state $|1\rangle$ at the position of resonance $|2\rangle$ yields $B_1(E_2)=\cos^2\phi_2$ since it is state $|2\rangle$ that determines the coupling angle $\theta=\phi_2$. The answer is very simple owing to the fact that on resonance the coefficients in the wavefunctions simplify (see eqs. 2). Near resonance $|2\rangle$, the sum over the two solutions reduces to a single term since excitation of character $|1\rangle$ couples to the $|+\rangle$ wavefunction. This result is just the complement of the normal branching ratio for state $|2\rangle$: $B_1(E_2)\approx 1-B_2(E_2)$.

In summary the following is found:

on resonance 1:	excite $\|1\rangle$	$B_1(E_1)\approx\cos^2\phi_1$
	excite $\|2\rangle$	$B_2(E_1)\approx\sin^2\phi_1$
on resonance 2:	excite $\|1\rangle$	$B_1(E_2)\approx\cos^2\phi_2$
	excite $\|2\rangle$	$B_2(E_2)\approx\sin^2\phi_2$ (5)

It is useful to ask when this phenomenon will be important. First, only a single character must be excited; thus, lines with a high Fano 'q' are required.[5] Second, the relative transition moments must favor the off resonant character. At $E=E_1$, with Γ_1,μ_1 and Γ_2,μ_2 the width and excitation transition moment of $|1\rangle$ and $|2\rangle$ respectively, primarily $|2\rangle$ character will be excited if

$$\frac{(\mu_2)^2}{(\mu_1)^2} \gg \frac{4(E_2-E_1)^2}{\Gamma_1\Gamma_2} \qquad (6)$$

If this condition is satisfied, the complement of the normal branching ratio for state $|2\rangle$ will be observed at $E=E_1$. If either width is small, this condition will not be met. If the off resonant state is narrow, then the excitation strength in the wings will be vanishingly small. If the resonant state is narrow, then even a small transition moment will lead to a large transition rate on resonance.

EXPERIMENT

We have demonstrated this phenomenon in autoionizing Rydberg states of barium. Three tunable dye lasers, pumped by the harmonics of a pulsed Nd:YAG laser, cross an effusive atomic beam of barium. The first two lasers excite one of the two valence electrons to a bound Rydberg state. The third laser is tuned around the remaining ionic core resonance. This is the

Isolated Core Excitation (ICE) method[8], and considers transitions of the form:

$$Ba(6sn\ell) + \hbar\omega \rightarrow Ba(6pn\ell)$$

In fig. 1, state $|1\rangle$ is the $6P_{3/2}18s$ J=1 state, and $|2\rangle$ is the $6P_{3/2}16d$ J=1 state. Since cross sections for two electron transitions and direct photoionization of the Rydberg electron are small, this technique is ideally suited for exciting only a single bound character. A given set of data was obtained by choosing an initial state (6s18s or 6s16d) and exciting it to the appropriate character ($6P_{3/2}18s$ or $6P_{3/2}16d$) of the autoionizing state. This method selects only one bound character for each set of data.

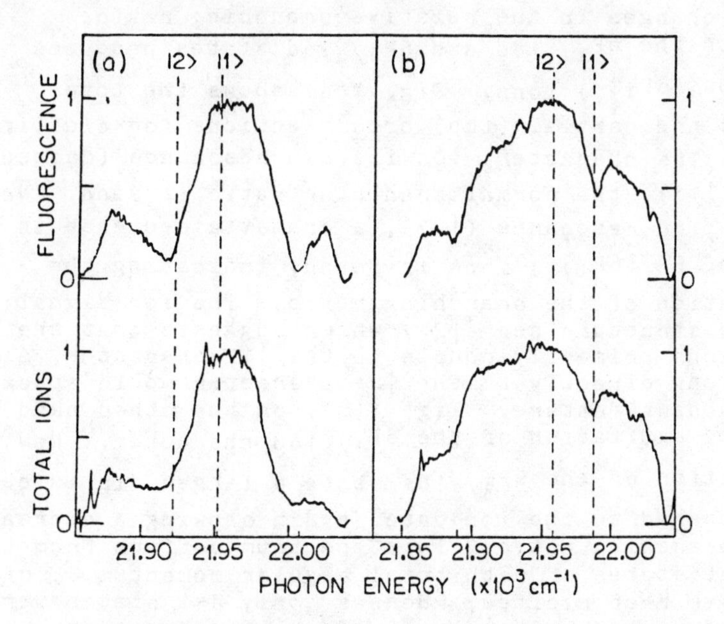

Fig. 1 The relative fluorescence yield (top) and relative ion yield (bottom) for exciting $6P_{3/2}n\ell$ states. In (a), the $6P_{3/2}ns$ character is excited; in (b), the $6P_{3/2}nd$ character is excited. The labels $|1\rangle$ and $|2\rangle$ show the locations of the $6P_{3/2}18s$ and $6P_{3/2}16d$ states, respectively. Note that in both (a) and (b), the relative fluorescence yield is lower than the relative ion yield at the satellite location. This shows that in both cases the branching ratio to produce excited $Ba^+(6P_{1/2})$ ions decreases at the satellite location.

In reality, there are more than two continua. The $Ba^+(6P_{1/2})$ continua may have two partial wave electrons. Further, both the $Ba^+(6s,5d)$ continua will have more partial waves available. Since we collect residual ion fluorescence, we cannot separate partial waves. In addition, we only monitor the fluorescence from the $6P_{1/2}$ ionic state. Thus, the experiment can only distinguish between two continua: i)the $Ba^+(6P_{1/2})$ continua, and ii)all other continua. The partial cross section was obtained by monitoring the residual ion fluorescence from the $Ba^+(6P_{1/2})$ state. The total cross section was found by collecting all of the ions produced. Although the detection system was not calibrated for absolute yields, the lineshape analysis illustrates the modification of the branching ratio by showing changes in the relative branching ratio.

Both the $6P_{3/2}18s$ and $6P_{3/2}16d$ states produce mostly $Ba^+(6P_{1/2})$ ions. Fig. 1(a) shows the total (bottom) and partial (top) cross sections for exciting the $6P_{3/2}18s$ character. On its own resonance (dashed line at $|1\rangle$), the normal branching ratio is seen. Near the $6P_{3/2}16d$ resonance ($|2\rangle$), a dramatic decrease in the number of $Ba^+(6P_{1/2})$ ions is found, indicating the modification of the branching ratio. The ion signal shows no structure near $|2\rangle$, which suggests that the transitions primarily couple to the $|1\rangle$ character, since transitions directly to the $|2\rangle$ character would appear as a resonant feature. Fig. 1(b), on the other hand, shows the excitation of the $6P_{3/2}16d$ character. Now, at the position of the $6P_{3/2}18s$ state a larger dip is seen when compared to the ion data, again showing a decrease in the branching ratio. This spectrum suffers from the fact that states of both total angular momentum, J, of 1 and 3 have been excited, whereas only J=1 states were populated in the left column. This introduces an additional background.

This data does not demonstrate perfect complementarity, but rather a dramatic modification of the normal branching ratio. This is due to the approximations made in our simple model: i)that the two bound states do not interact directly; ii)that only two continua exist; and iii)that the continua themselves do not interact with each other. A detailed analysis of these spectra, which discusses these conditions, will be presented elsewhere.[5]

CONCLUSION

We have shown that the excitation of an autoionizing state as a satellite to a second autoionizing state can produce the complement of the second states normal branching ratio. Although full complementarity may not routinely be achieved, dramatic modifications of the normal branching ratio have been seen. In any photoionization pumped laser scheme, the possibility of modified branching ratios must be considered.

ACKNOWLEDGEMENTS

This work was supported by the NSF under grant Phy85-00885.

REFERENCES

1 W.T. Silfvast, J.J. Macklin, and O.R. Wood,II. Opt. Lett. 8, 551 (1983).

2 S.E. Harris and R.G. Caro, Opt. Lett. 11, 10 (1986); R.G. Caro, P.J.K. Wisoff, G.Y. Yin, D.J. Walker, M.H. Sher, C.P.J. Barty, J.F. Young, and S.E. Harris; and W.T. Silfvast, O.R. Wood,II and D.Y. Al-Salameh, in Short Wavelength Coherent Radiation: Generation and Applications, J. Bokor and D. Atwood, ed. (AIP, New York, 1986).

3 J. Bokor, R.R. Freeman, and W.E. Cooke, Phys. Rev. Lett. 48, 1242 (1982).

4 H. Egger, T.S. Luk, W. Muller, H. Pummer, and C.K. Rhodes, in Laser Techniques in the Extreme Ultraviolet, S.E. Harris and T.B. Lucatorto, ed. (AIP, New York, 1984); T. Srinivasan, W. Muller, M. Shahdi, T.S. Luk, H. Egger, H. Pummer, and C.K. Rhodes, JOSA B, to be published.

5 U. Fano, Phys. Rev. 124, 1866 (1961)

6 C. Jugen and D. Dill, J. Chem. Phys. 73, 3338 (1980).

7 L.D. Van Woerkom and W.E. Cooke, to be published.

8 W.E. Cooke, T.F. Gallagher, S.A. Edelstein, and R.M. Hill, Phys. Rev. Lett. 40, 178 (1978); S.A. Bhatti and W.E. Cooke, Phys. Rev. A 28, 756 (1983).

AUTHOR INDEX

S

T

U

V

W

Y

Z

AIP Conference Proceedings

		L.C. Number	ISBN
No. 1	Feedback and Dynamic Control of Plasmas – 1970	70-141596	0-88318-100-2
No. 2	Particles and Fields – 1971 (Rochester)	71-184662	0-88318-101-0
No. 3	Thermal Expansion – 1971 (Corning)	72-76970	0-88318-102-9
No. 4	Superconductivity in *d*- and *f*-Band Metals (Rochester, 1971)	74-18879	0-88318-103-7
No. 5	Magnetism and Magnetic Materials – 1971 (2 parts) (Chicago)	59-2468	0-88318-104-5
No. 6	Particle Physics (Irvine, 1971)	72-81239	0-88318-105-3
No. 7	Exploring the History of Nuclear Physics – 1972	72-81883	0-88318-106-1
No. 8	Experimental Meson Spectroscopy –1972	72-88226	0-88318-107-X
No. 9	Cyclotrons – 1972 (Vancouver)	72-92798	0-88318-108-8
No. 10	Magnetism and Magnetic Materials – 1972	72-623469	0-88318-109-6
No. 11	Transport Phenomena – 1973 (Brown University Conference)	73-80682	0-88318-110-X
No. 12	Experiments on High Energy Particle Collisions – 1973 (Vanderbilt Conference)	73-81705	0-88318-111–8
No. 13	π-π Scattering – 1973 (Tallahassee Conference)	73-81704	0-88318-112-6
No. 14	Particles and Fields – 1973 (APS/DPF Berkeley)	73-91923	0-88318-113-4
No. 15	High Energy Collisions – 1973 (Stony Brook)	73-92324	0-88318-114-2
No. 16	Causality and Physical Theories (Wayne State University, 1973)	73-93420	0-88318-115-0
No. 17	Thermal Expansion – 1973 (Lake of the Ozarks)	73-94415	0-88318-116-9
No. 18	Magnetism and Magnetic Materials – 1973 (2 parts) (Boston)	59-2468	0-88318-117-7
No. 19	Physics and the Energy Problem – 1974 (APS Chicago)	73-94416	0-88318-118-5
No. 20	Tetrahedrally Bonded Amorphous Semiconductors (Yorktown Heights, 1974)	74-80145	0-88318-119-3
No. 21	Experimental Meson Spectroscopy – 1974 (Boston)	74-82628	0-88318-120-7
No. 22	Neutrinos – 1974 (Philadelphia)	74-82413	0-88318-121-5
No. 23	Particles and Fields – 1974 (APS/DPF Williamsburg)	74-27575	0-88318-122-3
No. 24	Magnetism and Magnetic Materials – 1974 (20th Annual Conference, San Francisco)	75-2647	0-88318-123-1

No. 25	Efficient Use of Energy (The APS Studies on the Technical Aspects of the More Efficient Use of Energy)	75-18227	0-88318-124-X
No. 26	High-Energy Physics and Nuclear Structure – 1975 (Santa Fe and Los Alamos)	75-26411	0-88318-125-8
No. 27	Topics in Statistical Mechanics and Biophysics: A Memorial to Julius L. Jackson (Wayne State University, 1975)	75-36309	0-88318-126-6
No. 28	Physics and Our World: A Symposium in Honor of Victor F. Weisskopf (M.I.T., 1974)	76-7207	0-88318-127-4
No. 29	Magnetism and Magnetic Materials – 1975 (21st Annual Conference, Philadelphia)	76-10931	0-88318-128-2
No. 30	Particle Searches and Discoveries – 1976 (Vanderbilt Conference)	76-19949	0-88318-129-0
No. 31	Structure and Excitations of Amorphous Solids (Williamsburg, VA, 1976)	76-22279	0-88318-130-4
No. 32	Materials Technology – 1976 (APS New York Meeting)	76-27967	0-88318-131-2
No. 33	Meson-Nuclear Physics – 1976 (Carnegie-Mellon Conference)	76-26811	0-88318-132-0
No. 34	Magnetism and Magnetic Materials – 1976 (Joint MMM-Intermag Conference, Pittsburgh)	76-47106	0-88318-133-9
No. 35	High Energy Physics with Polarized Beams and Targets (Argonne, 1976)	76-50181	0-88318-134-7
No. 36	Momentum Wave Functions – 1976 (Indiana University)	77-82145	0-88318-135-5
No. 37	Weak Interaction Physics – 1977 (Indiana University)	77-83344	0-88318-136-3
No. 38	Workshop on New Directions in Mossbauer Spectroscopy (Argonne, 1977)	77-90635	0-88318-137-1
No. 39	Physics Careers, Employment and Education (Penn State, 1977)	77-94053	0-88318-138-X
No. 40	Electrical Transport and Optical Properties of Inhomogeneous Media (Ohio State University, 1977)	78-54319	0-88318-139-8
No. 41	Nucleon-Nucleon Interactions – 1977 (Vancouver)	78-54249	0-88318-140-1
No. 42	Higher Energy Polarized Proton Beams (Ann Arbor, 1977)	78-55682	0-88318-141-X
No. 43	Particles and Fields – 1977 (APS/DPF, Argonne)	78-55683	0-88318-142-8
No. 44	Future Trends in Superconductive Electronics (Charlottesville, 1978)	77-9240	0-88318-143-6
No. 45	New Results in High Energy Physics – 1978 (Vanderbilt Conference)	78-67196	0-88318-144-4

No. 46	Topics in Nonlinear Dynamics (La Jolla Institute)	78-57870	0-88318-145-2
No. 47	Clustering Aspects of Nuclear Structure and Nuclear Reactions (Winnepeg, 1978)	78-64942	0-88318-146-0
No. 48	Current Trends in the Theory of Fields (Tallahassee, 1978)	78-72948	0-88318-147-9
No. 49	Cosmic Rays and Particle Physics – 1978 (Bartol Conference)	79-50489	0-88318-148-7
No. 50	Laser-Solid Interactions and Laser Processing – 1978 (Boston)	79-51564	0-88318-149-5
No. 51	High Energy Physics with Polarized Beams and Polarized Targets (Argonne, 1978)	79-64565	0-88318-150-9
No. 52	Long-Distance Neutrino Detection – 1978 (C.L. Cowan Memorial Symposium)	79-52078	0-88318-151-7
No. 53	Modulated Structures – 1979 (Kailua Kona, Hawaii)	79-53846	0-88318-152-5
No. 54	Meson-Nuclear Physics – 1979 (Houston)	79-53978	0-88318-153-3
No. 55	Quantum Chromodynamics (La Jolla, 1978)	79-54969	0-88318-154-1
No. 56	Particle Acceleration Mechanisms in Astrophysics (La Jolla, 1979)	79-55844	0-88318-155-X
No. 57	Nonlinear Dynamics and the Beam-Beam Interaction (Brookhaven, 1979)	79-57341	0-88318-156-8
No. 58	Inhomogeneous Superconductors – 1979 (Berkeley Springs, W.V.)	79-57620	0-88318-157-6
No. 59	Particles and Fields – 1979 (APS/DPF Montreal)	80-66631	0-88318-158-4
No. 60	History of the ZGS (Argonne, 1979)	80-67694	0-88318-159-2
No. 61	Aspects of the Kinetics and Dynamics of Surface Reactions (La Jolla Institute, 1979)	80-68004	0-88318-160-6
No. 62	High Energy e^+e^- Interactions (Vanderbilt, 1980)	80-53377	0-88318-161-4
No. 63	Supernovae Spectra (La Jolla, 1980)	80-70019	0-88318-162-2
No. 64	Laboratory EXAFS Facilities – 1980 (Univ. of Washington)	80-70579	0-88318-163-0
No. 65	Optics in Four Dimensions – 1980 (ICO, Ensenada)	80-70771	0-88318-164-9
No. 66	Physics in the Automotive Industry – 1980 (APS/AAPT Topical Conference)	80-70987	0-88318-165-7
No. 67	Experimental Meson Spectroscopy – 1980 (Sixth International Conference, Brookhaven)	80-71123	0-88318-166-5
No. 68	High Energy Physics – 1980 (XX International Conference, Madison)	81-65032	0-88318-167-3
No. 69	Polarization Phenomena in Nuclear Physics – 1980 (Fifth International Symposium, Santa Fe)	81-65107	0-88318-168-1

No. 70	Chemistry and Physics of Coal Utilization – 1980 (APS, Morgantown)	81-65106	0-88318-169-X
No. 71	Group Theory and its Applications in Physics – 1980 (Latin American School of Physics, Mexico City)	81-66132	0-88318-170-3
No. 72	Weak Interactions as a Probe of Unification (Virginia Polytechnic Institute – 1980)	81-67184	0-88318-171-1
No. 73	Tetrahedrally Bonded Amorphous Semiconductors (Carefree, Arizona, 1981)	81-67419	0-88318-172-X
No. 74	Perturbative Quantum Chromodynamics (Tallahassee, 1981)	81-70372	0-88318-173-8
No. 75	Low Energy X-Ray Diagnostics – 1981 (Monterey)	81-69841	0-88318-174-6
No. 76	Nonlinear Properties of Internal Waves (La Jolla Institute, 1981)	81-71062	0-88318-175-4
No. 77	Gamma Ray Transients and Related Astrophysical Phenomena (La Jolla Institute, 1981)	81-71543	0-88318-176-2
No. 78	Shock Waves in Condensed Mater – 1981 (Menlo Park)	82-70014	0-88318-177-0
No. 79	Pion Production and Absorption in Nuclei – 1981 (Indiana University Cyclotron Facility)	82-70678	0-88318-178-9
No. 80	Polarized Proton Ion Sources (Ann Arbor, 1981)	82-71025	0-88318-179-7
No. 81	Particles and Fields –1981: Testing the Standard Model (APS/DPF, Santa Cruz)	82-71156	0-88318-180-0
No. 82	Interpretation of Climate and Photochemical Models, Ozone and Temperature Measurements (La Jolla Institute, 1981)	82-71345	0-88318-181-9
No. 83	The Galactic Center (Cal. Inst. of Tech., 1982)	82-71635	0-88318-182-7
No. 84	Physics in the Steel Industry (APS/AISI, Lehigh University, 1981)	82-72033	0-88318-183-5
No. 85	Proton-Antiproton Collider Physics –1981 (Madison, Wisconsin)	82-72141	0-88318-184-3
No. 86	Momentum Wave Functions – 1982 (Adelaide, Australia)	82-72375	0-88318-185-1
No. 87	Physics of High Energy Particle Accelerators (Fermilab Summer School, 1981)	82-72421	0-88318-186-X
No. 88	Mathematical Methods in Hydrodynamics and Integrability in Dynamical Systems (La Jolla Institute, 1981)	82-72462	0-88318-187-8
No. 89	Neutron Scattering – 1981 (Argonne National Laboratory)	82-73094	0-88318-188-6
No. 90	Laser Techniques for Extreme Ultraviolt Spectroscopy (Boulder, 1982)	82-73205	0-88318-189-4

No. 91	Laser Acceleration of Particles (Los Alamos, 1982)	82-73361	0-88318-190-8
No. 92	The State of Particle Accelerators and High Energy Physics (Fermilab, 1981)	82-73861	0-88318-191-6
No. 93	Novel Results in Particle Physics (Vanderbilt, 1982)	82-73954	0-88318-192-4
No. 94	X-Ray and Atomic Inner-Shell Physics – 1982 (International Conference, U. of Oregon)	82-74075	0-88318-193-2
No. 95	High Energy Spin Physics – 1982 (Brookhaven National Laboratory)	83-70154	0-88318-194-0
No. 96	Science Underground (Los Alamos, 1982)	83-70377	0-88318-195-9
No. 97	The Interaction Between Medium Energy Nucleons in Nuclei – 1982 (Indiana University)	83-70649	0-88318-196-7
No. 98	Particles and Fields – 1982 (APS/DPF University of Maryland)	83-70807	0-88318-197-5
No. 99	Neutrino Mass and Gauge Structure of Weak Interactions (Telemark, 1982)	83-71072	0-88318-198-3
No. 100	Excimer Lasers – 1983 (OSA, Lake Tahoe, Nevada)	83-71437	0-88318-199-1
No. 101	Positron-Electron Pairs in Astrophysics (Goddard Space Flight Center, 1983)	83-71926	0-88318-200-9
No. 102	Intense Medium Energy Sources of Strangeness (UC-Sant Cruz, 1983)	83-72261	0-88318-201-7
No. 103	Quantum Fluids and Solids – 1983 (Sanibel Island, Florida)	83-72440	0-88318-202-5
No. 104	Physics, Technology and the Nuclear Arms Race (APS Baltimore –1983)	83-72533	0-88318-203-3
No. 105	Physics of High Energy Particle Accelerators (SLAC Summer School, 1982)	83-72986	0-88318-304-8
No. 106	Predictability of Fluid Motions (La Jolla Institute, 1983)	83-73641	0-88318-305-6
No. 107	Physics and Chemistry of Porous Media (Schlumberger-Doll Research, 1983)	83-73640	0-88318-306-4
No. 108	The Time Projection Chamber (TRIUMF, Vancouver, 1983)	83-83445	0-88318-307-2
No. 109	Random Walks and Their Applications in the Physical and Biological Sciences (NBS/La Jolla Institute, 1982)	84-70208	0-88318-308-0
No. 110	Hadron Substructure in Nuclear Physics (Indiana University, 1983)	84-70165	0-88318-309-9
No. 111	Production and Neutralization of Negative Ions and Beams (3rd Int'l Symposium, Brookhaven, 1983)	84-70379	0-88318-310-2

No. 112	Particles and Fields – 1983 (APS/DPF, Blacksburg, VA)	84-70378	0-88318-311-0
No. 113	Experimental Meson Spectroscopy – 1983 (Seventh International Conference, Brookhaven)	84-70910	0-88318-312-9
No. 114	Low Energy Tests of Conservation Laws in Particle Physics (Blacksburg, VA, 1983)	84-71157	0-88318-313-7
No. 115	High Energy Transients in Astrophysics (Santa Cruz, CA, 1983)	84-71205	0-88318-314-5
No. 116	Problems in Unification and Supergravity (La Jolla Institute, 1983)	84-71246	0-88318-315-3
No. 117	Polarized Proton Ion Sources (TRIUMF, Vancouver, 1983)	84-71235	0-88318-316-1
No. 118	Free Electron Generation of Extreme Ultraviolet Coherent Radiation (Brookhaven/OSA, 1983)	84-71539	0-88318-317-X
No. 119	Laser Techniques in the Extreme Ultraviolet (OSA, Boulder, Colorado, 1984)	84-72128	0-88318-318-8
No. 120	Optical Effects in Amorphous Semiconductors (Snowbird, Utah, 1984)	84-72419	0-88318-319-6
No. 121	High Energy e^+e^- Interactions (Vanderbilt, 1984)	84-72632	0-88318-320-X
No. 122	The Physics of VLSI (Xerox, Palo Alto, 1984)	84-72729	0-88318-321-8
No. 123	Intersections Between Particle and Nuclear Physics (Steamboat Springs, 1984)	84-72790	0-88318-322-6
No. 124	Neutron-Nucleus Collisions – A Probe of Nuclear Structure (Burr Oak State Park - 1984)	84-73216	0-88318-323-4
No. 125	Capture Gamma-Ray Spectroscopy and Related Topics – 1984 (Internat. Symposium, Knoxville)	84-73303	0-88318-324-2
No. 126	Solar Neutrinos and Neutrino Astronomy (Homestake, 1984)	84-63143	0-88318-325-0
No. 127	Physics of High Energy Particle Accelerators (BNL/SUNY Summer School, 1983)	85-70057	0-88318-326-9
No. 128	Nuclear Physics with Stored, Cooled Beams (McCormick's Creek State Park, Indiana, 1984)	85-71167	0-88318-327-7
No. 129	Radiofrequency Plasma Heating (Sixth Topical Conference, Callaway Gardens, GA, 1985)	85-48027	0-88318-328-5
No. 130	Laser Acceleration of Particles (Malibu, California, 1985)	85-48028	0-88318-329-3
No. 131	Workshop on Polarized ^{3}He Beams and Targets (Princeton, New Jersey, 1984)	85-48026	0-88318-330-7
No. 132	Hadron Spectroscopy–1985 (International Conference, Univ. of Maryland)	85-72537	0-88318-331-5

No. 133	Hadronic Probes and Nuclear Interactions (Arizona State University, 1985)	85-72638	0-88318-332-3
No. 134	The State of High Energy Physics (BNL/SUNY Summer School, 1983)	85-73170	0-88318-333-1
No. 135	Energy Sources: Conservation and Renewables (APS, Washington, DC, 1985)	85-73019	0-88318-334-X
No. 136	Atomic Theory Workshop on Relativistic and QED Effects in Heavy Atoms	85-73790	0-88318-335-8
No. 137	Polymer-Flow Interaction (La Jolla Institute, 1985)	85-73915	0-88318-336-6
No. 138	Frontiers in Electronic Materials and Processing (Houston, TX, 1985)	86-70108	0-88318-337-4
No. 139	High-Current, High-Brightness, and High-Duty Factor Ion Injectors (La Jolla Institute, 1985)	86-70245	0-88318-338-2
No. 140	Boron-Rich Solids (Albuquerque, NM, 1985)	86-70246	0-88318-339-0
No. 141	Gamma-Ray Bursts (Stanford, CA, 1984)	86-70761	0-88318-340-4
No. 142	Nuclear Structure at High Spin, Excitation, and Momentum Transfer (Indiana University, 1985)	86-70837	0-88318-341-2
No. 143	Mexican School of Particles and Fields (Oaxtepec, México, 1984)	86-81187	0-88318-342-0
No. 144	Magnetospheric Phenomena in Astrophysics (Los Alamos, 1984)	86-71149	0-88318-343-9
No. 145	Polarized Beams at SSC & Polarized Antiprotons (Ann Arbor, MI & Bodega Bay, CA, 1985)	86-71343	0-88318-344-7
No. 146	Advances in Laser Science–I (Dallas, TX, 1985)	86-71536	0-88318-345-5